MW01154224

VICTORIAN SENSATION

JAMES A. SECORD

VICTORIAN SENSATION

The Extraordinary Publication, Reception, and Secret Authorship of *Vestiges of the Natural History of Creation*

THE UNIVERSITY OF CHICAGO PRESS

Chicago and London

James A. Secord is Reader in History and Philosophy of Science at the University of Cambridge. He is author of *Controversy in Victorian Geology* (1986), editor of Robert Chambers's *Vestiges of the Natural History of Creation and Other Evolutionary Writings* (1994) and Charles Lyell's *Principles of Geology* (1997), and coeditor of *Cultures of Natural History* (1996, with Nick Jardine and Emma Spary).

The University of Chicago Press, Chicago 60637
The University of Chicago Press, Ltd., London

Printed in the United States of America
09 08 07 06 05 04 03 02 01 00 1 2 3 4 5

ISBN: 0-226-74410-8 (cloth)

Library of Congress Cataloging-in-Publication Data

Secord, James A.
Victorian sensation : the extraordinary publication, reception, and secret authorship of Vestiges of the natural history of creation / James A. Secord.
p. cm.
Includes bibliographical references (p.) and index.
ISBN 0-226-74410-8 (cloth : alk. paper)
1. Chambers, Robert, 1802–1871. Vestiges of the natural history of creation. I. Title.

QH363.S4 2000
576.8'0941'09034—dc21

00-009124

⊗ The paper used in this publication meets the minimum requirements of the American National Standard for Information Sciences—Permanence of Paper for Printed Library Materials, ANSI Z39.48-1992.

In memory of my parents

CONTENTS

ILLUSTRATIONS

ACKNOWLEDGMENTS

The writing and researching of this book have occupied a large part of my life, and it is impossible to acknowledge adequately the generous help of all those who have made it possible. My parents, Jane and John Secord, encouraged me in every way and fostered my own love of reading, not least during my childhood by putting a pile of library books by my bedside each week. My mother-in-law and much-missed friend Rita Goldhill and I talked about my hopes for the book on many occasions, and I will always remember the great day I was able to show her a completed typescript.

A number of institutions provided the support without which a big project of this kind cannot be contemplated. I am grateful to Churchill College, Cambridge, for appointing me to a Junior Research Fellowship from 1982 to 1985; to the Royal Society of London for travel funds in the early stages of my research; to a Wellcome Trust Research Leave Fellowship in 1989–90; and to the British Academy for a term's leave in 1998 that made it possible to complete the first draft.

The Department of History and Philosophy of Science at Cambridge is an exceptionally stimulating place to do interdisciplinary research. I wish especially to thank Nick Jardine, Martin Kusch, and Simon Schaffer for sharpening my analysis at crucial points in the writing; Soraya de Chadarevian, Andrew Cunningham, Marina Frasca-Spada, Silvia De Renzi, Patricia Fara, John Forrester, Nick Hopwood, Lauren Kassell, and Sachiko Kusukawa for probing questions and comments; and Peter Lipton, for creating an environment in which dialogue between historians and philosophers can flourish. Tamara Hug, David Allington, and John McWilliams have made the Department as efficient as it is lively and friendly. The approach taken in this book has been shaped by working in close proximity to the collections of the Whipple Museum under the successive curatorships of Jim Bennett and Liba Taub. Catriona West and Paul Webb, both of the museum staff, helped to prepare two especially recalcitrant illustrations. Rebecca Bertelloni Meli, Linda Washington, and Joanna Ball have made the Whipple Library a model of what a departmental collection should be.

I am much obliged to the many friends who have followed this book through its long gestation. My interest in the topic was stimulated by a copy of the third edition of *Vestiges* given to me by John Thackray, who provided much help before his untimely death. I am especially glad to thank Adrian Desmond for his encouragement, patience, and generosity. He performed the herculean task of reading the entire manuscript twice, and his perceptive comments transformed the structure of the finished book. Alison Winter has shared my enthusiasm for the early Victorian period and helped to change the way I think about it. In countless conversations, Jon Topham and Aileen Fyfe have given my successive drafts the benefit of their knowledge of early nineteenth-century publishing and religious history. Exceptionally helpful comments on the entire manuscript were also provided by Bernard Lightman, John van Wyhe, and the referees for the University of Chicago Press. Boyd Hilton, Jack Morrell, and Paul White cast a critical eye over several chapters, which have been much improved as a result. Rebecca Stott, Leah Price, Marilyn Butler, and Susan Bernstein provided invaluable guidance in literary history.

Many others commented on individual chapters, gave advice on special points, or provided important references. Among those whose help I wish to acknowledge are Katharine Anderson, Jean Archibald, William Ashworth, William J. Astore, Henry Atmore, Emm Barnes, Anne Barrett, Mary Bartley, Gillian Beer, John Beer, Michael Bott, Peter Bowler, William H. Brock, John Brooke, Iain Brown, Janet Browne, Jane Camerini, Geoffrey Cantor, James Chandler, Pamela Clark, Roger Cooter, Pietro Corsi, Robert Cox, John Creasey, Dennis Dean, Margaret De Mott, Susan Dench, Brian Dolan, Felix Driver, Simon Eliot, James Endersby, Sophie Forgan, Robert Fox, V. A. C. Gatrell, Patricia Gilhoulie, Jan Golinski, Lyubov Gurjeva, Beryl Hartley, Catherine Hemsley, Leslie Howsam, Frank James, Adrian Johns, Ludmilla Jordanova, Alice Beck Kehoe, David Knight, David Kohn, Kevin Knox, Trevor Levere, Eileen Groth Lyon, Sheila Mackenzie, Peter Mandler, James Moore, Iwan Rhys Morus, Ian Nelson, Christine North, Dorinda Outram, John Pickstone, Roy Porter, Theodore Porter, Philip F. Rehbock, Evelleen Richards, Marsha Richmond, Harriet Ritvo, Eugenia Roldan-Vera, Ulinka Rublack, Martin Rudwick, Nicolaas Rupke, Steven Shapin, Michael Shortland, Sujit Sivasundaram, Helen Smailes, Helen Small, Crosbie Smith, Emma Spary, David H. Staum, Joan Steigerwald, Ann Thwaite, Hugh Torrens, Sarah Wilmot, Carla Yanni, and Richard Yeo. Milton Millhauser's work on *Vestiges* and Sondra Miley Cooney's study of the Chambers publishing firm offered treasure troves of references. My students, both postgraduate and undergraduate, have contributed in many ways to this work, and have been remarkably tolerant of a supervisor who must often have seemed lost in the 1840s.

I have benefited greatly from invitations to present my ideas at conferences and seminars. At a critical stage of revision, several chapters were discussed in lively sessions at the Evolution Reading Group organized by Greg Radick at Cambridge. An abridged version of chapters 1 and 2 appeared in *Books and the Sciences in History,* ed. Marina Frasca-Spada and Nick Jardine (Cambridge University Press, 2000), and an early version of part of chapter 3 was published in *History, Humanity, and Evolution,* ed. James R. Moore (Cambridge University Press, 1989).

A project such as this one involves sifting through tens of thousands of printed, microfilm, and manuscript records, only a fraction of which yield relevant material. Inevitably, such a procedure places exceptional demands on librarians and archivists. I wish to thank in particular the two collections I have relied upon most: the National Library of Scotland, with its unrivaled collections of relevant manuscript material; and, above all, Cambridge University Library, the staff of which has been unfailingly helpful in dealing with a never-ending stream of request slips.

I owe a special thanks to Anthony S. Chambers, whose generous deposit of his family and business papers in the National Library of Scotland has made this important collection freely available to scholars. I am grateful to Mr. Chambers for permission to quote from these papers, his assistance with queries, and his warm hospitality. I am also extremely grateful to Sir Mark Norman for information on the contents of his collection of Chambers family papers, and for his generosity in allowing me to consult them.

Many other archives, libraries, and individuals have made it possible for me to use material in their possession. I especially wish to thank the American Philosophical Society; Billson of St. Andrews (for allowing me to photograph the illustration of St. Andrews showing Abbey Park); the Bodleian Library, Oxford; the British Geological Survey; the British Library; the British Museum, Department of Prints and Drawings; Calderdale District Archives, West Yorkshire Archive Service; Cambridge Central Library; the Department of Earth Sciences, University of Cambridge; the Cambridge Philosophical Society; City University Library; Cornell University Library; Edinburgh City Libraries; Edinburgh University Library; the Library of the Geological Society of London; the Guildhall Library; Castle Howard, North Yorkshire; Lord Howick of Howick Hall, Yorkshire; the Archives of Imperial College of Science, Technology and Medicine; Kirklees District Archives, West Yorkshire Archive Service; Liverpool University Library; the Local History Collection, Liverpool City Libraries; the London Library; the Mitchell Library, Glasgow; the Museum of the History of Science, Oxford; Manchester Central Library; the National Museum and Galleries of Wales; the Natural History Museum, London; Newcastle University Library; the Punch Library; the University of Reading; the Royal Commission on the Ancient and Historical Monuments of Scotland; the Royal Institution of Great Britain; the Royal Society of London; St. Andrews University Library; St. Bride Printing Library, London; Sir Ferrars Vyvyan, Trelowarren, Cornwall; the Wellcome Institute for the History of Medicine; Dr. Williams's Library; Wren Library, Trinity College, Cambridge; and Yale University Library.

My copyeditor, Michael Koplow, untangled bibliographical nightmares and saved me from scores of errors and inconsistencies. Martin White compiled a thorough and intelligent index. Working with the University of Chicago Press has been a great pleasure, and I wish to thank everyone involved, most especially my editor Susan Abrams. To a remarkable degree, she has combined patience in waiting for this book with faith in its (and my) potential. If anyone still needs convincing that books are not the sole product of their authors, this one could be used to demonstrate the case conclusively.

My greatest debt is to Anne.

ABBREVIATIONS

AI	Alexander Ireland
APS	American Philosophical Society
BJHS	*British Journal for the History of Science*
BL	British Library
CEJ	*Chambers's Edinburgh Journal*
CPS	Cambridge Philosophical Society
CUL	Cambridge University Library
DNB	*Dictionary of National Biography*
*E*1	[Robert Chambers], *Explanations, A Sequel,* 1st ed. (London: John Churchill, 1845); reprinted in Robert Chambers, *Vestiges of the Natural History of Creation and Other Evolutionary Writings,* ed. James A. Secord (Chicago: University of Chicago Press, 1994)
*E*2	[Chambers], *Explanations,* 2d ed. (London: John Churchill, 1846)
EUL	Edinburgh University Library
GSL	Geological Society of London
IC	Imperial College London
ILN	*Illustrated London News*
NHM	Natural History Museum, London
NLS	National Library of Scotland
NMW	National Museum of Wales
PT	*Pictorial Times*
RC	Robert Chambers
RSL	Royal Society of London
TC	Trinity College, Cambridge
*V*1	[Robert Chambers], *Vestiges of the Natural History of Creation,* 1st ed. (London: John Churchill, 1844); reprinted in Robert Chambers, *Vestiges of the Natural History of Creation and Other Evolutionary Writings,* ed. James A. Secord (Chicago: University of Chicago Press, 1994)
*V*2	[Chambers], *Vestiges,* 2d ed. (London: John Churchill, 1844)
*V*3	[Chambers], *Vestiges,* 3d ed. (London: John Churchill, 1845)
*V*4	[Chambers], *Vestiges,* 4th ed. (London: John Churchill, 1845)
*V*5	[Chambers], *Vestiges,* 5th ed. (London: John Churchill, 1846)
*V*6	[Chambers], *Vestiges,* 6th ed. (London: John Churchill, 1847)
*V*7	[Chambers], *Vestiges,* 7th ed. (London: John Churchill, 1847)
*V*10	[Chambers], *Vestiges,* 10th ed. (London: John Churchill, 1853)
*V*12	Chambers, *Vestiges,* 12th ed. (Edinburgh: W. & R. Chambers, 1884)

PROLOGUE

Devils or Angels

What a thing a book is! what power it has! It is a devil or an angel for power,—if a real, living book.

ELIZABETH BARRETT to Mary Russell Mitford, 1844

ELIZABETH HARRISON WAS SHOCKED by what she found in the parcel. The wife of a shopkeeper in a country village on the chalk downs of southern England, she loved novels, plays, and histories. She subscribed to a local library and encouraged her boys to read. But this little parcel, which her eldest son had left behind when he emigrated to the Australian gold diggings in 1852, contained books of a different kind. In it were books she had never read and never would. They were literary poison issued in the wake of the French Revolution, including Tom Paine's notorious *Age of Reason* and Count Volney's *Ruins of Empires*—dangerous books that attacked Christian truth as contrary to nature. Distressed and angry, Mrs. Harrison flung them into the fire.[1]

After much soul-searching and the entreaties of the family's youngest son, who at fourteen was already a keen naturalist and reader, one book was saved from the flames. That book was the celebrated *Vestiges of the Natural History of Creation.* As readable as a romance, based on the latest findings of science, *Vestiges* was an evolutionary epic that ranged from the formation of the solar system to reflections on the destiny of the human race. *Vestiges,* published in 1844, was more controversial than any other philosophical or scientific work of its time. In a hugely ambitious synthesis, it combined astronomy, geology, physiology, psychology, anthropology, and theology in a general theory of creation. It suggested that the planets had originated in a blazing Fire-mist, that life could be created in the laboratory, that humans had evolved from apes. Most intriguing of all, *Vestiges* was anonymous. No one seemed to know who the author was, or whether his or her references to a divine creator were just for show; the author's status, politics, and gender were a mystery.

Vestiges was a great sensation. Readers included aristocrats and handloom

1. Harrison 1928, 38–39, 28.

weavers; science writers and the wives of cotton manufacturers; evangelicals and militant freethinkers; as well as Queen Victoria, Alfred Tennyson, Florence Nightingale, Harriet Martineau, William Ewart Gladstone, Thomas De Quincey, Charles Darwin, Thomas Carlyle, and, on a conservative estimate, at least a hundred thousand other men and women across the spectrum of Victorian society. The present book is the story of this sensation.[2] To my knowledge, it offers the most comprehensive analysis of the reading of any book other than the Bible ever undertaken.

TODAY BOOKS ABOUT EVOLUTION are our devils and angels. It is through reading the successors of *Vestiges* that we make sense of our origins and potential futures. In best-sellers ranging from Stephen Hawking's *Brief History of Time* (1988) to Steven Pinker's *Language Instinct* (1994), readers trace stories that start from swirling clouds of cosmic dust and end with the emergence of mind and human culture. The literary agent who handles many of these books expresses what readers and producers usually take for granted: "The universe is changing in time, and it has evolved from something simpler to something more complex. That is the lesson to be learned from recent advances in evolutionary theory; the emergence of order has colored biology since Darwin and twentieth-century cosmology alike."[3]

How did evolution gain this pivotal role in the public arena? The answer turns out to have little to do with Darwinian biology or Big Bang astronomy. Instead, the critical period is the first half of the nineteenth century, and the turning point is the response of readers to *Vestiges*. The decades before its publication in the mid-1840s had witnessed the greatest transformation in human communication since the Renaissance. Mechanized presses, machine-made paper, railway distribution, improved education, and the penny post played a major part in opening the floodgates to a vastly increased reading public. Only now, with the advent of electronic communication, are we undergoing a period of equal change. "It is felt," John Stuart Mill wrote, "that men are henceforth to be held together by new ties, and separated by new barriers; for the ancient bonds will now no longer unite, nor the ancient boundaries confine."[4]

These transformations are manifest in the *Vestiges* sensation. Contemporaries called it the biggest literary phenomenon for decades, bigger perhaps than even

2. My aim is not, as some would have it, the "biography of a book," a description that is more appropriate for an account (for example, Darnton 1979) centered on production and authorship rather than reading. In any event, books do not have a "life" of their own independent from their use.

3. Brockman 1995, 33. Eger 1993 is helpful on issues concerning the genre of the evolutionary epic; Ruse 1996 discusses aspects of its history.

4. Mill 1986, 229.

Charles Dickens's early novels. The book was mentioned in thousands of letters and diaries, denounced and praised in pulpits, discussed on railway journeys, and annotated on an Alabama River steamboat. It was discussed at dinner parties, pubs, and soirées, reviewed in scores of periodicals and pamphlets, and in Britain alone sold fourteen editions and almost forty thousand copies.[5]

The remarkable story of *Vestiges* can be recovered through new approaches to reading and communication that are revolutionizing our interpretation of many aspects of the past. Reading has often been seen as a profoundly private experience, but it is better understood as comprehending all the diverse ways that books and other forms of printed works are appropriated and used. Taken in this sense, a history of reading becomes a study of cultural formation in action. My strategy will be to follow a single work in all its uses and manifestations—in conversation, solitude, authorship, learned debate, religious controversy, civic politics, and the making of knowledge.[6] We can then begin to understand the role of the printed word in forging new senses of identity in the industrial age. Rather unexpectedly, tracking a work like *Vestiges* proves to be especially revealing, for the handful of scientific books that became sensations have left more identifiable traces than comparable works of fiction, history, and poetry. References to fossil footprints and nebular Fire-mists have a specificity that makes their source relatively obvious. Because of this, a widely read scientific work is a good "cultural tracer": it can be followed in a greater variety of circumstances than almost any other kind of book.[7]

The most compelling attempts to view history from the perspective of reading have examined the period from the introduction of the codex in antiquity to the end of the eighteenth century. Reading itself proves to have a history, so that what it meant to read changes dramatically over the centuries. A fascinating variety of individual readers have begun to be located and studied. The creative use of scarce and often recalcitrant evidence—shifts in the use of words, a few pen marks in a margin—has contributed to new pictures of the origins of the

5. The *Vestiges* controversy has been discussed in many works, usually as a prelude to Darwin. The more significant accounts include Alexander Ireland's introduction to *V*12 (1884); Gillispie 1951, 149–83; Lovejoy 1959; Millhauser 1959; Bowler 1976; Ruse 1979, 94–131; Yeo 1984; Corsi 1988, 250–71; Desmond 1989, 176–80; and J. Secord 1994. Many other works analyze specific topics relevant to the debate.

For the reading of *Vestiges* on a steamboat, see the copy of an American edition (New York, 1845) in Special Collections, Syracuse University Library; information provided by David H. Staum, personal communication 5 Jan. 1998.

6. The cultural historian Roger Chartier (1988, 11–12) has called this an "object study." Among the most thorough accounts of a single book of which I am aware is Burke 1995, which follows the fortunes of Baldassare Castiglione's *Courtier* among readers from the Renaissance to the present.

7. For recent surveys of the history of science in relation to the history of the book, see Frasca-Spada and Jardine 2000, Johns 1998b, and Topham 2000.

novel, the transformation of seventeenth-century science, and the impact of illegal philosophical books on the French Revolution.[8]

The first half of the nineteenth century, when gossipy personal letters and private diaries coexisted with steam-printed books and cheap magazines posted by rail, is probably richer in sources for the history of reading than any other period. Yet this material has been quarried primarily for anecdotal color and to undercut stereotypes. Historians have only just begun to use it to challenge older narratives based on a limited canon of authors and the intellectual achievements of a few great minds. Although there are many pointers to what can be done, the potential of a history of reading in the machine age is largely unrealized.[9]

Nowhere is this more evident than in studies of the evolutionary debates, in which Darwin's *On the Origin of Species* (1859) acts both as the measure of a scientific approach as well as the interpretive guide by which earlier works are judged. In the most familiar version of this story *Vestiges* is dismissed as a "popular" work, a failed precursor of the *Origin.* Historians have found it difficult to escape what one biologist has called "Darwin's spectre,"[10] even as they have demonstrated that a Darwin-centered account is no longer credible. In recent years historical and literary studies have turned from the analysis of disembodied ideas toward an understanding of practices. Scientific theories, theological doctrines, and political ideologies are seen as forms of work, set in the context of everyday life. This book takes these approaches into territory that is only beginning to be explored. It suggests that the most abstract ideas about nature should be approached first and foremost as material objects of commerce and situated in specific settings for reading. Mundane considerations whose importance has long been recognized by librarians, bibliographers, and printers need to become the bedrock for literary and intellectual history. What once made sense as the "Darwinian Revolution" must be recast as an episode in the industrialization of communication and the transformation of reading audiences.

PLACING READING AT THE CENTER of a history opens up general possibilities for understanding what happens when we read. Reading always takes place in specific contexts of experiences and expectations. It unites an interpretation of

8. James Raven, Helen Small, and Naomi Tadmor provide an excellent survey in the introduction to an outstanding (1996) collection of essays. For examples and further references, see Blair 1997, Burke 1995, Darnton 1996, Davidson 1988, Jardine and Grafton 1990, Johns 1998a, Pearson 1999, and Sharpe 2000.

9. See Topham 2000 and Flint 1993, 187–247, both with good bibliographies. For examples of recent work, see Jackson 1999; Rose 1995; Small 1996, Topham 1998; and (among the older literature) the notoriously unreferenced Cruse 1935. A very general survey of French reading practices is in Allen 1991, esp. 225–320.

10. Rose 1998.

words on the page with an understanding of the physical appearance and genre of a work and the ways in which it is marketed and discussed.[11] Each of us implicitly addresses all of these issues every time we pick up a book or read a newspaper. Paradoxically, though, the study of reading is fragmented into a dozen different academic specialties, from economic analyses of the publishing industry to critical theories of reception and reader response.[12] The controversy over *Vestiges* offers an opportunity to bring these approaches together, not least because the author—the dominant figure in almost all literary and intellectual history—was hidden for nearly forty years. The text is a rich one, drawing on historical fiction, writings on science, and mass-market journalism. Evidence for the production, in its passage from manuscript to print and thence to different editions, is also unusually good.

But it is the evidence of readers that makes the case of *Vestiges* unique. By combining close study of the text with an understanding of contemporary reading practices, we can explore not only the origins of our current controversies, but the larger question of the role of reading in creating the first mass industrial society. This is a book about evolution for the people, and the evolving self-identity of "the people."[13] The profound reaction to *Vestiges* was a manifestation of the forces that led to the optimistic, imperial, professional, and relatively secular public culture of the second half of the nineteenth century. Reading about evolutionary progress offered common questions to bridge divides that threatened the nation's stability. Controversies about class and gender—among many potentially explosive issues—could thereby be subsumed into discussions of nature's progress. Hence the significance of the *Vestiges* sensation for new literary forms such as popular science and the realist novel, and its larger role in making "the people" a central category of the industrial order.

In archives, newspapers, and memoirs, there are thousands of traces of encounters with the book. These are not just records of ownership or borrowing, but substantial and often moving testimonies to the power of reading. Fourteen-year-old Ben Harrison discovered this in 1852, for his older brother and a

11. See esp. McKenzie 1986 and Chartier 1994; also Bourdieu 1977, 1984; Darnton 1990, 107–35.

12. Most writings on the history of reading recognize the need to combine such approaches, although surprisingly little work has been done to carry this out; see "First Steps toward a History of Reading," in Darnton 1990, 154–87. Garrett Stewart (1996, 8) is right to insist that there need be no contradiction between historical and literary approaches to reading; it is only disappointing that, despite locating his own study of nineteenth-century fiction at the border between the two, he goes on to assert the all-sufficiency of texts in "conscripting" readers, whose very power to resist is inscribed in the text. At the opposite extreme, Rose 1992 and 1995 effectively deny the utility of any analysis of the texts.

13. Important works on the changing meaning of "the people" include Joyce 1991, Vernon 1993, and Wahrman 1995.

skeptical friend had filled their copy of *Vestiges* with annotations debating the pros and cons of religious faith. The young boy had pleaded with his mother, who had fostered his passion for natural history, not to throw the book into the fire. Half a century later he could still recall how eagerly he opened it and began to read.[14] The book Elizabeth Harrison decided not to burn brought an evolutionary vision of the universe into the heart of everyday life.

14. Harrison 1928, 38–39, 28.

PART ONE

Romances of Creation

CHAPTER ONE

A Great Sensation

And what a sensation some books created!

The Autobiography of Mary Smith, Schoolmistress and Nonconformist, a Fragment of a Life (1892)

IN MID-NOVEMBER 1844 Alfred Lord Tennyson opened the latest issue of the *Examiner*, a weekly reform newspaper, and turned to the notices of books. The lead review, devoted to a just-published work called *Vestiges of the Natural History of Creation*, immediately caught his eye:

> In this small and unpretending volume we have found so many great results of knowledge and reflection, that we cannot too earnestly recommend it to the attention of thoughtful men. It is the first attempt that has been made to connect the natural sciences into a history of creation. An attempt which presupposes learning, extensive and various; but not the large and liberal wisdom, the profound philosophical suggestion, the lofty spirit of beneficence, and the exquisite grace of manner, which make up the charm of this extraordinary book.

Intrigued, Tennyson asked his bookseller to send him a copy, noting that the work "seems to contain many speculations with which I have been familiar for years, and on which I have written more than one poem."[1] In return Tennyson received a small volume bound in bright red cloth. Advertising bound inside showed that the publisher dealt in medical textbooks and monographs on obscure diseases; otherwise the origins and authorship were a mystery.

Tennyson was enthralled, "quite excited." As a contemporary remarked, "He reads all sorts of things, swallows and digests them like a great poetical boa-constrictor."[2] The book ranged from astronomy and geology to moral philosophy and the prospect of a future life, all drawn together in a gripping cosmological narrative. The early pages described a nebular hypothesis of the universe,

1. A. Tennyson to E. Moxon, [15 Nov. 1844], in Lang and Shannon 1982–90, 1:230. For the review, see *Examiner*, 9 Nov. 1844, 707–9, and Killham 1958.

2. William Makepeace Thackeray, quoted in Stott 1996, 1.

showing how stars, planets, and moons had evolved from a gaseous "Fire-mist." Tennyson then followed the book's story of geological progress from simple invertebrate animals up through fish, amphibians, reptiles, mammals, and man. These were ideas he knew well. God worked through a law that brought forth new species just as it did new worlds. Man's spiritual sense and reason were the products of development, part of what the unknown author called "the universal gestation of nature." There was, Tennyson later concluded, "nothing degrading in the theory."[3]

The *Examiner* had been one of the first to publish a review. Over five columns, Tennyson read of "the simplicity of the writer's manner, and the beauty of his style"; this was one of the great works of the age. The unknown author, someone who had "earnestly investigated Nature," had conducted his inquiry with "so much modesty and so much knowledge." There were no criticisms of mistakes or the wider philosophy. The evolution of new species, and even of human beings, although "a remarkable hypothesis," was described as worthy of consideration. In time, the author might even be able to throw off the mask of anonymity, for "there is now abroad in the world a certain rare disposition" to hear the truths of nature in "a beneficent spirit." The *Examiner* regretted only the author's failure to recognize Greek foreshadowings of its doctrines. "What are these," the reviewer asked, "but, in another and simpler shape, the noblest thoughts and the loftiest aspirations that have consoled and elevated the hopes of humanity in this world?"[4] Other works need only be borrowed; *Vestiges* was a book Tennyson wanted to buy.

Tennyson was fortunate to have ordered his copy. As his friend and fellow author Edward Fitzgerald reported, the *Examiner*'s eulogy sold out the first edition in a few days.[5] Extraordinary rumors began to circulate. A huge number of copies—perhaps most of the impression—appeared to have been given away.[6] The book seemed to emanate from the very center of English life: leading aristocrats, members of Parliament, and famous men of science were suggested as the author. As the novelist and politician Benjamin Disraeli wrote to his sister Sarah, *Vestiges* "is convulsing the world, anonymous" and from a publisher he had never heard of. As his wife Mary had told her: "Dizzy says it does & will cause the greatest sensation & confusion."[7]

3. Diary entry for 28 Mar. 1871, in Knies 1984, 63; see also p. 85.

4. *Examiner*, 9 Nov. 1844, 707–9.

5. E. Fitzgerald to B. Barton, 4 Jan. 1845, in Terhune and Terhune 1980, 1:471.

6. For these rumors, see [G. A. Mantell], "Vestiges of the Natural History of Creation," *American Journal of Science* 49 (Apr. 1845}: 191, which mentions two hundred copies being distributed for free.

7. B. Disraeli to S. Disraeli, [20 Jan. 1845]; M. Disraeli to S. Disraeli, 19 Jan. 1845, in Gunn 1982–, 4:154–55.

Mechanisms of Sensation

What did readers mean when they called *Vestiges* a "sensation"? "Sensation" needs to be our starting point because that is how readers first experienced the book. We might, from some perspectives, expect to begin with an author's life or a summary of the text. But neither of these strategies will do. Gossip, rumor, advertising, street hoardings, newspaper notices: these were the ways that word spread. The book was an event and needs to be seen as part of the changing history of how such events were constituted. So we will explore the meanings of *Vestiges* as "sensation." We begin with individuals—to see how reading engaged the passions and the senses—and then examine how these responses spread through society.

In the eighteenth century "sensation" had been part of the culture of sensibility; philosophical writings, most famously those of John Locke, stressed that mental states originated in the senses. Samuel Johnson's dictionary, for example, defined "sensation" as "perception by means of the senses."[8] In the most extreme versions of sensationalist psychology, all mental states were produced by the impact of corpuscles upon the brain. Reading involved nothing more than a series of physical shocks received from letters on a page and communicated from the eye to the brain, where they combined mechanically with other impulses to form ideas. This view of sensation gained notoriety as part of the philosophical underpinning for the French Revolution. Evolutionary narratives from an older classical tradition had been reshaped in the salons of enlightened Paris into materialist philosophical works such as those of Baron d'Holbach's *System of Nature.* In Britain, these books were blamed for the bloody horrors of the French Revolution, and any account that could be read as linking matter with mind through material causes became suspect. For fifty years after the Terror, such books were associated with revolutionary atheists, pornographers, radical medical men, and dissolute foreigners.

In reaction, "sensation" took on new meanings during the early nineteenth century. Everyday usage increasingly limited the term to the realm of immediate nervous stimuli, and defined the mind's consciousness of these stimuli as "perception." This distinction had been most explicitly developed in the eighteenth-century Scottish philosophy of common sense, which argued that sensations are the occasions rather than the materials of perception. Acceptance of the existence of an external world was part of "common sense," defined as the

8. Johnson 1755, 2: "Sensation." The meanings of the word before the 1860s would well repay further examination. My account is based on the *Oxford English Dictionary,* standard histories of sensationalist psychology, and a survey of electronic versions of eighteenth- and nineteenth-century prose. Related issues are discussed in Barker-Benfield 1992, Logan 1997, Poovey 1995, and Winter 1998a.

shared belief of rational men. The exact relation between the senses and the mind continued to be much debated. Many theories maintained a mediating role for an immaterial soul, with mental activities such as reading carried out under the guiding influence of a spiritual governor.

In cases of heightened feeling, however, the senses could overwhelm reason, contemplation, and the other faculties. Raw, unconsidered, animal passions could engulf not just individuals, but much larger groups. A "sensation" came to mean an excited or violent emotion felt by an entire community and produced by a common experience: the death of a monarch, a terrible accident, a shocking discovery, a public hanging, a remarkable book. The use of the word had changed.

As literacy increased and civil society seemed threatened, problems of social cohesion were portrayed in physiological terms, as disturbances in what the literary historian Mary Poovey has termed the "social body."[9] Matters of state were united with management of individual sensibility and public opinion. Society became a "mass"—undifferentiated and operating according to animal instinct rather than reason. "Sensation" became part of a language developed to diagnose this new social malaise; as the conservative *Quarterly Review* noted disapprovingly in 1817, it was "the phraseology of the present day."[10] "Sensation" did not always carry pejorative implications—a letter from a distant loved one could cause a "sensation" in a household—although in some circumstances ambivalence remained. "Sensation" could easily be linked with words such as "vulgar," "noisy," and "popular."

The language of sensation became ubiquitous. In his classic *Memoirs of Extraordinary Popular Delusions and the Madness of Crowds* (1841), the Scottish newspaperman Charles Mackay rewrote history as a series of delusional sensations:

> In reading the history of nations, we find that, like individuals, they have their whims and their peculiarities; their seasons of excitement and recklessness, when they care not what they do. We find that whole communities suddenly fix their minds upon one object, and go mad in its pursuit; that millions of people become simultaneously impressed with one delusion, and run after it, till their attention is caught by some new folly more captivating than the first.[11]

Mackay's book was almost entirely about the period before 1800, but its analysis in terms of "madness," "excitement," and "millions of people" is entirely

9. Poovey 1995.

10. [R. Southey], "Lord Holland's *Life and Writings of Lope de Vega*," *Quarterly Review* 18 (Oct. 1818): 1–46, at 10.

11. Mackay 1852, 1:vii.

characteristic of the 1840s. *Vestiges,* he later remembered, "excited a great sensation at the time. . . . highly praised by some, violently abused by others . . . before it finally 'blew over' and disappeared alike from public favour and animadversion."[12]

People varied in their susceptibility to sensation. Working-class readers were thought to be easily affected by sensual imagery, as their brains were assumed to associate words on the page with concrete, external objects. Cheap newspapers were dangerous because they brought the overt excitement of politics, murder, and other current events into ordinary cottages and working-class homes. Genteel readers, on the other hand, could remain aloof by rising to logical abstraction. Standard medical works explained that women could be subject to intense and rapidly changing sensations, which made them incapable of connected trains of reasoning. Sensation could be a disease of civilization, more easily affecting the refined nerves of upper-class women. As one Scottish weekly said in combating *Vestiges,* "It would almost appear . . . that the more civilized a society becomes, the more apt are visionary notions to spring up and flourish, just as we find hysterics and nervous vapours to prevail among fine ladies, while their robust maids are exempt from any thing of the kind."[13] Books suited to one kind of readers might be totally inappropriate for others. Who read what, and under what circumstances, mattered intensely.

Take the experience of Samuel Richard Bosanquet, a wealthy lawyer who read *Vestiges* at his Welsh country estate at Dingestow Court, Monmouthshire. A cultivated man of learning who had written several books, Bosanquet read *Vestiges* as a sign that the world was coming to an end. His *"Vestiges of the Natural History of Creation:" Its Argument Examined and Exposed,* a pamphlet published in two editions during 1845, railed against the "rapid circulation" and "very general approval" that the work had obtained. Bosanquet was a fervent evangelical in the Church of England, who identified the 1840s as the "last times" of apostasy that would precede Christ's second coming. As a premillennialist, he believed that God worked through "special" or "particular" acts involving the suspension of the laws of nature.[14] *Vestiges* was anathema because it denied special providence.

Bosanquet never doubted his ability to avoid temptation, but the serpent was poisoning the spiritual mind of the nation. Many readers, "especially the increasing class of female philosophers," were dangerously susceptible to the promise of a book of knowledge. Like Eve, they were led by "its most honied sweetness, to the most tasteful, and to the bitterest fruit." In *Vestiges,* the serpent

12. Mackay 1887, 1:177.

13. "The Delusions of the Day," *Torch,* 10 Jan. 1846, 21–23, at 21.

14. Bosanquet 1845, 1; Hilton 1988, 96–97.

(as in medieval images of the Edenic snake) "rears its head with human front and voice, and syren sweetness of address and invitation; while other idols exhibit their bestial foulness to only ordinary discernment."[15] The feminine was demonized by association with images of luxury, oriental corruption, and unthinking consumption.

In Bosanquet's reading, *Vestiges* was a temptress whose declarations of religious orthodoxy were carefully calculated lies. The bright red binding cloth of the book in his hands was the cloak of the whore of Babylon:

> We readily attribute to it all the graces of the accomplished harlot. Her song is like the syren for its melody and attractive sweetness; she is clothed in scarlet, and every kind of fancy work of dress and ornament; her step is grace, and lightness and life; her laughter light, her very motion musical. But she is a foul and filthy thing, whose touch is taint; whose breath is contamination; whose look, and words, and thoughts, will turn the spring of purity to a pest, of truth to lies, of life to death, of love to loathing. Such is philosophy without the maiden gem of truth and singleness of purpose; divorced from the sacred and ennobling rule and discipline of faith. Without this, philosophy is a wanton and deformed adultress.[16]

Behind the attractive cotton and gold-stamped spine all the tendencies of the work were bad: no special providence, no miracles, the Bible a fable, and human beings no better than beasts. *Vestiges* displayed the immodest attractions of an urban prostitute, female corruption disguised as feminine modesty.

The book, Bosanquet warned, was "the very romance of philosophy," characteristic of the last times of national degeneracy when reason had lost its proper place as a servant of religion. Now, even fictional stories had to appear in the garb of science: innocent children, who in previous ages had read fairy tales, were force-fed with supposedly "amusing and instructive" scientific philosophy. As a philosophical romance, *Vestiges* had "the due foundation on fact, and finishing of fancy"; but its loose reasonings and vague analogies reflected "a poetry, if not puerility, of mind." This was an age of cheap books and hasty reading; whoever he or she might be, the author was no Newton.[17] From Bosanquet's perspective, *Vestiges* could only be read as a poisonous infidel romance.

Yet faith could be secure even when the book could be described as offering the private, forbidden pleasures of a novel. Reason could tame sensation. Take the case of Mary Smith, whose love of reading was so great as to set her apart from friends and family as "an incomprehensible being." For as long as she could

15. Bosanquet 1845, 45.

16. Ibid., 3–4.

17. Ibid., 3, 10–12, 31. In contrast, Bosanquet's own previous work (1843) had been entitled *Principia.*

remember, literary pursuits had been "the inner cravings of my soul."[18] Smith read *Vestiges* in 1850 at the age of twenty-seven, while working without pay as a schoolteacher in the north of England in Carlisle. The daughter of an Oxfordshire boot- and shoemaker, she was a devout Methodist and hence a Dissenter from the Church of England. She was "intensely anxious" to read *Vestiges,* and recalled it as "the book that most excited the wonder and curiosity of the reading world." It was, as she remembered, the most sensational book of its day.

> Calvinism was a sober truth with millions of people up till then. There were many of all denominations who lived daily in the fear of hell; and scepticism of the archfiend's personal power was then considered equal in its wickedness to the doubt of a Deity and a future state. Judge then the alarm and head-shaking this book was received with in the religious world. Many of them read it clandestinely, and then silently waited for the comments and criticisms of the press and pulpit.

Night was the only time she could find for reading. When a copy came as a loan to her employer, she secretly borrowed it and "sat up till after daybreak, finishing its interesting pages by the first light of the morning, at my bedroom window."[19]

As Mary Smith was writing several decades after the event, her account may allude to Robert Martineau's famous painting, "The Last Chapter" (fig. 1.1), which shows a woman engrossed in one of the "sensation novels" by authors such as Mary Braddon and Wilkie Collins that became popular in the early 1860s. The psychology of "sensation" in this context had reference not to the enlightened culture of sensibility but to the newly discovered reflex reactions of scientific physiology. Her autobiography's implicit reference to these gripping fictional page-turners re-created the attractions of *Vestiges* for a later generation.[20]

Mary Smith always worried about her attraction to powerful narratives and sensuous language. She shared the ambivalence of many evangelicals to made-up stories and as a young girl had disobeyed her mother to read halfpenny versions of "Cinderella" and "Jack the Giant Killer" in secret. As an adult she condemned herself for novel reading, but could not give it up.[21]

The only way to control such a passionate response to books was to master them. Smith never took notes, but instead read intensely to make their contents

18. Smith 1892a, 192; for brief mentions of this reading, see Flint 1993, 233 and Astore 1995.

19. Smith 1892a, 161–62.

20. There is a large and sophisticated literature dealing with the "sensation novels" of this later period. See, e.g., Brantlinger 1998, 142–65; Flint 1993, 274–93; Hughes 1980; Welsh 1985; and Winter 1998a, 322–31.

21. Smith 1892a, 27, 61.

1.1 Reading as compulsive self-absorption: Robert Martineau, "The Last Chapter" (1863). The intense glow of the coal fire contrasts with the calm gray light of the dawn.

"my own." As she stressed, "every page I read I earnestly endeavoured to make myself sure of understanding."[22] The rational faculties, directed by God, could guide her approach to the printed page. She illustrated the dangers of superficial reading through the case of her employer at Carlisle, John Jones Osborn, whose exposure to skeptical works of philosophy and science had sapped his religious principles. It was through evening discussions in his circle—a group apparently otherwise composed solely of men—that she learned about the *Vestiges* sensation, and it was of course his copy that she had borrowed. Several years earlier Osborn had lost his position as a Baptist preacher through reading works of this kind. As Smith lamented, despite having authored manuals of logic and grammar he was "very partially educated," lacking "that stability and strength required to build up a truly wise and good man."[23] More generally, she felt sure

22. Ibid., 94.

23. Ibid., 110–11, 159. A copy of Osborn's chart of logic, published in Carlisle in 1848, is in the British Library (Tab. 597.c.3[79]); his identity is confirmed in the Post Office directory for 1858.

that *Vestiges* had reduced the widespread belief in the divine inspiration of the Bible. She, however, remained strong in faith, preferring to find "truth in something more godlike than logic."[24]

Thus the rising sun by which Mary Smith finished reading was, for her, a divine gift of light. Her eyes had not been seduced:

> On myself and my mode of thought, this book, and its successors in the same field, effected little change. Its arguments were to me much harder to believe than the dear old truths of the Bible, and the divine doctrines of the New Testament. These latter revive and quicken and inspire the spirit of man, thus proving their truth, as the organ of vision proves the light of day. Like Thomas Carlyle, my own early life owed its best and brightest influences to the devout Calvinism under which it was reared. Religion, I think, has little to fear from scientific inquiry, or its endlessly changing theories of nature and man.[25]

The young teacher, who later ran her own school, was open to new ideas, especially those of Carlyle and Emerson, whose essays she read from within her own Calvinist tradition. She campaigned for women's rights and refused to marry, arguing from her experience that women had the strength of character required for independence. But even as she pored far into the night over philosophical and scientific books at the tiny desk by her window, Mary Smith was drawn to the chimes of Carlisle cathedral, which she recalled in a poem published later that summer:

> Like voices long unregarded,
> Till, in some dark sad hour,
> They're heard; ah! then we wonder
> At their beauty and their power.[26]

The inner light of spiritual truth was clear as God's sunlight, far brighter than shifting visions of modern science.

Anonymous Power

For many readers, the most arresting feature of *Vestiges* was the lack of an author on the title page (fig. 1.2). More than anything else, this rendered it a sensation. Here was a work dealing with the most profound questions of existence, apparently in command of a dozen different sciences, but written by an unknown author. In a commercial society with an expanding population, in which people passed on the street in large cities without knowing one another, anonymity

24. Ibid., 160.
25. Ibid., 161–63.
26. "On Hearing the Chimes of Carlisle Cathedral at Midnight," in Smith 1892b, 29–30.

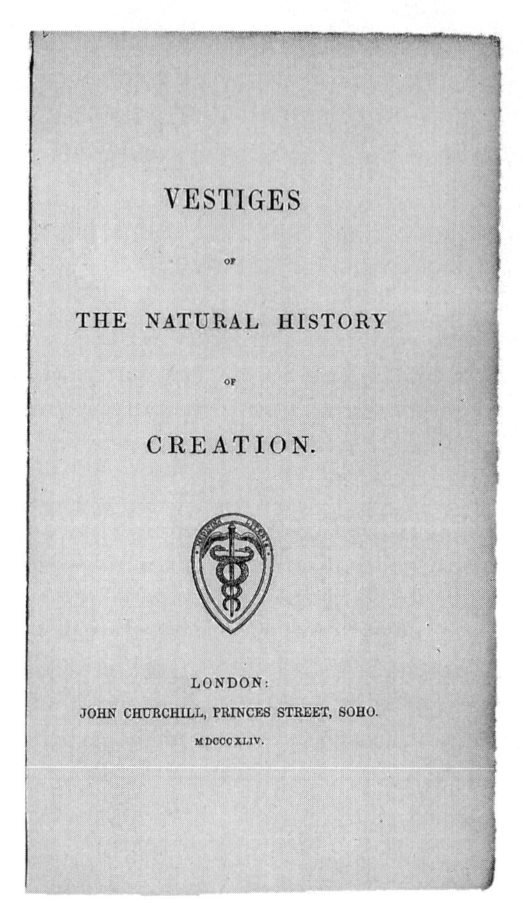

VESTIGES

OF

THE NATURAL HISTORY

OF

CREATION.

LONDON:

JOHN CHURCHILL, PRINCES STREET, SOHO.

MDCCCXLIV.

1.2 Title page of the first edition of *Vestiges of the Natural History of Creation* (London: John Churchill, 1844). Note the prominence of Churchill's name and device, which mark it as a medical or physiological work.

could raise anxieties about who might be pointing public debate in a potentially dangerous direction. The author's identity excited interest for months, in some quarters for decades. Speculations included reformers and reactionaries, women and men, aristocrats and working-class socialists, novelists, and celebrated naturalists. All the guesses had limited success. As one geologist wrote to an American friend, "A little volume of 390 pages, *anonymous* . . . has made a great sensation, chiefly I believe because the author cannot be detected."[27]

Nineteenth-century readers were, of course, far more familiar with anonymity than modern ones are. Almost all periodical journalism was anonymous, from the comic weekly *Punch* to the upmarket quarterlies, and many celebrated novels did not announce their author. *Vanity Fair, Mary Barton,* and *Yeast* (to name a few) were all unsigned; and *Jane Eyre, Adam Bede,* and *Wuthering Heights*

27. G. Mantell to B. Silliman, 28 Feb. 1845, Silliman Papers, Yale University Library, cited in Dean 1999, 209.

were issued under pseudonyms.[28] Famous poems, notably *In Memoriam,* also appeared anonymously. There were many reasons for avoiding identification. Women, including genteel ladies, did not want their literary reputations scrutinized too closely under the public eye; clerics, lawyers, or other professional men did not want to damage their prospects for advancement. An important class of political and theological works were anonymous, often to protect their authors from charges of heterodoxy. Anonymous periodical publication was widely defended as guaranteeing independence and freedom from personal bias; and even those, like the novelist Edward Bulwer-Lytton, who condemned the system ("anonymous power is irresponsible power"), did not extend their arguments to separately published books.[29]

Anonymity was so pervasive that most readers were little interested in cracking it and had no way to do so even if they were. Yet deep anonymity was unusual. Among those groups where knowing authors did matter—mainly among social and literary elites—books or articles that received any degree of celebrity were typically attributed within a few months. One Scottish newspaper could scarcely believe that the universal praise for *Vestiges* would not "drag the writer from his fancied obscurity into the brightness of the fame he has so nobly achieved."[30] To find so widely canvassed an unknown authorship, contemporaries had to look back to the early speculation about the identity of "the Author of Waverley," whose novels had begun to issue mysteriously from the press in 1814. Many names were proposed, although Sir Walter Scott quickly became the leading suspect. In fact the only close parallel was a full half-century before in the letters of "Junius," whose celebrated commentaries had rocked the eighteenth-century political world. "Since the days of Junius," one *Vestiges* reviewer noted, "few things have occurred to excite curiosity so much as the authorship of this extraordinary book."[31]

Anonymity was especially rare in history, biography, and science. The chief point of publication in science was to secure authorship of the facts of nature, so that anonymous scientific writings tended to be periodical essays and run-of-the mill textbook surveys. The implication was that unsigned works were unoriginal, part of the emerging genre of "popular science" that aimed to diffuse

28. Paradoxically, the *Wellesley Index* (Houghton et al. 1966–89), which identifies the authorship of thousands of otherwise unattributed articles, has made it possible to ignore the journals they originally appeared in and the context in which they were read: the greatest achievement of the study of Victorian periodicals has become a monument to the all-conquering power of the author. For stimulating pointers toward an alternative, see Beetham 1990 and Pykett 1990.

29. Bulwer-Lytton 1970, 239. The issues raised by anonymity are discussed more extensively in chapter 11.

30. *Scotch Reformers' Gazette,* 11 Jan. 1845, 4.

31. *Gardeners' Chronicle,* 4 Jan. 1845, 6–7, at 6.

known truths to the mass audience in useful knowledge tracts and newspapers. *Vestiges* failed to fit expectations. An anonymous book claiming conclusions at the highest theoretical level was a curiosity, and demanded an exceptional degree of trust from its readers. "Nothing," a reviewer wrote, ". . . can well be more out of the ordinary course of events than to find a writer of very extensive reading, high scientific attainments, and a perfect master of the arts of writing and reasoning, anxious to shroud himself in the most impenetrable mystery."[32]

The only other category of scientific works that appeared anonymously by convention were by the aristocracy, who might want knowledge of their authorship circulated only among a select few. Two names dominated gossip in fashionable society when the sensation was at its height: Ada, countess of Lovelace and Byron's only legitimate daughter; and Sir Richard Vyvyan, a leader of the opposition to the widening of the franchise in the 1832 Reform Bill. Both belonged to the hereditary aristocracy, which shows why the book was often read as emanating from the centers of metropolitan wealth and power. Both were strong possibilities, having written anonymously on the sciences before. In almost every other way, however, they could scarcely be more different, which shows the impossibility of tying *Vestiges* down to a single meaning.

About the only point on which most readers seemed to agree was that the book—despite its invocations of a deity—was too heterodox to have been written by a clergyman. The quality of the writing might be taken to indicate a journalist, novelist, or essayist. Some pointed to provincial authors of theologically liberal works, such as the young Francis Newman or Samuel Bailey of Sheffield, the author of *Essays on the Formation and Publication of Opinions and Other Subjects* (1821).[33] The eccentric, prolific Whig politician Henry Brougham was a common suspect.[34] Others pointed to the comic writer and journalist William Makepeace Thackeray, a Cambridge-educated man who had lost his family fortune. The author and political economist Harriet Martineau was certain that the phrenologist and botanical geographer Hewett Watson was the author. As she explained to a friend, Watson had just enough independent income not to depend on public favor. He was "safe in the respect, & satisfied in the love of his friends, & can brave (ie, disregard) the imputations of 'atheism' &c very comfortably."[35]

Martineau was herself a suspect, as was almost any other woman with scientific interests. Accusations of female authorship were used to undermine the work. For several months the Reverend Adam Sedgwick, a leading geologist,

32. Ibid.

33. *Literary Gazette,* 12 July 1845, 455; [Laing] 1846, 56.

34. For Brougham and Edward Bunbury, see Bunbury 1890–91, *Middle Life,* 1:37, 42 (journal for 19 Feb. and 5 Mar. 1845).

35. H. Martineau to H. C. Robinson, 24 June 1845, Dr. Williams's Library.

privately suspected that Ada Lovelace had written the "beastly book," which he condemned both in conversation on trips to London and in a widely read critique in the July 1845 number of the *Edinburgh Review.* Traces of feminine authorship could be found in the work's attractive style, popular appeal, and "ready boundings over the fences of the tree of knowledge." Most of all, it was "the sincerity of faith and love" with which the author adopted her chosen system.[36] It was on these grounds that Martineau was often pegged as the author. Her formidable reputation as a controversialist, mesmerist, and writer on political economy made her an obvious choice. Another common suggestion was Catherine Crowe, novelist and chronicler of the supernatural. Critics could attribute any weaknesses to the innate qualities of the female mind in such women: strong reasoning powers, but within a limited range. From this perspective, an impetuous longing after certainty made *Vestiges* just the sort of synthesis a woman might attempt.

Or perhaps *Vestiges* was written by a gentleman of science with wide-ranging interests. What about Andrew Crosse, a wealthy country squire famous for the insects that had emerged from his electrical experiments a few years earlier? These experiments played an important part in the book. Or how about Charles Babbage, the inventor of a calculating engine that also figured there? Other names put forward included those of Edward Forbes, the up-and-coming philosophical naturalist; Charles Lyell, author of the *Principles of Geology* (1830–33); and Charles Darwin, the invalid geologist and author of a round-the-world travel book.

Some people read *Vestiges* as the epitome of scientific expertise; others dismissed it as the product of a dilettante: it all depended on what one thought profound knowledge really was. Early in 1845, the most common suggestion of a recognized man of science was the Unitarian physiologist William Carpenter, who was known in aristocratic circles as tutor to Lord and Lady Lovelace's children. As the spring wore on, traces of dialect in the work began to be used to point to a Scottish voice, so that the moral philosopher Alexander Bain, the novelist Catherine Crowe, the phrenologist George Combe, and the astronomer John Pringle Nichol were sometimes suspected—usually on the basis of gossip from Edinburgh or Glasgow. Only after the first flush of interest in the book had subsided did suspicion begin to fall upon the Scottish journalist and publisher Robert Chambers, the cofounder of the largest mass-circulation publishing house in Britain.

The problem of anonymity is mentioned by everyone who writes about the period, but only to be forgotten. Text, book, and readers have been routinely collapsed into a single author, so that *Vestiges* becomes, paradoxically, "the anony-

36. [A. Sedgwick], "Natural History of Creation," *Edinburgh Review,* July 1845, 82:1–85, at 4.

1.3 [Horace Mayhew], "The Book That Goes A-begging," *Punch*, 11 Dec. 1847, 230.

mous work by Robert Chambers," and its meaning read off from his bourgeois sensibility, liberal politics, deistic religion, and status as a scientific outsider.[37] Such a facile equation was not available until the secret was formally revealed in 1884. Other than a handful of coconspirators, contemporary readers had to work hard to make such connections—and they could never be certain.

Three years after the book appeared, *Punch* could still make great play of the enigma. "The Book That Goes A-begging," in its number for 11 December 1847, sent up *Vestiges* as "'The Disowned' of literature," weeping unwanted at the entrance to a foundling hospital (fig. 1.3). No author would take it in. Lord Brougham (a standing *Punch* joke) had kept the title longer than most, but he threw it out, whence it ran about, "knocking at every scientific man's door," or calling on the "'Fast Man' of some light Review, or the editor of some heavy Quarterly." The presence of a strawberry leaf (the traditional mark on the coronet of a duke, a marquis, or an earl) would signal it as an aristocratic creation. Or perhaps Mr. Punch could be the author? The *Punch* staff were happy to

37. This problem is evident in all the previous literature on *Vestiges*, not least Secord 1989a, parts of which appear in a very different form as chapter 4 of this book.

shelter *Vestiges* for a while, as long as a "malevolent critic" did not repeat the attribution, to be copied into "every spiteful newspaper."

> Seriously, however, the destitution of this friendless little literary orphan is a most deserving case for the benevolent. We propose that a certain sum be subscribed in this wealthy metropolis, to pass it on to its own parish. But then again, there is this difficulty: which is its parish? for it does not know its father, and seemingly it never had a home. Heigho! we can only say that "It's a clever book that knows its own author!" Poor *Vestiges of Creation!* Hast thou no strawberry-leaf on thy frontispiece? no stain or blot about thee, by which thy parentage can be recognised? Unhappy foundling! Tied to every man's knocker, and taken in by nobody; thou shouldst go to Ireland!

Ireland was full of "kind fathers" who would be glad to take in a foundling (this was, to say the least, a feeble joke: during the later, terrible, stages of the potato famine, families received an extra allowance for every child they had to support). "It does not say much for the book," *Punch* joked, "or else the thing would have been claimed long ago, directly it had been known that the authorship of it was a profound mystery."[38]

Try as they might, contemporary readers could make no easy connection between author and text. The sense of "profound mystery" was fundamental to the experience of reading, and could pose acute moral dilemmas. Imagine you were a young gentlewoman who had ordered *Vestiges* from your bookseller on the strength of the early reviews. But what if the book was not lofty aristocratic philosophy, but a piece of infidel propaganda? The modest but handsome production, the declarations of faith, the range of knowledge: these might mask a sleazy hack or a crafty atheist. What if the anonymous author was not to be found in London's fashionable Pall Mall, but a few blocks away in the squalid dens of Holywell Street, notorious for atheism and pornography? What did you do when the Cambridge geology professor warned your brother-in-law that the volume sitting on your drawing room table was a "foul book," in which "[g]ross credulity and rank infidelity joined in unlawful marriage"?[39]

Guessing the author was not just a popular parlor game, but a desperate search with consequences for social cohesion and religious faith. Some sixty names were seriously proposed. The often entertaining controversy about the authorship demonstrates the power of readers as makers of meaning. Any conjecture expressed a reading of the work's politics, gender, religion, and expertise. The

38. [Horace Mayhew], "The Book That Goes A-begging," *Punch,* 11 Dec. 1847, 230.

39. A. Sedgwick to C. Lyell, 9 Apr. 1845, in Clark and Hughes 1890, 2:83. For Holywell Street, McCalman 1984; 1988, 217–18; and Winter 1993, 66–67, 214.

same is true for other once-anonymous or -pseudonymous Victorian works. Think how differently the novel *Mary Barton* (1848) was read when it was thought to be written by a male handloom weaver, rather than by Elizabeth Gaskell, a Unitarian minister's wife; or consider the effect of *In Memoriam* (1850) when read as the outpouring of a widow for her dead husband. Some suspected that "George Eliot" was a pseudonym for the famous comparative anatomist Richard Owen; this would have made *Scenes of Clerical Life* (1857) his first work of fiction.[40]

In the case of *Vestiges,* the failure to discover the author's name broke the usual bonds between narrative, narrator, author, and reader. Victorian readers expected a single, unequivocal meaning, which meant that the book could scarcely be "read" in the usual sense of the term. There was a fundamental difference between an attributed and an acknowledged work, and because the secret was kept so well and for so long, readings varied dramatically. Literary and aristocratic elites—who depended upon their personal knowledge of authors—could exert power over interpretation only with difficulty. Deep anonymity meant that *Vestiges* remained an "open" text for a very long time. As the ensuing chapters will suggest, different readings of the work were matched by different balances of probabilities attaching to the authorship. Names that seemed likely in Liverpool or Edinburgh were barely canvassed in Cambridge or Oxford; those that were common in London's fashionable West End were barely known in the Saint Giles rookeries only a few blocks away. The relationship that readers construct between the name on the title page and the rest of a work is always contingent. The radical uncertainty about *Vestiges* makes the case in an exceptionally powerful way.

The Industrial Revolution in Communication

"And what a sensation some books created!" Mary Smith's words are a reminder of how rapidly the sense of time and space in Britain was changing. The culture of sensation first emerged through the social and economic transformations in the wake of the French Revolution and the Napoleonic Wars. This was the age that saw the invention of the illustrated newspaper, the modern journalist, eye-catching street advertising, the international exhibition, and the paperback book. Thousands flocked to panoramic displays of the great events of the day: the Battle of the Nile, the fall of the Bourbons, the first railway journeys. Mary Smith was only one of the millions who traveled to London to see the Great Exhibition of 1851.[41]

The response to such events suggests that the meaning of economic trans-

40. Courtney 1908, 64.
41. Smith 1892a, 172–75.

THE RAILWAY JUGGERNAUT OF 1845.

1.4 [John Leech], "The Railway Juggernaut of 1845," *Punch*, 26 July 1845, 47.

formation was in perception as much as in output. The traditional picture of an industrial revolution focused on the late eighteenth century has been questioned in recent years. At least until the 1840s, factory production characterized only textiles, iron, and a few other sectors, and artisanal labor remained important even after parts of a process were mechanized. Industrialization is now seen as a long-term process focused on the restructuring of finance, labor practices, and patterns of consumption and extending back into the late seventeenth century.[42] Yet people who lived through the first half of the nineteenth century believed that they were witnessing unprecedented change. The consciousness of living in unique, historicized moments, which had come into being in response to the epochal events of the French Revolution and the Napoleonic Wars, was extended during the 1830s and 1840s into a sense that the material basis of human life was at a pivotal junction: the steam engine loomed over the intellec-

42. See Daunton 1995 and Price 1999 for reviews of the literature from this perspective, and Berg and Hudson 1992 for an important defense of the need to retain some notion of a dramatic economic transformation, especially in certain regions and sectors.

tual landscape as the chief symbol of a new age.[43] The machine dominated public debate partly because communication and transport were in the vanguard of the economic sectors undergoing industrialization. Perceptions of time became bound up with factory discipline and railway scheduling, as information traveled more quickly than ever before. New technologies, from steam presses to railway bookstalls, postage stamps and telegraph lines, were altering not only the terms of social interaction, but their fundamental forms. "Sensation" was one of the words that marked these changed perceptions.

Sensation in the individual body became wired up with sensation in the body politic; news seemed to be carried as quickly within the human frame as in the telegraph wires being strung up across the country. *Vestiges* suggested that sensation probably traveled as electrical impulses through the nervous system at the telegraph's transmission speed of 192,000 miles per second. "If mental action is electric," a footnote pointed out, "the proverbial quickness of thought—that is, the quickness of the transmission of sensation and will—may be presumed to have been brought to an exact measurement."[44]

At the time *Vestiges* appeared, the language of sensation was most consistently applied to the explosive growth of the railway network. The railways were compared to an alien "juggernaut," which turned the normally reserved English into blind worshippers of an all-conquering mechanical idol (fig. 1.4). The speculative "railway mania," which reached its peak in the mid- to late 1840s, became a focus for public debate about the effect of new forms of communication. The railways were the latest in a series of improvements in transport that had begun in the eighteenth century, from macadamized roads and canals to the high-speed horse carriage systems of the 1810s and 1820s. The first steam railway linked Manchester and Liverpool in 1830, and the network grew from about 500 miles in 1838, to 2,000 miles in 1844, to 7,500 miles in 1852. By the time *Vestiges* was published, the British Isles had effectively shrunk to between one-fifth and one-third of its former size, at a cost that made travel available to a much wider portion of the population (fig. 1.5).[45] Establishment of the rail network forever altered the relation between different parts of the country.

43. Chandler 1998 analyzes the emergence of a periodized historical consciousness; on the significance of the machine as an image of society, see Poovey 1995, 37–40; Berg 1980; and esp. [T. Carlyle], "Signs of the Times," *Edinburgh Review*, June 1829, 49:439–59, and J. S. Mill, "The Spirit of the Age," in Mill 1986, which appeared originally in 1831 as a series of articles by "A. B." in the *Examiner*.

44. *V1*, 335.

45. On the significance of the railways, see Freeman 1999, and also Daunton 1995, 285–317; Pollins 1971, 18–84; Freeman and Aldcroft 1985, 5–31.

1.5 The British railway system in 1845. For convenience, other places mentioned in this book are also included.

1.6 The sensation of reading on trains: "Lor dear! I've been and left 'The blood-stained Bandit', and all my 'Entertaining Knowledges' in the back Attic." *Punch's Pocket Book. For 1848.* . . . (London: Punch Office, 1847), facing p. 152.

Railways were progressive symbols of a technological age and the best contemporary example of a "sensation." William Wordsworth complained that rail travel disrupted the pastoral rhythms of country life, and others worried about the effects on the nervous frame of travelers. Reading on trains, which became a common practice in the 1840s, offered special cause for concern (fig. 1.6). It was feared that readers' emotions were being heightened by gripping narratives (*Vestiges* among them), at the same time that their bodies were being jostled and shaken in carriages packed with members of both sexes. The result could spell nervous collapse or the suspension of moral judgment.[46]

The Penny Post brought public sensation to doorsteps. Previously, sending letters had been expensive (unless you were lucky and had a friend in Parliament with franking privileges) and varied according to distance. The new system, proposed by Rowland Hill in 1837, set a prepaid flat rate for all letters under a certain weight, no matter what the distance. The experiment, which began

46. Schivelbusch 1986.

1.7 [William Newman], "The Post-Office Panic," *Punch,* 11 Oct. 1845, 159.

in January 1840 (stamps were issued in May), was a qualified success, although few among the working class used the system and it took until the mid-1850s to reach Hill's overoptimistic predictions. In the first two years, the number of chargeable letters increased from about 75 million to 196 million; and by 1849, the figure was 329 million. *Punch* joked about the Post Office "being suffocated, swamped, and smothered, by the enormous quantity of letters pouring in upon them at every chink, hole, corner, and cranny of the establishment" (fig. 1.7). Letter writing, previously a genteel art because of its high cost, had become quick, dependable, and cheap.[47]

The most significant changes in the mechanisms of sensation involved

47. "The Post-Office Panic," *Punch,* 11 Oct. 1845, 159. Vincent 1989, 32–52, argues that the main impact was in commerce and among the middle classes, and that the laboring poor began to take advantage of the Penny Post only later in the century. For those among the working classes with scientific interests the impact of the Penny Post was more immediately evident: A. Secord 1994b, 386–88.

printing and publishing. Innovations from paper-making machines to machine-stamped bindings, from the Penny Post to sales at railway bookstalls, transformed processes that had changed little since the sixteenth century. In terms of both output and innovation, the industry was of minor importance compared with textiles, iron, or the railways. But print culture, reporting on itself, occupied a central place in public awareness of industrial revolution.

The groundwork was laid for reading as a key part of modern mass culture. The steam-powered printing machine, machine-made paper, public libraries, cheap woodcuts, stereotyping, religious tracts, secular education, the postal system, telegraphy, and railway distribution played key parts in opening the floodgates to an increased reading public. The number of book titles published each year, as shown in fig. 1.8, reveals an impressive upward curve to the mid-1850s. Even more striking is the way that the price of books was declining. By the mid-1840s the number of medium-priced titles (3s. 7d. to 10s.) had already overtaken those at a higher price, and in the following decade the lead was taken by cheaper books (under 3s. 7d.). Like the other aspects of industrialization, these changes were not just technologically driven, but were part of wider developments—an explosion of the urban population, changes in the book trade, and a striking rise in literacy rates.[48]

Enterprising publishers and authors had begun to create what they ventured would become a mass readership. Edward Bulwer-Lytton deplored the effects in *England and the English* (1833):

> It is natural that writers should be ambitious of creating a sensation: a sensation is produced by gaining the ear, not of the few, but the many. . . . hence the profusion of amusing, familiar, and superficial writings. People complain of it, as if it were a proof of degeneracy in the knowledge of authors—it is a proof of the increased number of readers. The time is come when nobody will fit out a ship for the intellectual Columbus to discover new worlds, but when everybody will subscribe for his setting up a steam-boat between Calais and Dover.[49]

Glittering superficialities and familiar jocularities threatened original composition, just as steam was overtaking sail. The search for sensation in literature, as Bulwer-Lytton stressed, brought mechanization into the very texture of the language.

As in most economic sectors, the introduction of new technologies and new

48. Topham 2000 summarizes the debates on these issues. The figures are based on preliminary surveys in Eliot 1997–98; and S. Eliot 1994, esp. 59–65. For the effect of changing communication technologies in the American context, see Zboray 1993.

49. Bulwer-Lytton 1970, 294.

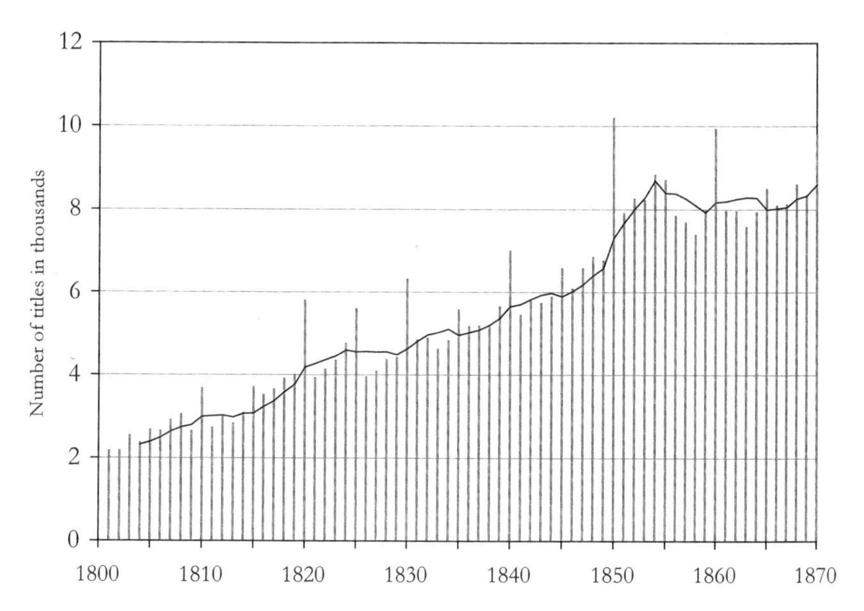

1.8 Number of titles per annum published in London, Oxford, Cambridge, Edinburgh, and Dublin, and listed in *Nineteenth-Century Short Title Catalogue,* 1801–70. Because the NSTC sorts undated items into the nearest year ending with a zero or a five, this graph has artificially high peaks at the start and in the middle of each decade. The superimposed line shows a five-year moving average. Based on Eliot 1997–98, 101.

forms of organization was gradual, so that books and newspapers continued to involve a combination of hand and machine work.[50] One of the early innovations had been the hand-operated iron-frame press, introduced at the start of the nineteenth century to replace the older wooden designs that had persisted almost unchanged since the introduction of printing. Although output was not much improved, it was exceptionally durable and could print small type with great delicacy. Costing about three times as much as the wooden models, iron-frame hand presses remained cheap enough to be bought by small proprietors, yet sturdy enough to withstand larger print runs. Artisan controlled, the hand press became a key symbol of the rights of the people (fig. 1.9); events such as the Queen Caroline affair, in which the prince regent's ill-treatment of his wife led to huge demonstrations, showed how public sensation could be used to further working-class radicalism. The freethinking publisher Richard Carlile, in his 1821 *Address to Men of Science,* proclaimed that the iron-frame press "has come

50. See Samuel 1977 and Berg 1985 on this characteristic combination, which as Samuel notes (7–16) should not be seen simply as involving the persistence of outdated hand labor processes in a machine age.

1.9 Dressed as a printer, a triumphant devil squashes the duke of Wellington, a bishop, and other characters in a hand press. From a parody of "The Devil's Walk" by Samuel Taylor Coleridge and Robert Southey. "The Printer's Devil's Walk," lithograph by Robert Seymour in *McLean's Monthly Sheet of Caricatures,* 1 May 1832.

like a true Messiah to emancipate the great family of mankind" from "kingly and priestly influence."[51]

The steam printing machine—large, fast, expensive—was a different matter. First introduced to break the power of the pressmen at the *Times* in 1814, it came into widespread use in the 1830s, after important patents had expired. Steam made publication possible on a massive scale and became especially important for the printing of newspapers. At the time *Vestiges* was published, the latest printing machines could print eight hundred sheets per hour, many times more than the fastest hand press. (Even then most specialized books continued to be hand-printed until midcentury.) The new technologies were ideal for entrepreneurs with large-scale financial backing, who could produce "cheap, amusing and instructive" publications for a penny or three halfpennies at a time. Radical publishers were in danger of having their wares drowned in the sea of useful knowledge pouring from the steam press factories.[52]

51. Reprinted in Simon 1972, 91–137, at 109; for the political significance of the press, see Hollis 1970 and Weiner 1969; for convenient surveys of developments in press technology, Feather 1988, 129–37, and Gaskell 1985, 198–200.

52. Hollis 1970; Curran, "Press History," in Curran and Seaton 1991, esp. 7–31, makes a cogent case for the effects of entrepreneurial control.

THE

PRINTING MACHINE:

A REVIEW FOR THE MANY.

No. 1.
SATURDAY,
FEBRUARY 15, 1834.
Price 4*d.*

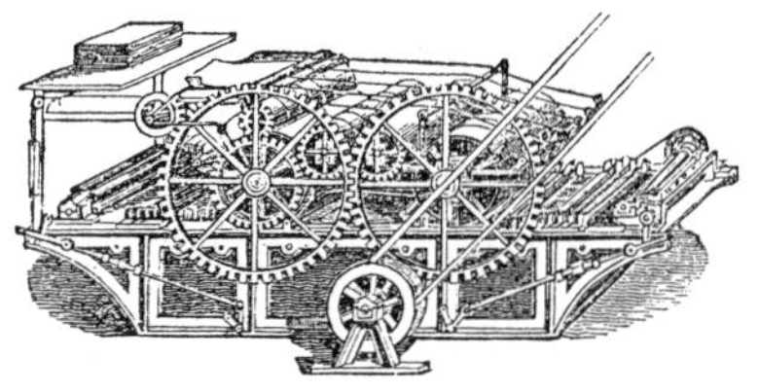

TO BE
CONTINUED
MONTHLY.

" What the PRINTING-PRESS did for the instruction of the masses in the fifteenth century, the PRINTING-MACHINE is doing in the nineteenth. Each represents an æra in the diffusion of knowledge; and each may be taken as a *symbol* of the intellectual character of the age of its employment."—*Penny Magazine.*

CONTENTS OF No. I.

1.10 Masthead for the first issue of Charles Knight's *The Printing Machine: A Review for the Many,* 15 Feb. 1834, 1. Note the illustration of a steam printing machine, which was used to print this journal as well as the *Penny Magazine* and other publications of the Society for the Diffusion of Useful Knowledge.

Publishing on an industrial scale thus proved effective in combating the radical hijacking of sensation. Cruder but more direct methods of censorship or taxation to keep political news from working people were breaking down anyway. In 1836 the newspaper tax fell from 4d. to 1d., thereby setting the stage for a 70 percent increase in press circulation over the next seven years. The number of dailies, weeklies, monthlies, and quarterlies reached new heights, creating the diversified topography of journalism that is so characteristic a feature of the Victorian period. The (ultimately successful) campaign of the late 1840s to remove the last of the "taxes on knowledge" argued for the need to create "a cheap press in the hands of men of good moral character, of respectability, and of capital."[53] Steam printing and railway distribution put publishing—though not reading—under the control of middle-class entrepreneurs. The chief symbol of the new order was embodied in the masthead for Charles Knight's monthly magazine, *The Printing Machine* (fig. 1.10). As one of the leading entrepreneurs of

53. Thomas Milner-Gibson, president of the Association for the Promotion of the Repeal of the Taxes on Knowledge, quoted in Curran and Seaton 1991, 29; the campaign in the 1840s and 1850s is outlined in Collet 1899, and its effects on restructuring the industry are sketched in Curran and Seaton 1991, 25–48.

print proclaimed, "Nothing, in our opinion, within the compass of British manufacturing industry, presents so stupendous a spectacle of moral power, working through inert mechanism, as that which is exhibited by the action of the steam-press."[54]

An Avalanche of Print

Today we tend to measure the impact of books by counting the number of editions and copies sold. However, the notion of a book as a "best-seller" came into common currency only a century ago, as one of the changes in the book trade at the start of the twentieth century. The best-seller emerged as part of a "middlebrow" literary culture dominated by mass-marketing and bourgeois consumption. Before that time, sales did of course matter to publishers and authors, but outside the trade they were less important than now. As one knowledgeable insider remarked, *Vestiges* "made a great sensation on its appearance, and several large editions were sold—two things which are not inseparable, for, as booksellers well know, a work may be praised in every newspaper, and discussed at every dinner-table, without having a great sale."[55] Taste in clothes, furniture, art, and books remained dominated by the aristocracy and urban gentry, so that in talking about a book too much stress on figures could be seen as vulgar. In that sense, no early Victorian books were best-sellers.

All the same, commerce did matter, and increasingly so. Multiple editions were one sign of continuing appeal, so that if a title was not selling, a publisher might issue it again (and again) with title pages announcing a "new" edition. Big sales, such as *Vestiges* was rumored to have, were evidence of "sensation." When Disraeli wrote to his sister in January 1845, he noted that the second edition was already sold out. There were two more editions that year, a total of ten after a decade, and fourteen in all during the nineteenth century. The publisher's account books show that first English edition had been only 750 copies, the second was 1,000, the third was 1,500, and the fourth was 2,000. By the end of 1860, 23,750 copies had been published in Britain, and by the end of the century the figure was just under 40,000.[56] These are large numbers. Publishers usually issued from 500 to 1,000 copies of new titles, and the great majority never went into a second edition. For some kinds of reading, however, they are not so striking. First editions of Dickens's novels regularly had print runs of ten thousand or more, as did Thomas Babington Macaulay's *History of England*. A

54. "Mechanism of Chambers's Journal," *CEJ*, 6 June 1835, 149–50, at 151.

55. Bertram 1893, 32. On middlebrow culture, see Radway 1997, 127–301, and Rubin 1992. This perspective underlies several of the classic studies of Victorian readers, notably Altick 1957 and Ellegård 1958.

56. For these figures, see chapter 4.

1.11 "Six O'Clock P.M.: The Newspaper Window at the General Post-Office." Sala 1859, 233.

book of advice or almanac might sell hundreds of thousands of copies but not qualify as a sensation.

Statistics, then, are not enough to explain why *Vestiges* was considered a sensation. We get a better picture when we recognize that knowing the number of editions of a book is only the starting point for understanding the spread of a work through advertisements, extracts, conversations, and notices in print. Then as now, if a book failed to make an impact in the first weeks after publication it was unlikely ever to do so. Even successful books were rarely fashionable for longer than a few months. And the pace of sensation was speeding up. During the 1830s and 1840s, quarterly periodicals such as the *Edinburgh, Quarterly,* and *Westminster* reviews—dominant in setting the literary agenda from the early 1800s through the 1820s—were supplanted as the most significant sites of debate by the monthlies and weeklies. "Magazine day," the first Monday of each month, became a major event on the publishing, bookselling, and Post Office calendar (fig. 1.11).

What would readers have found in the periodicals pouring out of the Post Office window toward the end of 1844? The reforming *Spectator* stressed the "power of popular exposition" in *Vestiges* and the "ingenuity" of its argument. The imperial *Atlas* had already praised the author's "extraordinary ability," "clearness of reasoning," and "the grandeur of the subjects of which he treats." It recognized, however, that the book would meet much opposition for its "wild hypothesis" that new species derived from existing ones. The leading medical weekly, the *Lancet,* hailed "a very remarkable book, calculated to make men think," valuable for revealing the connection between the different sciences. By the third week of November, the *Morning Chronicle*—the country's leading middle-class radical daily—tried to stop the swelling tide of acclaim. Because "some critics have expressed a much higher opinion," it feared the book was receiving unmerited celebrity.[57]

The excitement in the papers tied readings into the daily news. Take, for example, the debate about the repeal of the import tax on corn, probably the most vigorously fought political issue at the time the book appeared. A *Morning Chronicle* leading article—obviously written by someone more positively inclined toward *Vestiges* than its reviewer—brought out its relevance for the crisis of the industrial cities:

> It is no popular prejudice, that roast beef was at the bottom of our superiority over the French. Good food, and plenty of it, gives not only beautiful forms, but stout hearts, strong arms, and vigorous heads. How long the boasted superiority of England may continue, or how soon it may be numbered amongst those dreams of other times, if the legislature persist in narrowing the space whence supplies of food can be obtained, and enhancing the evils of a dense population, is not for us to conjecture; but we can safely say, according to the natural history we have quoted, that no system can be contrived more certain to destroy the supremacy of our beloved country . . . than to maintain a law which circumscribes the supply of food, and huddles the people together in comparative darkness and in stagnant air.[58]

These thoughts derived from a reading that made *Vestiges* relevant to ongoing parliamentary debates about whether nature's laws were opposed to import tariffs on corn. The book was not a political tract, the notice acknowledged, but had "a direct bearing on one at least of the most important legislative questions of the day."[59] In appealing directly to nature the book was all the more powerful as a political resource.

Vestiges also offered opportunities for the theatrical tactics of freethinking plebeian radicals. Even casual observers could not ignore the way in which the public sensation was being hijacked to further the cause of unbelief. Only a few minutes' walk from fashionable Bloomsbury, the windows of a freethought

57. *Spectator,* 9 Nov. 1844, 1072–73; *Atlas,* 2 Nov. 1844, 746; [E. Forbes], *Lancet* 2:265–66 (23 Nov. 1844); *Morning Chronicle,* 19 Nov. 1844, 3.

58. "Natural History and Legislation," *Morning Chronicle,* 1 Jan. 1845, 3.

59. Ibid.

bookshop had *Vestiges* prominently displayed.[60] A few weeks after Bosanquet's pamphlet went on sale, London was placarded with announcements that an atheist agitator with a prison record for blasphemy would be speaking on *Vestiges.* A year later, announcements promised an entire series of lectures on the book by the country's most notorious woman atheist.[61] The audience for such talks rarely exceeded one hundred, but handbills were plastered across the metropolis. Street advertising was, without question, vital to cementing the association of *Vestiges* with religious disbelief.

It was unprecedented for a book of science to attract so much attention. Most of the newspapers and periodicals that carried reviews sold in several thousand copies, each being read by several readers. The newspaper reviews that appeared in the first two months were gradually supplemented by those in the monthlies—especially the fast-growing religious press—in December and early January. There were discussions at public meetings in several of Britain's industrial and commercial cities, and consternation among gentlemanly men of science, alarmed to see a heterodox work advocating species evolution taking the public by storm. By then some of the more adventurous monthlies and quarterlies had published reviews, including supportive ones from parties of "advanced thought" in religion and medicine.

The prestigious quarterlies took longer. Two anonymous reviews in the summer of 1845, in the *Edinburgh Review* and the *North British Review,* attempted to puncture the sensation with a great show of authority. These reviews were widely hailed as the great refutations of *Vestiges,* but they also gave it more publicity and sales. There was an extensive correspondence in the *Times,* and further newspaper discussion. Only a handful of separately published pamphlets targeted the work, but dozens of scientific, theological, and literary works mentioned it in passing. As 1845 drew to a close, the still-unknown author replied to his critics in a small volume entitled *Explanations: A Sequel.* Although this had few of the literary attractions of the original, and sold only three thousand copies, it led to several dozen further reviews, which kept the sensation going. As a decade of debate drew to a close, virtually all the leading men of science had expressed an opinion, from the newspaperman and geologist Hugh Miller in his *Foot-prints of the Creator* (1849) to the master of Trinity College in Cambridge, William Whewell, in his *Indications of a Creator* (1845). *Vestiges* was dissected at public scientific meetings, condemned from pulpits and lecture platforms, borrowed from circulating libraries, and read.

Evolutionary theories became a common currency of conversation. Charac-

60. *Movement* 2 (29 Jan. 1845): 40; *Reasoner* 3 (7 July 1847): "The Utilitarian Record," 64.

61. For the lectures, *Movement,* 29 Jan. 1845; and *Reasoner* 1 (3 June 1846): 15; 1 (17 June 1846): 48; for the advertisement, *Reasoner* 3 (7 July 1847): "The Utilitarian Record," 64.

ters in fiction could be compared with *Vestiges,* as when one reader noted that the eponymous hero of *Sam Slick, the Clock-maker* was "reverting from any body he ever had into his primordial fire mist," a clear reference to the early history of creation in the evolutionary universe.[62] In Frank Smedley's comic novel *Harry Coverdale's Courtship* a half-drunken group of aristocrats and military men dine at a popular restaurant at Blackwall, consuming dishes in a developmental sequence that recapitulates the history of life as recounted in *Vestiges.* Fish, first served, transmute "into the higher forms of animal, into which the highest form of all—man—pitches cannibal-like, until the culinary cosmos is resolved into its pristine chaotic elements."[63] Judging from their constant appearance in letters and memoirs, such Vestigian analogies must have been ubiquitous in contemporary table talk.

Most books ceased to excite public comment after their first year. The annual social calendar continued to provide the most stable and significant chronology for literary fashion, which meant that most books lasted a single season.[64] Excitement over major deaths, battles, and discoveries typically lasted a few weeks or even days, and can best be traced in the newspapers. Sensations over books typically lasted longer, usually a few months; thus in this regard the decades-long *Vestiges* controversy proved an exception. Even in such a case, though, the most active phase of debate lasted for just twenty months, from October 1844 to June 1846. By the time Mary Smith had a chance to read *Vestiges* in Carlisle, it was no longer a sensation in London. Cheap editions had begun to appear from the English publisher in the spring of 1847, at a price that the middle classes could readily afford. These were rarely reviewed but were widely read on trains and in libraries. By this time *Vestiges* was no longer fashionable, but it was available to a far wider range of readers. The book had a substantial international impact too. It was translated twice into German and twice into Dutch,[65] and went through some twenty editions in the United States, where it sold more copies and had more readers than in Britain. A young lawyer and freethinker in Springfield, Illinois, Abraham Lincoln, read the work straight through—something he rarely did—and "became a warm advocate of the doctrine."[66]

62. J. Brown to J. T. Brown, [10 Feb. 1846], in Brown and Forrest 1907, 63.

63. Smedley 1855, 316. Further references are discussed below and in Millhauser 1959, 152–60.

64. Dawson 1979 and esp. Stein 1987 analyze the importance of the annual cycle of events as part of the "spirit of the age" in their discussions of the events of 1850 and 1837–38, respectively; in both cases the primary focus is on well-known works of literature.

65. The translations are listed in Chambers 1994, [220–21]; Rupke 2000 analyses the very different aims with which they were undertaken.

66. Herndon and Weik 1893, 2:147–48. There is no comprehensive study of the American reception of *Vestiges,* although most histories of science and religion in the United States discuss it briefly; see, e.g., Roberts 1988. Some relevant sources are in Millhauser 1959 and in Chambers 1994, [220, 222–29]; the latter lists most of the English-language reviews.

The Work of Sensation

Walter White, assistant secretary of the Royal Society of London and a former cabinet maker, described *Vestiges* as the greatest sensation since *Waverley*.[67] The book opened up possibilities for table talk; it could be employed in the early stages of courtship and to explore the possibilities of friendship; it could comfort the sick, or serve as a present; it provided an endless fund of party jokes; and it allowed specialists to display their knowledge without appearing pedantic. In a period when science was increasingly seen as a masculine arena and a highly technical one, *Vestiges* crossed boundaries of gender and expertise. No wonder it was so avidly discussed.

The encounters touched on so far juxtapose widely diverse sources—from autobiographies to private letters, religious pamphlets, newspapers, and reviews. They involve different localities, reading practices, and sorts of people. What they have in common is involvement in a sensation of a kind that was both novel and characteristic of the early industrial era. It was effectively impossible, only a few weeks after *Vestiges* appeared, to comment on it without being aware of sharing an experience with a wider national and even international community of readers. The title, as one American monthly noted in 1846, was "familiar to every reader."[68] *Vestiges* was the one book that all readers of the *Origin of Species* were assumed to have read. Sensation in the individual body, transmitted to others through the mechanisms of industrial communication, was seen to have electrified the body politic.

Victorian readers developed a variety of ways to report sensation; these are usually either highly individualized or involve "influences" that affect society en masse. Even today, it is easy to assume that the term "reading" applies best either to isolated individuals confronting words on a page, or to an undifferentiated public. Anecdotal reading experiences are juxtaposed against statistics of production and numbers of copies sold. To repeat these forms, as accounts of reading almost invariably do, is simply to reproduce the contradictions of the sensationalized social body of the nineteenth century.

How is it, then, that books exercise their power? What makes them, in Elizabeth Barrett's terms, devils or angels? Each of the present book's four parts tackles this question from a different perspective. The subsequent chapters of part one show how literary celebrity depended upon the combined action of readers, publishers, authors, and printers. The evolutionary narrative of *Vestiges*

67. Diary entry for 16 Oct. 1845 in White 1898, 65. White also wrote a letter, now lost, about the reception of *Vestiges* to his former employer Robert Chambers; see RC to AI, [Feb. 1845], NLS Dep. 341/110/175–76.

68. J. J. A[llen], *Christian Examiner and Religious Miscellany* (May 1840), 40:333–49, at 333.

emerged not from the margins of society, as is often assumed, but from the heart of the new industrial order, in distinctive formats targeted at a variety of readerships. Part two explores these readerships in relation to perceptions of place. Urban geography remained among the central parameters of intellectual life, so that the book was read in different ways in Edinburgh, Oxford, Cambridge, Liverpool, and aristocratic London. Reading in relation to individual identity is the subject of part three. The warfare between evangelicals and atheists brings this issue forward in its starkest form, which is then pursued in two cases: an eighteen-year-old apprentice, who kept amazingly detailed diaries, and the author, acting behind a mask of anonymity. The fourth part turns to readers, from leisured gentlemen to commercial hacks, who used *Vestiges* in debating the future role of the scientific practitioner. The emergence of paid scientific professionals, in conjunction with wider changes in journalism and publishing, led to the repackaging of progressive evolution as "Darwinism" after 1859. The foundations for a liberal polity had been laid in the preceding two decades: in new forms of communication, new kinds of urban spaces, new senses of individual identity, and new roles for expertise.

The next chapter begins to move us beyond the Victorian self-diagnosis of sensation. In reading a work, the first question to be decided is the genre to which it belongs. Assumptions about genre assist readers, publishers, and authors in creating stable conventions for interpretation. Locating a work within a particular genre makes some aspects appear central and others trivial. Achieving agreement about such issues was far from easy. The realm of the sciences, and of literature, theology, and politics, was hotly debated. To label *Vestiges* "the very romance of philosophy" could be praise for its novelistic merits or an accusation that it was infidel fiction flying under the false colors of fact. What kind of book was *Vestiges?*

CHAPTER TWO

Steam Reading

> Science had made great acquisitions, and it seemed desirable, if only for experiment sake, to see what kind of FRANKENSTEIN would result from the architectural union of her scattered limbs.
>
> *Atlas,* 20 December 1845

THE SPECTER OF FRANKENSTEIN haunted the *Vestiges* sensation. The book was a generic monster, the progeny of all the literary experiments that made reading so exciting. The early Victorians lived in an age when boundaries of genre and discipline were slippery, changing, and contested. A single work could transport readers between what seem to be distinct literary forms—epic, satire, sermon, romance, political tract, moral philosophy, natural history. New genres were created, too, ranging from the scientific monograph to the comic annual. Whatever else it was, *Vestiges* was a hybrid. As one critic said, the author could "create worlds with a dash of his wizard pen . . . animalise the dull lump of inorganic matter—and spiritualise, like another Frankenstein, the animal to which his fancy had given birth."[1]

Certain genres provided the context for *Vestiges,* especially the new kinds of systematic treatises, encyclopedic works, and introductions to knowledge that blossomed in the decade around 1830. These works have usually been discussed in connection with specific disciplines, such as induction in philosophy, uniformity in geology, the problem of "systems" in natural history. But contemporaries read them as contributing to debates about a vision of nature appropriate to the emerging order of the machine. They belong to new genres of literary production driven as much by publishing finances as by campaigns by men of science for authority over the emerging mass readership. Scientific systematizing emerged in a publishing climate engendered by reform agitation and industrialization.

1. "Vestiges of the Natural History of Creation," *Wade's London Review* 1 (1845): 382–94, at 383.

The Millennium of Useful Knowledge

The transformation of communication was implicated in a crisis of representation: an age of reflection, inward examination, national self-awareness. "What, for example," the *Edinburgh Review* commented in 1831, "is all this that we hear, for the last generation or two, about the Improvement of the Age, the Spirit of the Age, Destruction of Prejudice, Progress of the Species, and the March of Intellect, but an unhealthy state of self-sentience, self-survey; the precursor and prognostic of still worse health?"[2] From William Hazlitt to Robert Southey, from Henry Brougham to John Stuart Mill, authors diagnosed the "condition of England" and debated how to characterize the kaleidoscopic changes. These treatises were symptomatic of the novel idea that an "era" might have "characteristics," that the "age" might have a "spirit."[3]

During the second quarter of the nineteenth century, the controversies of the postrevolutionary period were transmuted into a national debate about the coming of an industrial society. As progressive forms of knowledge, political economy, geology, and nebular astronomy were at the heart of the response to the changes that were sweeping away the old order. The principles of the natural sciences could define the meaning of progress (fig. 2.1). The great debate was how to do this: and in line with contemporary political disputes, the problem was how to forge a viable program of reform after the French Revolution, especially as so many analytical methods and institutional forms in the sciences were Continental imports.[4]

The aim was to survey the principles of a science by introducing its history and defining its boundaries. A key work was the six-volume "supplement" to the fourth edition of the *Encyclopædia Britannica*, published in Edinburgh by "the Napoleon of the realms of print," Archibald Constable. This serves as a useful marker of how the encyclopedic enterprise had abandoned its unifying aims, as other Enlightenment genres—particularly the philosophic treatise in the mold of Adam Smith's *Wealth of Nations* (1776)—became the template for new genres of reflective works on the sciences. The concept of a general treatise prefacing a detailed account had been prominent in France during the late eighteenth and early nineteenth centuries. Notable works included the mathematician Pierre Simon Laplace on the evolution of the solar system, the naturalist Jean Baptiste

2. [T. Carlyle], "Characteristics," *Edinburgh Review* 54 (Dec. 1831): 351–83, at 365.

3. Chandler 1998 discusses the issues in the context of recent critical debates about literary historicism. For further evidence of the importance of "the glare of the present," see Altick 1991, 5–49.

4. Morrell and Thackray 1981, 2–34, Desmond 1989, Cooter 1984, Cannon 1978, Morus, Schaffer, and Secord 1992, and Yeo 1993, 28–48 survey some of the main controversies.

2.1 Henry De la Beche, "The Light of Science Dispelling the Darkness which Covered the World." NHM, General Library, Sherborn folder of broadsides.

Lamarck on the transformation of living beings, and the comparative anatomist Georges Cuvier on the revolutions of the earth.[5]

New notions of authorship accompanied the new genres. Original work in science depended on the credibility of the authors in a way that fiction and poetry did not. Discovery was not a democratic process, available to all through skill or practice; insights into nature came suddenly and to select individuals, those whom the chemist Humphry Davy had called "the sons of genius." Great men embodied in themselves the age's scientific spirit. As one author rhapsodized, "the mooned loveliness and divinity of Nature reveals itself only to the rapt dreamer upon lofty and remote places."[6] Publishers, more prosaically, encouraged a heroic role for the man of knowledge because familiar names sold books. Scientific portraits enjoyed brisk commercial sales (fig. 2.2).[7] While any gifted individual, whatever their background, could in principle contribute to science, the apotheosis of the author tended to imply that the production of knowledge should be kept out of the hands of all but a few "manly intellects" who could devote themselves to science as a vocation.[8]

5. Yeo 1991; for such works in France, see Outram 1984.

6. Bulwer-Lytton 1970, 329. For Davy, see Morus, Schaffer, and Secord 1992, 141.

7. Browne 1997 and Shortland 1996b discuss scientific portraiture; Prescott 1985 discusses its sale and collection.

8. On heroic authorship and the culture of print, see Kernan 1989, 91–117, which discusses the pivotal career of Samuel Johnson; Wilson 1989 penetratingly analyzes American authors in the late

2.2 The man of science as heroic author. This portrait of the prolific physiologist and textbook writer William Benjamin Carpenter was one of a series of about fifty lithographs drawn by Thomas H. Maguire for the Ipswich meeting of the British Association for the Advancement of Science and published in 1851.

Advocates of the new conception of science presented their case in a variety of ways. The British Association for the Advancement of Science, created in 1831, with its alphabetical sections for mathematical physics, geology, natural history, and so forth, represented an encyclopedic model of science in action. Created on a provincial impulse, it was soon led by the same coteries that controlled the metropolitan specialist societies like the Geological (f. 1807) and Astronomical (f. 1820) societies. The heroes of science gave presidential addresses and reports to further the ideals of specialist research. They also lectured, wrote in the quarterlies, and authored some of the key works of reflective science.[9]

According to contemporary stereotypes women were unsuited to scientific authorship.[10] A lithograph (fig. 2.3) shows just how unfeminine "a lady of scientific habits" was seen to be. A small *Craniology* volume tops her head, which is broadly based on a (French) *Encyclopédie;* her dress is *Pantologia,* a reference punning male trousers and the visionary schemes of utopian pantisocracy, and alluding to *Pantologia* (1808–13), a twelve-volume encyclopedia. Her feet are bound as Walker's *Tracts.* This woman is an author, as she holds a scroll under one arm ("Armstrong *On Slavery*") and quills under the other ("Handle, *Army*

eighteenth and the first half of the nineteenth centuries. On the vocation of science, see Outram 1984. For discussion of this vocation specifically in Britain, see Morrell and Thackray 1981; Porter 1978a; and Yeo 1993, 34–36. As shown in the previous chapter, Bulwer-Lytton diagnosed a decline of science and blamed it on an increase in the number of readers; [1833] 1970, 291–321.

9. Morrell and Thackray 1981. Disputes about leadership and the division of labor within science are discussed in Alborn 1996.

10. Self-presentations and literary styles for women were adapted to a notion of the writer as following a professional vocation: Benjamin 1991; Poovey 1984; Shteir 1996; Shteir 1997, 244–52.

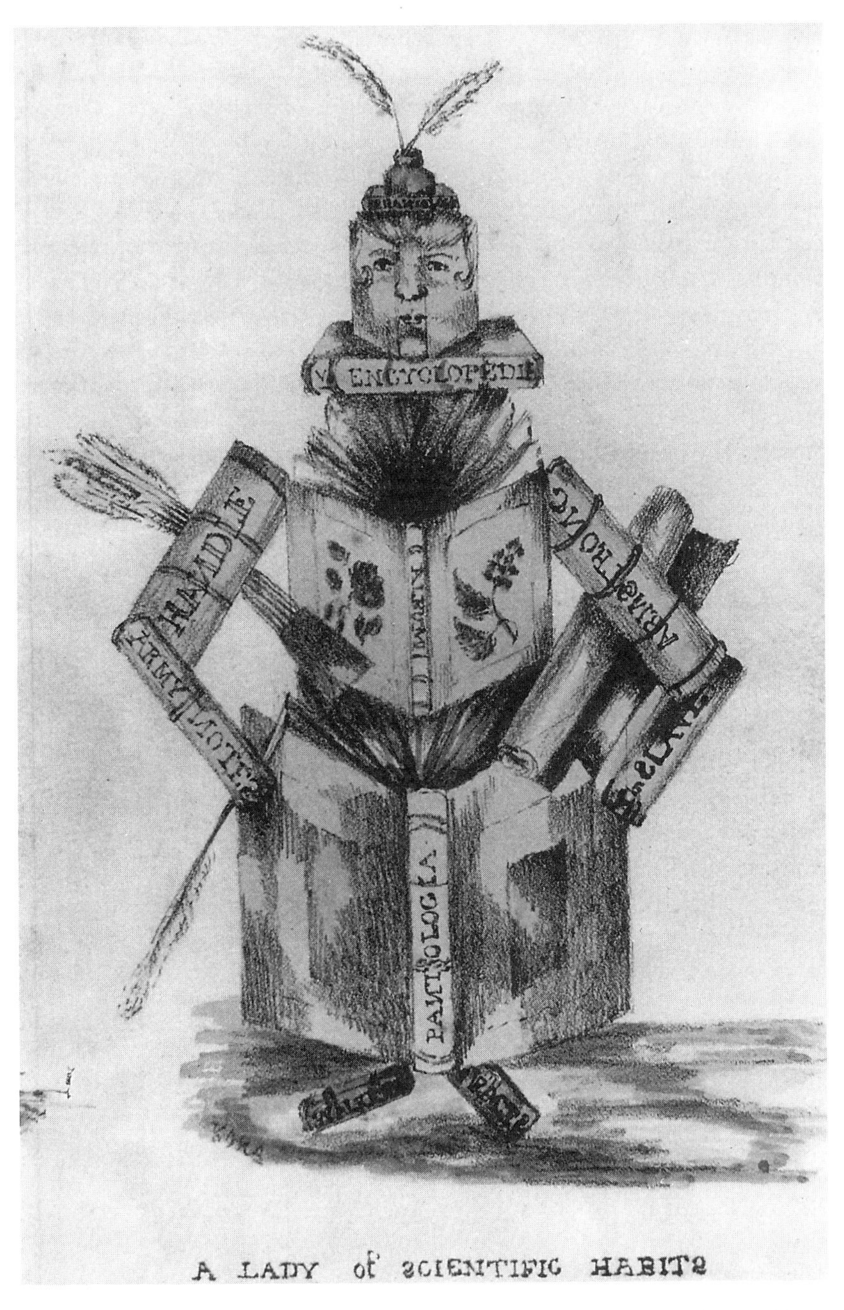

2.3 “A Lady of Scientific Habits.” Lithograph, early nineteenth century. The tradition of constructing scholars from their books goes back to Giuseppe Archimboldo, painter to the court of Rudolf II at Prague in the sixteenth century.

Notes"). Her garb of useful knowledge renders her deeply unfashionable; moreover, made of books, she has no body at all, and can give birth only to more books. The associations of knowledge with the masculine, the foreign, the rational, and the controversial could scarcely be clearer.

Women were caricatured, not necessarily because they were being edged out, but because their participation in cultural life was pervasive. The notion that science should be dominated by a clerisy developed as a strategic response to the increasing heterogeneity of the reading audience, both female and male. Without some select body of interpreters, Samuel Taylor Coleridge had feared the "*plebification*" of knowledge and the collapse of society.[11] Books needed to be appropriate for a public no longer limited to the genteel and commercial classes that had consumed natural philosophical systems during the eighteenth century. The wealthy London astronomer John Herschel agreed: without suitable reading matter the artisan population would be "dull boys," and as he recognized, "a community of 'dull boys' in this sense, is only another word for a society of ignorant, headlong, and ferocious men."[12]

An Army of Lilliputians

The readership for reflective surveys of knowledge expanded rapidly. As the London bibliophile Thomas Frognall Dibdin wrote in 1832, sales of ponderous quartos and classical folios were in the doldrums, but "there were 'brisk doings' below stairs," as smaller formats dominated the market (fig. 2.4).[13] The possibilities were canvassed in the early 1820s, most loudly in the *Edinburgh Review* and in *Practical Observations upon the Education of the People,* a pamphlet written by Henry Brougham that ran through twenty editions in 1825. The pamphlet sketched a utopian vision in which inexpensive readings on science, issued in parts at a few shillings, became available in even the poorest cottage.[14] In the same year, Archibald Constable announced plans for cheap editions of

11. "You begin, therefore, with the attempt to *popularize* science: but you will only effect its *plebification.*" Coleridge, *On the Constitution of Church and State* (1830), in Coleridge 1976, 69. Klancher 1987, 5–6, 164–70, and Schaffer 1991, 212–14, discuss Coleridge's clerisy as a response to changing reading audiences.

12. Herschel, "An Address to the Subscribers to the Windsor and Eton Public Library and Reading Room, Delivered . . . on Tuesday, Jan. 29, 1833," in Herschel 1857, 1–20, at 9. For fears of mass literacy, see Brantlinger 1998, Webb 1955, and the extensive literature on mechanics' institutes. For discussion of Herschel's address, see Altick 1957, 96, 105, 135, 139.

13. [Dibdin] 1832, 18–19.

14. [Brougham] 1825. Until 1971 the British used a nondecimal system based on the penny (abbreviated d.), shilling (s.), and pound (£). Twelve pence were worth one shilling, and twenty shillings were worth one pound. The guinea (worth twenty-one shillings) was not minted after 1813, but the term continued to be used for costly books and other luxury goods. Changes in relative values make it impossible to translate Victorian monetary values into modern ones. Housing,

2.4 "The March of Literature or the Rival Mag's." A dust cart driver and a chimney sweep ignore their duties while reading the *Penny Magazine* and *Saturday Magazine,* two of the leading penny weeklies of the 1830s. Behind them, ragged newspaper vendors hawk cheap publications that purport to supplant the classic authors they trample underfoot. Lithograph published 25 Oct. 1832 by Thomas McLean.

history, biography, and science. As he reportedly said, a monthly series of works at three shillings or half a crown "must and shall sell, not by thousands or tens of thousands, but by hundreds of thousands—ay, by millions!" A few months later Constable was bankrupt and living in squalor, a victim of financial mistakes and the general depression of trade.[15] The 1825–26 crash is often seen as a watershed, although few firms went under and the number of titles published soon recovered.[16]

---

for example, seems cheap by modern standards, while books appear exceptionally expensive. At 7s. 6d. the first edition of *Vestiges* cost one-third of the weekly income of a semiskilled print worker.

15. Lockhart 1837–38, 7:126–30, and for Constable's fate, Sutherland 1995, 298. As Sutherland (275–76) points out, there are reasons for being skeptical about the details of this drunken speech at Abbotsford in 1825.

16. The revisionist view of the 1826 "crash" was suggested in Sutherland 1987 and largely confirmed in S. Eliot 1994, 16–18.

Yet the crisis brought forward new trends important for the sciences. The paucity of well-known novelists between Sir Walter Scott and Charles Dickens is only partly a consequence of critical obsessions with the canon; it also reflects the difficulty of selling untried titles in the decade after 1826. Publishers became more cautious, less willing to take risks on new fiction and poetry. New novels were published as three-deckers priced at an extortionate thirty-one shillings and six pence. Literary publishers focused on reprints of fiction and poetry. Their key author was Scott, whose hugely successful "Magnum Opus" of 1829–33 used a five-shilling small octavo format—a handy, pocket size—which became standard for inexpensive book production.[17]

The most striking development was the rise of cheap nonfiction series, in which many of the new scientific works appeared. Constable's utopian vision was revived on a reduced scale as a miscellany, which began publication in 1827 in shilling numbers every week (whole volumes could be purchased for three shillings and six pence). The series did not reach so vast a readership as Constable had projected; as Knight later reflected, "The millions were not ready to buy such books at a shilling, nor even at six-pence."[18] But it did reveal an untapped market, resulting in a remarkable series of publishing experiments during the next decade (fig. 2.5). The Society for the Diffusion of Useful Knowledge (SDUK, f. 1826) forwarded Brougham's scheme to enlighten a wider range of readers through a Library of Useful Knowledge published by Baldwin and Cradock. The aim was, in part, to undermine political radicalism with rational information—as some working-class readers reportedly complained, "to stop our mouths with *kangaroos.*"[19] A six-penny number of the SDUK's Library of Useful Knowledge appeared every two weeks, in double columns of small type, with the aim of covering all the sciences. Brougham's captivating introduction to the series had sold 33,100 copies by the end of 1829, proclaiming that science "elevates the faculties above low pursuits, purifies and refines the passions, and helps our reason to assuage their violence."[20] A few of the other titles sold well, but Brougham's utopian hopes were not borne out. Cramming an amazing quantity of print into a small space, the treatises were often forbiddingly tech-

17. Millgate 1987 and Sutherland 1995, 329–33, on the "Magnum Opus." The relations between the publishing market and the novel have been extensively debated: see Altick 1957, 262–64; Feltes 1986; and Sutherland 1976. Although crudely economistic and limited to discussion of "literary" works, Erickson 1995 offers a stimulating account of the fortunes of various genres (poetry, fiction, the essay) in relation to trade fluctuations, changing business practices, and industrialization; there are clearly ways in which a similar analysis can be applied to the sciences.

18. Knight 1854, 243; Altick 1957, 268–69.

19. Quoted in Shapin and Barnes 1977, 56.

20. [Brougham], 1827, 2. Webb 1955, 69; Topham 1992, 413–19, on series publishing in useful knowledge.

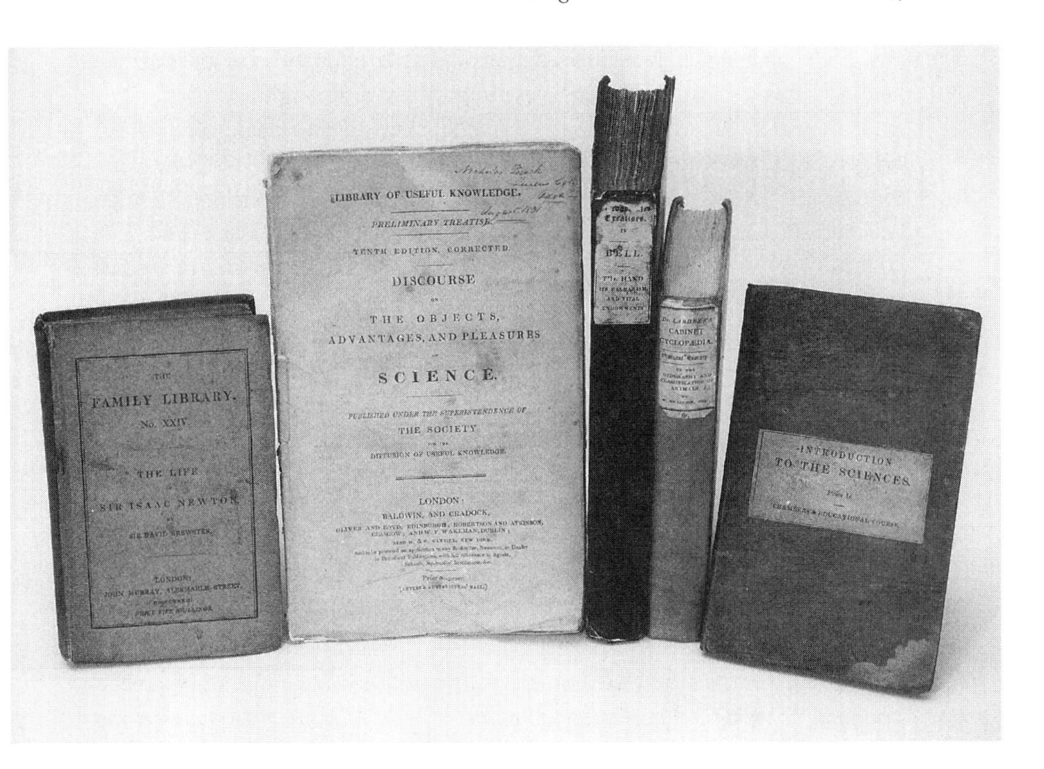

2.5 The binding of useful knowledge. The 1820s and 1830s were a period of great experimentation in publishers' bindings, as the old practice of providing books in boards was replaced by a variety of wrappers and cloth casings. Bindings were especially important in identifying books as part of a series. From left to right, a cloth-cased volume from Murray's Family Library (David Brewster, *The Life of Sir Isaac Newton* [London: John Murray, 1833]); a number in its original wrappers from the Library of Useful Knowledge ([Henry Brougham], *The Objects, Advantages, and Pleasures of Science* [London: Baldwin and Cradock, 1827]); a larger-format, cloth-cased Bridgewater Treatise (Charles Bell, *The Hand: Its Mechanism and Vital Endowments as Evincing Design* [London: William Pickering, 1834]); a volume from Lardner's Cabinet Cyclopedia (William Swainson, *A Treatise on the Geography and Classification of Animals* [London: Longman, 1835]); and from Chambers's Educational Course ([Robert Chambers], *Introduction to the Sciences* [Edinburgh: William and Robert Chambers, 1838]).

nical, stopping short only of calculus: an inexpensive form was at odds with the expectation of genteel learning. To counter this, Knight—also on behalf of the SDUK—began to issue the Library of Entertaining Knowledge at four shillings each, or six-penny weekly numbers. Knight went to great lengths to ensure that this series was more readable.[21]

21. Knight 1864, 2:151–54.

SDUK campaigns were noisy, public, and programmatic. The propaganda reflected a belief that the spread of useful knowledge among the working-class poor required philanthropic support analogous to that available through the Society for Promoting Christian Knowledge (f. 1698) and the British and Foreign Bible Society (f. 1804). Outside the realm of religious publishing, though, reliance on charitable donations and local committees proved a dead end. The SDUK's successes were due almost entirely to Knight's enthusiasm. Individual entrepreneurs—not just new men like Knight but gentlemanly publishers in the Regency mold—produced the innovations that became permanent features of science publishing.[22]

Those who wished to counter Whig domination of cheap publishing moved to meet the challenges of the new audience. John Murray of London's fashionable West End, publisher of the Tory *Quarterly,* issued fifty-three volumes of original nonfiction between 1829 and 1834 in a small octavo format. The series, called the Family Library, included David Brewster's *Life of Isaac Newton* (1833) and *Letters on Natural Magic* (1832). These works need to be seen not just as part of the author's intellectual trajectory, but within a commercial program of publishing. That is how the public saw them and how they were advertised and sold. Brewster's *Newton* rested on the same shelf as biographies of Napoleon and Nelson; *Natural Magic* was a follow-up to Scott's *Letters on Demonology and Witchcraft.* Even after Murray ceased commissioning for the Family Library, he continued to issue books in the same format. These included Humphry Davy's posthumous *Consolations in Travel* (1830), Mary Somerville's *On the Connexion of the Physical Sciences* (1834), and later editions of Charles Lyell's *Principles of Geology* (1834–40).[23] Touring the London booksellers, Dibdin recognized Murray as "the greatest 'FAMILY MAN' in Europe . . . surrounded by an extensive circle of *little ones.*"[24]

Murray's aim in adopting this form was counterrevolutionary, to bridge widening class divides.[25] Other publishers soon tackled the same market through the Edinburgh Cabinet Library, the Naturalist's Library, the Library of Sacred Literature, and similar series. Longman, the largest firm in the trade, issued the 133-volume Cabinet Cyclopædia edited by the Irishman Dionysius Lardner, including "preliminary discourses" on natural philosophy and natural history. These were prefatory to works on astronomy, botany, zoology, geology, optics, probability, and so forth that made up the rest of the set. The series format

22. Bennett 1982.

23. Secord 1997, xiv–xv, xxvii–xxviii, on the format of the *Principles.*

24. [Dibdin] 1832, 31. On the Family Library and its finances, see Bennett 1976.

25. Bennett 1976, 141.

meant that individual treatises could be bought for six shillings; purchasers need not be tied—as they had been with Longman's previous encyclopedia—to purchasing forty-five expensive quartos. As Dibdin put it, "A whole army of Lilliputians, headed by Dr. Lardner, was making glorious progress in the Republic of Literature."[26]

Even publishers skeptical about the virtues of cheapness were caught up in the changes. William Pickering, known for exquisitely printed volumes of literature, picked up the rights to the Bridgewater Treatises on "the Power, Wisdom, and Goodness of God, as manifested in the Creation," perhaps the most celebrated series of reflective treatises on science. The eccentric earl of Bridgewater had left £8,000 in his will for a work on natural theology, which a Royal Society committee distributed among eight specialist authors, each receiving the princely sum of £1,000 plus profits. Pickering expected the series to sell badly. He issued only 1,000 copies of the early titles, no more than required by the bequest, hand printed with fine illustrations, large margins, and widely spaced lines of type between which "the Earl of Bridgewater might almost have driven his cab!"[27] Only gradually did Pickering realize that the series "bid fair to traverse the whole civilized portion of the globe."[28]

At the height of the Reform agitation from 1828 to 1832, the cheaper format threatened to take over entirely. Murray predicted that in a few years "scarcely any other description of books will be published."[29] Expectations of success are revealed in ambitious print runs and razor-thin profit margins. For example, Murray printed 12,500 copies of Brewster's *Newton;* of these, just over half were sold and the rest had to be remaindered at a huge loss to Thomas Tegg, remaindering jackal of the Regency book trade.[30] Sales of the later numbers of the Library of Useful Knowledge also proved disappointing, and most appear to have been bought not by working people but by the middle classes. The same also appears to have been true of the SDUK's other productions. In that sense, Knight achieved only limited success in using cheap factory-based printing to replace the "foul trash" spewing forth from the pauper presses.

26. [Dibdin] 1832, 18; Peckham 1951 has further details about the Cyclopædia, and Hays 1981 about Lardner's career; Sheets-Pyenson 1981c and Altick 1957, 273–75, discuss other cheap series of the period.

27. The *Medico-Chirurgical Review* in 1835, quoted in Topham 1998, 245, which includes (pp. 237–49) a full account of the production of the Bridgewaters and their place in the publishing world of the 1830s. For more about the series, see Gillispie 1951, 209–16; Robson 1990; and esp. Topham 1993.

28. T. F. Dibdin in 1836, quoted in Topham 1998, 244.

29. J. Murray to C. Knight, Apr. 1829, in Smiles 1891, 2:296.

30. Bennett 1976, 161, 164.

The People and the March of Mind

Controversies about the future shape of the polity were embodied in the physical size and shape of printed materials. Many works reflected on the processes that had made their own production possible. At one end of the scale, the *Orchidaceæ of Mexico and Guatemala* (1837–43), among the largest books ever published, featured a caricature of itself being raised from the ground (fig. 2.6). A gentleman in a topcoat holds a speaking trumpet and directs a gang of working men, while (printer's) devils dance on the sidelines. Even works available for the libraries of the wealthy (the complete volume cost sixteen guineas, a year's wages for a servant) could exhibit an awareness of the labor relations involved in their production.

Cheap books explained how they had been produced at such a low price, and how that price could be even cheaper. The mathematician Charles Babbage's *On the Economy of Machinery and Manufactures* (1832) reflected on the role of machinery in industrial processes by referring to "objects of easy access to the reader."[31] Among these processes was book making, and Babbage asked readers to contemplate the physical object in their hands. Much to the consternation of booksellers, among whom such matters were trade secrets, the *Economy of Machinery* provided a full breakdown of costs for paper, printing, binding, payments to the author, and so forth. Babbage wished to show that the book's price had been determined not by production costs, but by taxes on knowledge and the restrictive practices of the book trade. The latter drove up "the price of the very pages which are now communicating information respecting it."[32] In citing pounds and pence, and allowing his book to "speak" directly to the public, Babbage was uniquely explicit about these issues—so much so that he had to switch publishers at the last moment.[33]

Babbage's concern with the political economy of print was not unusual, however, and his book typified the new reflective works. It was priced at six shillings, about half of what a similar text would have cost a decade earlier. The format was a small octavo, printed on large sheets that could go through steam presses. It had a case binding, which was coming into common use as a means for providing an attractive cloth cover for those who could not afford rebinding in calf. The work was issued in large numbers, with a first edition of three thousand

31. Preface to the second edition, included in the fourth edition, which I have used: Babbage 1835, vi; on Babbage's admission of his inability as an author to realize an ideal state of efficiency, see Poovey 1995, 39–40.

32. Babbage 1835, 314.

33. Ibid., vi–xi, 205–10, 314–33; Hyman 1984, 117–18, 203–4, briefly discusses this episode and its aftermath, as well as Babbage's long-standing interest in printing, an issue closely related to his plans for his calculating engines.

2.6 "The Librarian's Nightmare." Woodcut from a drawing by George Cruikshank, in J. Bateman, *The Orchidaceæ of Mexico and Guatemala* (London: R. Ackermann, 1837–43), 8.

copies selling out within two months, and three further editions soon following. This success, Babbage claimed, was the product of public demand for information about "the pursuits and interests of that portion of the people which has recently acquired so large an accession of political influence."[34] The publisher was Knight, rapidly gaining a reputation as a leading exponent of mass publishing. The physical character of the *Economy of Machinery* was integral to its message. Like Babbage's celebrated calculating engine,[35] the book was a demonstration device in political economy.

The intimate relation between cheap production and useful knowledge was hard to miss, even in books less insistent on their mode of production. Herschel's *A Preliminary Discourse on the Study of Natural Philosophy* was commissioned, written, printed, advertised, and stocked in shops as part of an encyclopedic series, the enterprise of Lardner and Longman. Spine, half title, and two of its three title pages give as much prominence to the Cabinet Cyclopædia as to the author (fig. 2.7). Purchasers were also made aware of cost: the six-shilling price could be seen from the paper label on the spine, and the size and cloth case binding identified it as a product of the latest innovations in machine

34. Babbage 1835, vi.

35. Ashworth 1996; Schaffer 1994, 1996.

THE

CABINET CYCLOPÆDIA.

CONDUCTED BY THE

REV. DIONYSIUS LARDNER, LL.D. F.R.S. L.&E.

M.R.I.A. F.R.S.Ast. F.L.S. F.Z.S. Hon. F.C.P.S &c. &c.

ASSISTED BY

EMINENT LITERARY AND SCIENTIFIC MEN

Natural Philosophy.

A PRELIMINARY DISCOURSE

ON

THE STUDY OF NATURAL PHILOSOPHY

BY

J. F. W. HERSCHEL, ESQ. M.A.

OF ST. JOHN'S COLLEGE, CAMBRIDGE.

LONDON:

PRINTED FOR

LONGMAN, REES, ORME, BROWN, GREEN, & LONGMAN,

PATERNOSTER-ROW;

AND JOHN TAYLOR,

UPPER GOWER STREET.

1832.

THE

CABINET

OF

NATURAL PHILOSOPHY.

CONDUCTED BY THE

REV. DIONYSIUS LARDNER, LL.D. F.R.S. L.&E

M.R.I.A. F.R.Ast.S. F.L.S. F.Z.S. Hon. F.C.P.S. &c. &c.

ASSISTED BY

EMINENT LITERARY AND SCIENTIFIC MEN.

A PRELIMINARY DISCOURSE

ON

THE STUDY OF NATURAL PHILOSOPHY.

BY

J. F. W. HERSCHEL, ESQ. M.A.

OF ST. JOHN'S COLLEGE, CAMBRIDGE.

LONDON:

PRINTED FOR

LONGMAN, REES, ORME, BROWN, GREEN, & LONGMAN,

PATERNOSTER-ROW;

AND JOHN TAYLOR,

UPPER GOWER STREET.

1832.

production. Opening the covers, the first thing the reader saw was a sixteen-page Longman's catalog, advertising other Cyclopædia volumes. The paper, as most readers knew, was made from old rags, one of the by-products of the transformation of the textile industry. The *Preliminary Discourse,* widely praised as the most high-minded scientific treatise ever published, thus bore the marks of industrial progress as clearly as a railway journey or a yard of factory-produced cotton.

Different formats carved out different markets. Five- or six-shilling reprints or series books, imbued with the aesthetics of the machine and the factory age, became ways of establishing a definition of the "people" broadly congruent with the beneficiaries of the 1832 Reform Act. But just as with political reform, such treatises were not exclusive to the middle class. Wealthy readers could have them rebound for genteel libraries; and self-improving artisans and clerks could consult them in mechanics' institute collections and pub libraries, or after making their way to street stalls. Longman or Murray advertised their respective series as suitable for families in the country who wished to build up a collection of books at a reasonable price; for emigrants who wished to take a library to the colonies or America; and for public institutions of education and learning.

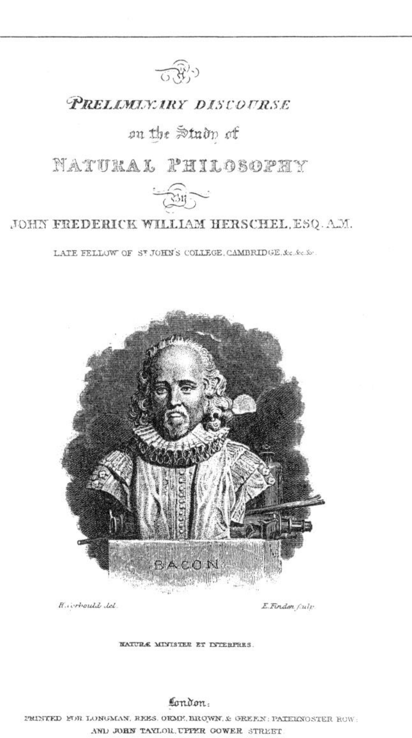
PRELIMINARY DISCOURSE
on the Study of
NATURAL PHILOSOPHY
By
JOHN FREDERICK WILLIAM HERSCHEL, ESQ. A.M.
LATE FELLOW OF ST JOHN'S COLLEGE, CAMBRIDGE, &c. &c. &c.

BACON

H. Corbould del. E. Finden sculp.

NATURÆ MINISTER ET INTERPRES.

London:
PRINTED FOR LONGMAN, REES, ORME, BROWN, & GREEN, PATERNOSTER ROW;
AND JOHN TAYLOR, UPPER GOWER STREET.
1830.

2.7 Each volume of Lardner's Cabinet Cyclopædia had three title pages, depending on whether the book was to be used on its own or as a volume in a comprehensive series. The engraved title page often included a portrait (in this case Francis Bacon), thereby underlining the genealogy of heroic authorship. In contrast, the spine label and the first two title pages both stress the editor and publisher more than the author. John Herschel, *A Preliminary Discourse on the Study of Natural Philosophy* (London: Longman, 1830).

Individual volumes were given as presents or as school prizes. The largest market was among the commercial bourgeoisie, so much so that purchasing books of this kind became a distinguishing feature of middle-class life.

Between Heaven and Earth

What held the sciences together in these new circumstances of production? Answering this question presented acute problems for authors who were carving up nature into new specialties and new genres. It was all very well to focus on strata hunting or stellar mapping within the meeting rooms of the Geological or Astronomical society; but this in no way met demands among those who bought five-shilling treatises. Such audiences demanded general concepts and simple laws. Defining these audiences and creating appropriate forms for addressing them were central to the making of the new sciences as authoritative, specialized activities in the first place. How was this to be done?

There was no agreed answer to this question. What an introductory treatise on science ought to look like was open to very different views. Price, format, length, illustrations, religious orientation, and demands on previous knowledge all varied dramatically. Such issues had a vital practical, pedagogical dimension

2.8 The geological record as a series of books. [Rennie] 1828.

that has often been ignored. Philosophical systems in science gained prominence because of a market-led demand for synthesis.

The new genres of reflective science were unified by a common debate about progress. Progress had been central to the encyclopedic tradition, but had become associated with materialism and human perfectibility, so that it needed to be used with care in presenting the specialist sciences. Geology, the newest and most controversial of the sciences, became identified with the progressive history of the earth and of life (fig. 2.8). Davy's *Consolations in Travel* offered dialogues on the meaning of science for metaphysics and religion. In the third dialogue, the "Unknown," a stranger, describes his vision of the geological past. It is a story of progress, beginning with the earth "in the first state in which the imagination can venture to consider it," cooling from original fluidity to become habitable. Tropical animals and plants of simple character are succeeded by shells, fish, reptiles, mammals, and finally human beings. In the early ages "there was no order of events similar to the present," for the crust was thin and the central fire close to the surface. Only gradually, as the planet cooled, did the world approximate its modern state.[36]

Similar progressive narratives provided the organizing principle for a wide range of geological works. Pickering sold ten thousand copies of William Buckland's Bridgewater Treatise, *Geology and Mineralogy* (1836), even though the book had nearly ninety plates and cost £1 15s.[37] Gideon Mantell's *Wonders*

36. Davy 1830, 133–37.

37. Topham 1993, table 5.1; Topham 1998, 249–61; Rupke 1983, 18–20.

of Geology (1838), based on lectures in fashionable Brighton, offered a cheaper alternative, as did works by commercial hacks. Mary Shelley proposed an elementary work on the science to Murray in the early 1830s, although she was unsure "how far such a history would be amusing."[38]

Progress became central to geology—although not to its practice in the field or museum—partly because it provided a narrative of a scope sufficient to counter literal readings of the Genesis story. The vast majority of the public continued to believe that the Creation, the Fall, and the Flood were defining moments in the physical history of the world. If geologists were to change this, some compelling account would have to take their place. Their findings challenged some interpretations of the Flood and the Creation, but could offer instead a divinely directed story of progress, preparing the earth for humans. Scripture and science were never locked in inherent conflict; had they been, introductions to geology would have been consigned to the gutter press. Rather, geologists (many of whom were clerics) wished to create a space for a science that was in danger of being reabsorbed into theological exegesis. Most often, they used a progressive sequence of "lost worlds" to fill the "gap" that was said to occur after the first verse of Genesis. On this reading, the six days referred only to the last creation, not the earlier ones studied by geologists.[39]

Progress was also invoked in astronomy. In 1833 Whewell's Bridgewater Treatise, *Astronomy and General Physics,* coined the term "the nebular hypothesis." This combined the suggestion of John Herschel's father, William Herschel, that nebulae might be new sidereal systems or stars being born, with the Laplacian theory of the formation of the solar system (fig. 2.9). It was the astronomer and political economist John Pringle Nichol's *Views of the Architecture of the Heavens,* however, that made the evolving nebulae into the symbol of astronomy as a progressive science. Issued in 1837 by the radical middle-class Edinburgh publisher William Tait, this work described in vivid language the evolution of the universe and the formation of galaxies and stars. It sold three thousand copies in its first two editions, with three further ones from the same publisher in the next decade.[40] Nichol nervously justified his project in a letter to John Herschel. He claimed to write something akin to a novel:

> Good introductions to Astronomical science are abundant, but it often occurred to me that the *moral results* of Astronomy might be made quite apparent to many minds, & quietly introduced among their modes of thought, by treatises, not

38. M. Shelley to J. Murray, 3 Sept. 1830, in Bennett 1980–88, 2:115.

39. These debates are discussed in Bowler 1976, Corsi 1988, Rudwick 1976 and 1986, Rupke 1983, and Yule 1976.

40. Nichol 1839, viii. The first five editions (1837–45) were followed by the eighth edition of 1850 from John W. Parker; no sixth or seventh edition was ever issued.

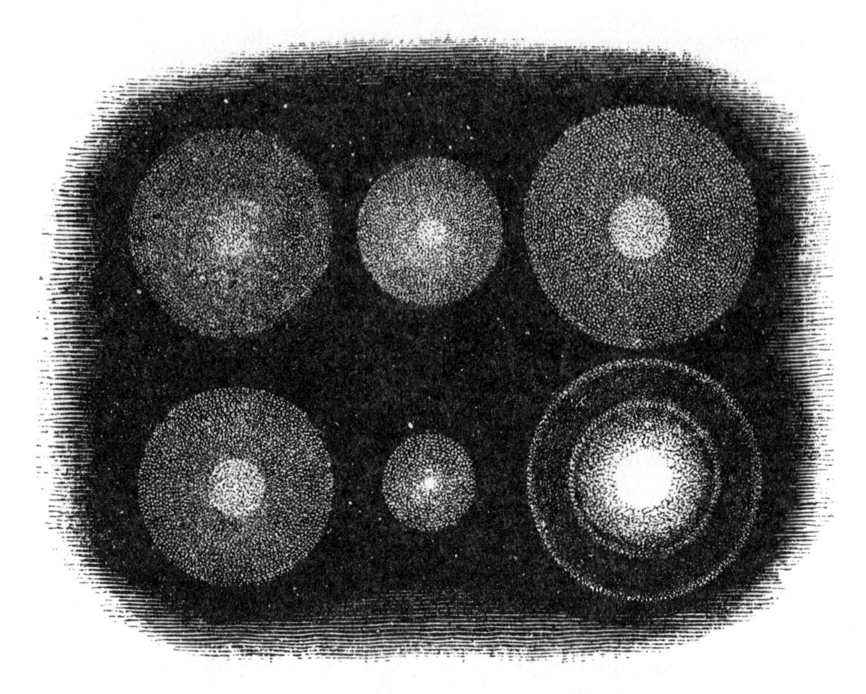

2.9 Progressive states of nebular condensation. From Mantell 1839, 1:24.

> without the shew of science, but without its form—having in that something of the form of Romance: and having time at my disposal I composed these works. . . . Under the guise of science I have endeavoured to teach what I esteem good bread upon the waters, & I doubt not what is good of it will not be lost.[41]

The *Architecture* presented itself as a paper instrument, with numerous fine plates "substituting for the want of powerful telescopes" and offering readers "something of the emotion" felt by expert astronomers. The nebular hypothesis showed that progress was written in "splendid hieroglyphics" across the sky.[42]

Nichol's nebular enthusiasm emerged from the context of radical middle-class journalism, at first in northeastern Scotland and then in London. He spent the 1830s writing on political economy for *Tait's Edinburgh Magazine* and John Stuart Mill's *London Review.* The nebular hypothesis proved an ideal way for Nichol to advance his fortunes as an author and lecturer; shortly before the *Architecture* appeared, he was appointed professor of practical astronomy in the University of Glasgow.[43] The nebular hypothesis became a staple of reflective and elementary books, many of them by Nichol himself. He advertised the

41. Nichol to J. Herschel, 4 Nov. 1838, RSL, Herschel Letters, vol. 13, f. 131.

42. Nichol 1837, vii, viii, 206–7.

43. Schaffer 1989.

Architecture as the first installment of a grand cosmological work on "the mechanism of Nature" in six volumes. The *Architecture* would come between volumes on the solar system and the earth, with three on life and the human race to follow.[44]

This plan required careful handling. Fears of association with Enlightenment cosmologies tempered enthusiasm for progress as a structuring device, for a narrative backed up by material causation could be read as making humans no more than better beasts. Moreover, a strong narrative line put scientific exposition uncomfortably close to novel writing. The emotional engagement of readers might lead to a suspension of judgment and unlicensed speculation. Introductory science books stressed the divine direction of their story lines, and that progression had nothing to do with the evolution of species or the origins of the human race. This was an important problem for the Bridgewater authors, who were not being paid to cast brute matter as the hero of romance. Only Buckland organized parts of his text around progress in nature, and it is not surprising that his treatise raised the most criticism.[45]

Bridgewater authors who discussed astronomy faced the dilemma that nebular condensation had been enlisted by Laplace in the service of a godless mechanical universe. However, they successfully purified the nebulae into signs of divine action. As Whewell said:

> Let it be supposed that the point to which this hypothesis leads us, is the ultimate point of physical science: that the farthest glimpse we can obtain of the material universe by our natural faculties, shows it to us occupied by a boundless abyss of luminous matter: still we ask, how space came to be thus occupied, how matter came to be thus luminous? If we establish by physical proofs, that the first fact which can be traced in the history of the world, is that "there was light;" we shall still be led, even by our natural reason, to suppose that before this could occur, "God said, let there be light."[46]

Whewell drew these conclusions for the sake of argument; if Laplace was right about the nebulae, it was important that his authority not be used to deny a first cause. Discussing astronomical progress in his Bridgewater volume on animal instinct, the Reverend William Kirby also expressed surprise that Laplace had failed to acknowledge that the nebular hypothesis required "an intelligent paramount central Being."[47]

In other publishing projects, there were sharp differences about whether

44. Nichol 1838, ix.
45. Rupke 1983, 209–18; Topham 1998, 249–61.
46. Whewell 1833, 191.
47. Kirby 1835, 1:xx.

progress applied to nature. Herschel, for example, denied that nebular hypotheses provided any basis for understanding astronomy. His Cabinet Cyclopædia volume had taken a more cautious view of his father's speculations on the formation of the universe than even Whewell had done. The nebulae furnished "an inexhaustible field of speculation and conjecture," but it was easy to become "bewildered and lost."[48] Herschel privately condemned the *Architecture* as an attention-grabbing potboiler, which led him to take "a very humble opinion" of Nichol's astronomical abilities. The book was bombastic and overwrought, "a deliberate attempt to *burst the English language by inflation.*"[49]

Readers should not despair, Herschel stressed, in reaching the limits of valid induction. The *Preliminary Discourse* compared the lessons to be learned from analyzing the mechanism of the heavens with those to be gained from analyzing a printing machine. Both could be understood without agreement on the source of power. During factory tours with Babbage and their friends, Herschel had seen presses in action, presumably including those of Andrew Spottiswoode, who printed the Cabinet Cyclopædia:

> We might frequent printing-houses, and form a theory of printing, and having worked our way up to the point where the mechanical action commenced (the boiler of the steam-engine), and verified it by taking to pieces, and putting together again, the train of wheels and the presses, and by sound theoretical examination of all the transfers of motion from one part to another; we should, at length, pronounce our theory good, and declare that we understood printing thoroughly.

Readers might even go on to apply the theory in building new machines for other purposes. Yet "not without a show of reason," various hypotheses to explain the engine's action might be proposed: on the evidence of its heat, breathing noises, and need for fuel, the boiler might be "the den of some powerful unknown animal." Doubt as to the correct explanation, however, should not lessen our sense of knowledge, for we could understand a process even though its ultimate cause remained uncertain.[50]

The *Preliminary Discourse* located progress within human understanding itself, so that the reader's emotions were mobilized in admiration of scientific genius. This view was widely shared by those who wished to transform natural philosophy by importing Continental analytical mathematics. As Mary Somerville wrote, the physical sciences were united "by the common bond of analysis, which is daily extending its empire, and will ultimately embrace almost

48. Herschel 1833, 406.

49. J. F. W. Herschel to W. H. Smyth, 20 Feb. 1846, RSL, Herschel Letters, vol. 22, f. 268; see Hoskin 1987 and Schaffer 1989, 136–38.

50. Herschel 1830, 192–95.

every subject in nature in its formulae." Mathematical formulas were not susceptible to common understanding, which meant that what Somerville called the "connexion of the physical sciences" could be comprehended only by the few. There was a wide distinction, she stressed, "between the degree of mathematical acquirement necessary for making discoveries, and that which is requisite for understanding what others have done."[51] The same approach is evident in the astronomy volume of the Library of Useful Knowledge, which despite being packed with trigonometry did not claim to be "of a purely scientific character."[52]

The geologist Charles Lyell based his challenge to progress on an analogous concern for the foundations of knowledge. He compared the experience of reading Lamarck's evolutionary views to the pleasures of light fiction, and his *Principles of Geology* argued that the science would advance only when geologists stopped assuming that the earth had a progressive history. Instead, progress should center on the human observer; drawing on Dugald Stewart and Scottish common sense philosophy, Lyell argued for a reform of the earth sciences that depended on the witnessing of visible causes. Pursued in this way, geology would join the other sciences in exemplifying the progressive character not of nature, but of human reason.[53]

The Laws of Life

Progress dominated the image of geology and astronomy because of its potential for addressing a broad readership. Books on physiology, anatomy, and the forms of plants and animals also had to tie together the sprawling mass of encyclopedic knowledge. Such works were usually written by medical men, who took advantage of commercial demand to supplement their incomes. Even more than works by geologists and astronomers, treatises by medical men were often based on lectures; some served both as student texts and as reflective works for a general audience. The science of "life" or vitality became the most widely used scheme for structuring works of this kind. The central concept, adapted from French naturalists and especially from Étienne Geoffroy Saint-Hilaire, emphasized the unity of composition of all living beings from the lowliest invertebrates to humans.[54]

The most prolific contributor to this genre was the Whig physician Peter Mark Roget, one of the founders of the SDUK. Roget, later famous for his thesaurus, contributed to numerous encyclopedias. He wrote on electricity and magnetism for the Library of Useful Knowledge, authored book-length

51. Somerville 1834, 5.
52. [Malkin] [1829], 2.
53. Secord 1997; Bartholomew 1973; Laudan 1982.
54. Desmond 1989; Jacyna 1983, 1984; Rehbock 1983.

accounts of physiology for the *Encyclopædia Britannica,* negotiated with Lardner about two Cabinet Cyclopædia volumes (but pulled out of the project because of copyright problems), and accepted the irresistible offer of a thousand pounds to write a Bridgewater Treatise. The latter work showed just how effective philosophical anatomy could be for a reflective treatise on the laws of life in relation to natural theology.[55]

Doctrines of the unity of type were also marketed in less expensive works. John W. Parker published Perceval Lord's *Popular Physiology* (1834) in a series sponsored by the Society for Promoting Christian Knowledge. Using his lecture notes from Edinburgh, Lord reported the findings of the Continental anatomists such as Friedrich Tiedemann, Étienne Serres, and Geoffroy, but in a Christian (and strongly antiphrenological) context. At seven shillings and sixpence for a small octavo, Lord's *Physiology* cost less than a quarter of the price of Roget's Bridgewater, and was explicitly "adapted for general readers."[56]

Many of these new works drew on the lectures of Robert Edmond Grant, lecturer on philosophical anatomy first in Edinburgh and from 1828 in London. Grant's approach was based on study of common plans in the style of Geoffroy, and although an evolutionary descent was a consequence of such views they were not the main focus of his research. Grant believed that the new anatomy belonged in the hands of a trained elite and campaigned passionately for reform in training and research. Fellow gentlemanly naturalists and geologists usually identified Grant not as a radical or a religious skeptic but as the nation's leading expert on sponges.[57] Grant's priorities were clear in his detailed and highly professional lecture course at the University of London. The course was not published separately in English, although Hippolyte Baillière began to issue Grant's *Outlines of Comparative Anatomy* in parts in 1841. However, the *Outlines* was never completed: it was too dry, too abstruse, too unwilling to address any but aspiring anatomical professionals.[58]

The Edinburgh medical lecturer John Fletcher, like Roget, was more concerned to sell science to a broad audience, to replace the "inflammatory political trash" from the radical press and show reactionaries that physiology led to the true Presbyterian religion.[59] His posthumous *Rudiments of Physiology* (1835–37),

55. Desmond 1989, 222–32.

56. Lord 1834, title page and passim.

57. Grant's father, like Lockhart and Scott, was a successful writer to the Signet, the highest grade of solicitor in the Scottish courts.

58. Desmond 1989 is the best source on Grant; for the professional exclusivity of the program of Thomas Wakley and his circle (an issue that Desmond sometimes underplays), see Richardson 1987. "Biographical Sketch of Robert Edmond Grant," *Lancet* 2 (1850): 686–95 is the most extensive contemporary account.

59. Fletcher 1836.

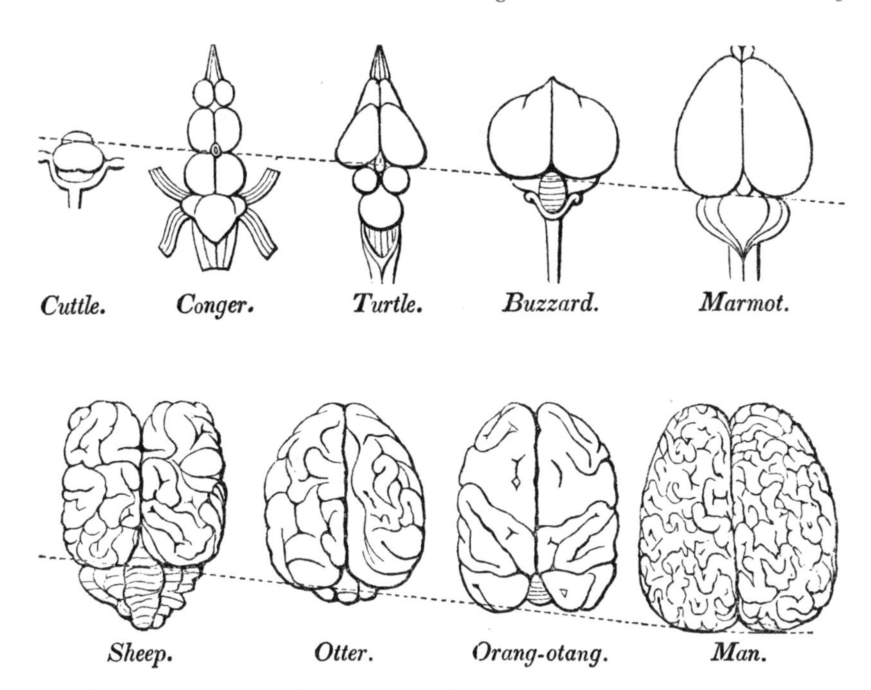

2.10 Comparative views of the brains of adult organisms in John Fletcher's introductory anatomy, ranging from the cuttlefish to man. The dotted line marks the posterior boundary of the cerebral hemispheres. Fletcher 1835–37, 47–48.

published by John Carfrae in Edinburgh (and Longman in London), was based on illustrated lectures delivered to a large mixed audience. The *Rudiments* stressed the unity of structure exhibited by series of individual organs—for example, of the heart or brain (fig. 2.10). It listed the animal sequence in order of increasing complexity, described the theory of recapitulation at length, paralleled the embryological and geological sequences, and argued for serial continuity from fungus to pine and from monad to man. But it denied that any of this supported the evolution of species. The functional integrity and perfection of each species could not be maintained if fish became reptiles, reptiles mammals, and so forth. Embryological recapitulation did not hold for whole organisms, but only within individual organs. The heart progressed through the sequence at a different rate from the brain, and the brain at a different rate from the liver. The geological record also supported "the generally immutable character of each tribe."[60]

By the time the *Rudiments* appeared, its emphasis on unity of plan and the animal series was becoming outdated. The new trends are evident in William Carpenter's *Principles of Physiology*, issued in 1839 by the London medical publisher

60. Fletcher 1835–37, 14; also Desmond 1989, 71–73.

John Churchill. Carpenter, son of a famous Unitarian minister, had attended Grant's lectures in London, and studied in Edinburgh in the mid-1830s. By giving his *Principles* a unified basis in a law of progress, Carpenter hoped to reach a combined audience of students and serious general readers, doing for the life sciences what Herschel had done for natural philosophy. Imposed on matter by the Creator, this law "should harmonise and blend together all the innumerable multitude of these actions." Carpenter drew on work by the Estonian naturalist Karl Ernst von Baer, who disputed the transcendental anatomists' claim that the embryological sequence repeated the history of the species. Rather than claiming that the embryo successively resembled a fish, a reptile, and a mammal, von Baer's version of development pictured general forms being replaced by more specialized ones.[61]

The dangers of being tarred with the brush of speculation were most apparent in presenting a systematic account of living beings. Any treatise that discussed the mechanism of creation was in danger of being accused of infidelity—what one evangelical Tory condemned as "the absurd cry of Atheism."[62] "Materialism" was not a coherent doctrine, but a term of abuse. Systematic treatises downplayed or rejected the evolutionary implications of philosophical anatomy because it could be so readily associated with infidelity, pornography, or the unstamped press. Such associations were fatal to any campaign to establish a safe science of life, or a medical profession based on technical expertise and science. A causal account of generation of higher species was, even in the most liberal medical circles, simply too speculative to be an available subject for general treatises, regular research, or a lecture course. Instead, "unity of type" was the battle cry of the earlier reformers, not the origin of new species. To see the situation otherwise is to obscure contemporary priorities in the everyday practice of research, teaching, and compiling systematic works. Similarly, von Baer's developmental embryology had been targeted against what he saw as vague claims of transcendentalist nature philosophy, not against species transmutations. The context in which it was discussed in Britain was not evolutionary either.[63]

The force of reaction meant that physiologists had to distance themselves from any hint of materialism. In this context the theological pronouncements of Grant, one of the most uncompromising philosophical anatomists, are revealing. Grant was no fiery atheist, but a providentialist working for professional

61. On Carpenter and von Baer, see Desmond 1989, 210–22; Gould 1977, 108–14; Ospovat 1976; Richards 1987, 134–37. Carpenter 1839, 463–64 details his notion of law.

62. J. G. Children to W. Swainson, 11 July 1831, quoted in Desmond 1985, 167.

63. Appel 1987 (on France) and Lenoir 1989 (on Germany). The point is also evident in Desmond 1989, which however does tend to stress the link with transmutation.

medical reform. As Grant proclaimed in his inaugural address at the University of London, zoology enhanced "our views of the wonderful harmony which everywhere pervades the economy of the universe" and elevated "our conceptions of the infinite wisdom, power, and goodness of the great Author of Nature, as displayed in his minutest works." The new anatomy would "lay the most rational and lasting foundations of piety and virtue, and strengthen the best principles of morality and religion."[64] Grant's later lecture to the Liverpool Mechanics' Institution showed how geological progress could be buttressed by scriptural authority. He emphasized "how perfectly does this order of development of living beings upon our globe accord with the Mosaic account of their successive creation."[65] Such parallels were neither ironic parodies nor avowals of evangelical literalism; rather, they were ways of explaining progression to audiences more familiar with the Bible than with the Bridgewaters. As Fletcher wrote, in the adaptation of organic life "it is surely no cant to acknowledge here the finger of God, or to say that an undevout anatomist must be a maniac."[66]

Works engaged in disseminating philosophical anatomy judiciously limited the application of the new views. As Fletcher said, any attempts to postulate the generation of human beings from the lower animals were "wild speculations." The doctrine had "nothing but the most vague and rambling presumptions in its favour."[67] Roget's Bridgewater decried the "remote and often fanciful analogies," the "seductive speculations" that pretended to explain the universe by a few general laws; Lamarck could be condemned for claiming that the inherent perfectibility of matter obviated the need for a Creator. Roget could then reformulate creative progress in terms amenable to Christianity:

> It is impossible, however, to conceive that this enormous expenditure of power, this vast accumulation of contrivances and of machinery . . . can thus, from age to age, be prodigally lavished, without some ulterior end. Is Man . . . formed but to perish with the wreck of his bodily frame? Are generations after generations of his race doomed to follow in endless succession, rolling darkly down the stream of time, and leaving no track in its pathless ocean? May we not discern, in the spiritual constitution of man the traces of higher powers, to which those he now possesses are but preparatory; some embryo faculties which raise us above this earthly habitation?[68]

64. Grant 1828, 5, 7.

65. R. Grant, "Antediluvian Remains at Stourton Quarry," *Liverpool Mercury*, 24 Apr. 1838; also in *Magazine of Natural History* 3 (1839): 43–48, but without the remarks on progress; see Desmond 1989, 311.

66. Fletcher 1835–37, 69, 62.

67. Ibid., 14, 62.

68. Roget 1834, 2:640.

The huge "machinery" of nature would be justified by the advance of humanity beyond it, when progress would occur on a spiritual plane. Roget's skill in discussing such sensitive issues made him an ideal systematizer and showed why he was so sought after as a commercial writer.

Avoiding being read as an infidel required subtle strategies. The most notorious case occurred when a reviewer attacked Carpenter's *Principles* for selling a machine world able to run without divine interference. His career in the balance, Carpenter fought back. In his defense, he could cite passages carefully planted in his text, among them a definition of natural law that undercut readings of his work as godless materialism:

> To imagine that the plan of the Universe, once established with a definite end, could require alteration, during the continuance of its existence, is at once to deny the perfection of the Divine attributes; whilst, on the other hand, to suppose, as some have done, that the properties first impressed upon matter would *of themselves* continue its actions, is to deny all that Revelation teaches us regarding our continued dependence on the Creator. Let it be borne in mind, then, that when a *law* of Physics or of Vitality is mentioned, nothing more is really implied than a simple expression of the *mode* in which the Creator is *constantly* operating on inorganic matter, or on organised structures.[69]

Passages like this offered an exquisitely balanced account of developmental theories of life, perfectly calculated for use in controversy.

In wording key passages with such care, Carpenter negotiated the limits and opportunities for debate that had been marked out by his patrons. Because of this, he could accompany his defense with supporting letters from men of science, medicine, and religion, many of whom were celebrated authors of systematic overviews and reflective treatises. Their letters show how Carpenter's care in constructing the *Principles* had paid off. Herschel thought the reading of the crucial passages was a matter of "common sense, and logical interpretation of words." The royal physician Henry Holland thought Carpenter had "never exceeded the authorized bounds of physical research, as pursued by the most eminent physiologists."[70] The reply and testimonials (which Carpenter had printed locally in Bristol) were sent out with Churchill's *British and Foreign Medical Review* to clear both the publisher and his most promising new author of charges of infidelity.[71]

The risks Carpenter faced can be summed up by the fate of two books with-

69. Carpenter cites his 1839 text in his 1840, 3.

70. Carpenter 1840, 6–7, 8. For detailed accounts of Carpenter's dilemma, see Desmond 1989, 210–22, and Winter 1997, 35–43.

71. Carpenter's 1840 pamphlet is typically found bound at the back of the April 1840 number of the *British and Foreign Medical Review;* see CUL Q.300.c.104.9.

drawn from circulation at the height of the Reform crisis. One was Richard Vyvyan's *Psychology*, an anonymous argument for the immortality of the animal soul. This was released by the antiquarian publisher John Bowyer Nichols in 1831 but called in immediately afterward.[72] The other and more celebrated case involved Thomas Hope's *Essay on the Origin and Prospects of Man*, issued posthumously in three volumes in 1831—the only proevolution book published by John Murray before the *Origin of Species* nearly three decades later. Hope was already famed as a novelist and advocate of neoclassical furniture design when he withdrew from society to write a cosmological masterpiece. His aim was to unite "the first creation of matter" with "the final destination of man," spiritualizing matter so that belief in an afterlife could be based on the inexorable progress of nature. By arguing that matter was imperishable, Hope believed that he could advocate the origin of mind in matter (and the return of the dead to dust) without suffering "the indelible odium" associated with materialism.[73] Murray and Hope's executors recognized that for most readers this was philosophical dynamite. Only 250 copies were printed, and at thirty-six shillings, the *Essay* cost more than a three-decker novel. This strategy failed to stem public alarm, and the limited run was withdrawn. To foil the suppression, the weekly six-penny *Literary Gazette* printed long extracts from what it hailed as "one of the most extraordinary productions of the age," predicting that Hope's analysis of the progress and future of the universe would "engage the pens of the ablest philosophers of the present and subsequent times."[74] Other reviews were less positive, the *Edinburgh* condemning it as the dreary culmination of the mechanistic spirit of the age.[75]

Constituting the People

On the eve of Reform enterprising publishers attempted to extend the market for reflective science. The working and lower middle classes had previously read chapbooks, astrological almanacs, religious tracts, bibles, and radical pamphlets and periodicals. Now the aim was to sell them products of the steam press factories.

Among the first in this field was the Scotsman William Chambers, who began publishing *Chambers' Edinburgh Journal* in 1832.[76] He was soon joined by his brother Robert, who had written many historical books in Constable's

72. Boase and Courtney 1878–82, 2:840.

73. Hope 1831, 1:26, 2:236–37.

74. *Literary Gazette,* 18 June 1831, 393; 25 June 1831, 408–9; 2 July 1831, 423–5; 9 July 1831, 439–41. For the withdrawal see Bulwer-Lytton 1970, 317.

75. [T. Carlyle], "Characteristics," *Edinburgh Review* 54 (Dec. 1831): 351–83.

76. William Chambers founded it as *Chambers' Edinburgh Journal.* After Robert joined him, *Chambers's* with an *s* after the apostrophe was usually (but not always) used.

Miscellany. Their journal offered the same quantity of words as a six-penny number of the SDUK's Library of Useful Knowledge, but at a quarter of the price. The opening editorial promised to feed "the universal appetite for instruction" shared by *"every man in the British dominions."* Early issues featured a weekly essay; biographies of great men (including a substantial life of Scott); long extracts from Herschel's *Preliminary Discourse* and other reflective works on science; and stories and an occasional poem.[77]

Never before had there been a mass market for knowledge, and so predicting sales involved much uncertainty. The only publications that had approached such a broad audience were working-class political papers; these had news but were unstamped and therefore illegal. For entrepreneurial publishers, defining readerships required an astute combination of financial and political judgment. In this respect, W. & R. Chambers were cannier operators than Knight, whose expectations of the polity for cheap knowledge verged on the millennial. Two months after the launch of *Chambers's Journal* Knight commenced the *Penny Magazine* on behalf of the SDUK. It was cheaper and illustrated with woodcuts, although without fiction and poetry. Both periodicals were risky, depending on an unprecedented level of sales. Both had a huge success, the *Penny Magazine* peaking with a circulation of two hundred thousand in 1832. But the underlying finances were very different. The *Penny Magazine* had to sell nearly five times as many copies in order just to break even, and although its initial figures were far beyond those for *Chambers's,* these proved impossible to maintain.[78]

Knight's enterprise failed because the reading public it had helped to bring into being proved unstable. Above all, the crisis of expectations associated with reform made the *Penny Magazine* seem at first the dawn of a new age of knowledge, but soon after dull, fact-ridden, and Whiggish. The "people" it addressed came to be recognized as excluding most of the working classes; yet a wider readership was needed to break even. The Chamberses also recognized that their audience was not primarily among individual male factory operatives, but from a financial point of view this mattered far less. The marriage of fiction and fact, even for an extra halfpenny and without pictures, proved more salable among a readership whose core was solidly middle class. By 1834 circulation had reached fifty thousand and a decade later it stood at nearly ninety thousand.[79]

77. [W. Chambers], "The Editor's Address to his Readers," *CEJ,* 11 Feb. 1832, 1–2. The standard work on the firm is Cooney 1970, to which I am much indebted. For the financing of *Chambers's Journal,* see Bennett 1982; for its politics and aims, Scholnick 1999.

78. Bennett 1982, 240–45, although his argument tends to underplay the significance of the Chamberses in creating a mass audience; see also Bennett 1984 and Anderson 1991. For literary strategies in addressing the crowd, see Klancher 1987, 85–88.

79. Bennett 1982, 236; Cooney 1970, 96; the middle-class readership is stressed in [W. and R. Chambers], "Address of the Editors," *CEJ,* 25 Jan. 1840, 8, which notes that "[a] fatal mistake is

Chambers's Journal became the basis for the firm's preeminent reputation as "publishers for the people." Unlike the established gentlemanly entrepreneurs in London, who contracted out printing and binding, the Chamberses were simultaneously publishers, producers, and authors. Their state-of-the-art four-story factory at 19 Waterloo Place, with ten steam presses in constant activity, became one of Edinburgh's wonders, brazenly set within the religious heart of the old town. It was considered a model of industrial management, with a savings bank, evening classes, and a nine-hundred-volume library for its employees (there were eighty by 1843). *Chambers's Journal* aimed at a family readership among the middle and working classes, a new polity of consumers around which the firm developed a whole range of factory-produced printed materials. Within a year or two it expanded operations into cheap books, encyclopedias, and school texts. From 1833 the Information for the People offered folded sheets on a hundred different disciplines, "each a distinct branch of knowledge" for three halfpence. By the early 1840s some seven hundred thousand sheets had been sold, about fifteen tons of paper.[80]

From 1835 W. & R. Chambers began to apply its huge economies of scale to produce what it called People's Editions. These aimed to bring full-length works to readers unable to afford the Library of Useful Knowledge, Family Library, or similar series. People's Editions cost one shilling sixpence for an entire book, or more depending on length. They were printed with double columns of closely printed type on cheap paper, and bound in thin card like a magazine. The firm made money on volume sales: runs were at least two thousand and often greater than that by a factor of ten. Titles included an eclectic range of classic works that had previously been available only in editions for the wealthy: Bacon's *Essays,* Adam Smith's *Wealth of Nations,* David Hume's *History of England,* Benjamin Franklin's *Life* and *Miscellaneous Writings.* They included the first English translation of Adolphe Quetelet's *Treatise on Man,* a pioneering work in applying statistical laws to human life.[81]

The possibilities of production on this scale first became apparent when W. & R. Chambers began to publish the Edinburgh phrenologist George Combe's *The Constitution of Man Considered in Relation to External Objects*

committed in the notion that the lower classes read." The appeal of *Chambers's Journal* to the better-off is evident from the fact that by 1844 almost half the total circulation was in monthly numbers, not weekly ones.

80. *CEJ,* 8 July 1843, 197–99, at 198; on the factory, "A Few Words to Our Readers," *CEJ,* 4 Jan. 1845, 1–3.

81. [W. and R. Chambers], "Address of the Editors," *CEJ,* 25 Jan. 1840, 8, listing thirty-two works published up to that date. Quetelet's work was translated in 1841 by Robert Knox, the anatomist disgraced in the Burke and Hare resurrectionist scandals of the 1820s (Lonsdale 1870, 257).

from 1835. This became the most widely read of all the reflective treatises in the Reform era and the single most important work for debates about natural law and progress in Britain and the United States. The mutual relations of industrial production and the sciences is shown nowhere more clearly than in the People's Editions of the *Constitution.*

Combe was convinced that nature's laws could not be understood without a science of the human mind; and phrenology was that science. As developed in the writings of Franz Joseph Gall and Johann Gaspar Spurzheim, phrenology located specific mental functions in specific organs, which together composed the brain (fig. 2.11). During the 1820s Combe had begun to apply phrenology's lessons to all areas of human activity, a project that culminated in the *Constitution.* He developed the theme in an address of 1826 to the Edinburgh Phrenological Society, which he circulated first in a privately printed pamphlet and then as a book two years later.[82] This argued that proper conduct involved exercising the full range of faculties under an enlightened comprehension of nature's laws. Combe identified three classes of law: physical, organic, and moral. Disobey any one, and bad consequences were sure to follow. The 1825–26 trade depression, he argued, had resulted from "excessive activity of Acquisitiveness, and a general ascendancy of the animal and selfish faculties over the moral and intellectual powers." Infringement of the moral law had led to "mental anguish." Under the organic law, this in turn produced "a morbid nervous influence" and "bodily disease," leading in turn to a generation of children who "will inherit weak bodies, with feeble and irritable minds, a hereditary chastisement of their father's transgressions."[83]

Such disastrous consequences could be avoided through exercise of the appropriate faculties. Self-help could then proceed on a rational basis. The results would be progressive: individual improvement would maximize happiness and a sense of the Creator's benevolence. Obeying natural laws, however, would only make the best of the situation and would not lead to human perfectibility or a socialist utopia. "The system of sublunary creation," Combe said, "does not appear to be one of optimism."[84] The book carefully stressed that an understanding of law was no substitute for salvation, which could be achieved only through reading the revealed Word.

When first published in 1828, the *Constitution* appeared in the usual six-shilling format, in an edition of fifteen hundred copies from John Anderson, the Edinburgh bookseller who had issued most of Combe's earlier works; in

82. On Combe and the *Constitution,* Cooter 1984, 101–33; Gibbon 1878, 2:180–84; and work in progress by John van Wyhe of the University of Cambridge.

83. Combe 1828, 281.

84. Ibid., 16–17.

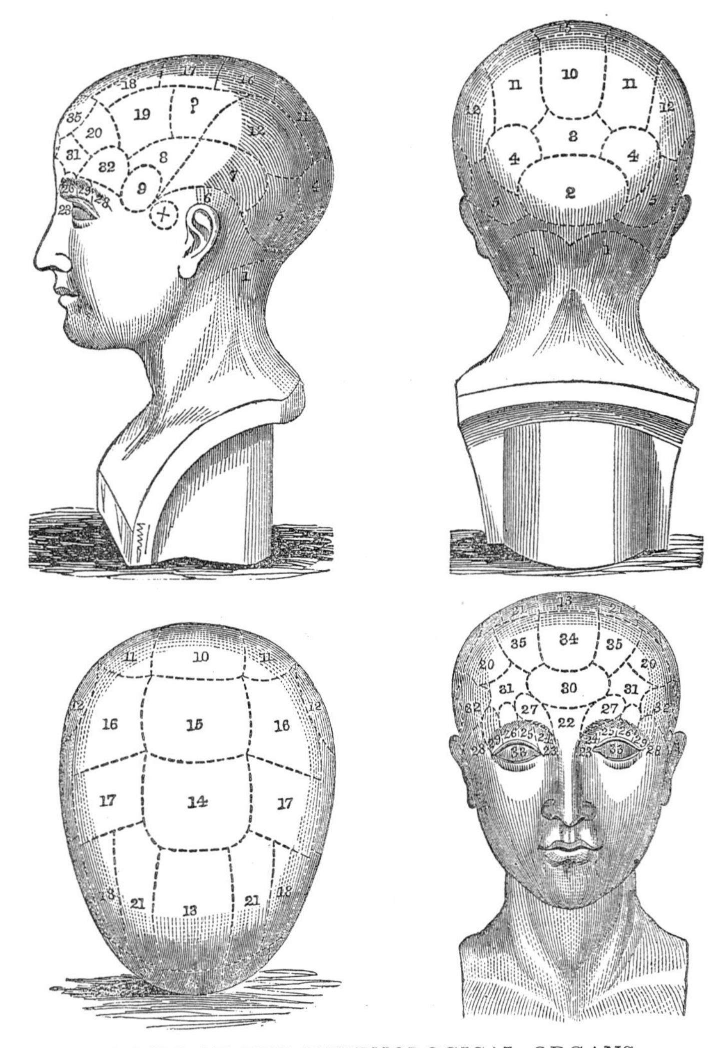

NAMES OF THE PHRENOLOGICAL ORGANS,

REFERRING TO THE FIGURES INDICATING THEIR RELATIVE POSITION.

AFFECTIVE.		INTELLECTUAL.	
I. PROPENSITIES.	II. SENTIMENTS.	I. PERCEPTIVE.	II. REFLECTIVE.
1 Amativeness *P.* 8	10 Self-Esteem *P.* 12	22 Individuality *P.* 18	34 Comparison *P.* 23
2 Philoprogenitiveness, 8	11 Love of Approbation 12	23 Form 19	35 Causality 24
3 Concentrativeness 8	12 Cautiousness 13	24 Size 19	
4 Adhesiveness 9	13 Benevolence 13	25 Weight 19	Modes of Activity 24
5 Combativeness 9	14 Veneration 14	26 Colouring 19	Practical Directions 28
6 Destructiveness 9	15 Firmness 15	27 Locality 20	Combinations 29
+ Alimentiveness 10	16 Conscientiousness 15	28 Number 20	Materialism 31
7 Secretiveness 10	17 Hope 16	29 Order 20	
8 Acquisitiveness 11	18 Wonder 16	30 Eventuality 21	
9 Constructiveness 11	19 Ideality 16	31 Time 21	
	? Unascertained.	32 Tune 21	
	20 Wit or Mirthfulness 17	33 Language 22	
	21 Imitation 17		

2.11 "Phrenological bust. Names of the mental faculties, the positions of the organs which are marked upon the bust." Combe's identification of the organs, based on that of Johann Gaspar Spurzheim, was widely used in Britain. Combe 1836c, 1: frontispiece.

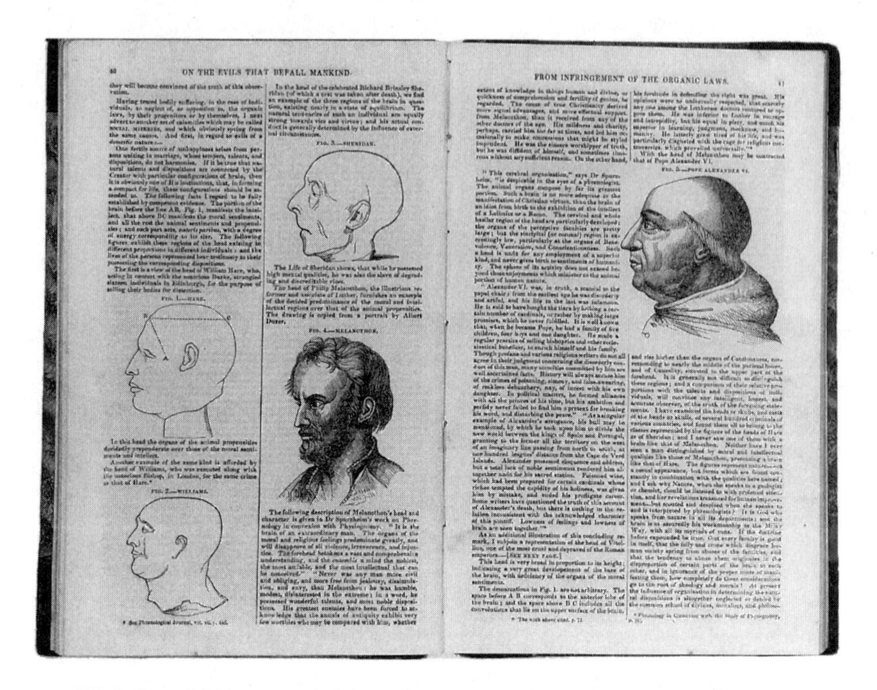
40 ON THE EVILS THAT BEFALL MANKIND.

FROM INFRINGEMENT OF THE ORGANIC LAWS. 41

2.12 With their double columns of small type, People's Editions of the mid-1830s brought books to a wider range of readers. Here, illustrations of the phrenological characteristics of famous men could be provided at a low cost because the wood blocks could be borrowed from the original publisher. Compare the predominance of "animal propensities" in the profiles of two murderers and the pope, with the high forehead of the Protestant reformer Melanchthon. Combe 1836a, 40–41.

London it was sold by Longman. Even at this price, it took nearly seven years to sell out, although this was not unusual for a work of this kind. The status of the work began to change in March 1835, when Anderson sold a second edition of three thousand in just four months. The bargain price of two shillings sixpence was subsidized from a bequest by a wealthy benefactor of phrenology.[85]

Intrigued by sales of the cheap edition, the Chambers brothers—committed phrenologists and friends of Combe—tried an experiment. In September 1835 they issued the fourth edition of the *Constitution* in a new double-column format, at the sensationally low price of one shilling sixpence (fig. 2.12). This did not rely on a subsidy, but rather on a bold publishing move that the book itself proudly announced. Multiple copies of the plates were made by the new

85. "Advertisement to the Fourth Edition," in Combe 1836a, iii, which states categorically that no funds from the Henderson Bequest were used to subsidize the People's Editions. A third edition of the *Constitution,* with a limited subsidy and a four-shilling price, sold more slowly. All quotes from the People's Edition are from the sixth impression of the fourth edition.

process of stereotyping, and printed on the same steam press that produced *Chambers's Journal* and the Information for the People. A preface by Combe explained the principles behind the production. "Operatives," he claimed, failed to buy books on "moral and intellectual science" not through lack of interest, but because they were priced far too high. Books like the Bridgewater Treatises were wasted opportunities.[86]

The cheap editions transformed the *Constitution of Man.* From an interesting but minor contribution to phrenology, it became one of the most popular books in the English language, a new bible for natural law. The first two thousand copies of the People's Edition sold in ten days; the next thirty thousand within a year. By 1848, eighty thousand copies had been sold in Britain alone.[87] Although the shilling-and-a-half price meant that most purchasers were not factory operatives,[88] the *Constitution* could be found in households that possessed no other books than the Bible, *Pilgrim's Progress,* and Paine's *Rights of Man.*

The effect on the meaning of Combe's text to readers was, if anything, even more striking. Few of the words had changed—if anything, the text was very careful and conciliatory. The first edition had already left out controversial chapters on lower forms of life and on human responsibility. But mass distribution radically altered the import of what Combe had to say. The privately circulated essay and first edition had already produced major schisms within Edinburgh phrenology. Combe had been deluged with letters from friends and attacked in newspapers by the evangelical wing of the Presbyterian Church. Establishing natural law as a basis for moral law seemed to lead to determinism and a denial of free will. Leading clerics withdrew their support.[89]

These internal schisms, however, differed in character from the onslaught after 1835. The *Constitution* became a major public target for evangelicals, and the debate about it a defining moment in the relations between science and faith. What had been seen as misguided (and, from some perspectives, a bit silly) when limited to a genteel, educated audience was dangerous in the hands of the people at large. Thus the evangelical phrenologist William Scott attacked the book in public only after 1836, when it had begun to penetrate into working-class homes in the industrial districts. He feared that "many thousands of readers not best fitted to detect its fallacies," clerks and factory operatives faced with an insufficient provision of churches, would slide down the

86. "Henderson Bequest," in Combe 1836a, v–vi. Comparisons between Combe's work and the Bridgewater Treatises were often made: Topham 1992, 402.

87. Combe 1836a, iii; Gibbon 1878, 2:274.

88. A point stressed, in opposition to critics, in "Advertisement to the Fourth Edition," Combe 1836a, iii.

89. Gibbon 1878, 1:181–86, 1:211–13; de Giustino 1975, 104–35; Cooter 1984, 129–33.

slippery slope that led to atheism.[90] Sermons, pamphlets, and pulpit orations proclaimed the dangers of the book and railed against its diffusion.[91]

Publication for the people led to readings of the *Constitution* as an evolutionary book. Scott's pamphlet argued that Combe was now open to attack on the origin of organic beings. All the editions of the *Constitution* drew on standard presentations of geology to show that the world was arranged "on the principle of gradual and progressive improvement" and that the introduction of the moral and intellectual faculties was part of a progressive system.[92] Scott argued that these passages could be read as verging on heterodox evolutionary views. Whatever his failings in the eyes of the faithful, however, Combe never saw himself as a transmutationist: the *Constitution* explicitly denounced "the French revolutionary ruffians" and their program of "fraud, robbery, blasphemy, and murder."[93] The botanist and phrenologist Hewett Watson also sprang to Combe's defense, after learning that an evangelical publisher planned a people's edition of Scott's attack. Watson identified passages from the *Constitution* that denied the spontaneous evolution of matter without a creator. Watson's pamphlet was privately published, although it was discreetly available for sale from Longman. That Watson avowed his proevolutionary views for the first time in this debate testifies to the impact of cheap printing.[94] A woman touring the Chamberses' factory had been horrified to see a steam press printing Combe's book. She threw down the sheet, still wet with ink, "as if it had been a serpent, and exclaimed—'Oh, Mr Chambers, how *can* you print that abominable book?'"[95]

Defining "the people" meant identifying customers for a true knowledge of nature. From aristocratic palaces to the humblest cottage, the market was potentially unlimited. An annual festival in the Chamberses' factory celebrated this inclusive, industrial ideal of the people. The compositors' hall was decked out with garlands, and the employees and Edinburgh worthies gathered to savor lemonade, fruit, cake, and speeches (fig. 2.13). This was a temperance version of the printers' traditional alcohol-soaked "wayzgoose feast."[96]

The meetings, William Chambers told one such gathering, brought employer and employed together "on the broad principles of humanity" during a period of industrial unrest, "to produce harmony between the elements of the social machine." His brother went on to tell the workers about the iron rules of political economy, laws as unchanging "as those which regulate the movement

90. Scott 1836, preface.

91. For examples, see Cooter 1984, 130.

92. Combe 1835, 5–7; Combe 1836a, 2.

93. Combe 1828, 301; Combe 1836a, 100.

94. Watson 1836, 4, 22–30.

95. Letter to an unnamed correspondent from G. Combe, 16 July 1836, in Gibbon 1878, 1:331.

96. McKenzie 1960; Johns 1998a, 95.

2.13 Employers and employed: factory workers, guests, and Edinburgh's leading literary men listen to a speech by William Chambers in the typographical gallery of the publishing firm, decked with garlands for the annual temperance soirée. Note the presence of women, both in the audience and near the head table; the girls in the foreground are probably some of Robert's daughters. "Messrs. Chambers's Soiree," *PT,* 23 June 1843, 236.

of the planets." "We cannot," he said, ". . . expect that the working-classes are all at once to become careful of their earnings, and consequently infant capitalists—that must be the work of time."[97] Knowledge would be the key to responsible political action; the "people" would be defined as the all-inclusive readership that would understand nature's laws.

At the end of the 1830s, five years after cheap editions of the *Constitution* went on sale, Carpenter would face down the onslaught on the *Principles of Physiology.* By that time, reflective treatises by Herschel, Lyell, Nichol, and others had shown how natural laws might be discussed without outraging public sensibilities. Outside the shadowy underworld of freethought, the slightest opportunity for accusations of materialism had to be blocked, especially when human

97. "Messrs Chambers's Soirée," *CEJ,* 8 July 1843, 197–99, at 198. For debates about definitions of "the people" in relation to the middle class, see Joyce 1991, Vernon 1993, and (for the Reform era) Wahrman 1995, esp. 321–22. The significance of the political language focused on the constitution of the polity was first brought out in "Rethinking Chartism" in Stedman Jones 1983, 90–178. For the factory as an image of society, see the discussion of Babbage in Poovey 1995, 37–40.

origins were at stake. To go further than the *Constitution of Man* in making natural law the basis of a general work would require the most consummate judgment. It was a telling sign of the times that after the fiascoes surrounding Hope's and Vyvyan's books in 1831, no respectable British publisher issued a cosmology extending an evolutionary law from the stars to plants, animals, and human beings until 1844.[98]

98. Even Powell 1838, which addressed itself to a learned, cultivated audience, went no further than to argue that Lamarckian transmutation was a legitimate topic for philosophical contemplation. If there is an exception, it would be Tiedemann 1834, a translation from a German textbook, made by two medical phrenologists; Desmond 1989, 175. However, this work had a limited circulation; it focused on physiology and did not extend to astronomy or geology.

CHAPTER THREE

Evolution for the People

But Secretiveness is not merely useful in the arts,—it is of essential service in the invention of plots, both in reality and in works of fiction.

WILLIAM SCOTT,
in *Transactions of the Phrenological Society* (1824)

ROBERT CHAMBERS HAD A LARGE ORGAN of Secretiveness. We now know his greatest secret, the authorship of *Vestiges,* in a way that most of his contemporaries never could. Yet although no longer an anonym, Chambers remains little known. This chapter analyzes the circumstances out of which the text of *Vestiges* emerged. It asks why Chambers wrote it and how the writing created an appropriate voice for drawing in so wide a spectrum of readers.

Friendship offering, atheist propaganda, new gospel, target for fashionable wit: how could a single work be open to such different uses? Understanding the range (and limits) of readings requires examining the rhetorical conventions within which the writing was situated, and especially how it addressed potential readers. Readers needed a place as imagined partners in the process of composition. They had to be "written in"—"conscripted," as a recent critic has put it—to the very fabric of the text.[1] At every turn of the argument, in every phrase, in each choice of word, *Vestiges* anticipated fears, doubts, difficulties, assumptions, and prejudices on the part of implied readers. It had to project what Hans-Georg Gadamer has called a "horizon of expectations," questions defining the anticipated relation between narrator and reader. This kind of analysis is rarely applied to scientific writings. As Chambers said of Jane Austen, "She was too natural for them. . . . They did not consider that the highest triumph of art

1. Stewart 1996. Important works include, among a large literature, Fish 1980; Iser 1978; Suleiman and Crosman 1980; Tomkins 1980. The most accessible entry point remains "Literary History as a Challenge to Literary Theory," in Jauss 1982, 3–45.

consists in its concealment; and here the art was so little perceptible, that they believed there was none."[2]

Close analysis is especially rewarding in this case because the process of composition drew on an experience in addressing contemporary audiences surpassed only by Dickens. Literary practices from journalism, from new sciences of progress, and from Sir Walter Scott's fiction made it possible to recast the secular Enlightenment cosmologies as evolutionary science. *Vestiges* brought together these practices and created new possibilities for prose, what its concluding pages identified as "the first attempt to connect the natural sciences into a history of creation."[3] The evolutionary narrative, I hope to show, emerged as part of the same processes that produced the classic form of the European historical novel.

In one sense, this chapter unveils the secret story about *Vestiges* that the Victorians most wanted to discover. Like the Victorians, we are inheritors of what Jerome McGann has described as the "romantic ideology," which assumes that the meaning of a work is definitively revealed in private documents from the author's hand.[4] My aim here is not to cast Chambers as an unsung literary genius, nor to explicate his intentions, but to show how the writing of *Vestiges* drew on experience at the center of early Victorian literary production.[5]

The Dark Ages

As the only reader who was also the author, Chambers commented on the meaning of *Vestiges* in ways that were both idiosyncratic and atypical. He defined it as the great work of his life, expressing his deepest feelings as author, father, husband, and reformer. Written as a kind of moral therapy at a moment of conscious withdrawal and contemplation, *Vestiges* was at once an attempt to reestablish his equilibrium and a carefully crafted intervention in local politics and national debate. That writing could do all these things was part of the heroic image of the man of letters that Chambers himself had actively encouraged. This image can be seen in early portraits (fig. 3.1), which show him with a pen in one hand and a volume of *Chambers's Journal* in the other.

By the early 1840s Chambers was among the most prolific authors in Britain, with over thirty books and the larger part of the weekly *Chambers's Journal* to his credit.[6] Much of this work was written in the factory at Waterloo Place, with

2. Gadamer 1975, 245–73. Chambers 1843, 2:572.

3. *V*1, 388.

4. McGann 1983; for more recent discussions, see Chandler 1998, 33 n. 66.

5. My account differs from Hodge 1972b, which is explicitly devoted to reconstructing the authorial intentions behind the writing of *Vestiges*.

6. The principal biographical sources on Chambers are Chambers 1872, Cooney 1970, Layman 1990, and Millhauser 1959. Of these, Cooney's outstanding account is dependably based on the manuscripts, most of which are on deposit in the National Library of Scotland in Edinburgh.

3.1 Robert Chambers as man of letters. Steel engraving by T. Brown from an oil portrait by Sir J. Watson Gordon. Frontispiece to *V*12.

3.2 Abbey Park, the country villa on the outskirts of St. Andrews rented by Anne and Robert Chambers in the early 1840s. Detail from John Heaviside Clark, "The Town of St Andrews" (London: Smith & Elder, 1827).

the noise of ten steam press "type-smashers" sounding in his ears. The strain had been too much. His wife, Anne Chambers, noted that "incessant mental exertion" and "intense mental action" had left him with a mind in an "unsound, or partly in a *diseased* state."[7] He felt overwhelmed and undervalued and that his cold-blooded brother William, whom he hated, had no conception of what literary production on this scale entailed. To escape, the family moved in 1842 to a villa, Abbey Park (fig. 3.2), on the outskirts of the seaside university town of St. Andrews, where Chambers took up golfing and participated in the local literary culture. In this quiet family retreat, removed from controversy and the pressure of business, Chambers wrote the work that became *Vestiges.* Such pastoral isolation perfectly embodied the ideal setting for philosophical contemplation. At Abbey Park, his daughter recalled, he "could work at his secret with all the security of a criminal unrecognized in the midst of the police."[8]

Chambers saw his mental illness in phrenological terms. He had disobeyed the laws of his bodily constitution, disregarding the natural play of his faculties through "constant and intense mental action."[9] As he told William, not only his mind but their factory needed reorganizing; they had to become "more publishers and less authors." Perhaps William should carry on independently, or *Chambers's Journal* should fold. Robert recalled the fate of Brewster, who during the final terrible stages of compiling the *Edinburgh Encyclopedia* had been unable to face people on the street.[10] Nervous exhaustion also brought about tensions in the marriage, as Anne Chambers found her husband irritable and unpredictable. Pressures of work were multiplied by domestic tragedy. Scarlet fever took a two-and-a-half-year-old daughter in March 1842, and another daughter (just three weeks old) died six months later. "A child," an early number of *Chambers's Journal* had noted, "is apt to warp itself into the vitals of our very soul; so that, when God rends it away, the whole mental fabric is shattered."[11]

Most significantly, Chambers's mother died in 1843. A "noble-minded woman," Jean Chambers had nurtured her son's literary talents and followed every step of his advancement. "Of all friends," Chambers wrote in an early

7. A. Chambers to RC, 23 Nov. 1842, NLS Dep. 341/82/32. For the usual story, which is that Chambers left Edinburgh for secrecy, see Millhauser 1959, 29–30. Conditions at Waterloo place are described in "A Few Words to Our Readers," *CEJ*, 4 Jan. 1845, 1–2.

8. Priestley 1908, 43; also RC to A. Chambers, 7 June 1843, Norman mss. Grant, McCutcheon, and Sanders 1927, 67, 70–71, discusses the history of the building, which is now part of Saint Leonards School. Chambers became active in the local Literary and Philosophical Society, to which Anne donated an "Indian Harp": entry for 7 Nov. 1842, Literary and Philosophical Society of St. Andrews, Minute Book, St. Andrews University Library, Ms. UY8525/1, f. 72v.

9. RC to A. Chambers, 23 Nov. 1842, NLS Dep. 341/82/32.

10. RC to W. Chambers, [30 Nov. 1842], NLS Dep. 341/82/29.

11. *CEJ*, 22 Sept. 1832, 133–34; RC to D. R. Rankine, 14 Mar. 1842, NLS Dep. 341/109/3.

letter, "a mother is the most sympathising . . . there is nothing could make her so happy as the honourable distinction of her son and nothing so miserable as his debasement."[12] Her death, peaceful and honored, marked the end of a family drama of decline and recovery.

Throughout his early life, Chambers (born in 1802) had defined his ambition as filling the blank left by his father's improvidence. James Chambers had been a successful manager in the Scottish handloom industry in the small border town of Peebles, with over a hundred looms under his control. But bad debts, drink, and a failure to adjust to the emerging factory system led the family into decline. On one occasion he was found on the road after a drinking bout, beaten up and robbed. Jean Chambers held the family together through these "dark ages," keeping a tavern and later a hotel. She made sure that Robert received a classical education and was alarmed when at the age of sixteen he followed in William's footsteps by taking a few pounds worth of books to set up a stall on Leith Walk, a cheap commercial thoroughfare connecting Edinburgh with its port.[13]

Robert and William witnessed extremes of political debate among the artisan classes. One of their closest friends, a grocer's apprentice named John Denovan, was a "violent *radical*" who worshipped the champion of the illegal unstamped press in England, Richard Carlile. In 1819, the two brothers assisted Denovan in publishing the radical twopenny *Patriot,* one of scores of short-lived street periodicals that sprouted up after the Peterloo Massacre in Manchester.[14] Robert and William later claimed not to have shared the politics of the *Patriot,* although Robert's first verses appeared there, and they underwrote it financially. They soon realized the dangers of linking their names with advocacy of universal suffrage as the only remedy to a corrupt and tyrannous regime.

As his fortunes began to improve, Chambers became increasingly sensitive about his social status. His meticulously researched *Traditions of Edinburgh* (1824), with its vast fund of anecdotes about the customs and manners of the city, brought him to the attention of the literary world of Regency Edinburgh. As he told a friend, "I am indeed getting along wonderfully just now, and expect soon to go daft with excessive popularity."[15] By 1830 Chambers owned a respectable bookshop and circulating library in Edinburgh's fashionable New

12. RC to A. Constable, 25 Feb. 1822, NLS mss. 670, ff. 136–37. Her death is mentioned in Chambers 1872, 262.

13. Chambers's early life is best described in the reminiscences he compiled in 1833; these are partly printed in Layman 1990; see also Chambers 1872 and Cooney 1970.

14. Eight weekly numbers appeared, from 9 Oct. to 27 Nov. 1819; the only known copy is in the Mitchell Library, Glasgow. For background to the agitation, see Thompson 1968 and Wiener 1983.

15. RC to W. Wilson, Sept. 1825, quoted in James Grant Wilson, "Robert Chambers," *Lippincott's Magazine* 8 (1871): 17–26, at 20.

Town; he had married Anne Kirkwood, daughter of a well-known engraver and clockmaker; he had established himself as a leading author of literary and topographical works about Scotland; and he had a powerful patron in Sir Walter Scott, who introduced him to customers and encouraged his antiquarian researches.

Chambers signed himself "Young Waverley" and devoted his first book to identifying the originals for characters in the then-anonymous Waverley novels, which like most people he suspected were written by "the Prince of modern authors." He drafted part of a historical novel of his own, and met "the great inspirer" for the first time in 1822. "I have seen and spoken to Sir Walter Scott," he told Constable, "and like the Comet who travels to the Sun once in a thousand years and lays in such a stock of heat and blazing glories as serves him in all his wanderings through the coldest bounds of his orbit, I have received so much reflected greatness from my own near approach to this centre of the Literary System, that the experience of a century of mere, common, *prose* life could scarcely expend it."[16]

In political terms the connection with Scott was no coincidence, for at this time Chambers was a confirmed Tory. For two years during the Reform agitation he edited a Tory newspaper, and his publications, he assured the conservative Edinburgh publisher William Blackwood, took "a vigorous stand... against the prevailing tide of radicalism."[17] The Reform Bill agitation was hurrying society toward mob rule, as he confessed to Scott:

> The present political excitement, which I could see at the devil, puts polite literature and all its interests at the wall; and some periodical publications which lately circulated to the amount of 10,000, are now down to a fourth of the amount. This fervour is as fatal to literature as the irruption of the goths. Nor do I think it near an end: it is rather at a beginning. People formerly had a maxim, which history in all its ages showed to be good, that the great object of informed and civilised society was to keep the mob in check; but now the maxim is, that the government must reside in the mob. . . . The fiend, say I, take all the fools who are now hurrying us on to revolution and vandalism. But really this way madness lies; I must to business.[18]

In these charged circumstances, antiquarian chronicles were read in terms of current politics. In his history of mid-seventeenth-century Scotland, Chambers

16. RC to A. Constable, 25 Feb. 1822, NLS mss. 670, ff. 136–37. The plan to write a novel is mentioned in Layman 1990, 85–86. For "Young Waverley," see James Grant Wilson, "Robert Chambers," *Lippincott's Magazine* 8 (1871): 17–26, at 19.

17. RC to William Blackwood, [1830–32], NLS mss. 4714, ff. 183–84.

18. RC to Scott, 30 Mar. 1831, NLS mss. 3917 ff. 156–57.

confessed to being unable to offer an even-handed treatment; "good sense and good feeling" demanded that he "give his countenance almost exclusively to the royal cause" rather than "the rude cause of the populace."[19] Like Scott's own works, such histories celebrated a world of deferential paternalism that seemed to be slipping away under the force of progress.

Even at this early date, Chambers thought his great work would be on science. In Peebles, he had read through most of the fourth edition of the *Encyclopædia Britannica* after finding a copy in an old attic trunk. "I plunged into it," he later remembered, "I roamed through it like a bee." With William, he conducted electrical experiments in the company of an old porter, Jamie Alexander, whose Edinburgh garret "resembled the den of an alchemist or magician."[20] Robert became captivated by astronomy, especially through the account of a plurality of worlds by the great preacher the Reverend Thomas Chalmers.[21] In 1821 the two brothers started the *Kaleidoscope*—named after Brewster's recently invented optical toy—which Robert largely wrote and William printed on a tiny hand press. Robert's first published writing in science, "Vindication of the World and of Providence," appeared in this journal, and shows his adherence to the nostalgic politics of the Waverley novels:

> If we compare the whole creations of Providence to a vast and intricate Machine, in which Wheels and Parts may represent Men and Societies, and in which, however delicately formed each separate proportion, there is such a *laxity of fitness*, that though a wheel or even a part be accidentally abstracted or worn out of use, it causes no detriment to the whole, which still works on its immense purposes, uninjured: if we could comprehend a general idea of this vast and interminable machine, we could then be only amazed at the intricate proportions yet perfect beauty of Providence.

It is a cosmology underlining the stability and equilibrium of the moral, political, and natural world. Radicals and reforming Whigs are dismissed as "whiners" who long for perpetual improvement and development.[22]

Chambers's attitudes, however, soon changed. In 1832 Scott died and Parliament passed the Reform Act; Robert celebrated his thirtieth birthday and joined William as a proprietor and principal author of *Chambers's Journal.* He announced that *Reekiana* (1833) was his "last contribution" to Scottish antiquities, as his attention would be engaged in "literary objects of more extensive

19. Chambers 1828, 1:10–11.

20. From Layman 1990, 58–59, 84; Chambers 1872, 62, 106–9.

21. Chambers quoted in Layman 1990, 85; on Chalmers, see Crowe 1986, 182–90.

22. Chambers, "Vindication of the World and of Providence" (1822), in Chambers 1994, [199–203].

utility, and requiring a greater exertion of moral reflection."[23] He later explained the shift in concerns:

> I loved the old tales and legends of my native country with the most passionate ardor, and delighted to gather up every little trait of bygone times. [My early] works were an effluence of mental youth, analogous to a green phase of the studious mind of England at the present day, which shows itself in a love of patristic reading and of Gothic architecture. The mind, in progressive men, passes out of such affections at thirty, and the national mind will pass out of them when the time comes for its exercising its higher faculties.[24]

Although Chambers did write other historical works, his interests had shifted to themes of progress, utility, improvement, and modernity that mark *Chambers's Journal.* In politics he became a liberal Whig, emphasizing the need for educational reform, free trade, and harmony between the classes.

A related change involved Chambers's attitude toward religion. Throughout the early years of their marriage, Robert and Anne Chambers had attended the Presbyterian church of Saint Cuthbert's in Edinburgh, where they held fashionable pews in the first row of the gallery. Chambers's daughter Eliza records the incident that precipitated their departure into the less evangelical fold of the Scottish Episcopalians. Waving a copy of the first issue of *Chambers's Journal* from the pulpit, the minister had denounced its secular tone and neutrality on religion.[25] The preacher was almost certainly the Reverend David Dickson, the senior parish minister. His concern extended back at least two decades, as evidenced by a sermon delivered before the Society for Propagating Christian Knowledge. As Dickson had said, "Great as are the advantages of knowledge and learning, they are wholly unavailing . . . if separated from the spirit and influence of vital, personal godliness." "Unsanctified knowledge," he continued, "instead of profiting us, will only increase our final doom; for however useful we may be as members of society, we shall be found guilty of having diverted our talents from promoting the glory of God."[26] Precisely what the Chamberses heard at Saint Cuthbert's is not known, but it must have been similar in tone, for they never returned.

The sermon damning *Chambers's Journal* precipitated a shift that was probably inevitable. Chambers had moved into the prosperous Anglicizing middle

23. [Chambers] [1833], vii. This work, *Minor Antiquities of Edinburgh,* was called "Reekiana" on the half title and throughout the text.

24. General preface, in Chambers 1847a, 1:iii–iv.

25. Priestley 1908, 30. Chambers told a similar story to Charles Kingsley and A. K. H. Boyd in 1867; see [Boyd] 1890, 1:86, which connects the incident with the first issue of *Chambers's Journal* in March 1832.

26. Dickson 1814, 69–71. For Dickson, see [Anderson] 1832.

classes, nouveaux riches who provided the main social basis for the Episcopalians in Scotland.[27] His departure from Saint Cuthbert's would thus have seemed natural. His historical writings had already condemned the seventeenth-century Covenanters (blessed martyrs in Presbyterian eyes) as "an oligarchy of mingled demagogues and fanatics." He took a similar view of their modern representatives, especially those who in 1843 joined the breakaway Free Church, which he saw as an obstructing group of ignorant zealots. Religion was a drag to reform.[28]

In characterizing his views in private letters and journals, Chambers maintained his belief in God only in a distant and abstract sense. He had only perfunctory contacts with the formal church; as Anne Chambers noted, he could not be accused of being "very guilty of church going," although she and the children seem to have attended with some regularity. When asked why he kept two pews, each in a different church, he jokingly replied that "when I am not in the one, it will always be concluded by the charitable that I am in the other."[29] As a moderate deist, Chambers hated evangelical enthusiasm and doctrinal controversy. Religion occupied much the same place in his personal life that it did in *Chambers's Journal:* a minor one. Because of this, it is misleading to speak of Chambers's inquiries into natural law as an attempt to construct a "theodicy."[30] By the 1830s justifying God's ways to man was not central to Chambers's sense of self-identity.

The shifts toward anticlericalism, Whiggery, and religious indifference form the broad background to *Vestiges.* The setting for these changes lies in Edinburgh phrenology. From the early 1830s, Chambers's interest in natural law, begun through his reading of Thomas Chalmers, Adam Ferguson, and the Scottish encyclopedists, developed through close contacts with George Combe's phrenological coterie. The circle included Chambers's best friend Robert Cox, Combe's nephew and editor of the *Phrenological Journal* from 1830 to 1837; Cox's wife Ann; Combe's brother Andrew; and Combe's wife Cecilia, daughter of the celebrated actress Sarah Siddons. Other friends with liberal leanings included the novelist Catherine Crowe; the London physician Neil Arnott; the Glasgow professor of astronomy John Pringle Nichol; and the Manchester-based journalist Alexan-

27. Brown 1987, 49–50.

28. RC to AI, n.d., NLS Dep. 341/110/243–44. For a similar case involving a move away from a Presbyterian background, see the sensitive account of George Combe's career in Cooter 1984, 101–33. An account of the Moderate faction in the Scottish church in which Chambers was raised is provided in Drummond and Bulloch 1975, esp. 224–33, and Hoeveler 1981, 3–32.

29. Payn 1884, 142. Drummond and Bulloch (1975) claim that Chambers continued his allegiance to the Presbyterian church in his parish, although the source they cite does not support this contention. For Anne Chambers's comment, see A. Chambers to RC, 5 Sept. 1836, NLS Dep. 341/90/19.

30. Hodge 1972b, 131–33; Moore 1979, 344; Moore 1990, 156; Ruse 1979, 112; Young 1985.

der Ireland, who served as Chambers's confidential intermediary in the publishing of *Vestiges*.[31]

If *Vestiges* is central to the introduction of an evolutionary narrative, as everyone who has written on the subject agrees, then historians must devote at least as much attention to the phrenological movement from which the work sprang. Phrenology was as much a way of life as a specific doctrine of mind, and these wider applications appealed most to Chambers. From the early 1830s, his long-standing interest in natural science was expressed in the phrenologically inspired educational programs in publishing undertaken by the family firm. In 1835, he and William launched a series of textbooks for young children, Chambers's Educational Course, implicitly based on Combe's doctrines. Robert's most successful contribution was also his first book on science, a general *Introduction to the Sciences*, published in 1836. The book began with chapters on the "extent of the material world" and "the stars," and ended with "man—his mental nature." This book, anonymous like others in the series, was widely used in schools. It had sold over 120,000 copies by 1849, even more than the People's Editions of Combe's *Constitution*.[32]

The *Introduction*, like other pre-Vestigian works by Chambers, drew on phrenology's lessons for living in accordance with natural laws—what he termed "the philosophy of phrenology." It did not refer to its "organology"—the physiological and anatomical basis of the doctrine. And although its discussion of the relation between mind and matter could have been written by Combe, the word "phrenology" was not mentioned.[33] This was to be expected in a schoolbook, for many parents would equate phrenology with materialism and a denial of man's immortal soul. For similar reasons, almost no references to phrenology are to be found in *Chambers's Journal*.[34]

In this strategy Chambers differed from Combe, who argued that drawing on phrenology without advocating the physiological views underpinning it was neither candid nor fair. As a commercial publisher Chambers knew that revealing his debt to phrenology would harm the circulation of *Chambers's Journal* and the Educational Course. The money-losing *Phrenological Journal* was not a

31. Surprisingly, one phrenological enthusiast who was not an intimate of the Chambers's household was Robert's brother William. From the late 1830s onward their relationship was confined to business matters. On a personal level, they were not close. The rift has hitherto remained a private matter masked by public displays of brotherly goodwill. See RC to W. Chambers, 28 Mar. 1859, NLS Dep. 341/93, ff. 50–59, and Norman mss., for evidence of long-standing antipathy.

32. For these points, and a discussion of the Educational Course, see Cooney 1970, 153–213; see also [Chambers] 1836.

33. "In what manner the material and immaterial things are connected during life, no one can tell" ([Chambers] 1836, 97.

34. Hutchinson 1980, 93–94.

model to be followed. The basic message could be communicated, and received by readers "as if it was a new gospel," without alienating them with the word "phrenology":

> The process is analogous to that recommended by Duff for converting the heathen. First convince their understandings that a correct system of mind has been discovered, and there will then be no obstacle to the reception of its fundamental truths except what may still be presented by the habit of venerating former systems. When we reflect that some of the forms of heathenism survived in Scotland till the close of the last century, perhaps eight hundred years after Christianity had been acknowledged to all intents and purposes, we must not fret at the slow progress which phrenology makes.[35]

Similarly, Chambers maintained that his friend failed to make enough concessions to religion. Combe, he wrote in a letter, "is too materialistic."[36] The basic differences between the two men, however, involved strategy rather than substance. Chambers accepted the tenets of phrenology and the application of natural law to human life. "Phrenology," he remarked in recommending Combe for Edinburgh University's logic chair, "appears to bear the same relation to the doctrines of even the most recent metaphysicians, which the Copernican astronomy bears to the system of Ptolemy."[37]

Cosmic Waverley

Vestiges thus grew directly from phrenological soil. Its author's self-image, however, continued to be dominated by Sir Walter Scott. Chambers had always defined his relationship to his literary father in nationalistic terms, contending that his status (after Scott) as Scotland's "favourite national writer of the day" deserved more recognition; as he told Scott himself in 1831, "to moralize a great nation" was "a noble thing." After the death of Scott in the following year, *Chambers's Journal* became a vehicle for forwarding his legacy as a universal man of the people.[38] However, Chambers's early goal of emulating Scott by writing a great work—especially a treatise on science—proved elusive. Editorial work and business concerns gave time only for compiling short books in odd moments. These had ranged from *Notices of the Most Remarkable Fires which have Occurred in Edinburgh* (1824), written in three days to catch public interest after

35. RC to G. Combe, 25 Nov. 1835, NLS mss. 7234, ff. 140–41. Alexander Duff (1806–78) was a leading Presbyterian missionary on the Indian subcontinent.

36. RC to AI, [1845], NLS, Dep. 341/110/88–9. For further comments on Combe and materialism, see "George Combe," in Chambers 1864, 2:213–14.

37. See Combe 1836b, 55.

38. RC to W. Scott, 1 Mar. 1831, NLS mss. 3917, ff. 3–4.

a devastating conflagration, to *Scottish Jests and Anecdotes* (1832), a foolhardy attempt to prove the Scots a "jocular race." By 1829 Scott (who was scribbling himself to death) was already worried that Chambers was "a clever fellow who hurts himself by too much haste."[39]

Scott's son-in-law, John Gibson Lockhart, took a dim view of these efforts to hijack the great man's memory to the cause of mass journalism and progressive improvement. Sharp-tongued, snobbish, and satirical, he disdained the "offensive vulgarity" of Chambers's works, which "will never obtain a place in the English library." His seven-volume life of Scott quoted criticisms of Chambers from a private diary, and in the *Quarterly Review* Lockhart condemned *Chambers's Journal* as the "ephemeral compilation" of a "worthy trafficker":

> All's well that ends well—for Mr. Chambers's till. We are well pleased that his till thrives. He started in life, however, with ambition of a different sort from that which has been thus copiously gratified. The performances of his juvenile pen afforded promise of distinction in the historical and antiquarian departments of literature. We see for what pursuits these early favourites have been cast aside. . . . It was, however, though singularly successful, an abandonment of exactly the same species which we recognize in a very great proportion of Mr. Chambers's contemporaries—men who have continued to be authors, but dropped by degrees, as the experience of life grew on them, the high aspirations which apparently animated their youth, and, in place of setting and keeping before them some great plan not to be fulfilled without a life-long devotion, have bestowed their ink upon those easier kinds of literature which furnish amusement sufficient for the hour, and for which the pay of the hour is sufficiently liberal.[40]

Lockhart scorned Chambers as an upstart tradesman, whose "amusements" allied authorship with money-grubbing commercialism and the effeminate world of the lower-middle-class clerk. As *Quarterly* editor, Lockhart wished to make authorship a calling suitable for gentlemen. Comments like these—and Lockhart was not the only one to make them—demanded that Chambers define himself as an author and a man by writing a substantial antiquarian treatise.

Chambers, in contrast, had long maintained that the modern spirit of the age could best be realized in a scientific work. He argued that the sciences were fast overtaking the kind of researches advocated by Lockhart—just as higher forms had succeeded insects, worms, and reptiles in the record of life. *Vestiges* noted that Scott's brain, like Shakespeare's, had been marked by its balance and universality, so that in the appropriate circumstances he might have been a man

39. Diary entry for 24 Feb. 1828, quoted in Lockhart 1837–38, 7:179.

40. [Lockhart], "The Copyright Question," *Quarterly Review* 69 (Oct. 1841): 186–227, at 199; also [Lockhart], "Mahon's *History of England*," *Quarterly Review* 63 (Jan. 1839): 151–65, at 153.

of science: "not only the poet, but the warrior, the statesman, and the philosopher"[41] (fig. 3.3). As Chambers had written in 1842, science was now the way forward: "physiology alone could throw more light upon the origin and progress of nations, within the bounds of one small volume, than could be done by a whole library of political history, or the united labours of a score of archaiological societies."[42]

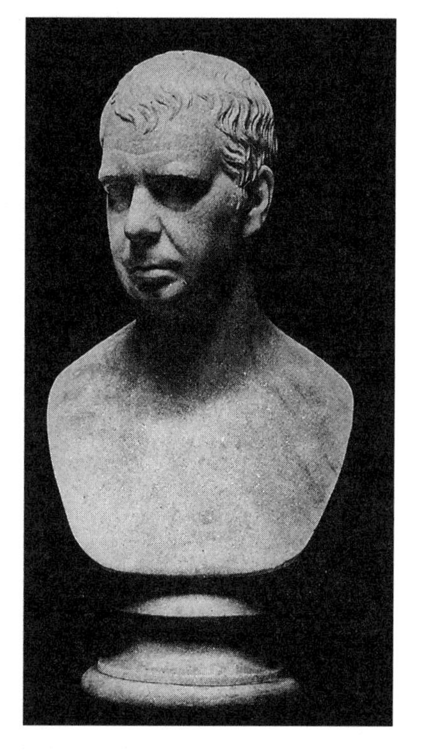

3.3 Phrenologically informed marble bust of Sir Walter Scott. Based on careful measurements with calipers, the sculpture was modeled in clay by Lawrence Macdonald in 1831 while Scott dictated a romance to his amanuensis. Chambers would have often seen the bust at the house of George Combe, who owned it for many years. Photograph in Stirling Maxwell 1872, plate 4.

Vestiges applied Scott's methods on a cosmic scale. As a young man in the 1820s, Chambers had begun writing a historical novel based on the story of the earl of Montrose; and although this was never completed, his account of the progress of nature is underpinned by the conventions of the genre. The historical novel of the early nineteenth century was characterized by its reconstruction of past worlds in relation to specific circumstances, so that history was seen as a shaping force in the most minute aspects of everyday life. In Scott's novels, the force behind all history is the onward progress of commercial society. The Waverley narratives are an attempt to preserve the memory of an older feudal social order that is, on the testimony of the novels themselves, inevitably going to disappear.[43] In this way, the novels rely upon the philosophical history of nature and society given currency during the Scottish

41. *V*1, 351–52.

42. [Chambers], "Thoughts on Nations and Civilisation," *CEJ,* 21 May 1842, 137–38, at 138.

43. These points have been extensively discussed in the literature on Scott, but not connected with the recasting of evolutionary narratives in the sciences. See Bann 1984, Brown 1979, Forbes 1953, Garside 1975 for discussion of Scott and the philosophical historians of the Enlightenment. The classic account remains Lukács 1962; in defining the generic conventions of the historical novel, I have found Shaw 1983 especially useful.

Enlightenment. Chambers was familiar with this grand vision not only through fiction, but also through key Enlightenment texts, such as Adam Ferguson's writings, the *Encyclopædia Britannica,* and David Hume and Tobias Smollett's *History of England.*[44] Such works had sympathy neither for religious enthusiasm nor for divine intervention. Human life could be viewed with the eye of reason, seen as part of the time-bound world of the everyday; this made possible the construction of a science of man based on history.

Vestiges stripped the Waverley novels to their essentials in nature's laws. By retaining traces of the generic conventions of historical fiction, however, the evolutionary cosmology of the Enlightenment was recast in a form appropriate for a Victorian readership. A story that had been deemed too dangerous for polite conversation could be brought to the center of public debate. Hence the widespread acknowledgment among contemporary readers that *Vestiges* read like a novel.

Chambers began to see the possibilities in such a project soon after he joined the phrenological party of reform. As he began to view nature's laws from the phrenological perspective of Combe's *Constitution,* he praised Scott as a great practical phrenologist—though he recognized that his hero was not a believer in the doctrine. In 1835, as the People's Edition of Combe's book was pouring from the Waterloo Place factory, Chambers described his scientific *chef d'oeuvre* as a treatise on "the philosophy of phrenology"—"a system of mind on the phrenological basis." He spent his spare hours researching the project, corresponding with friends and borrowing books and periodicals, including a run of the *Phrenological Journal.*[45]

The scope of the planned work widened to include the entire natural world after Chambers read Nichol's *Views of the Architecture of the Heavens* (1837),[46] a work that was itself indebted to Scott's historical fictions. He had long been aware of nebular speculations, and as a young boy discussed William Herschel's ideas on the structure of the universe with his father. But Nichol's influence was critically important. The pages of *Chambers's Journal* indicate an intense enthusiasm for the *Architecture,* for "high and rational Wonder has never been so delightfully associated with moral feeling."[47] In 1837, the year the *Architecture* was published, Chambers, Nichol, Alexander Ireland, and their phrenological

44. Chambers 1872, 62–63; Layman 1990, 84.

45. RC to AI, [Nov. 1837], NLS Dep. 341/110/9–10; also RC to AI, [1845], NLS Dep. 341/113/164–5; and J. G. Wilson, "Robert Chambers," 22.

46. *V*10, v; see also Ogilvie 1975.

47. [Chambers], "Professor Nichol's Views of the Architecture of the Heavens," *CEJ,* 29 July 1837, 210–11. On the early science discussions, [Chambers] 1872, 45; and Chambers in Layman 1990, 85.

friends toured Ireland together for sixteen days;[48] Anne Chambers jokingly warned her husband not to be carried away by astronomical enthusiasm.

Once the nebular hypothesis had led to the application of a law of progress to the whole realm of nature, Chambers began to explore other relevant sciences. Thus the day after returning from Ireland, he told a friend that he was "in the commencement of a geology fever, and extremely anxious to make up a little collection of the appropriate objects."[49] He studied zoology and botany as well. Most pieces of the puzzle are evident in *Chambers's Journal.* There were essays on monstrosities, the progressive nature of the fossil record, the habits and instincts of animals, the learning abilities of dogs and pigs, the spontaneous generation of insects through electricity, and the effects of diet and exercise upon health. Articles on races, nations, languages, and civilizations elaborated a developmental model almost identical to the *Vestiges* chapter on the "early history of mankind." In "Gossip about Golf," Chambers even applied the model to his favorite sport, which he argued was an inevitable result of the "existence of a certain peculiar waste ground called links." Similarly, cricket was said to be a natural outcome of village greens in England.[50] The ease with which these topics could be accommodated within a family periodical underlines the precedents for constructing narratives in which human life is the product of circumstance.

In *Vestiges,* as in Scott's fiction, picturesque scene settings seem to be digressive excursions but prove essential to an underlying vision of history as the product of minute particulars. *Vestiges* characterized the habits and appearance of extinct trilobites and ammonites with the same care that Scott used to recreate the world of Rob Roy and Edie Ochiltree. When early fish are described as having "a large crescent-shaped head, somewhat like a saddler's cutting-knife," this is both a homely detail and evidence for their anatomical connection with earlier, crustacean forms of life.[51] As in the Waverley novels, different epochs have highly specific characteristics and occur in a definite succession. Similar borrowings from Scott—though in more limited domains—are evident in Nichol's *Architecture* and Hugh Miller's *Old Red Sandstone* (1841), among other works; these not only provided models for *Vestiges,* but made its direct debt to fiction less obvious.

By grounding his narratives in universal history, Scott had transformed the novel from a genre associated with languorous femininity (and female authors)

48. [Chambers] "A Few More Days in Ireland," *CEJ,* 7 Oct.–11 Nov. 1837, 289–90, 301–2, 309–10, 317–18, 325–26, 333–34; see also RC to A. Chambers, 23 Aug. 1837, Norman mss.

49. RC to D. R. Rankine, 3 Sept. 1837, NLS Dep. 341/109/1.

50. [Chambers], "Gossip about Golf," *CEJ,* 8 Oct. 1842, 297–98; also in Chambers 1847a, 2:313–24.

51. *V1,* 69.

to one of moral improvement and masculine intellectual vigor. Chambers also attempted to make a threatening kind of writing safe and manly, by reversing the process—recasting Enlightenment cosmologies into literary forms drawn from Scott.[52] He had no wish to produce a work that could be associated with the feminizing qualities of light fiction or French novels by the likes of Balzac or Eugene Sue. In this Chambers looked to what the author of *Waverley* had announced as the source of all historical truth: "It is from the great book of Nature, the same through a thousand editions, whether of black-letter, or wire-wove and hot-pressed, that I have venturously essayed to read a chapter to the public."[53]

The Great Mystery

How did Chambers come to give the origin of new organic beings such an important place in his argument? He knew that Herschel had proclaimed the law of life as "the mystery of mysteries"; but unlike the great astronomer, he claimed it as ripe for solution, an appropriate topic for a bold philosophical work. Even the most liberal authorities had avoided this issue: Carpenter's textbook discussed the origin of new species only in the context of degeneration, though without dismissing this as "absurd or untenable";[54] Lyell's *Principles* implied that some natural cause (not transmutation) would be found; Nichol's *Architecture* had spoken of "the germs, the elements of that LIFE, which in coming ages will bud and blossom," but left the extension of natural law into the organic world to be inferred.[55] In the excitement of reading the *Architecture,* Chambers saw that Nichol's goal of uniting "the mystical evolution of firmamental matter with the destinies of man" could became his own; he would carry the story further, into the living world of plants, animals, and human origins.[56]

The problem was, for Chambers, that transmutation was advocated by atheists and Frenchmen, and had few if any supporters in middle-class phrenological circles. The years on Leith Walk had inoculated Chambers against the underworld of ultraradical materialism. In the early 1840s Edinburgh was awash

52. RC to G. Combe, 17 Dec. 1840, NLS mss. 7254, ff. 1–6. On the masculine character of Scott's fictions in relation to other contemporary novels, see Ferris 1991.

53. Scott 1986, 5.

54. Carpenter 1839, 396; J. F. W. Herschel to C. Lyell, 20 Feb. 1836, in Babbage 1838, 225–26, which famously stated that the Creator "operates through a series of intermediate causes, and that in consequence the origination of fresh species, could it ever come under our cognizance, would be found to be a natural in contradistinction to a miraculous process—although we perceive no indications of any process actually in progress which is likely to issue in such a result." Recognition of "the Great Mystery" by men of science is stressed in *V*10, vi.

55. Nichol 1837, 127. On Lyell, see Secord 1997.

56. *V*10, v.

with freethought propaganda and handbills, while the local newspapers reported blasphemy trials and denounced itinerant infidel lecturers. The northern outpost of the atheist mission had been raided by the police in 1841, who found it filled with infidel pamphlets and pornography. As an editor who kept a finger on the pulse of working-class debate, Chambers might even have come across a series on "regular gradation" in one of the freethought periodicals that enjoyed public notoriety at this time. Transmutation also featured in Henry Hetherington's *Free-thinker's Information for the People* (1843), which threw down the gauntlet to W. & R. Chambers. The convicted blasphemer George Jacob Holyoake sent *Paley Refuted in his own Words* to Chambers from prison, and dedicated it to the firm.[57]

Chambers's early writings responded by denouncing materialism and transmutation; as early as 1825 the *Traditions of Edinburgh* had mocked Lord Monboddo's "fantastic theory of human tails."[58] Ten years later *Chambers's Journal* expressed amazement that "some very eminent philosophers" had claimed that "man himself, Socrates, Shakespeare, and Newton, were merely zoophytes in a state of high improvement and cultivation!" The limits of variability, the sterility of mules, and a host of other evidences precluded any possibility of change between one species and another. Lyell's demolition of Lamarck was "so satisfactory as to require us to say nothing in addition." This continued to be *Chambers's Journal*'s policy through 1838, when an editorial dismissal of those "most absurd notions" prefaced an account of the orangutan.[59] This stance was of a piece with advocacy of self-help, Whig political economy, and Malthusian restraint among the working class.

By building the narrative around the "universal gestation of nature," Chambers attempted to defuse the radical threat by incorporating it. If natural law were to apply with full force to the organic world and human mind, he wanted it to triumph under the banner of liberal reform rather than revolutionary freethought. Moreover, he hoped that the universal application of natural law would confound the Presbyterian "saints" whose domination of Scottish intellectual life he saw as a disaster. He began his project aiming to rid science of miracles. As he told Ireland, "every effort is made that reason and common sense would at all admit of to keep smooth with the sticklers—though I daresay I shall not succeed with the extreme ones."[60]

57. Holyoake [1847], v. More generally, see Desmond 1987 and chapter 9 below.

58. Chambers 1825, 2:177.

59. "Natural History: Animals with a Backbone," *CEJ*, 24 Nov. 1832, 337–38; "Popular Information on Science: Transmutation of Species," *CEJ*, 26 Sept. 1835, 273–74; "Sketches in Natural History: Monkeys," *CEJ*, 1 Sept. 1838, 251–52.

60. RC to AI, 30 June 1844, NLS Dep. 341/110/32–33.

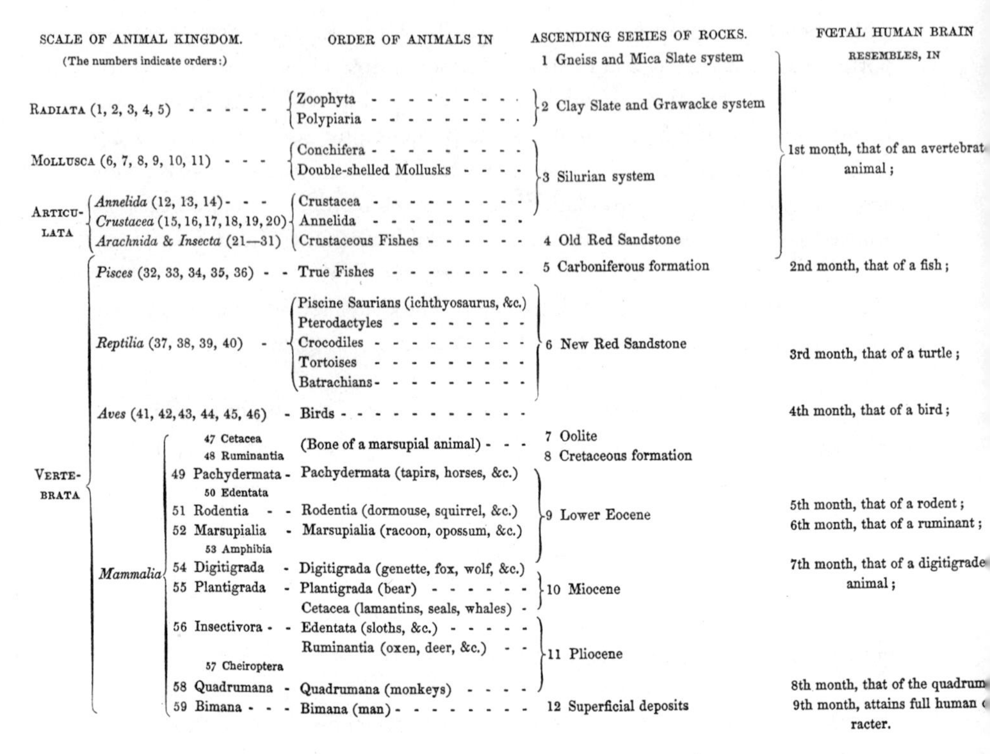

SCALE OF ANIMAL KINGDOM. (The numbers indicate orders:)		ORDER OF ANIMALS IN	ASCENDING SERIES OF ROCKS.	FŒTAL HUMAN BRAIN RESEMBLES, IN
			1 Gneiss and Mica Slate system	
RADIATA (1, 2, 3, 4, 5)		Zoophyta Polypiaria	2 Clay Slate and Grawacke system	
MOLLUSCA (6, 7, 8, 9, 10, 11)		Conchifera Double-shelled Mollusks	3 Silurian system	1st month, that of an avertebrat animal;
ARTICULATA	*Annelida* (12, 13, 14)	Crustacea		
	Crustacea (15, 16, 17, 18, 19, 20)	Annelida		
	Arachnida & *Insecta* (21—31)	Crustaceous Fishes	4 Old Red Sandstone	
VERTEBRATA	*Pisces* (32, 33, 34, 35, 36)	True Fishes	5 Carboniferous formation	2nd month, that of a fish;
	Reptilia (37, 38, 39, 40)	Piscine Saurians (ichthyosaurus, &c.) Pterodactyles Crocodiles Tortoises Batrachians	6 New Red Sandstone	3rd month, that of a turtle;
	Aves (41, 42, 43, 44, 45, 46)	Birds		4th month, that of a bird;
	Mammalia: 47 Cetacea 48 Ruminantia	(Bone of a marsupial animal)	7 Oolite 8 Cretaceous formation	
	49 Pachydermata	Pachydermata (tapirs, horses, &c.)	9 Lower Eocene	
	50 Edentata			
	51 Rodentia	Rodentia (dormouse, squirrel, &c.)		5th month, that of a rodent;
	52 Marsupialia	Marsupialia (racoon, opossum, &c.)		6th month, that of a ruminant;
	53 Amphibia			
	54 Digitigrada	Digitigrada (genette, fox, wolf, &c.)	10 Miocene	7th month, that of a digitigrade animal;
	55 Plantigrada	Plantigrada (bear)		
		Cetacea (lamantins, seals, whales)		
	56 Insectivora	Edentata (sloths, &c.)	11 Pliocene	
		Ruminantia (oxen, deer, &c.)		
	57 Cheiroptera			
	58 Quadrumana	Quadrumana (monkeys)		8th month, that of the quadrum
	59 Bimana	Bimana (man)	12 Superficial deposits	9th month, attains full human racter.

3.4 The three-fold parallel between the scale of the animal kingdom, the fossil record, and the development of the brain of the human fetus. Adapted from *V*1, 226–27.

From the late 1830s Chambers had come as close as he dared to discussing his changed ideas about species in *Chambers's Journal,* but possibilities were limited. Within a few years the hints became bolder. As he noted in an essay on "The Educability of Animals," "We become more confident in the improvability of our own species, when we find that even the lower animals are capable of being improved, through a succession of generations, by the constant presence of a meliorating agency."[61] Studies of race underlined the point. Instances of "white negroes" demonstrated "that the rise of the white races of men out of the black is within the range of possibility."[62] Given the constraints on publishing on sex and generation in a family periodical, Chambers never spoke more directly. Of all the issues "behind the veil," this was the most darkly hidden.

61. [Chambers], "Educability of Animals," *CEJ,* 16 Apr. 1842, 97–98; also in Chambers 1847a, 4:154–62.

62. [Chambers], "Popular Information on Science: Effects of Climate, &c. on Human Beings," *CEJ,* 18 Nov. 1843, 346–47.

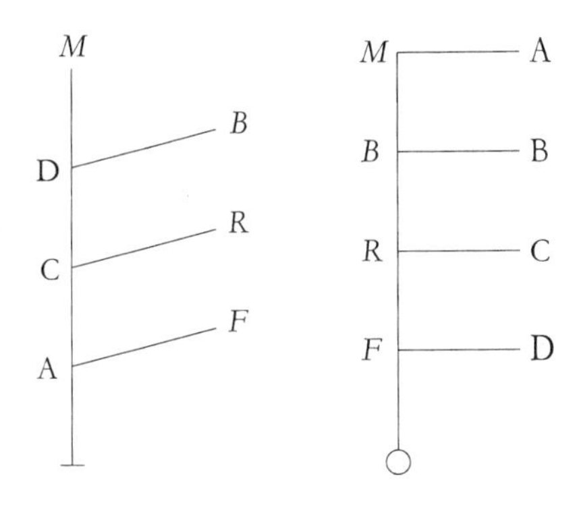

3.5 On the left, diagram illustrating progressive changes of type in the fetus according to the development theory. Starting from the bottom, the embryo of the fish advances to stage A, and then diverges to F; that of the reptile progresses to C and then to R; while the bird embryo diverges at D toward B. Mammals follow the full path of development through A, C, and D to M. The diagram shows that the adult forms (M, B, R, F) would differ from each other and from their fetal ancestors. Redrawn from *V1*, 212. The original diagram in Carpenter 1841, 197 is on the right.

In medical and physiological books, however, theories of embryological development were common currency; Chambers relied on three works by authors who had taught or been trained in Edinburgh. (His awareness of Continental writings was at second hand.) Probably the first book Chambers consulted was Perceval Lord's elementary antiphrenological *Popular Physiology* (1834). Lord explained how the human embryo passed through stages resembling a fish, a reptile, a bird, and a mammal; it then "successively represents the characters with which it is found in the Negro, Malay, American, and Mongolian nations." Chambers turned this recapitulation doctrine to an evolutionary end, giving his own work orthodox credentials.[63] His second source was John Fletcher's more sophisticated textbook, which discussed recapitulation at length and also rejected phrenology. Although Fletcher denied transmutation and any simple Lamarckian series, he did range the entire animal kingdom in ascending order of complexity. Chambers found this list very suggestive, especially when juxtaposed with the appearance of various species in the geological record. A chart illustrating this comparison became an important part of his effort to enhance the visual impact of his argument (fig. 3.4).[64]

Finally, Chambers knew of the newly imported, rival embryology that argued that embryos do not recapitulate stages resembling the adult forms of simpler organisms. Instead, the process was one of differentiation (fig. 3.5), with the embryo starting as a generalized form and gradually exhibiting the special characters of the adult species. As *Vestiges* explained, "the resemblance is not to

63. *V1*, 200–201, quoting (with minor alterations) from Lord 1834, 347–48. The reference to Tiedemann in *V1*, 201, is taken directly from Lord, as is the reference to Serres (*V1*, 306–7). His comments on Lamarck (*V1*, 230–31) imply no further knowledge than is available from Lyell's *Principles*.

64. *V1*, 226–27.

3.6 Silhouette of Anne Chambers playing the harp, with eight of her nine children, c. 1844: James (age 3), Amelia (6), Anne (9), Mary (11) with William (1), the twins Janet and Eliza (8), and Nina (14). The eldest son, Robert (12) is not shown; three other children had died in infancy.

the adult fish or the adult reptile, but to the fish and reptile at a certain point in their foetal progress."[65] This idea, designated "von Baer's law," was known to him chiefly through the same work that had received backing against the onslaught of the Edinburgh "saints," Carpenter's *Principles.* A summary of Carpenter's presentation and a version of his diagram appeared in *Vestiges,* though without acknowledgment—presumably so as not to stir up further trouble.[66]

More important than technical treatments of embryology, however, was Chambers's familiarity with everyday processes of birth and development. Having been born with six fingers and toes, his own body was marked with the truth of his theory, signs that he knew could be passed to his offspring. An enthusiastic father, Chambers had nine children (six girls and three boys) in the family during the years at St. Andrews, and watching them mature was a fundamental part of the experience that went into the making of the book (fig. 3.6).[67] As he later wrote, "the ordinary phenomenon of reproduction" had impressed him as "the key to the genesis of species."[68] Chambers had been profoundly

65. As Gould 1977, 110–12, points out (p. 110), Chambers "read the vertical line of fish-reptile-mammal-human as a foreordained path of evolutionary advance."

66. *V1*, 212. It is worth noting that the source was explicitly credited from the fifth edition of 1846 (*V5*, 220), when Carpenter himself was actively contributing to the revision of *Vestiges.*

67. For the family, see Norman mss., and Priestley 1908, 28–49.

68. *V10*, vii.

moved by the birth of his first son, Robert, and published a signed poem, "To a Little Boy," in *Blackwood's Magazine* for July 1835. "The feelings there expressed," he told a friend two years later, "have suffered no change, for he himself, in advancing out of childhood, has lost none of its endearing qualities. What a history, though an incommunicable one, resides in my mind respecting the aspects, talk, doings, and traits of progressive intelligence of this dear boy in the course of his brief existence."[69] This "incommunicable history" was finally expressed in a vision of all nature engaged in gestation and development.

The Voice of Nature

Reintroducing a causal account of all natural phenomena demanded the creation of an appropriate literary form. Any direct debts to the historical novel had to be muted, lest controversial claims be dismissed as imaginative fantasies (which some readers did anyway). Univocal modes of address typical of law, mathematics, or physics would be equally inappropriate. Chambers needed to create not only a new vision of creation, but a narrative voice through which to speak to potential readers.

There were many resources to hand for defining such a voice. An omnivorous reader and critic, Chambers had a good understanding of classical rhetoric and an unrivaled familiarity with literature in English. His two-volume *Cyclopædia of English Literature,* written with the assistance of Robert Carruthers of Inverness and published in 1844, became a standard work on both sides of the Atlantic. Widely used in university teaching and in print for nearly a century, this critical anthology was of major importance in defining the "Pantheon of our national authors."[70] Chambers was also instrumental in developing the Scottish literary canon, especially through his editions of Robert Burns. A keen follower of scientific developments, he read all the major works of reflective science, and had most of them in his large private library at St. Andrews.[71]

Finding an appropriate voice had always been critical to the commercial success of *Chambers's Journal.* By 1847 Chambers had written nearly four hundred leading essays, together with a far larger number of shorter pieces. "My design from the first," he said, "was to be the essayist of the middle class."[72] In these contributions, he assumed the character of an "intimate acquaintance or friend"

69. RC to D. R. Rankine, 3 Sept. 1837, NLS Dep. 341/109/1. See also [Chambers], "To a Little Boy," *Blackwood's Edinburgh Magazine* 38 (July 1835): 70.

70. "A Few Words to our Readers," *CEJ,* 4 Jan. 1845, 1; Chambers 1843.

71. The auction catalogue for Chambers's library (Chapman 1873) is obviously of a much later date, but many of the titles were presumably purchased at this time as he was a keen book collector throughout his life.

72. Chambers 1847a, "General Preface," 1:iv. Chambers's editorial policies are discussed in Cooney 1970.

on a weekly visit to his readers, sitting at the fireside for a friendly chat. This consistency of tone gave the periodical "a strong and abiding character," which could never be obtained by a miscellaneous assemblage of papers. "When a periodical has such a presiding spirit," Chambers told Combe, "it may be said to form a friendship with every one of its readers."[73]

The narrator in *Vestiges* is very close to the "presiding spirit" of anonymous early Victorian periodical journalism. The reader is implicitly led by an earnest companion and guide, rather like a bourgeois Virgil in Dante's *Divine Comedy* or the archangel Raphael in the seventh book of Milton's *Paradise Lost.* Like the "presiding spirit" of mass journalism, this narrator is aware of the divisions within industrial society and among its readership, but draws the people together by being itself without class, age, or gender. The narrator is a spiritual *flâneur,* moving in and out of a crowd of phenomena that are seen simultaneously en masse and according to a highly individualized typology.[74] This "universality" of voice was in Chambers's view the key to Scott's status as an author whose understanding of human nature came second only to that of Shakespeare. It was also the key to the narrative voice for *Vestiges.*

Implied narrator and implied reader explore the book of nature together, and the relationship runs as a subplot throughout the work. The narrative voice is patient, modest, with few of the clearly marked gender or class characteristics found in Thackeray or Edward Bulwer-Lytton's novels.[75] There is no hint of the dogmatic certainty that had characterized Enlightenment materialism. Take the brutal declaratives that had opened D'Holbach's *System of Nature:* "Man always deceives himself when he abandons experience to follow imaginary systems.—He is the work of nature.—He exists in nature.—He is submitted to her laws."[76]

An entirely different relationship with the reader is established in the first paragraph of *Vestiges:*

> It is familiar knowledge that the earth which we inhabit is a globe of somewhat less than 8000 miles in diameter, being one of a series of eleven which revolve at different distances around the sun, and some of which have satellites in like manner revolving around them.[77] The sun, planets, and satellites, with the less intelligi-

73. RC to G. Combe, 17 Dec. 1840, NLS mss. 7254, ff. 1–6; also in Gibbon 1878, 2:126–27.

74. Klancher 1987, 76–97; for the spiritualized quality of this voice, see Winter 1995 on Harriet Martineau.

75. Flint 1996.

76. D'Holbach 1834, 1:9.

77. In the nineteenth century, the term "planet" referred not only to Mercury, Venus, Earth, Mars, Jupiter, Saturn, and Uranus (Neptune was discovered in 1846) but also to a number of what today would be termed "asteroids."

> ble orbs termed comets, are comprehensively called the solar system, and if we take as the uttermost bounds of this system the orbit of Uranus (though the comets actually have a wider range), we shall find that it occupies a portion of space not less than three thousand six hundred millions of miles in extent. The mind fails to form an exact notion of a portion of space so immense; but some faint idea of it may be obtained from the fact, that, if the swiftest race-horse ever known had begun to traverse it, at full speed, at the time of the birth of Moses, he would only as yet have accomplished half his journey.[78]

The emphasis is on ascertained fact shared between reader and narrator. The opening words juxtapose the declarative "It is" with the reassuring "familiar knowledge," prefiguring the theme that an understanding of nature is accessible to all. From the very beginning, the new cosmology takes its place in the domestic, "family," circle.

The relationship between reader and narrator in *Vestiges* is initially subdued. This is in contrast not only with most novels of the 1840s, but also with the openings of reflective works in the sciences. Nichol addresses the reader directly in the first of the letters that make up his *Architecture;* Buckland's Bridgewater presents itself as an imaginary travel book; Miller starts with advice to workingmen and then engages the reader in the story of his life in the *Old Red Sandstone.*[79]

Vestiges begins with facts. But the possibility of greater intimacy between author and reader is hinted at from the start. Narrative voice and prospective readers are immediately united through the pronoun "we," which appears three times in the first two sentences. This imperial "we" implicitly includes all those who inhabit the earth. The closest parallels in the tradition of reflective science are in Mary Somerville's *Connexion of the Physical Sciences,* which opens by invoking the elevated character of the search for the laws of nature; in Combe's *Constitution of Man,* where "we discover" by "surveying the external world"; and in Herschel's *Preliminary Discourse,* which situates "man on the globe he inhabits."[80] Compared to these works, though, the opening of *Vestiges* moves quickly from such transcendent concerns to concrete particulars and problems—a shift characteristic of the historical novel.

Combining an address to all humanity with an emphasis on facts required a finely tuned understanding of what readers might be expected to know or might find boring. A blunt declamation (e.g., "The earth is 7,902 miles in diameter") would have indicated a didactic work appropriate to children, an improving lecture, or the *Penny Magazine,* which was deeply unfashionable by the early

78. *V*1, 1–2.

79. Nichol 1837, 3; Buckland 1836, 1; Miller 1841, 1–4.

80. Somerville 1834, 1; Combe 1835, 1 (changed from the first edition of 1828); Herschel 1830, 1.

1840s.[81] "It is familiar knowledge," announces a work that hopes to address a wide audience, without being condescending or dull. (More subtle are echoes of celebrated fictional openings, perhaps most especially that of *Pride and Prejudice.*) The opening also acknowledges fears of didacticism by establishing a hierarchy of significance in the information provided; thus important facts about the sun, planets, and satellites are given priority over details about the "less intelligible" comets (whose status is consistently parenthetical). All numbers are only as precise as needed for the purpose at hand; thus the earth's diameter and the solar system's dimensions are only approximations.

The dimensions of the solar system emerge in the second sentence of the opening paragraph, which engages reader and narrator in a joint attempt to solve a problem of mutual interest. "If we take" certain parameters, "we shall find": after the familiarity of the first sentence, the quasi-mathematical nature of this problem—and the announcement of the fact that the solar system is "not less than three thousand six hundred millions of miles in extent"—creates a dramatic tension, in which issues of didacticism and difficulty arise again. This tension is resolved by acknowledging that "the mind fails to form an exact notion of a portion of space so immense." Any failure of comprehension is thereby attributed not to the individual reader's stupidity but to characteristics of the universal "mind." Even by the end of the first paragraph, readers have been enlisted in a common quest to understand the most sublime phenomena of nature.

The formation of this "mind"—whose development will proceed throughout the work—begins here through an act of measurement, a shared comprehension of human scales of space and time. These scales draw on precisely the kind of "familiar knowledge" that Chambers anticipated all his readers to possess: a speeding racehorse, traveling from Moses' birth, could give at least "some faint idea" of the immensity of the heavens. The same strategy is used in opening his anonymous *Introduction to the Sciences,* where a mile is defined for the young reader as "about as much as he can walk at once without being tired."[82]

The emphasis on "familiar knowledge" continues through the first third of *Vestiges,* which is plotted as a narrative of origins taken from astronomy, chemistry, and especially geology. The main organizing metaphor is the ancient one of the "book of creation," so that readers will be drawn into the text as a direct transcript of nature. Even the physical act of turning pages becomes a metaphor for scientific exploration: the description of the strata series is presented as "the leaves of the *Stone Book,*" and subdivisions in the work are the "record of this period," "great natural transactions," "a new chapter in this marvellous history,"

81. The figure is stated this way, for example, in "Geography," in Chambers and Chambers 1842, 1:33–38, at 33.

82. [Chambers] 1836, 1.

or the "next volume of the rock series."[83] Causal language and temporal sequence, combined with the forward-directed process of page turning, brings out the force of progress in nature, so that the act of reading affirms progressive development.

The image of nature as a book was associated not only with geological and other scientific writings but with the Bible. The divine authorship of the "book of creation" is explicitly brought out for the first time at the end of the astronomical chapter:

> But it is impossible for an intelligent mind to stop there. We advance from law to the cause of law, and ask, What is that? Whence have come all these beautiful regulations? Here science leaves us, but only to conclude, from other grounds, that there is a First Cause to which all others are secondary and ministrative, a primitive almighty will, of which these laws are merely the mandates. That great Being, who shall say where is his dwelling-place, or what his history! Man pauses breathless at the contemplation of a subject so much above his finite faculties, and only can wonder and adore![84]

Here the language echoes the ecstatic prose of *Architecture of the Heavens,* to defuse critics who might condemn *Vestiges* as an attempt to distance God from the creation. Yet this and other passages open up the possibility that the scientific study of nature's laws says nothing about a First Cause, which must be inferred "from other grounds," and their fervent tone contrasts with the rest of the argument. Readers of *Vestiges* were thus effectively given a choice of devotional languages. In his secret correspondence Chambers stressed the strategic character of his invocations of the deity. "I am happy to say," Chambers wrote to Ireland in sending a final batch of manuscript, "that I have been able at the end to introduce some views about religion which will help greatly, I think, to keep the book on tolerable terms with the public, without compromising any important doctrine."[85]

To avoid accusations of undermining faith, nature's story is presented as unthreatening, uncoercive, and implicitly narrated within a domestic scene. (In a passage added to the sixth edition, the reader of creation is literally compared to a child sitting "at a mother's knee" asking "of the things which passed before we were born.")[86] Familiar images of birth, childhood, the family, and the home—the staples of Victorian mass journalism—are embedded throughout. In

83. *V1*, 57, 116, 144, 66, 94.

84. Ibid., 25–26.

85. RC to AI, [1844], NLS Dep. 341/110/157.

86. *V6*, 17. In the helpful terminology of Greg Myers (1990, 195–203), the early chapters of *Vestiges* perfectly exemplify the "narrative of nature."

explaining the evidence for the nebular hypothesis, the narrator appeals to an analogy with details of human growth. Stars are visible in every stage of development; "[i]t may be presumed that all these are but stages in a progress, just as if, seeing a child, a boy, a youth, a middle-aged, and an old man together, we might presume that the whole were only variations of one being."[87]

The domestic model is then extended to the solar system. A long passage is quoted from Herschel's *Astronomy,* describing how the planets are joined in "a true *family* likeness . . . interwoven in one web of mutual relation and harmonious agreement."[88] This appearance of happy domesticity could be taken literally, *Vestiges* suggests, for the nebular hypothesis showed that the planets were "children of the sun," generated from the stellar body according to the same universal law that had produced the galaxies. Readers could be expected to be familiar with language of this kind from Nichol's *Architecture,* which spoke of "nebulous parentage" and the "inexhaustible womb of the future."[89]

Explanations of order and structure were accorded most power when attached to numbers.[90] Following a position argued in John Stuart Mill's *System of Logic* (1843), *Vestiges* refers to mathematical arguments to produce conviction on empiricist grounds, not because they embody any necessary truth.[91] In *Vestiges,* numbers, lines, and points are expected to gain the reader's assent because they are visible. The regularity of planetary distances is visually striking; Babbage's calculating engine provides a convincing explanation of miracles because it can be watched in operation.

The uncontroversial, factual character of the early chapters of *Vestiges* is enhanced by the subdued persona adopted by the narrator, who cites over eighty different authorities and presents the work as a summation of the scientific findings of the preceding two decades. Famous names are marshaled both in the text and notes: Friedrich Argelander, Thomas Henderson, William Herschel, Pierre Simon Laplace, Ottaviano Fabrizio Mossotti, and other astronomers back up the nebular hypothesis, even against the evidence of "our eyes." Later passages speak of "men eminent in science," note that "mechanical philosophy informs us," and report "the information which chemistry gives us." Geology is

87. *V*1, 8.

88. Ibid., 11–12, quoting (with slight modifications) from Herschel 1833, 264. See also the comments on this passage in Beer 1983, 169.

89. Nichol 1837, 194, 195.

90. A good example of this can be found in an exchange of letters in the *Manchester Guardian* concerning Argelander's calculations of the motion of the solar system in space. See "J. G. S." [Chambers], "Motion of our Solar System," *Manchester Guardian,* 8 Feb. 1845, 5, in reply to H. H. Jones, "Motion of Our Solar System," *Manchester Guardian,* 29 Jan. 1845, 8.

91. Mill 1973–74, 1:224–61; Holyoake [1847]; for account of Mill's views, see Bloor 1991, 87–92, and Richards 1988.

"calculated to excite our admiration" and exalts "the dignity of science, as a product of man's industry and his reason."[92] If creation can be read as a book, simple enough for a child to understand, it is because of the interpretative work of the astronomers, naturalists, and geologists, who are credited—more regularly than God—as authors of nature's story.

This deferential perspective toward science is typical of the useful knowledge movement of Charles Knight, the SDUK, and the *Penny Magazine.* Its emphasis in the early chapters is strategic, to gain the reader's confidence and allies among men of science. Chambers had been greatly heartened by their public defense of Carpenter after his *Principles of Physiology* had been condemned as tending to atheism.[93] Carpenter's supporters had included John Herschel; the Cambridge physiologist William Clark; the Congregationalist leader John Pye Smith; the Oxford natural philosopher Baden Powell; and the royal physician, Henry Holland. It was an impressive list. Chambers was equally encouraged by the meeting of the British Association for the Advancement of Science at York in September 1844, which he attended "in the pure simplicity of a love of science and of science's cultivators," just two weeks before *Vestiges* appeared. Among the heroes of the report in *Chambers's Journal* was the Reverend Adam Sedgwick, who defended modern geology against "the whimsicality" of the dean of York's scriptural version of the science.[94] All this augured well for the reception of *Vestiges,* which Chambers hoped would lead readers—and especially men of science—to still bolder views.

Toward this end, the narrator does not just report established findings, but maintains independent positions almost from the very beginning. For example, the regularity of planetary distances from the sun is said to be "a most surprising proof of the unity which I am claiming for the solar system." When connecting the origin of life and the appearance of limestones in the geological record, reference is made to "my hypothesis."[95] Such statements are accompanied by modest disclaimers, but they invite the reader to go beyond established knowledge by drawing out larger consequences and general laws.

These passages exhibit a tension between the authority granted to men of science (both as readers and as experts) and criticism of their timidity. This can be seen in vacillations over what to call the book—an issue of vital importance when so little else was going to be on the title page. In August 1844 Ireland

92. *V*1, 2, 179, 13, 35, 145–46.

93. RC to AI, 30 June 1844, NLS Dep. 341/110/32–33. Crucially, Carpenter must have sent Chambers a proof copy of his 1840 defense, as this is in the papers of W. & R. Chambers (NLS Dep. 341/637).

94. [Chambers], "The Scientific Meeting at York," *CEJ,* 23 Nov. 1884, 321–24.

95. *V*1, 11, 57.

objected to Chambers's working title, "The Natural History of Creation," so that "Vestiges of" was added only weeks before publication.[96] His reasons are not known, but the change had advantages, for "The Natural History of Creation" was ambiguous. "Creation" could be taken either as a static noun, implying a completed whole; or it could be taken as an active process, focused on the very act of creation itself.[97] Similarly, a "Natural History" had traditionally been a descriptive work, contrasted with natural philosophies that dealt with causes; but this distinction was beginning to break down. "The Natural History of Creation" might announce the definitive attempt to explain the origin of all things (and hence challenge the literal truth of Genesis), but it might be a static description of existing nature. The first meaning was too bold, the second too tame. "Vestiges" derived from the Latin *vestigium,* or footprint, and had previously been used for the titles of antiquarian works, such as the classic *Vestiges of Ancient Manners and Customs Discoverable in Modern Italy and Sicily* (1825) by the Reverend John James Blunt. It suggested traces, hints, fragments—as in the just-published *Eöthen, or Traces of Travel Brought Home from the East,* which everyone in publishing circles knew was causing a sensation in London. "Vestiges of the Natural History of Creation" promised empirical modesty and an active role for readers in assembling fragments and completing the narrative. At the same time, the title posed a question of trust, for the "vestiges" were records of a process that could not be witnessed directly.

The Great Plot

The key moment in the development of the reader's relation to the narrator comes in the chapter on "general considerations respecting the origin of the animated tribes." When Chambers sent a fresh batch of manuscript and noted that "the great plot here comes out" it surely dealt with this transition, the most difficult in the entire work.[98] The reader is drawn from "the wondrous chapter of the earth's history which is told by geology" to the problem of giving a causal explanation for that history, and finally to the origin of species:

> A candid consideration of all these circumstances can scarcely fail to introduce into our minds a somewhat different idea of organic creation from what has hitherto been generally entertained. That God created animated beings, as well as the terraqueous theatre of their being, is a fact so powerfully evidenced, and so universally

96. RC to AI, "Saturday" [24 Aug. 1844], NLS Dep. 341/110/129, which can be dated by its reference to [D. Jerrold], "The Burns' Festival.—'Repentant Scotland,'" *Punch,* 24 Aug. 1844, 81–82; and for a reference to the change in title, RC to AI, [early Sept. 1844], NLS Dep. 341/113/144.

97. For the latter reading, and an analysis of Darwin's *Origin* to which I am generally indebted, see Beer 1983, esp. 64–65.

98. RC to AI, [pmk 30 July 1844], NLS Dep. 341/109/10.

> received, that I at once take it for granted. But in the particulars of this so highly supported idea, we surely here see cause for some re-consideration. It may now be inquired,—In what way was the creation of animated beings effected?[99]

The narrator, having established a claim to be believed, now gives explicit reassurance in the first person on only one point: God's role as the creator ("I at once take it for granted"). The reader is then ready to be invited on a new quest, to give a "candid consideration" of ideas that—for the very first time—are announced as being different from those "generally entertained." The question asked at the end of the passage is, crucially, both an acknowledgment of the reader's implicit willingness to proceed ("It may now be inquired") and a statement of the problem ("—In what way was the creation of animated beings effected?"). Everything in *Vestiges* up to this point has led up to establishing the legitimacy of this question.

In one sense, then, the remaining 250 pages follow out the narrative line, drawing the reader to extend the law of progress from astronomy and geology to living beings, to civilization, and to the destiny of the human race. These themes, however, cannot be developed without overcoming resistance. The narrator has to anticipate objections, engage in dialogue with the implied reader by making a case and providing evidence. The carefully cultivated relationship between narrator and reader means that this analytical voice can be adopted without seeming dogmatic, egotistical, or overly speculative: the reader, enlisted as a friend, is invited to share in confidential discussion.

This growing intimacy can be traced in the increasing frequency of the first-person singular in the work's second half, especially through the use of the rhetorical techniques of direct address ("I cannot here but remind the reader"). The strength of the narrative voice, and the corresponding demands on the bond with the reader, are especially striking in the pivotal chapter on the "hypothesis of the development of the vegetable and animal kingdoms":

> The idea, then, which I form of the progress of organic life upon the globe—and the hypothesis is applicable to all similar theatres of vital being—is, *that the simplest and most primitive type, under a law to which that of like-production is subordinate, gave birth to the type next above it, that this again produced the next higher, and so on to the very highest,* the stages of advance being in all cases very small—namely, form one species only to another; so that the phenomenon has always been of a simple and modest character.[100]

Italics signal this as a major claim to novelty. In announcing something new, the

99. *V1*, 152–53.
100. Ibid., 389, 222.

work projects a horizon of expectations that implicitly acknowledges that many readers will dismiss any discussion of organic progress as a doomed effort to revive the transmutation theories of the past. Chambers, having ridiculed Lamarck in *Chambers's Journal* on many occasions, knew all too well that this would be a common response. To obtain the reader's tacit assent to continue, the narrator becomes very explicit about the novelty of the theory being put forward:

> I take existing natural means, and shew them to have been capable of producing all the existing organisms, with the simple and easily conceivable aid of a higher generative law, which we perhaps still see operating on a limited scale. I also go beyond the French philosopher to a very important point, the original Divine conception of all the forms of being which these natural laws were only instruments in working out and realizing. The actuality of such a conception I hold to be strikingly demonstrated by the discoveries of Macleay, Vigors, and Swainson, with respect to the affinities and analogies of animal (and by implication vegetable) organisms.[101]

Novelty is vital to credibility. At the same time, though, claims to original discovery have the potential for being read as brash egotism, which would disrupt the developing relationship between reader and narrator. To minimize this possibility, a range of responses are acknowledged as appropriate, from outright conviction to suspended disbelief: "I do not indeed present these ideas as furnishing the true explanation of the progress of organic creation; they are merely thrown out as hints towards the formation of a just hypothesis."[102]

To make transmutation read as a transcript from nature, techniques from popular journalism are used to avoid the complexities of developmental embryology. The reader is introduced to the generalized version of recapitulation through a long extract from Lord's *Popular Physiology*, a "safe" work published by the Society for Promoting Christian Knowledge. Only then is the more technical treatment in von Baer's law of differentiation introduced as a qualification, and its source—Carpenter's controverted *Principles*—is not mentioned at all. The juxtaposition of two different developmental models—which had actually emerged in opposition to one another—has often been taken to demonstrate Chambers's lack of scientific sophistication, but from an expository point of view there were advantages. Von Baer's law was not well known in Britain, so that relying solely on it would mean explaining a novelty with a novelty. The latest results are stated clearly, but in a modest and uncoercive setting.

If new species come about in this way, why don't we see the process happening in the ordinary course of nature? Almost immediately after the extract

101. Ibid., 231–32.
102. Ibid., 230.

from Lord's *Popular Physiology,* the reader's attention is diverted from such questions, and the dangerous territory of sex and generation, by a visit to the celebrated calculating engine of the mathematician and inventor Charles Babbage. This is the most authoritative site in the country for demonstrating the lawlike character of miracles. Here, the narrator averts a potential crisis in trust by asking the reader to see the problem from a very different and less threatening perspective:

> The reader is requested to suppose himself seated before the calculating engine, and observing it.

The narrator then withdraws, allowing Babbage to take over. A quotation from the *Ninth Bridgewater Treatise*—the longest in the entire book—begins by reiterating the direct address, but this time transferred to the voice of the acknowledged authority:

> 'Now, reader,' says Mr. Babbage, 'let me ask you how long you will have counted before you are firmly convinced that the engine has been so adjusted, that it will continue, while its motion is maintained, to produce the same series of natural numbers?'[103]

Such a series might continue as a regular succession—say up to one hundred million and one; but at that point another law might come into action, so that the sequence of numbers from that point onward would advance in a different and unexpected way. "It is not difficult," the narrator in *Vestiges* resumes, "to apply the philosophy of this passage to the question under consideration."[104] As Babbage had said many times to visitors, what would appear a miracle is simply the outcome of a preprogrammed "higher law." The reputation of Babbage is crucial here, and it is his voice that makes the visit to the engine such an effective strategy for tackling the origin of new species.

Fears about transmutation are also addressed by rehearsing images of domesticity and the family. Readers are asked what it would be like to be "acquainted for the first time with the circumstances attending the production of an individual of our race." Faced by the facts of sex and reproduction, "we might equally think them degrading, and be eager to deny them, and exclude them from the admitted truths of nature." But these facts could not be denied by "a healthy and natural mind." Neither, by implication, could the analogous processes of gestation that gave birth to new species.[105] The production of new species "has never been anything more than a new stage of progress in gestation, an event as simply natural, and attended as little by any circumstances of

103. Ibid., 206, quoting Babbage 1838.
104. *V*1, 210.
105. Ibid., 233–35.

a wonderful or startling kind, as the silent advance of an ordinary mother from one week to another of her pregnancy."[106] The birth of a new species is neither more nor less to be feared than the birth of a child.

As the story unfolds, the reader is encouraged to view nature from a dizzying variety of perspectives: not only as an expectant mother, but as an uneducated rustic, an ephemeron fly, a sanitary inspector, an insect in a garden, a scientific expert, and a primitive savage. Life on other planets seems improbable until we realize that our position on earth is like that of the people in ancient times who make their first boat, and on going out to sea observed a fleet of other ships—"a set of objects they had never before seen." "Precisely in this manner," the reader is told, "we can speculate on the inhabitants of remote spheres."[107] Such analogies, in linking the familiar and fantastic, stress the need to transcend the limits of ordinary human sense.

Even passages that invoke familiar domestic settings shift perspective rapidly. Within the space of a few sentences, the narrator reports that "I have been told" how Manchester factory children develop new language; notes that "I have seen" private vocabularies develop in families; and calls on readers' experience of what occurs in families generally: "I believe I am running little risk of contradiction when I say that there is scarcely a family, even amongst the middle classes of this country, who have not some peculiarities of pronunciation and syntax." Like the "presiding spirit" of mass journalism, the narrator here assumes familiarity with middle-class family life but in no way ties the reader to it. The very next sentence concludes that the three or four thousand known human languages all have a common origin. With the evidence provided, this "is easy to understand."[108]

The aim is to give readers a growing sense of power, the ability to see things from any point of view; subjects that would have come across in an opening paragraph as outrageous speculations or undiluted materialism can be introduced as self-evident. "It is hardly necessary to say," the narrator casually remarks in the penultimate chapter, "much less to argue, that mental action, being proved to be under law, passes at once into the category of natural things. . . . the distinction usually taken between physical and moral is annulled."[109] The highest qualities of mind—numeracy, sagacity, wonder, imitation, love—first appeared in animals and are, most likely, electrical impulses. These were exceptionally controversial claims, but they can be offered for consideration because the character of the narrator as a spiritual guide has been established.

106. Ibid., 223.
107. Ibid., 162.
108. Ibid., 317–18.
109. Ibid., 331–32.

Parables of Reception

Vestiges, as we have seen, reinscribed Enlightenment cosmology as evolutionary narrative by using techniques from Scott, mass journalism, and the tradition of reflective writing on the sciences. The most touchy subject it treated, the creation of new species by transmutation, had never been allowed into polite conversation and middle-class homes. Building its generative model around familiar images of pregnancy, childhood, the family, and the hearth minimized the fears of potential readers, as did a carefully managed narrative structure. The anonymous persona of the Victorian weekly periodical—the voice that was everywhere and nowhere simultaneously—developed in the course of the narrative into a friendly guide, a spirit hovering over creation.

The book ends, like *Waverley,* with a "note conclusory" that under other circumstances could have been a preface. The narrator expects neither to respond to critics nor to forward the cause of the work, although these possibilities are not ruled out. "Thus ends a book, composed in solitude, and almost without the cognizance of a single human being." Written for the "benefit" of "my fellow-creatures"—an explicit echo of an earlier chapter's description of the role of genius in the development of civilization—the work "goes forth to take its chance of instant oblivion, or of a long and active course of usefulness in the world."[110]

The narrator's hopes for the reception had already been signposted, in parables of reading that show how unconventional ideas ought to be received. Skepticism about the creation of life through electricity is compared with the response to Copernicus; the rejection of phrenology is attributed to philosophical prejudice; geologists are shown to have struggled against fears of conflict with Scripture.[111] In other contexts Chambers acknowledged the importance of the geological parable. As he reassured Ireland, "There is nothing in it of a worse character than geology is when you consider its inconsistency with the Mosaic record."[112] Within *Chambers's Journal,* articles on credulity, trust, and the reception of new truths were aimed at preparing a wider audience for the forthcoming bombshell.

The final words of *Vestiges* rehearse these parables of reading in the context of an even more famous book of creation:

> But may not the sacred text, on a liberal interpretation, or with the benefit of new light reflected from nature, or derived from learning, be shewn to be as much in

110. Ibid., 387–88; for studies of images of reading within the texts (in rhetorical terms, "intradiegetic parables") in Victorian fiction, see Stewart 1996.

111. *V*1, 187–88, 350, 389.

112. RC to AI, 30 June 1844, NLS Dep. 341/110/32–3.

> harmony with the novelties of this volume as it has been with geology and natural philosophy? What is there in the laws of organic creation more startling to the candid theologian than in the Copernican system or the natural formation of strata? . . . Thus we give, as is meet, a respectful reception to what is revealed through the medium of nature, at the same time that we fully reserve our reverence for all we have been accustomed to hold sacred, not one tittle of which it may ultimately be found necessary to alter.[113]

Even the final "tittle" recalls a scriptural reference to the laws of God. Bible, strata, heavens, and book: all will be in harmony if given the same "liberal interpretation" and "respectful reception."

Having been guided this far, the reader is invited to take over, to press forward the development of humanity. The future progress of civilization reinforces the need for a favorable reception of the work in hand. The British nation, even through imperial aggression, will diffuse "light over the adjacent regions of barbarism." Readers are implicitly encouraged to support the "original, inventive, and aspiring minds" who "strike out new ideas for the benefit of their fellow-creatures."[114] The fate of the work will depend on harbingers of "a nobler type of humanity, which shall . . . realize some of the dreams of the purest spirits of the present race."[115] Reading *Vestiges* is part of the universal gestation of nature. Narrator and reader, on the moment of parting, are joined in a union projected into the future.

113. *V*1, 390.
114. Ibid., 321–22.
115. Ibid., 276.

CHAPTER FOUR

Marketing Speculation

> Among all the manufactures which—for the mental and mechanical skill required in their prosecution, the remarkable steps by which they have attained their present rank, and the influence which they exert on society generally—claim our attention and admiration, none perhaps is more striking than the *manufacture of a book.*
>
> GEORGE DODD, *Days at the Factories* (1843)

ALL BOOKS ARE MATERIAL COMMODITIES—printed, bound, advertised, sold, excerpted, reprinted, reviewed, read, and revised. They are part of a capitalist process of literary production that was at a critical juncture in the 1840s. As a manuscript, *Vestiges* would have circulated among a few friends in Edinburgh. As a book, it served an extraordinary range of readers and gained a huge variety of new meanings. The manufacture of the first edition reveals only a small part of this process. Further editions appeared in different forms, and these were multiplied by reviews, excerpts, advertisements, editorials, and translations. The work was not a stable entity, but the sum total of an expanding array of representations.

The outlines of this story will be familiar to some readers, although the implications have never been followed out for a nonfiction book in the industrial era.[1] *Vestiges* offers a unique opportunity to see what happened when a leading specialist publisher joined forces with an anonymous author who happened to be the nation's most successful publisher for "the people."

Making a Book

The medical publisher John Churchill (fig. 4.1) first heard of a proposed volume on the natural history of creation in a letter, dated June 1844, from

1. Surprisingly, Patten 1978, Shillingsburg 1992, and esp. Dooley 1992 are among the first studies that have done this for nineteenth-century literary works. For the importance of such studies, see Sutherland 1988.

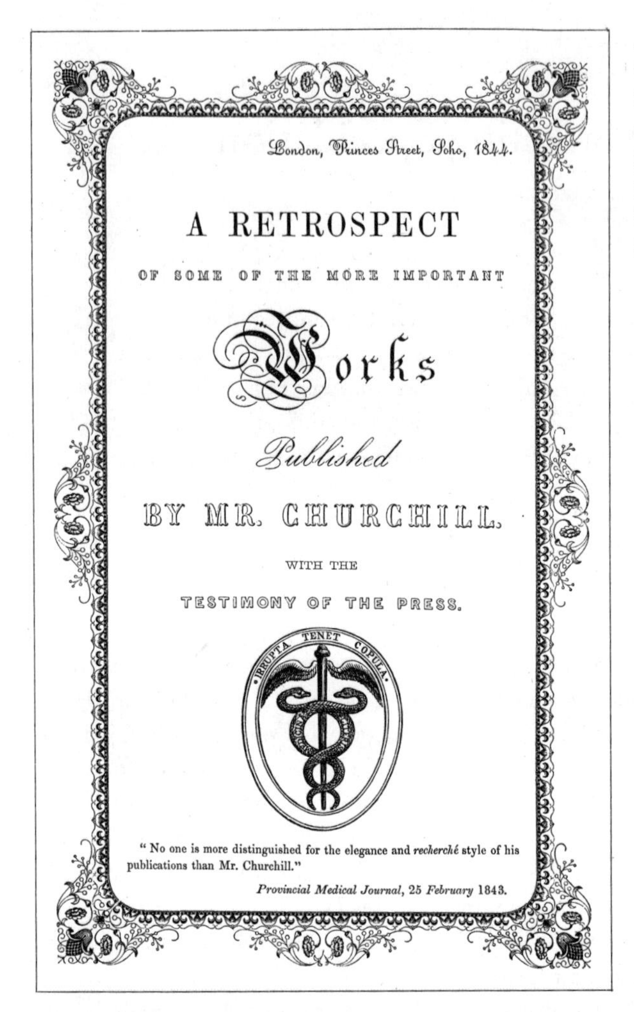

4.1 Cover of John Churchill's catalogue for 1844. The Latin motto ("an unbroken bond unites") comes from Horace, *Odes,* bk. 1, ode 13, line 18. (CUL Q 300.c.85.14.)

Alexander Ireland on behalf of an anonymous friend.[2] Churchill had been born in 1801, the third son of a well-to-do Dissenting minister. Intended for the medical profession, he instead became apprenticed in 1816 to Cox and Son, medical booksellers in Southwark in London. For several years he was employed at Longman, where he learned the wholesaling trade. In 1832, with assistance from his wife's fortune, he bought out the established firm of Callow and Wilson and set up in their premises at 46 Princes Street, Soho, just north of Leicester Square.[3]

2. J. Churchill to AI, 27 June 1844, Reading University Library, mss. 1393/381, p. 135. This is the only letter from this correspondence to have been preserved; the contents of the others have been inferred on the evidence of replies from Chambers to Ireland.

3. R. D., "John Churchill," *Medical Times and Gazette,* 14 Aug. 1873, 197–200. There is no good modern history of Churchill; among the few such studies of other specialist medical and scientific

PRINCES STREET, SOHO, October 1844.

BOOKS ON THE EVE OF PUBLICATION,

BY MR. CHURCHILL.

1.
DR. MARSHALL HALL, F.R.S.
PRACTICAL OBSERVATIONS AND SUGGESTIONS IN MEDICINE.
Post 8vo.—*On November 1.*

2.
MR. FOWNES, PH.D.
A MANUAL OF CHEMISTRY.
With numerous Engravings on Wood.
Fcp. 8vo. 12s. 6d.—*Now ready.*

3.
DR. GOLDING BIRD.
THE DIAGNOSIS, PATHOLOGICAL INDICATIONS, AND TREATMENT OF URINARY DEPOSITS.
Post 8vo.—*On the 1st of November.*

4.
DR. CARPENTER, F.R.S.
PRINCIPLES OF HUMAN PHYSIOLOGY.
Second Edition, revised; with Additions.
8vo. 20s.—*Now ready.*

5.
DR. RAMSBOTHAM.
PRINCIPLES AND PRACTICE OF OBSTETRIC MEDICINE AND SURGERY.
Second Edition, revised; with Additions.
8vo.—*On the 1st of November.*

6.
DR. RANKING.
RESEARCHES & OBSERVATIONS ON SCROFULOUS DISEASES.
BY J. G. LUGOL.
Translated; with an Appendix. 8vo.

7.
DR. PROUT, F.R.S.
CHEMISTRY, METEOROLOGY, AND THE FUNCTION OF DIGESTION:
Being a Third Edition of the Bridgewater Treatise.
Revised; with Additions. 8vo.

8.
MR. NEWNHAM.
HUMAN MAGNETISM VINDICATED;
Or, its Claims to dispassionate Inquiry:
Being an Attempt to shew the Utility of its Application for the Relief of Human Suffering.
Post 8vo.

9.
DR. BINNS.
THE ANATOMY OF SLEEP.
Second Edition, revised; with Additions.
Post 8vo.

10.
MR. EDWARD SHAW.
THE MEDICAL REMEMBRANCER.
Concisely pointing out the Treatment to be adopted in Medical and Surgical Emergencies.
Second Edition, revised; with Additions. 32mo.

11.
DR. EVANS RIADORE.
ON ELECTRICITY AND GALVANISM.
Their Application for restoring the Healthy Functions of the Nervous System.
Post 8vo.

12.
MR. GRANTHAM.
FACTS AND OBSERVATIONS IN MEDICINE AND SURGERY.
8vo. 7s. 6d.—*Now ready.*

13.
VESTIGES OF THE NATURAL HISTORY OF CREATION.
Post 8vo. 7s. 6d.—*Now ready.*

14.
DR. BUDD, M.D. F.R.S.
ON DISEASES OF THE LIVER.
8vo. (1051)

4.2 From the *Publishers' Circular*, 15 Oct. 1844, 303, with Churchill's advertisement for the first edition of *Vestiges.*

Churchill's shop soon became an informal center for London medical men, with a circulating library, facilities for reading newspapers and periodicals, and a wide range of new and secondhand works for sale. His success was based on providing students with the texts of lectures in the London medical schools. He was intimate with the leading medical authors, from the patrician physicians of Guy's and Saint Thomas's hospitals to the reformers of the nearby Gerrard Street anatomy school. From 1838 Churchill became the publisher of the

publishers are Brock and Meadows 1998 (on Taylor and Francis) and McKitterick 1998 (on Cambridge University Press). Topham 2000 offers a thorough survey of the existing literature, and James 2000 briefly surveys trends in science publishing. Brief accounts of some of the major firms, mostly devoted to fiction and poetry, are available in Anderson and Rose 1991 and Curwen 1873.

prestigious *British and Foreign Medical Review*, edited by John Forbes, and from 1842 to 1847 he issued Thomas Wakley's weekly *Lancet*.

Churchill was a religious man, who attended chapel twice on Sundays and kept the Sabbath.[4] Although his sympathies lay with reform, his list included only a few publications advocating mesmerism, phrenology, and other disputed doctrines of mind; whatever his own views, such works would have gained him little favor with some of his medical friends. Characteristic titles were Golding Bird's *Elements of Natural Philosophy* (1839), William Prout's *On the Nature and Treatment of Stomach and Renal Diseases* (1840), W. J. Erasmus Wilson's *Anatomist's Vade-mecum* (1840), and others shown in the advertisement from the *Publishers' Circular* (fig. 4.2). William Carpenter was his most promising and prolific new author.

Churchill's list reveals a willingness to be associated with a variety of "advanced" medical views based on continental philosophies, ranging from Samuel Taylor Coleridge's posthumous *Hints towards the Formation of a More Comprehensive Theory of Life* (1848) to "heretical" works such as a translation of Friedrich Tiedemann's *A Systematic Treatise on Comparative Physiology* (1834), a work that advocated serial transformism. The *Lancet* had to be checked for possible libel actions. An experienced veteran of sectarian attacks, Churchill had demonstrated his ability to hold fast during the storm over Carpenter's *Principles*. This is why the *Vestiges* manuscript was forwarded to Princes Street. As its author had noted to Ireland a few months before publication, "I do not think Churchill is likely to boggle, for publishers of that class are a little used to such things." Even so, he wondered later that summer if Churchill—whom he had not met—might suffer "a sudden attack of squeamishness." In that case, the natural history publisher John Van Voorst was a possible alternative, and failing that, the book distributors and general publishers Simpkin and Marshall. The author anticipated, however, "that Churchill will stand good—London booksellers having after all more acquisitiveness than veneration."[5]

Churchill set high store on his ability to keep secrets, but found that this mysterious author took no chances. All dealings came through correspondence with Alexander Ireland in Manchester—whose involvement in the production was also supposed to remain secret. Even when in London, Ireland kept away from the Princes Street shop lest some suspicious adversary decided to do a bit of de-

4. See RC to AI, 29 Aug. 1847, NLS Dep. 341/110/251–252, which speaks of "a London publisher whom you know"; in the context, this is likely (but not certain) to be Churchill.

5. RC to AI, 30 June 1844, NLS Dep. 341/110/32–33; RC to AI, 12 Sept. 1844, NLS Dep. 341/110/153–154. For Acquisitiveness and Veneration, see fig. 2.11, faculties 8 and 14. Several years later, Chambers still had not met Churchill, and it is likely that they never did: see RC to AI, 29 Aug. 1847, NLS Dep. 341/110/251–52. For the *Lancet* and controversial character of medical publishing, see Desmond 1989.

tective work. Through two decades of negotiations, Churchill seems to have met Ireland just once, and he never was told the name of his most celebrated author.

Churchill replied to Ireland's initial inquiry in polite and businesslike terms, qualities for which he was celebrated in the trade. "I thank you for the favor of your communication and beg to state, that if intrusted with the publication of the book referred to, it should receive my best attention."[6] He agreed to handle accounting and sales for a commission of 10 percent of the copies sold at the trade price; all expenses were to be paid by the author, who took all risk and potential profit. These were standard terms for publishing on commission.[7] Like most specialist medical, religious, and legal publishers, Churchill issued many works at the authors' own risk; medical orations and prize essays were typically printed in small numbers (between 100 and 250) for limited circulation. Of all the medical speeches he ever published, only one paid its expenses.[8] Initially, Churchill was told to have 1,000 copies of the mysterious work printed, but almost immediately was asked to reduce the run to 750.[9] That way, if the book failed there would be no embarrassing pile of copies to remainder or pulp. Publishing at the author's risk guaranteed Churchill a small profit no matter what happened.

Churchill received the manuscript, copied in a fine Italianate hand (apparently that of a woman) in batches during July and August 1844. Revisions to important issues—ranging from the title to the concluding reflections on religion—were made at the last moment. In several cases, scraps arrived in a different hand (that of the unknown author, in fact). Churchill may have recognized this, as it was among the best-known in the industry; later, he would have heard rumors about the authorship, but kept any suspicions to himself. Most people could scarcely believe that the publisher didn't know the secret, and within a few months, his silence became a byword for propriety.[10]

Churchill's good character was vital to the book's success. His name, after all, was the only one on the title page. Publication by Churchill meant that the work might be seen to emanate from reputable medical or scientific circles. If published by John Van Voorst, it might appear too closely related to popular natural history, as he had a strong list in this area; if issued by Simpkin and Marshall,

6. Churchill to AI, 27 June 1844, Reading University Library, mss. 1393/381, p. 135.

7. Howsam 1993, 64–65.

8. R. D., "John Churchill," 199.

9. RC to AI, dated Friday [July 1844], NLS Dep. 341/110/114–15; RC to AI, [3 Aug. 1844], NLS Dep. 341/110/78–79.

10. "Chapman was prepared to be as mysterious as Churchill on the 'Vestiges' question; when he found Mr. Kingsley himself had told everybody, and that all his fibs were falsehoods, thrown away." M. R. Mitford to Mrs. Ouvry, 1849, in Chorley 1872, 2:119.

it would come from a general firm without any reputation in science. Anonymous cosmologies had a way of falling dead from the presses, and Churchill's imprint had many advantages.

As the manuscript arrived, Churchill forwarded it to the printer a few blocks to the east at Saint Martin's Lane in Charing Cross. This was one of the largest printing firms in London, owned by Thomas Choate Savill. By midcentury, printing was ceasing to be characterized by small shops using artisanal labor and was dominated by a few large firms working on an industrial scale. Savill was noted as an efficient printer of books and some half dozen periodicals, including the *Lancet.* He had joined the firm of Thomas Lawrence Harjette in 1828, and took over six years later.[11] As a master printer, Savill was an artisan rather than a gentleman and Churchill treated him accordingly.

Churchill was an important customer. Just as his firm and others like it were transforming the publishing of specialist works in medicine and science, so too were Savill and others revolutionizing the printing of quality books. Savill produced books by a combination of technologies as old as Gutenberg and as new as the steam press. As a physical object, *Vestiges* was a hybrid product, the result of work from artisans and machines. This becomes evident if the production process is traced in its passage through the printing house.

The model Savill was told to use for printing was a recent volume on Churchill's list, the chemist George Fownes's *Chemistry, As Exemplifying the Wisdom and Beneficence of God* (1844), an Actonian Prize Essay printed by Savill in June when Churchill and Ireland were negotiating the arrangements. Fownes's essay was only 184 pages long, but it offered a suitable exemplar in terms of page size and type.[12] From the printer's point of view it was a duodecimo, with each sheet of paper producing twelve leaves or twenty-four pages; from the outside, it had the appearance of an octavo (in which case each sheet would have provided eight leaves or sixteen pages).

Savill would have sent the packets of manuscript immediately on to the compositors' room. By 1855 there were over one hundred compositors in his employ, and in the summer of 1844 (before they moved in the following March to larger premises at Chandos Street, Covent Garden) there must have been several dozen.[13] The process was entirely manual, as composing machines did not come into common use until large-scale production of the Linotype in the

11. The best source is Howe and Waite 1948, 148, 153. For the firm's history, Todd 1972, 90, 169; Brown 1982, 83, 170.

12. Churchill to AI, 27 June 1844, Reading University Library, mss. 1393/381, p. 135. Twyman 1998 is a marvelously illustrated overview of printing in this period.

13. Howe and Waite 1948, 148, 153. As I have been unable to find an account of the specific procedures followed at Savill's, the following paragraphs are based upon those for other large London printers of the period, esp. Tomlinson 1854, 2:471–508; Dodd 1843, 326–60.

1890s. Suitable machines had been invented—at least one of Churchill's books had been composed on them by another printer earlier in the decade—but militant unions kept them out of common use. As shown in figure 4.3, the machines could be worked by female labor and by children, which threatened typesetting as a skilled activity pursued by better-paid men.[14]

Each compositor worked at a frame, which was built to hold two cases, each holding all the letters of the alphabet, together with spaces, punctuation, numerals, and other characters in both roman and italic type (fig. 4.4). The font for the main text of *Vestiges* was a standard small pica in the "machine-age vernacular of modern-face roman," with a vertical stress and mechanically regular cut (fig. 4.5).[15] The biggest headache was the plethora of species names and technical terms, unfamiliar to the compositors, to the printer's proofreaders, and (as became obvious) to the mysterious author.[16]

Over 150 different characters were in the compositor's cases, and those most used were nearest to hand: speed was essential as payment was for 1,000 ens (an en being half the body size, or em, of the type) set and corrected; Savill's had a major strike over this issue in the late 1840s.[17] The compositor worked by setting individual pieces of a line of type into a composing stick, corrected any errors, and adjusted spacing to justify the right margin. The type was then put into a galley of the length of a page. The 396 pages of the first edition of *Vestiges* averaged twenty-six lines of type, with spacing maintained by interlinear leading. These pages of type were then wedged together into a forme for printing one side of a sheet. As a large duodecimo, *Vestiges* had twelve pages blocked together in this way, and the resulting formes could be carried around, stored, and set into the press for printing. Once the compositors completed enough pages to make up a forme, proofs were taken and sent to Churchill, to Ireland, and through him to the author. The process was fast. Savill returned the first sheets less than a month after agreeing terms with Churchill, although there were subsequent delays.[18] By September Savill had all the corrected proof in hand and was ready to go to press.

If the process of composition had remained largely unchanged since the Renaissance, the paper *Vestiges* was printed on was a product of the new industrial

14. Moran 1965, 23–28; the employment issues are also raised in the contemporary literature, e.g., "Young and Delcambre's Type-Composing Machine," *Mechanics' Magazine,* 25 June 1842, 497–500; "Young's Composing Machine," *Magazine of Science, and School of Arts,* 2 July 1842, 105–8. The work published by Churchill and composed on the machine was the first edition of Edward Binns's *Anatomy of Sleep* (1841).

15. Gaskell 1985, 210.

16. RC to AI, dated Tuesday [1844], NLS Dep. 341/110/157.

17. Howe and Waite 1948, 166.

18. RC to AI, [1844], NLS Dep. 341/110/129.

Mechanics' Magazine,

MUSEUM, REGISTER, JOURNAL, AND GAZETTE.

No. 985.] SATURDAY, JUNE 25, 1842. [Price 3*d.*

Edited, Printed and Published by J. C. Robertson, No. 166, Fleet-street.

YOUNG AND DELCAMBRE'S TYPE-COMPOSING MACHINE.

VOL. XXXVI. K K

4.3 Machines for composing type, based on the design of the upright piano, were produced in the early 1840s, but their development was blocked by unionized compositors. The machine divided the process into separate tasks carried out by cheap female or child labor. A woman played the keys and another justified the type; one child operated the eccentric mechanism, while four others kept the machine supplied with type and running smoothly. "Young and Delcambre's Type-Composing Machine," *Mechanics' Magazine,* 25 June 1842, 497.

4.4 Compositor at his frame, setting a manuscript into type with his composing stick. This was highly skilled and relatively well paid work for men. Dodd 1843, 334.

IT is familiar knowledge

4.5 The standard, modern-face roman used in the first edition of *Vestiges:* vertical stress, unbracketed serifs, and machine-made uniformity. *V*1, 1.

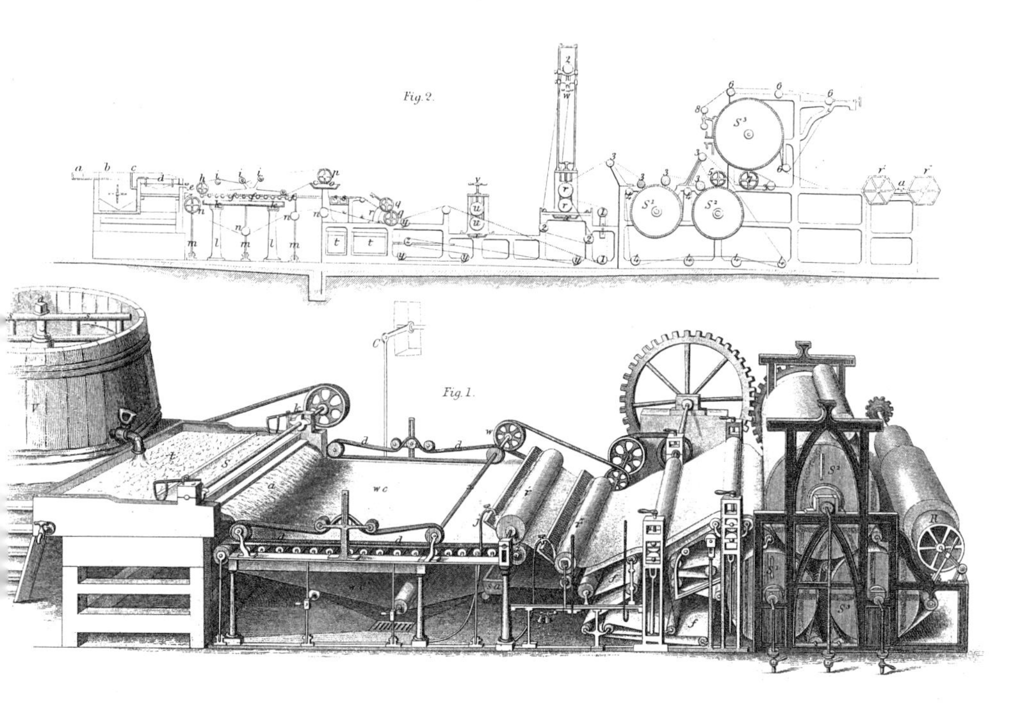

4.6 Fourdrinier paper-making machine. The process was one of the few in book production in which the raw material passed to finished product with minimal human intervention. Stuff (from linen and cotton rags) flowed from the vat on the left onto a rotating belt of woven wire. Water was drained away, the web passed through various rollers, was pressed to give strength, and finally emerged from the machine. Tomlinson 1854, 2: facing p. 365.

order. In the eighteenth century, paper had been expensive and hand-made, produced in frames one sheet at a time in a highly controlled labor-intensive craft enterprise. This all changed as a result of the astonishing productivity of the Fourdrinier paper-making machine, which laid pulp onto a continuous web of woven wire, removed excess water, and dried and smoothed it (fig. 4.6). The paper specified by Churchill was a good-quality rag. Cheaper, acidic wood-based papers were introduced only later in the century and were used on the twelfth (1884) and later editions. For all the editions published by Churchill, the paper underwent a further finishing process called coldpressing; used in the better kinds of books, this involved pressing the paper between glazed boards to prepare it to take the delicate impression of fine printing.[19]

By the 1840s the printing of books was reaping the benefits of the steam technology introduced for newspaper publication earlier in the century. This was true even for the limited press runs required by specialist scientific and medical

19. Dodd 1843, 357; on papermaking, see Gaskell 1985, 214–30; Tomlinson 1854, 2:357–74.

publishers. Almost all books by this time were printed by steam, although for very small editions the hand frame steel press could still be more economical. According to a guide published in 1855, Savill was famed for his "impressive machine installations" and use of new technologies.[20] He normally used small-cylinder machines run by steam, even for press runs of five hundred; the early editions of *Vestiges* were probably printed in this way (fig. 4.7). Such machines were run by one skilled worker, with a boy to lay the paper in the machine. Reform agitation against child labor often focused on the printing houses. Typical machines could print eight hundred sheets in an hour.

After printing, the sheets were sent to Remnant and Edmonds of Lovell's Court, just off Paternoster Row. The firm had been founded in 1837 through the partnership of Frederick Remnant and Jacob Edmonds and quickly became one of the three largest binderies in London. By the early 1850s the firm was doing up some thirty thousand volumes every week.[21] The process was standardized and involved extensive division of labor. The set of sheets for each copy needed first of all to be folded, so that the twenty-four pages from each sheet would be in their proper positions. The best folders could do five hundred sheets an hour, a rate possible even with duodecimos when working with machine-made paper. The gatherings were then sent to collators, who checked for correct order and folding, using the signature marks printed at the bottom of each sheet. Each book of loose sheets was then placed in a rolling press, a machine that compressed them prior to binding.

After a second collation, the sheets were then passed to the sewer, who sewed some two or three thousand sheets every day. Calculated at such a rate, the entire first edition could have been hand sewn with a week's labor. Errors at any stage could be disastrous. For example, the later editions had several "cancels": when mistakes were discovered after a sheet was printed, an individual page or sheet would be reset and reprinted. On one occasion, both the original canceled page—slashed through for destruction—and its replacement were left in by mistake. Such a copy would be a gift to a hostile reviewer.

The labor of folding, collating, and sewing was carried out entirely by women (fig. 4.8). Working by hand, women carried out repetitive tasks that required a high degree of accuracy: sheets aligned, pages in exact order. All work had to be carried out with extraordinary speed, for payment was by piecework. Because bookbinding involved sewing and other activities associated with the domestic

20. Howe and Waite 1948, 153.

21. Unless otherwise mentioned, details of the binding process are taken from Tomlinson 1854, 1:153–62, which is based on visits to Remnant, Edmonds and Remnant (as it was by then called). The evidence that the firm bound *Vestiges* is based on binder's tags; for an example of the first edition, see the copy (For QH363.C4) in the St. Andrews University Library.

4.7 The Napier Gripper machine, made and sold by the Scottish inventor David Napier. It is not known on what kind of press *Vestiges* was printed, although a small perfecting machine (which printed both sides of the paper) is a possibility. One skilled worker operated the press, the other (usually a boy) fed the sheets of paper. Wilson 1879, 74.

4.8 Women were extensively employed in the "female department" at Remnant and Edmonds, and other binderies. On the left, a woman uses a folding stick to fold the large octavo sheets; in the right-hand figure, a woman sews the assembled signatures of a book together along the two stretched strings or bands. Once a book was completed the next one would be begun using the same bands, until the sewing press was full. Tomlinson 1854, 1:155, 156.

sphere, it was seen by middle-class reformers and employers to be a form of labor appropriate for women. In a large establishment at the peak of the season, two hundred women "who might have no other resources to fly to" could find employment; income averaged from ten to eighteen shillings a week.[22]

The heavier tasks of binding, including all those associated with machines, were undertaken by men (fig. 4.9). Until the early 1830s, most books had been sold in boards, in the anticipation that purchasers would have them rebound in leather. As the market for print expanded into the middle classes, almost all books were sold in cloth. Wealthy purchasers could have them rebound, while others had a serviceable cover.[23] In a large establishment such as Remnant's, an entire edition of *Vestiges* could be done up in cloth in a day or two. The first six editions were cased in stiff pasteboards covered by a bright red cotton (fig. 4.10) used on many of Churchill's books. Remnant's obtained it from dealers who dyed and finished cloth specially manufactured in Manchester for bookbinding. The back of each book as received from the sewers was first glued and rounded, and the edges of the pages left uncut. To prepare each case, the boards had to be cut to the right size with a blade, the cloth attached, and a decorative pattern and rectangular framed border blind-stamped upon it, with title and publisher in gold on the spine. There was, of course, no author shown. The case then had to be secured to the book and placed in a standing press for a few hours. The whole process, as one visitor to Remnant's noted, "is indeed an extraordinary example of the power of numbers of skilful workpeople, and the effect of a refined system of division of labour."[24]

The mathematician Augustus De Morgan wittily expressed the close fit between the book's physical qualities and its developmental message:

> Its form is a case of the theory: the book is an undeniable duodecimo, but the size of its paper gives it the look of not the smallest of octavos. Does not this illustrate the law of development, the gradation of families, the transference of species, and so on? If so, I claim the discovery of this esoteric testimony of the book to its own contents; I defy any one to point out the reviewer who has mentioned it.[25]

De Morgan punned on the ambiguities of medium and message. "Form" has both its general meaning and its technical one in printing—the body of type secured together for printing a sheet. "Case" similarly has both its general meaning and the binders' reference to a binding to enclose a book. Poised between a duodecimo and an octavo, *Vestiges* illustrated its own law of develop-

22. Dodd 1843, 371.
23. Sadleir 1930 and Carter 1935 discuss the early history of cloth bindings.
24. Tomlinson 1854, 1:152–62, at 159.
25. De Morgan 1872, 210.

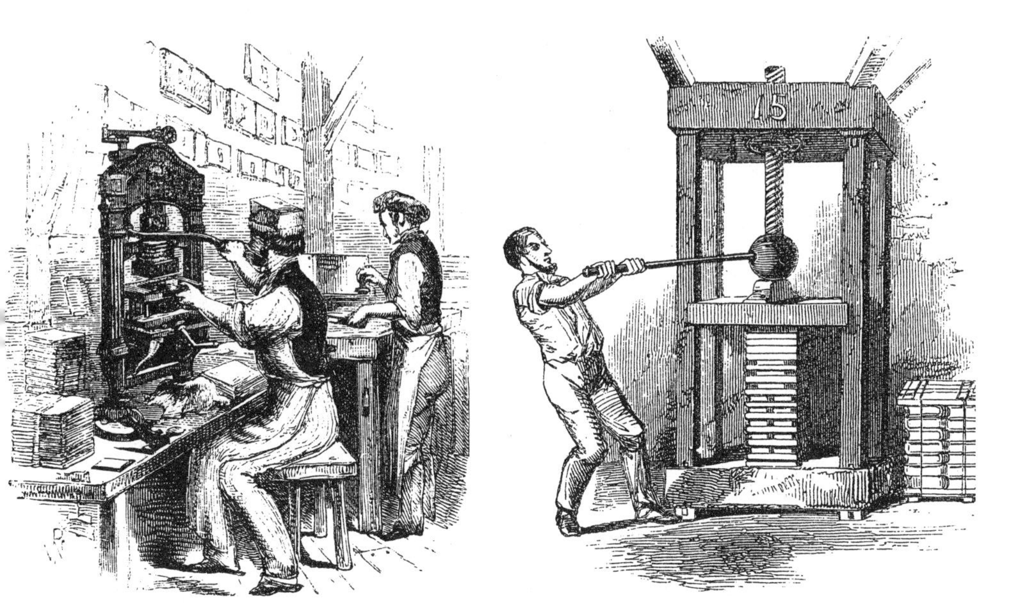

4.9 Examples of work carried out in the "male department" of Remnant and Edmonds. On the left, a workman uses a small blocking press to fix gold lettering on the spine, before it is attached to the book. On the right, the books have been placed in a standing press, where they will be compressed for a few hours and then packed in crates to be sent to the publisher. Tomlinson 1854, 1:157, 159.

ment. From the outside, it looked like a gentlemanly octavo, but collation revealed it to be an upstart duodecimo.

The attractions of this format were substantial. It was small enough not to be too expensive, thereby keeping costs and risks of publication low. At the same time, the clear modern typeface, attractive cloth binding, generous margins and wide line spacing provided the requisite air of gentility. It was small enough to be held in a lady's hand, while substantial enough not to appear out of place in a gentleman's library. The address to the reader in the text—spiritualized yet without ostentation—was thus embodied in the aesthetics of production.

In agreeing to publish, Churchill had quoted £5 16s. per sheet of twenty-four pages for an edition of 1,000 copies; binding on the model of Fownes's *Chemistry* would be 7d. a copy, or £21 17s. 6d. for the edition, with advertising not less than £20. Not including corrections to proof sheets, this suggested that an edition of 1,000 would cost at least £137 11s. 6d. The author, in assuming the entire risk of publication, did not want to lose more than about £100, a substantial sum by contemporary standards. This, too, may have been a factor in reducing the

4.10 Binding types used on *Vestiges of the Natural History of Creation* (1844–90). The first six editions, together with both editions of *Explanations,* appeared in a red cloth (the fifth of *Vestiges* and the second of *Explanations* are shown here). The cheap editions (seventh through ninth) appeared in paper wrappers; like modern paperbacks, these often wore out, as can be seen from the badly chipped spine of the seventh edition. Almost all surviving copies of these people's editions were rebound. The large illustrated tenth edition of 1853 was in a light brown cloth casing; the binding for the eleventh (1860) occurs in both green and red variants. In the final decades of the century, highly decorated cloth bindings became common, a fashion followed by the 1884 and subsequent editions. (Seventh ed., Dr. Williams's Library, (W)C.2.4.)

print run, for an edition of 750 copies brought any potential loss closer to a reasonable level.[26]

The actual costs of production, insofar as they can be determined, are shown in figure 4.11a. Clearly, Churchill was not dealing with someone who intended to make a lot of money. This is best shown by the low retail price of 7s. 6d., a real bargain for a volume of this size and quality. It was just about as low as the price could be without arousing suspicion, and was just enough to cover costs.

26. RC to AI, [3 Aug. 1844], NLS Dep. 341/110/78–79; Churchill to AI, 27 June 1844, Reading University Library, mss. 1393/381, p. 135.

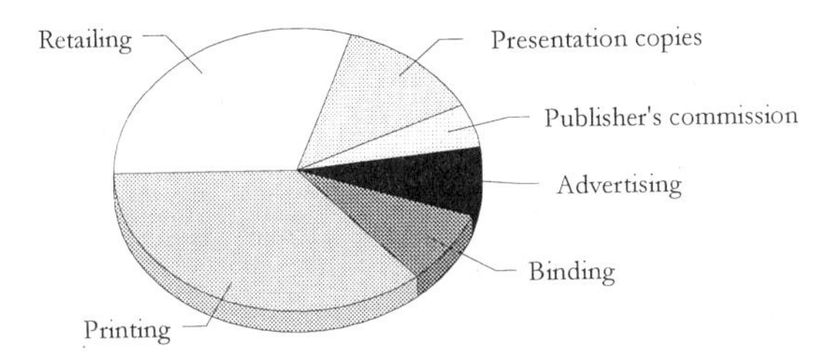

(a) First edition (1844), 7s. 6d.

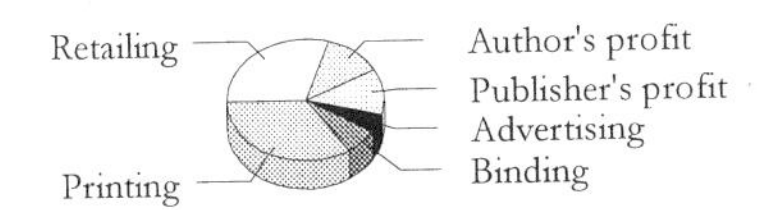

(b) Eighth edition (1850), 2s. 6d.

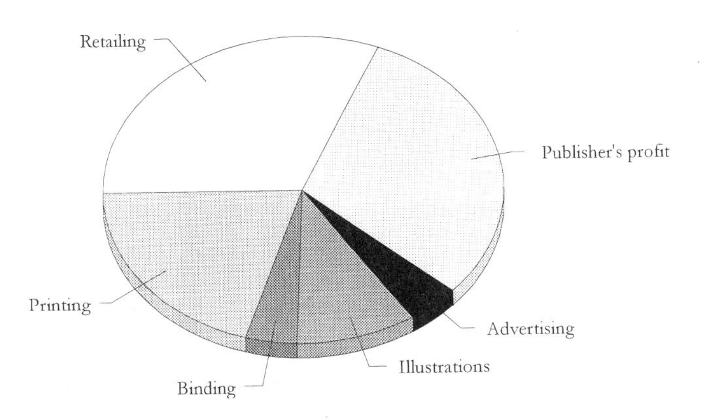

(c) Tenth edition (1853), 12s. 6d.

4.11 The overall dimensions of these pie charts illustrate the dramatically different costs to the consumer of a single copy of *Vestiges* in three characteristic formats of mid–nineteenth century publishing: (a) the first "gentlemanly" edition; (b) a people's edition for the middle classes; (c) a large-format illustrated edition. The relative component costs are shown as slices of the pies. The publisher made between 3d. and 4d. profit on every copy of the people's edition, but more than ten times as much (3s. 9d.) on each sale of the illustrated version. Based on data from John Churchill's ledgers, University of Reading Library, Mss. 1393/385/253, 1393/386/1737, and 1393/386/347–48.

Copies were sold to the trade at 5s. 4d., discounted by one-third from retail; and as was the custom, 25 copies were sold as 24. Five copies had to be deposited in Stationers' Hall for copyright. If all the remaining 745 copies had been sold to the trade, they would have brought in £190 15s. 4d., leaving a possible surplus (after Churchill took his cut) of about £30. In fact, the author gave this potential income away, by having Churchill distribute 150 free copies.

The first edition, with its gentlemanly format, tempting price, and huge free distribution, went on sale late in the second week of October 1844. Crafted to intervene in the key market with a maximum impact, it was the first step in a campaign for exploiting the print culture of Victorian Britain.

Literary Replication

The importance of *Vestiges,* and indeed any book, lies not just in the words on the page. The work had to be advertised, carried on trains, placed in shops, talked about, excerpted, and reviewed. This process, in an analogy with experimental practices, can be called "literary replication." The replication of a scientific experiment used to be thought of as a mechanical process in which an identical object was produced. Historical and sociological work over the past two decades has shown that this is not the case, and that replication is an accomplishment, achieved through agreement that two experiments are in fact "the same." Similarly, historians of printing and publishing have shown that textual stability, even within a single edition, has been difficult to achieve. This problem of stability extends far more widely, for attempts to reproduce the work also extended beyond the original bookseller, printer, or publisher—from anthologies of reprinted extracts issued by other publishers to brief quotations in newspapers.[27]

Following the historian Robert Darnton, this might be considered in the context of a "circuit of communication," which remains the most widely used model in studies of book history. In its simplest form, the circuit moves from author to publisher, printer, shipper, bookseller, readers, and back again to the author.[28] However, by turning the process in on itself, Darnton's cycle tends to limit attention to those elements shared among all those in the circuit, thereby making the history of readers an extended version of publishing and printing history. The notion of literary replication is more open-ended. It does not stress unduly the elements of feedback in the system, which makes readers important only insofar as their responses cycle back to the author. Instead, textual stability and reproducibility are examined where they mattered most immediately, in the actual business of producing and distributing print.

Like many other books in this period, *Vestiges* went through a range of different formats from the publisher and different publication arrangements. These illustrate some of the fundamental issues involved in mechanized book production, and their role in defining readerships and meanings.

As a gentlemanly production, the first edition had to make its way in a crowded market. To make it less likely that *Vestiges* would be ignored, Churchill was instructed to give away 150 copies, a large part of the impression. Copies were sent to many periodicals; to the leading men of science in London, Oxford, and Cambridge; to the major libraries of universities, mechanics' institutes, and literary and philosophical institutions; and to educational writers, politicians, and authors.[29] Distributing free copies was a useful way of intervening in the mar-

27. Johns 1998a.

28. Darnton 1990, 108–35.

29. RC to AI, 12 Sept. 1844, NLS Dep. 341/110/153–54.

ket, although unusual on this scale. Churchill did not receive a commission on free distributions, and could have reasonably complained. Adding 47 new names to 70 or so already sent, Ireland explained that "the author adopts this as one of the best modes of advertising," and hoped Churchill would "not grudge the trouble."[30] Churchill also gave the book additional publicity by including it in the catalogue of recent titles bound into every book he published; all his editions of *Vestiges* included such a list.

The advertising campaign can be reconstructed from the ledgers (fig. 4.12). For the first edition, Churchill targeted some two dozen titles in the daily and weekly metropolitan newspaper press. As well as the main London newspapers, such as the *Times, Globe, Morning Herald,* and Whig *Morning Chronicle,* there were also general intellectual weeklies, especially the *Literary Gazette* and the *Athenaeum.* Politics made little difference, with advertisements appearing everywhere from the reactionary *John Bull* to the liberal *Examiner,* which Tennyson read. Churchill placed advertisements across the religious spectrum, including the Nonconformist *Patriot* and *Watchman,* the ultraevangelical *Record,* the High Anglican *English Churchman,* and the *Witness,* newspaper of the Scottish evangelical Presbyterians. There were imperial papers like the *Atlas* and the *Times of India,* both popular among the commercial classes. One of the earliest advertisements appeared in the *Lancet,* which Churchill published; and a month later one came out in the *Pharmaceutical Journal.* An insertion in any one of these papers cost anywhere from 5s. 6d. to 8s.[31]

Newspapers were heavily taxed and therefore expensive—regularly affordable only by those who might spend 7s. 6d. on a book. The layout of contemporary newspapers, with no display headlines and only occasional woodcuts (to symbolize the departure of a ship, for example, or to mark the beginning of the editorial section) suggests that most people scanned the paper with care. Newspapers were organized as much spatially as visually, with material usually appearing in the same place in every issue. Because of this, advertisers could feel confident that the book notices—however inconspicuous to twentieth-century eyes—would be seen. Many of the earliest newspaper advertisements for *Vestiges* carried a one-sentence abstract of its argument.[32] This was simultaneously a way to attract potential purchasers, and to encourage them to approach the book in the right spirit.

30. Chambers instructed Ireland to write this to Churchill in RC to AI, 12 Sept. 1844, NLS Dep. 341/110/153–54.

31. Reading University Library, authors' ledger of J. Churchill, mss. 1393/385, p. 254.

32. "This work, after connecting the Nebular Hypothesis with the Discoveries of Geology, suggests a Physiological Explanation of the Development of the Vegetable and Animal Kingdoms; leading to the conclusion that the Designs of Creative Wisdom were entirely effected by the intervention of Natural Law." This version of the advertisement appeared in, e.g., *Athenaeum,* 12 Oct. 1844, 916, *Examiner,* 5 Oct. 1844, 640 and *Times,* 12 Oct. 1844, 6.

254.

First Edition.

		£	s	d	£	s	d
1844 Sept 4	By Cash				20	—	—
6th	Lancet 110/- 3/- Chron Slips	—	18	6			
	Carriage Manchester 3/8 York 3/- Liverpool 3/- Edinburgh 4/6	—	14	2			
	Times 5/6 Mail 5/- Examiner 5/- Herald 5/6	1	3	—			
	Globe 5/6 Standard 5/6 Post 5/6 Atlas 7/- Lit Gaz 7/-	1	13	6			
	Athenaeum 7/- Pic Jour 6/- 20 38 Gazette 880. 82.	1	11	—			
	Bell 8/- Record 7/- Patriot 7/- Watchman 7/- Evan Mag 10/7	1	16	—			
	Weekly 2/6 Times 5/6 Chronicle 5/6	1	0	6			
	Carriage Smith 1/- Conolly 8/- Booking 1/6 4d	—	3	6			
	Chronicle Slips 7/- Examiner 6/- Athenaeum 7/-	—	18				
	Record 7/- Atlas 7/- Spectator 7/- Times 5/6	1	4	6			
Nov	Blackwood 15/- Britannia 7/- Post 5/6 Standard 5/6	1	13	—			
	St Jas Chron 5/6 E Churchman 5/6 Lit Gaz 6/- Mail 6/-	1	3				
	Times 5/6 Standard 5/- Britannia 7/- Pic Times 7/-	1	2	6			
	Pic News 7/- Lit Gazette 7/- Examiner 7/- Globe 5/7	1	3	—			
	Sun 5/- Post 7/- Athenaeum 7/- Times 5/- Johnson 82 4/-	1	6	—			
	Phar Jour 7/6 31 Bell 25 10/6 Friend India 6/-	1	4				
	Chron 5/6 Slips 1/- 4 Papers 2/- ditto 1/-	—	9	6			
	Examiner 7/- Herald 5/- Bell 7/- Observer 7/- Atlas 7/-	1	11				
	Standard 5/6 Spectator 7/- Times 5/6 & Mail 5/6	1	3	6			
	St James Post 5/-	—	11				
19	Pd Mr Clarke, Dean St.	—	10	6			
18	To Simony Books, & 20th Nov.	1	3	6			
	Boarding 750 Copies, 8d	25	0	0			
		49	3	2	20	0	0

4.12 The cost of sensation: payments for advertising the first edition of *Vestiges.* Near the bottom of this page from Churchill's ledger is the fee paid to the agent, for various books, and for binding ("boarding") 750 copies. (From John Churchill's ledgers, University of Reading Library, mss. 1893/385/254.)

Newspapers and other periodicals (how many is not known) were also sent review copies, which editors used in a variety of ways. Many extracted short passages of fine writing from the early chapters that described striking phenomena of nature—volcanoes, fossils, comets, nebulae. Excerpting of this kind had been a standard way of filling space since the eighteenth century, although by the 1840s such "scissors and paste" journalism was acceptable only for making up the odd column. The book thereby gained additional publicity and in a setting that implied that it was a standard work. Such extracts sometimes appeared in surprising places, as when the *Witness*, a newspaper supporting the Scottish evangelicals, included a brief quotation from *Vestiges* on the age of mountains.[33]

By the 1840s newspaper reviews were the most effective route to establishing a book (especially one by an unknown author) on the public stage. As interest gathered momentum, an unusually large number of weeklies and dailies took the opportunity for a review. Many were effectively compilations of extracts, although in other cases they involved substantial analyses of several thousand words. Early reviews appeared in the *Atlas* (2 November), the *Examiner* (9 November), the *Spectator* (9 November), the *Morning Chronicle* (19 November), and the *Scotsman* (16 November). Robert Jameson, editor of the monthly *Edinburgh New Philosophical Journal*, just slipped a notice into his October issue, but otherwise the earliest reviews came out three weeks after publication. By the time any general-interest monthly or quarterly publication could notice the book, the first edition had sold out.[34]

Behind-the-scenes contacts could be crucial in obtaining reviews. We know little of how this was managed, although the distribution of free copies among potentially friendly journalists and editors was clearly important. The most significant of the early reviews, in the *Examiner*, almost certainly had something to do with the subeditor Philip Harwood having been sent a free copy. Harwood, born in Bristol, had been a student at the University of Edinburgh, a Unitarian preacher, and assistant to the Reverend W. J. Fox at London's South Place Chapel. In 1843 he had become the *Examiner*'s subeditor and was in a good position to ensure a positive review.[35]

The newspaper notices established *Vestiges* as more than just another cosmology from a specialist publisher. Impressed, Churchill offered to assume the risk of all future editions. His author did not take this up, but did agree that the swelling demand should be met by the quick release of further copies.[36] Savill

33. *Witness*, 22 Nov. 1845.

34. A comprehensive list of English-language newspaper, periodical, and pamphlet reviews of *Vestiges* and *Explanations* is in Chambers 1994, [222–29].

35. *DNB*; RC to AI, [Sept. 1844], NLS Dep. 341/110/58–59 for the free copy.

36. RC to AI, [Dec. 1844], NLS Dep. 341/110/171–72.

had not kept the work in type, as this locked up capital and was done only when an immediate reprint was thought likely. To save time, though, only those sheets with extensive corrections were to be sent to the author for checking. A thousand copies were rushed through the presses, with eight hundred copies sold to the trade on publication in mid-December, and the remaining two hundred nearly gone by the end of the year.[37] This edition sold out faster than Churchill expected, so that the book was out of print throughout most of January—frustratingly, just as many of the most important reviews began to appear. The third edition of fifteen hundred copies then sold out on the day of publication in mid-February.[38] This was advertised (though not on the title page) as "greatly amended." Only in April, with two thousand copies of the fourth edition, did the immediate need for copies appear to be met (see table opposite).

Writing of all kinds, especially in science and medicine, was increasingly tied to the schedules of periodical publishing. Not only were expensive monographs and important surveys issued in parts, but book reviewing had become a major literary industry. The second decade of the nineteenth century was characterized by the expansion of what was known as "class journalism"—periodicals aimed to supply information relevant to a special group. A host of journals catered to the needs of medical men, covering an extraordinary range of perspectives. *Vestiges,* based on physiological theories of generation and issued by a medical publisher, was widely reviewed. The weekly *Lancet* and the quarterly *British and Foreign Medical Review,* as might be expected from journals in Churchill's stable, were highly positive; although both offered criticisms on particular points, they both recommended the book to their readers. The weekly *Medico-Chirurgical Review,* another reformist organ, identified the author as "a man of great information and reflection," whose "doctrines have come out a century before their time."[39] On the other hand, the conservative weekly *London Medical Gazette* could only condemn "the ingenious sophistry of this anonymous pseudo-philosopher."[40] What it called "the conflict of the 'Vestiges' with

37. On proof checking for the second edition, RC to AI, [Dec. 1844], NLS Dep. 341/110/22–23; RC to AI, 8 Dec. 1844, NLS Dep. 341/110/165–66; RC to AI, [Jan. or early Feb. 1845], NLS Dep. 341/110/143–46.

38. *Publishers' Circular,* 15 Feb. 1845, 53.

39. "Vestiges of the Natural History of Creation," *Medico-Chirurgical Review,* 1 Jan. 1845, 145–57, at 157; [Forbes], Review of *Vestiges, Lancet,* 23 Nov. 1844, 265–66; [Carpenter], "Natural History of Creation," *British and Foreign Medical Review* 19 (Jan. 1845): 155–81. On medical journalism, see Bynum, Lock, and Porter 1992, and more specifically, Desmond 1989, esp. 195, 209, on the *Vestiges* reviews.

40. Review of *Explanations, London Medical Gazette,* 12 June 1846, 1044–45, Desmond 1989, 16. *Vestiges* was also condemned by the London-based physician and journalist J. Stevenson Bushnan, a disciple of Fletcher; see Bushnan 1851, 86–107.

Publishing History of *Vestiges* and *Explanations*

Vestiges of the Natural History of Creation

Edition	Date	Length	Price	Printed (Given Away)	Publisher	Notes
1	Oct. 1844	vi + 390	7s. 6d.	750 (150)	Churchill	
2	Dec. 1844	vi + 394	7s. 6d.	1000 (8)	Churchill	
3	Feb. 1845	iv + 384	7s. 6d.	1500 (18)	Churchill	
4	May 1845	iv + 408	7s. 6d.	2000 (25)	Churchill	
5	Jan. 1846	iv + 423	7s. 6d.	1500 (23)	Churchill	
6	Mar. 1847	iv + 512	9s.	1000 (46)	Churchill	
[7]	May 1847	iv + 294	2s. 6d.	5000 (3)	Churchill	people's edn
8	July 1850	vi + 319	2s. 6d.	3000 (5)	Churchill	people's edn
9	July 1851	iv + 316	2s. 6d.	3000 (5)	Churchill	people's edn
10	June 1853	xii + 325 + lxvii	12s. 6d.	2500 (45)	Churchill	illustrated edn
11	Dec. 1860	iv + 286 + lxiv	7s. 6d.	2500 (28)	Churchill	1st post-*Origin* edn
12	Apr. 1884	vi + 418 + lxxxii	5s. 6d.	5000	Chambers	reveals author (in print through 1930s)
[13]	Jan. 1887	286	1s.	8000	Routledge	Morley's Univ. Lib. (text of 2nd edn)
[14]	July 1890	286	1s.	2000	Routledge	Morley's Univ. Lib. (text of 2nd edn)
Explanations, A Sequel						
1	Dec. 1845	vii + 198	5s.	1500 (56)	Churchill	reply to critics
2	July 1846	vii + 205	5s.	1500 (17)	Churchill	

Note: Edition numbers in brackets were not identified as such on their title pages. Publication dates and prices are from the *Athenaeum* and *Publishers' Circular.* Information about copies printed and given away is compiled from J. Churchill, authors' ledgers (1842–71, mss. 1393/385; 1848–61, mss. 1393/86; 1854–83, mss. 1393/388; W. & R. Chambers, publication ledger no. 4, p. 380 (NLS Dep. 341/27); and George Routledge & Co., publication book no. 10, 1881–89, p. 747, and No. 11, 1889–1902, p. 616 (Routledge Archives Microfilm). Five further copies of each edition were deposited by law in Stationers' Hall. Extended and corrected from J. Secord 1994, [xxvii].

the 'Anti-Vestiges' party" was especially characteristic of the metropolitan medical world.

It was to the periodicals that theologically inclined readers also looked for guidance in assessing new books. A religious faction or sect was scarcely thought to exist without a weekly, monthly, and quarterly issuing from one of the religious publishing houses on Paternoster Row. The evangelical newspapers—notably the Anglican *Record* and the Dissenting *Patriot*—were too busy denouncing popery to have much to say about *Vestiges;* and like most newspapers, they noticed few books anyway. The religious monthlies and quarterlies, on the other hand, featured review sections in every issue. Many of these stressed the unholy origins of progressive development. These included the *Baptist Magazine* (Particular Baptist), *Christian Observer* (moderate Anglican evangelical), *Eclectic Review* (Nonconformist, especially Congregationalist), *Evangelical Magazine* (nondenominational evangelical), and *Wesleyan-Methodist Magazine.* A new venture, the moderate Nonconformist *British Quarterly Review,* reviewed *Vestiges* in May 1845, in its second number, and *Explanations* in the following year.

"Infidel" books like *Vestiges* needed careful handling in the religious press. Few editors accorded it the dignity of a direct review, focusing instead on Bosanquet or another effective pamphlet antidote. Arguments for and against a deity were not to be considered on an equal basis; the only reason for discussing unbelief was to refute it. (This is why preachers often accepted challenges for debate with atheists, either in freethought journals or on public platforms.) Most evangelicals believed that God gave man the printing press to reinforce the message of salvation through his Word, not to induce questioning or doubts.[41]

Periodicals created unity and a sense of purpose among the various religious parties, through reference to shared doctrines and points of difference with other groups. Reviews thus reinforced the boundary between the saved and the damned. In the secular metropolis, individual believers might feel beleaguered, isolated, and unsure; but congregations who worshipped on Sundays could be held together during the rest of the week as readerships. Hence each denomination went to great lengths to provide a full range of periodical publications, even when these required subsidies. Too expensive to be regularly read by the artisanal and lower middle classes, these periodicals (and the pamphlets) involved in the *Vestiges* debate targeted wealthier believers and clergymen. In turn, these readers could use them for sermons or to arm domestic missionaries for battle in working- and lower-middle-class homes.

By the spring of 1845 a curious feature of the sensation began to emerge. Few of the prestige quarterlies, whether religious or not, were featuring reviews. The Unitarian quarterly *Prospective* had a powerfully supportive notice in January, but the *Edinburgh, Quarterly,* and other most widely circulated periodicals were silent. Newspapers and monthlies could bring books to notice, but the quarterlies remained an authoritative court of judgment. While no longer as politically influential as had been the case, they defined the position of books over the

41. On salvation through the press, see Pearson 1853, 473–512. Altholz 1989 provides a helpful guide to religious periodicals.

4.13 The editor as gentleman: John Gibson Lockhart of the *Quarterly.* Bates 1873, facing p. 12.

longer term, solidly and sometimes stolidly. To be ignored by the quarterlies—the fate of most books—was worse than being torn apart.

As talk about *Vestiges* swept through London clubs and drawing rooms, the editors of the major quarterlies had begun to search for reviewers. This was not easy, for authors able to speak authoritatively about science were difficult to find. William Hickson, editor of the radical *Westminster,* solved the problem in the easiest way, by having a notice of *Vestiges* tacked on to a previously prepared notice of Alexander von Humboldt's *Cosmos.* The result was unsatisfactory, not only because of slapdash editorial interventions, but because the anonymous reviewer—John Crosse, Andrew Crosse's eldest son—had little status in metropolitan science. The problem of attracting reviewers who would command respect in scientific circles had dogged the *Westminster* since its foundation.[42]

At John Murray's the situation was rather different, for Lockhart, the *Quarterly's* editor, had many connections in London science (fig. 4.13). "Have you read 'Vestiges of Creation'?" he asked the geologist Roderick Murchison. "This is a very clever man—God knows whether mad or not I mean madder than all clever men seem to be." Lockhart resolved to have the work reviewed. He passed word of his interest to Churchill, and told the *Quarterly's* crusty stalwart, John Wilson Croker, that it was an extraordinary concoction of infidelity

42. [J. Crosse], "The Vestiges, etc.," *Westminster Review* 44 (Sept. 1845): 152–203; J. Crosse to A. Lovelace, [Sept. 1845], BL Add. mss. 37192, f. 234; Rosenberg 1982.

and expertise. Croker hated *Vestiges,* but agreed it must be dealt with. "I can make nothing of the book," he replied, "but a curious collection of facts spoiled for either instruction or amusement by being pressed into the service of an odious, disgusting, revolting and irrational theory."[43] Their first choice was John Herschel, who had written brilliantly upon Whewell's *History.* Mantell, Owen, Whewell, Sedgwick, and John Fleming were then approached on Lockhart's behalf by Murchison.[44] All turned him down.

Months went by and no review ever appeared. Lockhart was reduced to asking the reviewer of another work to praise one of the pamphlet replies in a footnote; and after a few years, an essay on human origins included a few perfunctory comments.[45] Meanwhile, though, there was another way to bring *Vestiges* to attention of the *Quarterly*'s readers. On several occasions, Churchill's agent inserted paid announcements, sometimes costing a guinea or more, in the hefty advertising supplements included with every number (fig. 4.14). These "advertisers" were a major forum for literary news and a feature of all the leading journals, but have been ignored because they are missing from bound copies in libraries.

All the same, the *Quarterly Literary Advertiser* was a poor substitute for the *Quarterly Review.* Only in the late spring did the major quarterlies begin to issue substantial reviews in reaction to the growing furor. Unlike the early newspaper notices, these were almost uniformly negative. A few early pamphlet replies, including Bosanquet's, sparked reviews in the religious press. The storm of criticism reached a peak in the summer, with an especially savage notice in the *Edinburgh* in July. In the mid-1840s the *Edinburgh* cost six shillings and had a circulation of seven thousand, although it was read by an audience several times that large. At the time it was published, more people had read that account than the original book, which had sold something under five thousand copies. These weighty antidotes then sparked off further discussions in print, where new issues of periodicals were regularly surveyed. This led in turn to further sales; the review in the *Edinburgh* sold two hundred copies of the fourth

43. J. G. Lockhart to R. Murchison, 14 Nov. [1844], BL Add. mss. 4127, f. 93. Lockhart's interest is noted in RC to AI, 18 Nov. 1844, NLS Dep. 341/110/43–44. Paston 1932, 49, prints the correspondence with Croker. On Lockhart as editor, see Shattock 1989.

44. G. Mantell, diary entry for 24 Feb. 1845, in Curwen 1940, 191; R. Murchison to R. Owen, 2 Apr. 1845, in Owen 1894, 1:254; J. Romilly, journal for 6 Apr. 1845, CUL Add. mss. 6823, pp. 22–23; J. Fleming to R. Murchison, 14 Apr. 1845, GSL M/F8/1.

45. [Holland], "Natural History of Man," *Quarterly Review* 86 (Dec. 1849): 1–40, at 14. J. D. Forbes to J. G. Lockhart, 13 Nov. 1845, St. Andrews University Library, J. D. Forbes Papers, letter book 4, pp. 27–28. This letter shows how reluctant Forbes was to recommend Mason's tract in [Forbes], "Humboldt's *Cosmos,*" *Quarterly Review* 77 (Dec. 1845): 154–91, at 167.

QUARTERLY
LITERARY ADVERTISER.

LONDON, JUNE, 1849.

MR. CHURCHILL'S PUBLICATIONS
On Science and Medicine.

MR. COLERIDGE.

THE IDEA OF LIFE.

Edited by SETH B. WATSON, M.D. Post 8vo., cloth, 4*s*.

'This book is one of the finest of the late Mr. Coleridge's philosophical essays. The internal evidence is sufficient to establish its authorship. Both in matter and form it is indubitably Coleridgean. The work demands and deserves the studious and earnest perusal of the philosophic reader.'—*Athenæum.*

VESTIGES OF THE NATURAL HISTORY OF CREATION.

Sixth Edition, post 8vo., cloth, 9*s*.

By the same Author.

EXPLANATIONS;

A Sequel. Second Edition, post 8vo., cloth, 5*s*.

THE
NATURE AND ELEMENTS OF THE EXTERNAL WORLD.

8vo., cloth, 10*s*.

ELEMENTS OF NATURAL PHILOSOPHY;

Being an Experimental Introduction to the Study of the Physical Sciences.
By GOLDING BIRD, M.D., F.R.S.
Third Edition, with numerous Engravings on Wood. Foolscap 8vo., cloth, 12*s*. 6*d*.

'A volume of useful and beautiful instruction for the young.'—*Literary Gazette.*

'We should like to know that Dr. Bird's book was associated with every boys' and girls' school throughout the kingdom.'—*Medical Gazette.*

'This work marks an advance which has long been wanting in our system of instruction. Dr. Bird has succeeded in producing an elementary work of great merit.'—*Athenæum.*

THE UNDERCLIFF, ISLE OF WIGHT;

Its Climate, History, and Natural Productions.
By GEORGE A. MARTIN, M.D.
Post 8vo., cloth, 10*s*. 6*d*., with Panoramic View.

'We regard Dr. Martin's work as very superior to a mere guide-book; it is an interesting addition to the medical and general topography of the British Islands. To those who visit the Undercliff with natural history objects in view the book will be found a useful guide.'—*Athenæum.*

'While its pages contain everything requisite for the purposes of the tourist and valetudinarian visitor, they are replete with matter of interest and value for the scientific reader.'—*Morning Chronicle.*

CYCLOPEDIA OF PRACTICAL RECEIPTS IN ALL THE USEFUL AND DOMESTIC ARTS;

Being a complete Book of Reference for the Manufacturer, Tradesman, and Amateur.
By ARNOLD JAMES COOLEY.
Second Edition, 8vo., cloth, 14*s*.

This work embraces all the latest improvements in science and art, which the author has been assiduous in collecting for many years. It consists of several thousand practical receipts.

LONDON: JOHN CHURCHILL, PRINCES STREET, SOHO.

Q. Rev., No. 169. *B*

4.14 *Quarterly Literary Advertiser* for June 1849 (no. 169), bound at the front of every copy of the corresponding issue of the *Quarterly Review.* Note the advertisement for the sixth edition of *Vestiges* and the second of *Explanations;* for this Churchill was charged fifteen shillings. The extended range of subjects on this list suggests that Churchill's business was expanding, partly through the success of *Vestiges.*

edition[46] and led to an extensive correspondence in the *Times* that brought further notoriety.

One of the most important ways that *Vestiges* became available to a wider audience was through a lengthy supplement to the *Atlas* newspaper, published on 30 August 1845. So popular did this become that extra impressions of the issue were required, and in 1846 an updated version was separately published as a seventy-page pamphlet. This gave an "expository outline" of *Vestiges* and a critical but respectful review of the controversy and its "bearing on the moral and religious interests of the community." At a price of four pence, the *Atlas*'s potted piracy dramatically widened participation in the debate, for a detailed summary of the work could be bought for a fraction of the original price. The green cloth binding, with the title stamped on the front cover (fig. 4.15), also made this an attractive purchase for polite readers who wanted the gist of the debate for use in conversation. The anonymous redaction was by Samuel Laing, a leading liberal politician and railway proprietor.

Besides the *Atlas* reprint, most of the other separately published discussions were religious pamphlets, a mainstay of Protestant polemic since the Reformation. Some religious disputes spawned dozens or even hundreds of tracts; debates that did tended to involve central questions of doctrine or church policy, such as the government grant to the Catholic college of Maynooth in Ireland. Given its public notoriety, it may seem surprising that *Vestiges* sparked only a small pamphlet literature. John W. Parker, who published many religious works from his office on West Strand, had two anti-*Vestiges* works on his list. The first, at 5s. 6d., was authored by the master of Trinity College in Cambridge, William Whewell; the other, at 5s., was by the Irish balloonist and theologian Thomas Monck Mason. Whewell's book was widely used among Anglicans, and was popular among both Nonconformists and the Low Church Establishment. The evangelical publisher on Piccadilly, John Hatchard, published Samuel Bosanquet's punchy millenarian attack in the same week in February 1845 as Whewell's book; both went through two editions. Jackson and Walford of Saint Paul's Churchyard issued the Anabaptist lay preacher John Sheppard's *Lecture on the Arguments for Christian Theism,* originally delivered at the Literary and Philosophical Society of Frome in Somerset. Most of these pamphlets were intended as antidotes for readers who might be misled by *Vestiges.* Printed with the same page size and format, they could easily be bound with the book to counteract its effects.

The number of pamphlets was small because publishers and authors recognized that books like *Vestiges* could most effectively be dealt with in more general works. Direct attacks just provided free advertising to the enemy. The number of works that included a brief reference or reply is enormous; explicit

46. RC to AI, 26 July 1845, NLS Dep. 341/111/29–30.

4.15 Progressive development in inexpensive dress. The *Expository Outline* of *Vestiges,* shown here in the lower right corner, was issued by the *Atlas* newspaper in 1846. At only four pence, this sold well among all classes and in its embossed cloth binding was suitable for the hands of a lady. In May 1847 the seventh edition of *Vestiges* was published in cheap card covers, as shown here in the lower left corner. Priced at only 2s. 6d., this format—an early example of the modern paperback book—made the complete work accessible to the middle and working classes. Wealthier purchasers often had this inexpensive "people's edition" rebound to make it more permanent or attractive. The three upright copies shown here range from paper-covered boards to full and half leather. (Seventh ed., Dr. Williams's Library, (W)C.2.4.)

discussions of *Vestiges* are typically found among the sects of Old Dissent that tended to emphasize learned argument. Congregationalist leaders such as John Harris and John Pye Smith sniped at *Vestiges* in footnotes. The evangelical publisher William Collins of Glasgow reissued several American works that discussed *Vestiges,* notably *The Religion of Geology* by the Presbyterian theologian and geologist Edward Hitchcock, which had a chapter on the subject and went through many editions. The relatively low cost of typesetting meant that existing refutations of freethought could be easily updated.[47] There were, as we will see in later chapters, many other works—from novels and scientific articles to philosophical surveys and poems—that drew on *Vestiges* in different ways.

What are we to make of all these redactions, reviews, advertisements, and expositions? It is easy to retreat into thinking that the first edition of a book—or, even better, a holograph manuscript—is all that really matters, that everything after that is diffusion, debasement, and summary. But valorizing first editions makes sense only in literary interpretations grounded in authorial intentions. Once readers become fundamental to the making of meaning, then the ink on the 750 copies of the first edition of *Vestiges* becomes merely one step—and in some ways not a particularly important one—in a much wider process of literary replication.

Circulating Science

What the steam printing machine was to the reproduction of literary works, the steam railway was to their distribution. The railway mania of the 1840s, near its height when *Vestiges* was published, extended lines to all parts of the United Kingdom, giving firms like Churchill's access to a national market at low cost. At the same time, the Penny Post made ordering and payment quick and simple. Quick distribution was vital, for only the specific agents for a book could be expected to hold it in stock; otherwise, copies had to be ordered. A gentleman or lady who wished to purchase would make a request to their bookseller, who would in turn obtain the volume from a provincial agent, a wholesaler, or direct from Churchill. Because Churchill was a specialist publisher, the role of middlemen was particularly significant; Simpkin and Marshall, the leading wholesaler to the provincial trade, took six hundred copies of the third edition alone—two-fifths of the entire run.[48]

It is hard to be more specific about patterns of distribution for *Vestiges*, or for any book of the period, as most of the relevant records have been destroyed. What does seem to be clear is that sales figures are only a partial guide to the actual number of readers, many of whom—probably most—read the work in borrowed copies. Though the early editions were "filling at the price" as one reader said,[49] 7s. 6d. was still a substantial sum for a casual read, even within the best circles or for those with a special interest in the subject.

47. See also Taylor 1855 and Crofton 1855; and the additions in Godwin 1853, 208.

48. "Literary News," *Publishers' Circular*, 15 Feb. 1845, 53.

49. J. D. Hooker to C. Darwin, 30 Dec. 1844, in Burkhardt et al. 1985–99, 3:103.

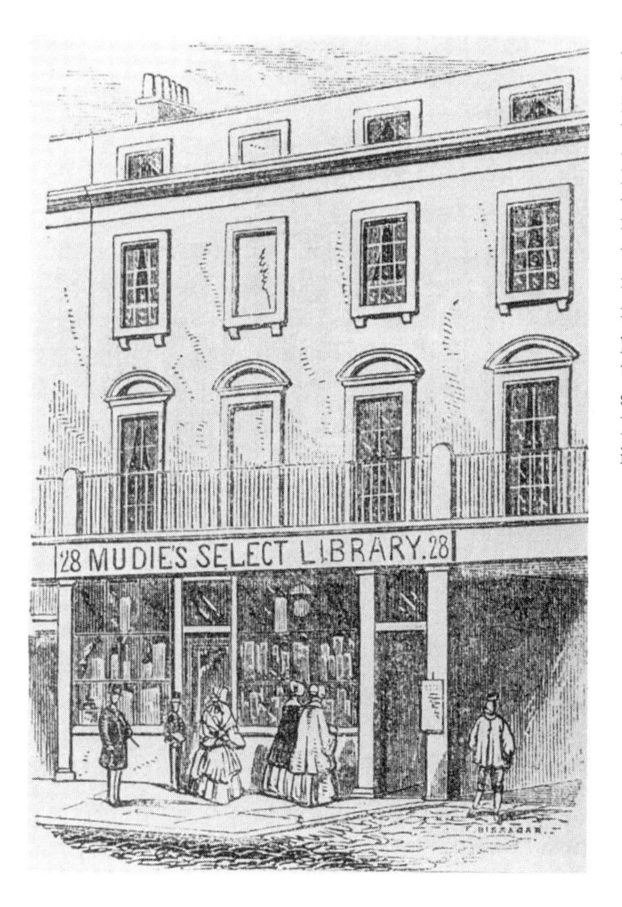

4.16 George Mudie's circulating library in its first premises on Upper King Street in Bloomsbury. Although Mudie became famous for vetting the nation's taste in three-decker novels, his original reputation was as a purveyor of advanced thought in philosophy, science, and history. From Cruse 1935, facing p. 320.

An elaborate system of private circulating libraries, clubs, and reading rooms had developed to circumvent the high prices of books.[50] Circulating libraries had begun in the provincial towns, especially at the seaside and at spa resorts like Bath, but in the 1840s they were becoming national institutions. In London in 1844, there were several dozen general circulating libraries, not to mention more specialized ones such as Churchill's in his Princes Street shop. The fastest-growing circulating library had been started in 1843 by the Bloomsbury stationer and newsagent, Charles Edward Mudie. From his premises on Upper King Street (fig. 4.16), Mudie offered a wide selection of newspapers and books to subscribers, who paid a guinea a year to have a single volume out at a time. The shop was tiny, only about twenty by fifteen feet, with several hundred volumes on two or three shelves running the length of the shop. Mudie, his sleeves

50. Black 1997 surveys the literature on Victorian libraries.

rolled up, served customers from behind the counter.[51] Within a decade, with premises on Museum Street and Oxford Street, he had become a major force within the publishing industry, ordering several thousand copies each of popular works. It is largely through his influence that the three-decker novel at 31s. 6d. lasted until the end of the century.[52]

Mudie's undoubted significance for the novel has obscured the fact that over half the books in his lists were nonfiction, with strengths in travel, biography, and science. Liberal in theology and radical in politics, Mudie was noted in the 1840s for offering "a little library of books of a progressive kind," including writings by the American transcendentalists.[53] The first of his surviving catalogues, from 1848, shows that readers could have borrowed not only *Vestiges* and its sequel, but also works by Carlyle, Combe, Emerson, Fry, Humboldt, Lyell, Nichol, and the Bridgewater authors. Mudie carried almost no specialized scientific or medical treatises; he avoided religious polemics, and had none of the anti-Vestigian theological pamphlets. Later catalogues show that he kept up with the debate, so that Hugh Miller's *Testimony of the Rocks* was added in 1849, Darwin's *Origin* in 1859, and the new edition of *Vestiges* in 1860.[54]

The significance of reading in borrowed copies is evident if we examine another collection established in the early 1840s, the London Library. Unlike Mudie's, which aimed to meet the demand for the most popular books—no matter how many copies might be needed—the London Library was established to provide an in-depth general collection of "standard" philosophical and literary works. Located at Saint James's Square in the heart of fashionable Pall Mall, the subscription was not cheap: a six-pound entrance fee and two pounds annually, which allowed ten books to be taken at a time, fifteen for those outside the metropolis. The 1847 catalogue lists a comprehensive selection of works relevant to the *Vestiges* debate, including all the pamphlets and book replies, and even a considerable volume of related scientific literature, though any intention of competing with specialist libraries (such as the Royal Society or the Royal College of Surgeons) was disavowed.[55]

The London Library's issue books show that *Vestiges* was among the most popular titles in the collection. The library was sent a presentation copy, which was first checked out on 7 October 1844 by a subscriber in Leamington Spa in

51. Crosland 1898, 227.

52. Griest 1970 provides an account of Mudie's focused on the novel.

53. Espinasse 1893, 27.

54. A copy of the 1848 catalogue, with a supplement for 1849, is in the Guildhall Library; later ones (1857, 1861, 1865, et seq.) are in the British Library.

55. Cochrane 1847, including a list of subscribers; the regulations are published on pp. xiv–xxxi. Baker 1992 discusses the library's early history.

the English Midlands. A few days later a Miss Everett of Clapham, among the library's most active readers of science, borrowed another copy. As a woman, she was barred from all the specialist libraries. A few additional copies (probably half a dozen in all) were purchased, so that the title was issued at least forty times in the next six months or so, and at least forty times again in the twelve months after that. During the rest of the 1840s *Vestiges* was borrowed about twenty times each year, often together with *Explanations* and occasionally with the replies by Bosanquet or Whewell. Copies were sent out with crates of summer reading to the seaside, to country houses, and with parcels of specialist books to learned authors. At the peak of interest, demand was clearly outrunning supply.[56]

Readable and widely discussed, unavailable for two and a half years in a cheap edition: *Vestiges* was ideally suited to the circulating libraries. Like the railways and the postal system, circulating libraries were part of the system for multiplying the places in which a work could be found. Like periodical reviews and literary salons, they provided a social mechanism for coping with the large number of titles being published. The significance of being able to borrow books, as we have seen, was by no means limited to fiction. They made it possible for people (especially women) to read more scientific books, even if these were borrowed by husbands or fathers. However, circulating libraries had little impact on widening access among different social classes, for they were expensive to join and catered for a wealthy urban elite. Other, less exclusive libraries—including a number of mechanics' institutes—also had early editions of *Vestiges*. But even these collections were used primarily by the middle classes. What were regularly referred to as the "gentlemen's" editions were just that—targeted at a readership in the professional classes and urban gentry. The price was not expensive for a 396-page book, but the complete work remained relatively inaccessible to those who could afford only the penny periodicals.

Explanations and the People

Churchill soon realized that his unknown author wanted *Vestiges* to reach a broader readership: this was not just another speculator happy to see his philos-

56. From 1844 through mid-1846, the London Library kept detailed record books organized by subscriber (issue books no. 5 and no. 8). Total borrowings for an individual title can be calculated, as here, but are inevitably approximations based on examining nearly two thousand pages of manuscript. From 1846 to 1849 records were kept by title, which makes the calculation much easier (book of issues no. 9). A further issue book contains scattered records for 1856 through 1858 (register book of issues). No other records for the relevant period appear to have been preserved. For the earliest borrowings of *Vestiges,* see issue book no. 5, pp. 104, 196. My estimate of six copies is based on a "No. 6" marked by the book in the 17 July 1845 entry for H. W. Bevan, issue book no. 8, p. 62.

ophy in print. A month before publication of the first edition, Ireland wrote him that "the author intends, whether the book succeeds in its present form or not, to publish ere long a *people's edition,* which he has no doubt would meet an extensive sale."[57] True, the compositors at Savill's were the first sample of readers among this audience, and their typesetting blunders suggested that *Vestiges* might be pitched at too high a level.[58] Despite these doubts, a people's edition remained high on the agenda.

After the enthusiastic reviews and brisk sales, plans for a cheap edition began to move ahead in December 1844. At his author's request, Churchill prepared estimates for producing five thousand copies of a cheap edition.[59] As the new year began, however, publisher and author began to view the market in different ways. The book was clearly a sensation, but also a target for criticism. After publication, Churchill seems to have been surprised to discover that his author had scant firsthand knowledge of natural history and had been poorly placed to correct proof, unlike the specialist medical men he was used to dealing with. Setting the table of affinities (fig. 3.4) presented so many problems with brackets and names that it finally had to be dropped from the fifth and subsequent editions.

Early criticisms of mistakes and misspellings were followed beginning in January by damning reviews in the *Athenaeum, Literary Gazette,* and *Gardeners' Chronicle.* As a specialist publisher, Churchill depended absolutely on the credibility of his list. Although willing to press forward with further editions for the higher classes, he insisted on extensive revisions before the market could be flooded with an inexpensive reprint. He was concerned that a cheap edition would undercut the sales of expensive ones and that his reputation would be compromised if the book was released to a wider public while full of errors.[60] The notice in the *Athenaeum,* the most authoritative scientific and literary weekly, appears to have been especially alarming. The review, as Churchill would have heard, was by the physician Edwin Lankester, who reviewed almost all works in physiological and natural history there. The *Lancet* review, which pointed to numerous mistakes, also had a powerful impact on Churchill and the author.[61]

Churchill was in a good position to ask naturalists and medical men in his circle to take on revision work for a set fee. When the second edition was being rushed through the press in December 1844, he had contacted Lankester, who

57. Ireland was told to say this in RC to AI, 12 Sept. 1844, NLS Dep. 341/110/153–54.

58. RC to AI, dated Tuesday [1844], NLS Dep. 341/110/157.

59. RC to AI, [Dec. 1844], NLS Dep. 341/110/39–40; [1846], NLS Dep. 341/112/84–85; RC to AI, [Jan. 1845], NLS Dep. 341/110/45–46.

60. Churchill's concerns can be inferred from RC to AI, 31 July 1845, NLS Dep. 341/110/141–42.

61. Wilson and Geikie 1861, 383.

supported himself by literary odd jobs of this kind. The author was anxious that the mistakes in nomenclature should be corrected as soon as possible.[62] Lankester made further changes for the heavily revised third edition, as did George Fownes, pharmaceutical chemist and author of the book that had been used as a model in the production.[63] The author seemed to have further advisers of his own.[64]

Churchill also asked Carpenter, the rising star of his list, to improve the work. Carpenter had followed up his review essay in the *British and Foreign Medical Review* with a letter to the anonymous author, outlining suggestions and corrections. In thanking him in a letter (recopied by Ireland for Churchill to forward), the anonymous author outlined an ambitious plan for improving the fourth edition:

> Restrained however as he is by the necessity of keeping his secret, he requires a scientific assistant for the proposed revision of the book. He begs to know if Dr Carpenter will for any moderate fee, left to his own appointment, undertake to get the volume thoroughly revised. He will himself be able to do all that is necessary in some departments, particularly the physiology. For the rest he will employ other persons, such a man for instance as Edward Forbes in the natural history, and some equally able person for the geology. Perhaps no more would be required. These persons he will pay, looking to the author for reimbursement. . . . The author, however, wishes him to understand that the book must remain in its present general form, which the author is confident is the most attractive, and that corrigenda on special points, or commentaries, are mainly what he desires.[65]

Carpenter regularly did editorial work of this kind, but turned down this unusual proposal to lead a multidisciplinary team. He did, however, comment on the chapters on classification and geographical distribution. Although only indirectly his doing, details of the revisions were indebted to him, and Ireland arranged that he was paid for his trouble.

After the *Edinburgh*'s onslaught in July, Churchill was less eager than ever for a cheap edition. He learned from Ireland that the author hoped that turning to the people would be the best way of answering the attacks. Churchill's experience as a publisher, as before, was highly valued. He was asked "to state what he thinks would be the feeling of the aristocratic, the religious or any other

62. RC to AI, [Nov.–Dec. 1844], NLS Dep. 341/112/70–71.

63. RC to AI, [Jan. 1845], NLS Dep. 341/110/125–26; RC to AI, [Jan.–Feb. 1845], NLS Dep. 341/110/143–46.

64. RC to AI, [1844], NLS Dep. 341/113/27–28 (on Cox); RC to AI, [Feb. 1845], NLS Dep. 341/110/112–13 (on Page).

65. RC to Churchill, quoted from RC to AI, 1 Mar. 1845, NLS Dep. 341/110/155–56.

world, in the event of a cheap popular edition."[66] Churchill must have responded in no uncertain terms, for the cheap edition was delayed again. Through Ireland, he recommended extensive revision before making it available to the people at large.

Such caution led to the adoption of a very different strategy. In the summer of 1845 Churchill received another of the intriguingly anonymous letters, written as usual in the third person and sent through Ireland:

> The author &c. is obliged to you for your advice respecting the proposed cheap edition. He will deliberate upon it, and take no rash step in that direction. It certainly ought to be a first object with him to make the book as nearly unexceptionable in respect of scientific details as possible, before making it more accessible to the uneducated classes than it is; but still it is part of the general aim which he has in view, to diffuse a knowledge of the natural system of the world; and he hopes you will not fail to second him in that object when the proper time comes. In the meantime, he is writing a defence of the book, with particular reference to the coarse attack of Mr Sedgwick, and he wishes to consult you regarding its publication.

The author envisioned publishing this defense as a series of letters to the *Times* (because of an extensive correspondence that was appearing there in the wake of the *Edinburgh* critique), and then as a separate pamphlet of a hundred pages or so. The direct target would be the *Edinburgh* reviewer, who was widely known to have been Adam Sedgwick of Cambridge. The combination of newspaper and pamphlet publication, a common one for political and religious controversy, would secure maximum interest, and "keep public attention alive" by occasioning further reviews and a readership over a longer term. Churchill was asked to give his advice. "Would you say if you think the Times would give it admission, or if you would recommend the pamphlet form first in preference?"[67]

It is not known precisely how Churchill replied, but his concerns were instrumental in determining the course of events. The newspaper plan was dropped, and the reply broadened into a subtle attempt to meet specialist criticism on its own ground. During the autumn of 1845, Savill set up and printed fifteen hundred copies of *Explanations*, a 206-page work with the same paper, page size, type, and binding cloth as the original work. "This mode of publication will, I think, be most dignified," the author admitted to Ireland, "and the character of this new attempt demands such consideration, for it is far more powerful in argument than the opus itself, although not so popular in its nature."[68] What had begun as a newspaper polemic against the *Edinburgh* reviewer

66. 31 July 1845, NLS Dep. 341/110/141–42.
67. RC to AI, [July–Aug. 1845], NLS Dep. 341/113/48–49.
68. RC to AI, [Sept. 1845], NLS, Dep. 341/110/55.

was now, thanks in large part to Churchill, a "forcible and argumentative" work, aimed at "convincing stern-minded men."[69]

Churchill published *Explanations* at the end of 1845, setting the price at five shillings. He was impressed with the work, which would maintain sales by providing a peg for further reviews. Both he and the author recognized the importance of early publicity, and Churchill sent advance copies to the *Athenaeum* and the *Literary Gazette.*[70] These duly appeared just as *Explanations* became available for sale. The trade subscription, however, was relatively limited, and a slightly altered issue of fifteen hundred copies (called the "second edition") was postponed until June 1846.[71] Not only did *Explanations* rely on specialist evidence and analysis, but page for page it was more expensive than the original work. The author suggested reducing the price to four shillings, although Churchill did not do this. The second edition sold slowly (only two hundred copies in the first six months or so) but steadily, well into the 1850s.[72] In January 1846 a fifth edition of *Vestiges* was ready, incorporating further revisions. It was usually advertised and often reviewed with *Explanations,* as in figure 4.14, and Churchill had suggested that the two works be issued bound together. His anonymous author, however, pointed out that the fifth edition already incorporated all the relevant arguments, and "the smaller a book is, it is the smaller evil to the public and the greater gain to the author and publisher."[73]

These exchanges have a significance beyond the *Vestiges* debate. They show how the staging of intellectual controversy was a collaborative act involving both publisher and author. Churchill knew little about "people's editions" and selling science to the "uneducated"; but he was familiar with the medical men who frequented his shop in Soho. Because of this, he served as an intermediary between author and market, advising on how to convince an audience to whom the first edition had seemed amateurish and ill conceived. A cheap edition in 1845 would have produced a very different reception. Instead, through Churchill's contacts in the medical and scientific world, the advice of the best specialists was applied, so that later editions grew in both length and sophistication. Most tellingly, *Explanations*—which eventually transformed the terms of the debate—would never have been written without the publisher's recognition of the need to counter accusations of incompetence. Only after the process of improvement was complete could a cheap edition for the masses begin

69. RC to AI, [1845], NLS Dep. 341/110/108–9.

70. RC to AI, [Nov. 1845], NLS Dep. 341/113/128.

71. RC to AI, 12 Jan. 1846, NLS Dep. 341/110/199–200.

72. Authors' ledger of J. Churchill, Reading University Library, mss. 1393/385, pp. 405, 477 (2d ed.); mss. 1393/386, p. 473 (2d ed.). For the two hundred copies, see RC to AI, n.d., NLS Dep. 341/110/211–12.

73. RC to AI, [1845], NLS Dep. 341/112/62–63.

to take shape. *Explanations* was not just a response to the *Edinburgh* reviewer, but to Churchill. The invisible author recognized Churchill as an ally who would "not fail to second him" in diffusing "a knowledge of the natural system of the world" among the people at large.[74]

The Truth in Wrappers

With *Explanations* and the fifth edition out of the way, Churchill learned at the beginning of 1846 that the author planned one more "edition for the higher classes and for libraries," to appear almost simultaneously with the long-postponed people's edition.[75] This would correct the most serious problem that had emerged in the intervening period, the celebrated resolution of the Orion nebula by Lord Rosse's gigantic telescope; the mechanism for transmutation and revised scheme for organic affinities could also be detailed in as authoritative a manner as possible. The attempt to meet the criteria of science, as mediated by the publisher, was nearly complete: the message could go to the people at last.[76]

More than any other author in Britain, George Combe knew the difference a people's edition could make. As he had told the American feminist and reformer Lucretia Mott, a cheap printing of *Vestiges* "would be another battery erected against superstition."[77] He underlined the advantages in a letter to the author, whom he addressed through Churchill after receiving the first edition:

> It will make its way into circulation; but allow me to suggest that you would benefit the thinking community greatly by publishing a cheap edition of it. This would not injure the sale of the present edition, but in my opinion promote it; for many wealthy persons would buy it, first in its cheap form, ascertain its merits, and then place it in their libraries in its present elegant shape. You will be abused by all orthodox prints, probably ridiculed as a visionary [by] many men of science, & openly supported by few of the oracles of the press: your resource against all these disadvantages lies in a cheap edition.[78]

The only reason for delay, Combe continued, was "the chance of criticism enabling you to improve your next impression." The author of *Vestiges,* as Combe's own publisher, was of course fully aware of these advantages.

By the mid-1840s, the market for inexpensive books was much better known than it had been a decade earlier. There were several options. The close-set

74. RC to AI, 31 July 1845, NLS Dep. 341/110/141–42.
75. RC to Churchill, [1846], NLS Dep. 341/113/84–85.
76. RC to AI, [June–July 1846], NLS Dep. 341/110/201.
77. G. Combe to L. Mott, 10 Oct. 1844, in Gibbon 1878, 2:18.
78. G. Combe to the author of *Vestiges,* 30 Oct. 1844, NLS mss. 7388, ff. 780–81.

double-columned format used for the *Constitution of Man* was no longer selling. Most publishers were issuing elegant, single-column, small-type books like the weekly or monthly volumes of Charles Knight, the Chambers firm, and the Religious Tract Society. John Murray of London, whose Family Library had failed in the early 1830s, tried again a decade later with the Colonial and Home Library, which aimed to take advantage of changes in international copyright.[79] The market for mass-produced nonfiction had proved to be not among working people but the middle class.

The text of the first inexpensive edition and that of the sixth gentlemanly edition were identical; but the differences between the volumes were immensely significant. Although both were probably printed on steam presses, a big firm like Savill would have employed a larger machine on the longer print run: five thousand of the people's edition against one thousand of the sixth. The type on the former was considerably smaller (bourgeois rather than small pica) and more closely spaced: it had thirty-eight printed lines of text on each page, compared with twenty-six for its pricier counterpart (fig. 4.17). In consequence, what occupied 516 pages in the one volume was condensed into 298 pages in the other. The paper was still coldpressed and of good quality.

An obvious difference was the size of the pages. The sixth edition is technically a duodecimo, the people's edition an octavo; but the people's edition is a much smaller book (105 mm by 175 mm against 123 mm by 196 mm). Perhaps most striking to contemporaries would have been the difference in binding (figs. 4.10, 4.15). The sixth edition, like its predecessors, sported a high-quality red cotton over boards; the cheap reprint, on the other hand, was sold in stiff paper wrappers of the kind used for magazines. The binding, however, remained sewn; a new and cheaper option, which involved cutting off the spine folds and attaching the leaves to a coating of flexible rubber, was rejected.[80] Still, the wrappers were quite flimsy, so that virtually all copies of the cheap edition that were not thrown away were rebound in cloth or leather.

Issuing books in what would today be called a paperback was a new development in the 1840s, related to the popularity of reading on trains. Most other characteristics of the people's edition were typical of cheap nonfiction books from the mid-1820s through the 1850s. The format was not often used by Churchill—his usual productions were large medical textbooks and monographs—but it would have been familiar to him. In 1846 he was asked by the

79. On formats, see RC to G. Combe, 8 Mar. 1847, NLS mss. 7283, ff. 123–26. On Murray, see Fraser 1997. As Murray's advertising said, "the Works are issued at a rate which places them within the means not only of Colonists, but also of a large portion of the less wealthy classes at home." *Quarterly Literary Advertiser*, no. 150 (Mar. 1845): 36.

80. Gaskell 1985, 234.

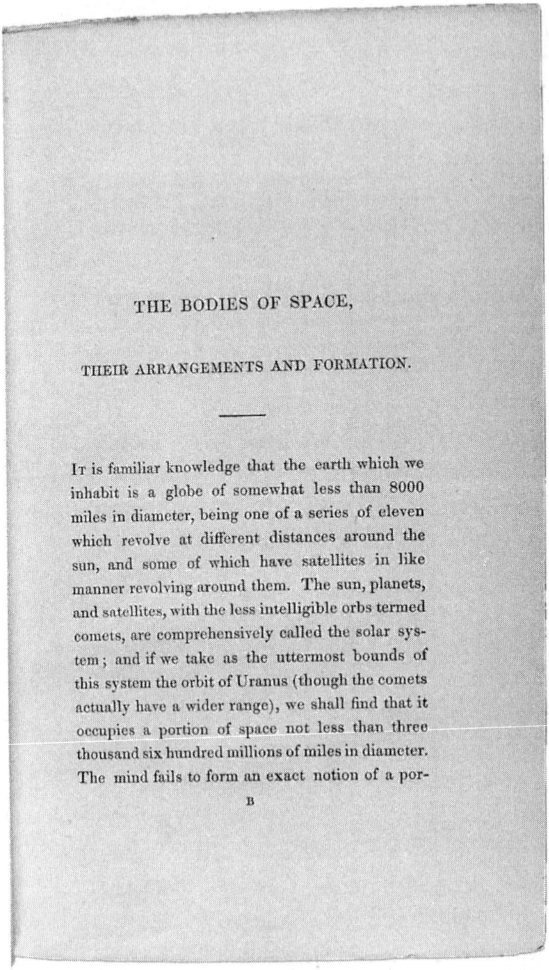

THE BODIES OF SPACE,

THEIR ARRANGEMENTS AND FORMATION.

It is familiar knowledge that the earth which we inhabit is a globe of somewhat less than 8000 miles in diameter, being one of a series of eleven which revolve at different distances around the sun, and some of which have satellites in like manner revolving around them. The sun, planets, and satellites, with the less intelligible orbs termed comets, are comprehensively called the solar system; and if we take as the uttermost bounds of this system the orbit of Uranus (though the comets actually have a wider range), we shall find that it occupies a portion of space not less than three thousand six hundred millions of miles in diameter. The mind fails to form an exact notion of a por-

B

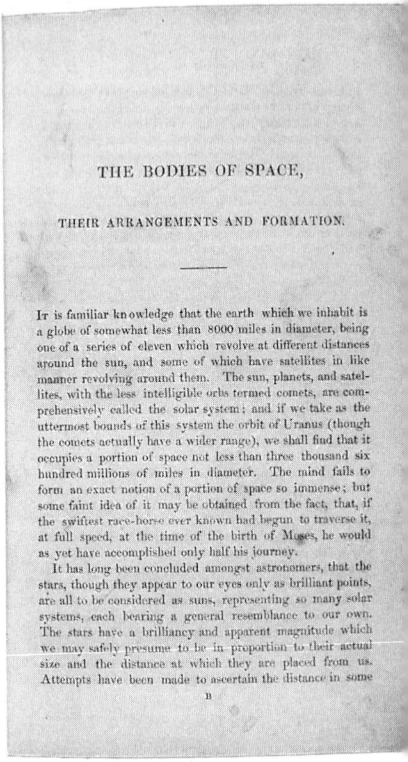

THE BODIES OF SPACE,

THEIR ARRANGEMENTS AND FORMATION.

It is familiar knowledge that the earth which we inhabit is a globe of somewhat less than 8000 miles in diameter, being one of a series of eleven which revolve at different distances around the sun, and some of which have satellites in like manner revolving around them. The sun, planets, and satellites, with the less intelligible orbs termed comets, are comprehensively called the solar system; and if we take as the uttermost bounds of this system the orbit of Uranus (though the comets actually have a wider range), we shall find that it occupies a portion of space not less than three thousand six hundred millions of miles in diameter. The mind fails to form an exact notion of a portion of space so immense; but some faint idea of it may be obtained from the fact, that, if the swiftest race-horse ever known had begun to traverse it, at full speed, at the time of the birth of Moses, he would as yet have accomplished only half his journey.

It has long been concluded amongst astronomers, that the stars, though they appear to our eyes only as brilliant points, are all to be considered as suns, representing so many solar systems, each bearing a general resemblance to our own. The stars have a brilliancy and apparent magnitude which we may safely presume to be in proportion to their actual size and the distance at which they are placed from us. Attempts have been made to ascertain the distance in some

B

4.17 Publishers used different line spacing and type size for different classes of readers, as shown by comparing a page from the sixth edition of *Vestiges* (1847) with one from the cheap people's edition (also 1847).

author for his "opinion as to the doing up of the cheap edition," specifying exactly the physical form appropriate to spreading knowledge about universal progress to the mass audience. "The author," the inquiry noted, "who all along has thought more of the dissemination of what he believes to be the truth of nature than anything else, leans to a sewed form, with printed cover, as in Murray's Colonial Library."[81] The aim was to mass-produce a consumer object that was inexpensive, but not obviously so.

81. RC to Churchill, [1846], NLS Dep. 341/113/84–85.

The greatest difference between gentlemen's and people's editions involved price. By 1847 revisions and more generous spacing had extended the number of pages substantially, so that the gentlemen's sixth edition was 20 percent longer than the fifth and required almost four extra sheets. The 7s. 6d. price had to be raised to 9s., and the one thousand copies printed sold rather slowly. By contrast, the people's edition was only 2s. 6d., an extraordinary bargain. Five thousand copies were issued, almost as many as the first four editions combined, and sales were brisk. As the author noted, "The sale of this new edition strikes me as wonderfully good—only some three of the dullest months of the year, yet half the edition gone."[82]

The intended audiences for the two books were very dissimilar. Churchill was told to hold back the already printed people's edition until after the gentlemanly one appeared. That way, the title page of the one for the mass audience could read "Reprint of Sixth Edition," implying that it reprinted a text already in circulation among gentlemen, rather than the other way around.

Publishing costs differed too. Because of the difficulty of reconstructing Savill's charges for paper, composition, and printing, the comparison is clearest if the accounts for producing the first edition of *Vestiges* in 1844 are compared with those for the second people's edition (the "eighth edition") of the work in 1850. Even leaving aside the difference in the lengths of the two texts (the word count had grown by over a third in the interval), it is easy to see how much less the people's edition cost to produce (fig. 4.11b). Particular savings were made on binding, paper, and above all on the printing. Notably, the costs involved for producing a book in this format had changed little from the early 1830s, as the expenses closely match those calculated for Murray's Family Library.[83]

Above all, the figures show that cheap publishing continued to demand extensive capital resources, with the sale of thousands of copies securing only a tiny profit. Copies of the people's editions sold to the trade at 1s. 10d. a copy, twenty-five copies for the price of twenty-four. The estimates Churchill sent for the eighth edition showed that the profit on two thousand copies would be just £2 8s. more than the author would be paying out, even if advertising were kept to a minimum.[84] When Churchill received in response an offer to sell the copyright, he replied by offering to publish an edition at his own risk, with the profits to be divided equally. The main point, from the author's point of view, was to keep the price low, to "bind him down as to a price per copy to the public, beyond which he should not go."[85] Thus total profit on the second people's edition of 1850

82. RC to AI, 16 Aug. 1847, NLS Dep. 341/110/245–46.

83. Bennett 1976.

84. RC to AI, 28 Apr. 1849, NLS Dep. 341/112/40–41.

85. Ibid.

was £91 15s. 8d., or a little over seven pence a copy. About the same amount was made on an edition of three thousand a year later. These were respectable sums, even after being divided between publisher and author: but scarcely enough to keep either in business. Churchill must have found this distinctly odd.

A situation more characteristic of the trade is revealed by the two editions issued in 1853 and 1860 at the publisher's risk. These are notable as the first to have illustrations. Churchill, who like most scientific publishers was used to large numbers of woodcuts and plates, had pressed for *Vestiges* to be illustrated from 1846. In contrast, the author—whose success in the publishing business was based on avoiding expensive woodcuts—had always refused on the grounds that it "would be dreadfully troublesome, and endanger privacy."[86] By 1851 he had changed his mind, as long as publication (and profits) were entirely at Churchill's risk, and the task of choosing figures deputed to a qualified person. Churchill once again hired Carpenter, who drew material largely from his physiological textbooks, most of them woodcuts engraved by William G. T. Bagg, a skilled medical illustrator who had worked for the firm for years. Because of this, the selection of figures reflected the contents of *Vestiges* only in part (lots of fossils and embryos, no nebulae or electrically generated mites). Effectively, the book became more physiological and anatomical. Although ultimate responsibility remained with the author, Carpenter also commented on the geological and physiological chapters, and saw the work through the press. His fee was seventy-five pounds, of which the author paid thirty-five pounds for the revisions, a guinea for every sheet examined.[87]

Both of the resulting editions (the tenth and eleventh) were highly profitable. Physically, they looked like the "standard works" for which Churchill was renowned; and they made money in much the same way. Churchill netted £469 11s. 3d. on the twenty-five hundred copies of the tenth edition of 1853, which was in print right through the end of 1859. At 12s. 6d. retail, it was five times the price of the people's editions, but realized more profit than all the other editions combined (fig. 4.11c). The eleventh edition of 1860, which was priced at 7s. 6d. like the first edition, could take advantage of lower production costs (the paper tax, for example, had been removed) and a print run of twenty-five hundred. In print until at least 1874, it made a profit of £250 6s. 0d., even after £100 was paid to the author, presumably for a transfer of copyright.

The Philanthropic Commodity

The various editions of *Vestiges* exploited a market that was well developed and clearly differentiated by the 1840s. The author made several hundred pounds

86. RC to AI, [1846], NLS Dep. 341/110/223–24.
87. RC to AI, 8 July 1853, NLS Dep. 341/112/107–8.

(without the printer's full accounts, it is difficult to be sure exactly how much), but this was a far smaller sum than a book that sold so many copies would normally have earned. Profits on the gentlemen's editions from 1844 to 1847, as we have seen, would have been much higher had the book been priced closer to the usual rate. The situation was compounded with the people's editions: not only was the price kept artificially low, but the profit margin on publications of this kind tended to be perilously small in any event. In effect, the literary labor that went into *Vestiges* became almost as invisible as the author.

Publishing the first nine editions of *Vestiges* at such low prices was, in effect, a form of liberal philanthropy. The evangelicals, with their cheap tracts and bibles, had pioneered the use of the latest technologies of print to effect a social transformation.[88] By the 1830s and 1840s the opponents of the evangelicals were fighting back with the weapons of commerce. "Advanced thinkers" like Combe and Mill supported the loss-making *Westminster Review* or the *Phrenological Journal,* but publishers like W. & R. Chambers were using their knowledge of the market to make money by selling science at a price the middle classes could afford. The *Vestiges* sensation thus benefited from a confidence in razor-thin profit margins and a willingness to forgo the full financial rewards of authorship. Faith in the power of cheap print was crucial to the effective intervention of *Vestiges* in the market.

At the same time, the anonymous author was only one participant in the process of bringing *Vestiges* into the public sphere. Churchill, as an experienced specialist medical and scientific publisher, was central to the success of the enterprise. He piloted the book through metropolitan waters that were almost entirely unfamiliar to the author, whose contacts with specialist men of science were limited. The delaying of the people's edition, the extensive program of revisions, the issuing of *Explanations* and the subsequent transformation of the narrative persona: Churchill played a crucial part in all these actions, which were critically important for the reputation of the book and its message.

The role of the publisher should be obvious from Churchill's prominence on the title page. But several dozen other people were involved, from Churchill's accountant to the boys at Savill who fed the steam presses, from the women who folded and sewed the sheets, to the papermakers and the laborers who stamped the binding. Moreover, the work of reproduction only began with the first edition. As we have seen, the text became available to most readers through advertisements, excerpts, reviews in the burgeoning periodical literature, and the sequence of revised editions.

Inexpensive labor meant that authorial changes and proof corrections were easy to make, so that *Vestiges* could be dramatically transformed from edition to

88. Howsam 1991.

edition. In this way, readers literally could become partners in composition; as implied readers became actual readers, reactions were recycled in refining and rewriting. New editions engaged in continuous dialogue with reviews, conversation, correspondence, and reactions from publisher and printer. The resulting edition sequences are typical of nineteenth-century books, especially in nonfiction and poetry. A vast literature has grown up around the related question of the effects of part-publication on novels, but the far more widespread issues posed by edition sequences has yet to be studied.[89] Books like Carpenter's *Principles of Physiology,* Lyell's *Principles of Geology,* Mills's *Logic,* and Darwin's *Origin* are usually studied only in the static form of their first editions. They are better seen as serial publications, part of a process of constant rereading and revision.[90] The political economy of print—a flourishing periodical press, gendered labor regimes, unionized and skilled typesetting, factory production, and the steam press—provided the material foundation for intellectual debate.

The irony revealed by this production history is that the cheapening of books and periodicals was made possible by the emergence of a labor force that was, by and large, unable to afford them. Like all books during the Industrial Revolution, *Vestiges* represented an intricate combination of machine and hand production, work by women and men, intellectual and physical labor—so that its text and physical form reinforced a particular political economy of class and gender. Contemporary readers were sensitive to these issues, which featured largely in debates about the role of machinery in the emerging industrial economy. They read the same articles, books, and encyclopedia essays that have been used to construct this chapter. We need not look to individual passages of prose or the setting of particular lines, to trace the effects of the production process on meaning. The entire volume, from its binding cloth to its type size, spoke directly to such issues, not least through the final arbiter of all literary commodities, accessibility to potential readers.

89. On serial production, see Hughes and Lund 1991, Feltes 1986, and several of the essays in Jordan and Patten 1995.

90. In relation to *Vestiges,* the empirical foundations for this work have been laid in Ogilvie 1973.

PART TWO

Geographies of Reading

CHAPTER FIVE

Conversations on Creation

Reading has charms—and they perceived the charm
Was much increased by reading arm in arm;
And found their scientific walk so sweet,
The more they met, the more they wish'd to meet.

"Love and a Cottage," in T. HAYNES BAYLY,
Songs, Ballads, and Other Poems (1844)

EARLY IN JANUARY 1845, Lady Amelia Murray—artist, philanthropist, and maid of honor to Queen Victoria—was invited to a party given by her old friend, Lady Noel Byron. The scene at such a gathering is familiar to us from countless film and television adaptations of contemporary novels—for our culture has inherited the Victorians' fascination with high Society.[1] Gentlemen are wearing fitted black tailcoats, ladies are corseted in dresses of exquisite embroidered silk, taffeta, and ribbons. The colors in fashion are light—delicate shades of rose, sky blues, and pale golds—and the candle-lit rooms airy and uncluttered, with guests conversing in animated groups. Music and the mingled scents of wax, punch, and perfume fill the air. The conversation, however, is a surprise. Lady Murray and the other guests are talking about science—about cosmic evolution. Throughout fashionable London, from Buckingham Palace to intimate parties like those at Lady Byron's, the most powerful men and women in the country would be discussing *Vestiges*.

The early editions appeared to target this select readership, not the lower middle classes with which the book later became associated. Copies were presented to leading gentlemen's clubs and to progressive-minded politicians and genteel men of science. The first reviews appeared just as the literary season was getting under way in the late autumn, and during the next four months celebrity was secured. Yet until mid-February 1845 only 1,750 copies were in existence,

1. Following Davidoff (1973, 103 n. 5), I use the uppercase to distinguish high Society from the more general society.

5.1 The clubs and aristocratic mansions of London's fashionable West End, viewed from the top of the duke of York's column looking north up Regent Street. This giant folding panorama (81.5 cm × 180 cm), of which this is only half, was compiled from a mosaic of daguerreotypes and engraved on sixty woodblocks. Supplement to the *ILN*, 7 Jan. 1842.

which makes it possible to infer the broad outlines of the readership and its use of the book. Observers of the London scene had parties like this one very much in mind when they reported that *Vestiges* was "much talked of at present."[2]

Because historians so often view London as the national stage, a local perspective is applied only to events that happened elsewhere. Yet its streets were just as idiosyncratic, with their own sights and smells, as those of any other town. The *Vestiges* controversy affords a prime opportunity for exploring the geographies of reading, both in London itself and throughout the British Isles. Victorian towns and cities were defined through the character of their literary life, which was in turn shaped by industrial structure, class, population size, and tradition. Metropolis differed from province, north from south, York from Leeds, Manchester from Liverpool. Some cities were celebrated for scientific institutions and the eminence of their savants. Others failed to develop any continuous intellectual tradition, worshipping instead what the art critic John Ruskin called "the great Goddess of 'getting on.'"[3]

Reading and the culture of print occupied a key place in civic definition.

2. C. Bunbury, diary entry for 19 Feb. 1845, in Bunbury 1890, 1:37. For Lady Murray, see *DNB;* for the party, Carpenter 1888, 34.

3. Quoted in Morrell 1985, 2.

Most towns of any size had at least one newspaper, often more, together with bookshops, circulating libraries, and literary societies where new publications could be debated and discussed. What happened after parcels of *Vestiges* arrived at the rail stations that were springing up across the country? What role did the book have in the ferocious religious disputes of Edinburgh or Liverpool, or in polite conversation at London soirées? How could it be vilified at Cambridge, but read as supporting a new kind of science at Oxford? Even perceptions of what it might mean to participate in a national controversy differed from place to place. "One ought to try and see the local districts," the traveler Hippolyte Taine wrote, "for it is not possible to understand the social fabric properly until one has studied three or four of its component threads in detail."[4]

To address these issues, the history of reading can best begin as local history. We start in the West End of London (fig. 5.1), celebrated for its role in initiating fashion.

"Books, Balls, Bonnets and Metaphysics"

London was the largest city the world had ever known—a sprawling, gigantic, unruly Babylon, with a population of two million. Yet the social elite centered on several hundred families in the aristocracy and urban gentry, who

4. Quoted in Briggs 1959, 1.

possessed an influence in judging books far beyond their numbers. An account of the reading of *Vestiges* can therefore begin in the great mansions and town houses around Mayfair and Pall Mall, and newer areas such as Belgravia. This was a world of birth and breeding, focused (at least symbolically) on the court of Queen Victoria. Life in fashionable Society revolved around the months from November to July, when aristocrats left their country houses for the annual session of Parliament. The calendar of learned and scientific meetings matched these months. New books appeared, professional men were at their chambers, and most private entertaining took place. The height of the social calendar, the Season, ran from May through July. There were garden parties, court receptions, and the grand debutantes' ball where eligible young ladies "came out" onto the marriage market.

The beau monde created intellectual fashion through conversation and letter writing. The upper classes, and especially the aristocracy, maintained a distance from the literary marketplace. They wrote novels, philosophical works, poems, and essays for the quarterlies or the odd letter to the newspapers, but tended to leave the rough-and-tumble of newspaper reviewing to commercial writers. For that reason, very little evidence of how *Vestiges* became the talk of the town comes from published sources, and none from reviews. Rather, traces of it appear in diaries and letters.

Surprisingly few attempts have been made to reconstruct the practice of fashionable conversation. There are no histories of the soirée, no analyses of male conversation in clubs, no accounts of the techniques of celebrated talkers, no surveys of the centrality of talk to intellectual debates. There is no history of silence, and the study of boredom has just begun. The neglect is especially remarkable given the uniquely rich sources: conversations in London Society during the first months of a book's existence really mattered, and contemporaries recorded them in detail.[5]

An idealized vision of the place of books in genteel society is shown in figure 5.2. At a soirée, new books and prints provide topics of conversation for the fashionably dressed gentlemen and ladies scattered in small groups around the room. Moving from left to right, the woodcut illustrates a progressive sequence from a single woman reading a book, to mixed groups discussing books and prints, to social interactions of many different kinds. The centrality of print to the smooth functioning of Society is clear.

5. Through the combined influence of gender studies, ethnomethodology, and the anthropology of everyday life, the assumption that conversation is trivial—say, in comparison with publication—has been widely questioned in recent years: see, for example, the lively account of gossip in Spacks 1986 and the analysis of the London Season in Davidoff 1973. Among historians, Burke 1993, Goldgar 1995, Shapin 1994, Terrall 1996, and Walters 1997 focus on the early modern period.

5.2 "Soirée at the British and Foreign Institute." Note the new books on the tables, the woman reading a volume in the corner, and illustrated prints serving as topics for conversation between men and women. *PT,* 23 Mar. 1844, 177.

Books, like everything from clothes to political news, were part of a culture based on fashion. Tastes in reading changed as quickly as the latest Paris hats, and it became as unthinkable to talk about last year's books as to wear last year's cravat or cut of sleeve (fig. 5.3). At lavish soirées in Mayfair and Belgravia, new books were laid out on tables to serve as conversation pieces. Besides *Vestiges,* the major titles in 1844–45 included the anonymous *Eöthen, or Traces of Travel Brought Home from the East,* a subtle account of a refined sensibility in a region of diplomatic and religious importance; Charles Lyell's *Travels in North America,* which mixed scientific discussions with impressions of society, manners, and education; and Arthur Penrhyn Stanley's biography of the educator Thomas Arnold. Poetry readers welcomed Richard Monckton Milnes's *Palm Leaves,* like *Eöthen* appealing to the rising tide of orientalism, and Elizabeth Barrett's *Poems.* It was a poor season for novels, with *Martin Chuzzlewit* by general agreement having been a less powerful performance than Dickens's earlier works. The same author's *Chimes,* however, was the big Christmas book. The most discussed novel was Disraeli's political *Sybil, or the Two Nations,* which went through three editions in the year. Many of these works were concerned with defining the bonds that might (or might not) hold industrial soci-

5.3 Fashions for August 1846. The book being read aloud is unlikely to be *Vestiges*, which was no longer fashionable twenty months after publication. *The London and Paris Ladies' Magazine*, Aug. 1846, plate 4.

ety together. By the 1840s, reading was typically a silent, solitary act, but the object of reading was social—to maintain relations through civil conversation.

Within weeks of publication, *Vestiges* had become a major talking point. As one diplomat wrote from the Foreign Office in November 1844, "We have a topic here that brings up many lively discussions."[6] Taking just those who borrowed the book from the London Library, which was conveniently located in the West End, the registers show the names of Sir Herbert Compton, Lord De l'Isle, Sir Edmund Head, Major General Sir Charles Pasley, Sir Harry Verney, and dozens of others from the upper strata of metropolitan Society. The duke of Somerset and Lord Morpeth marveled at the "noise" the book had made; Sir John Cam Hobhouse noted that "the book is very remarkable & has produced & must produce a great sensation."[7]

Just how effectively *Vestiges* could serve as a counter in fashionable conversation was shown at a dinner party at Sir William Heathcote's country residence in February 1845. This was a glittering affair, with fancy plate, many courses, and a servant for each guest. The dinner introduced the family of William Edward Nightingale to Alexander Baring, the first Lord Ashburton, and his wife. Baron Ashburton was a leading Tory diplomat; his American wife, Anne Louisa, was from a prominent Philadelphia family. She sat next to the twenty-four-year-old daughter of the Nightingales, Florence, and despite differences in age and status, the two found much to talk about. Nightingale reported the conversation in a self-consciously "clever" letter to one of her regular correspondents, probably her sister.

Nightingale and Lady Ashburton began by talking about Boston, upon which they "swore eternal friendship." Since Nightingale had never crossed the Atlantic, this implied a one-sided conversation: "I having, you know, much curious information to give *her* on that city and its inhabitants." She pretended to be fascinated instead by "a raspberry-tart of diamonds" (raspberries being a slangy reference to the aristocracy) on Lady Ashburton's forehead. Their talk then turned to mesmerism, a topic in the newspapers since the celebrated invalid Harriet Martineau announced that she had been cured in a trance. Mesmerism opened up possibilities for discussing the boundaries of consciousness and knowledge; it also placed women in unprecedented positions of authority, an issue that became central to Nightingale's campaign to create new feminine roles through the manipulation of domestic stereotypes.[8]

6. G. W. Featherstonhaugh to A. Sedgwick, 16 Nov. 1844, CUL Add. mss. 7652.I.E.105.

7. John Cam Hobhouse, diary entry for 18 May 1845, BL Add. mss. 43747, f. 112v; issue books 5 and 8, London Library; duke of Somerset to E. Seymour, 20 Jan. 1845, in Mallock and Ramsden 1893, 268; Lord Morpeth, journal entry for 26 Mar. 1845, Castle Howard Archives, J19/8/6.

8. F. Nightingale to [?P. Nightingale], [Feb. 1845], in Cook 1913, 1:37; see also Goldie 1983, fiche 1, G2, 216. On stereotypes, see Poovey 1988, 164–98. On mesmerism, see Winter 1998a.

Mesmerism led to *Vestiges.* The powers of mind revealed by mesmerism might be one indication that what the book had said was true—that the present race might be succeeded, by "a nobler type of humanity."[9] As Nightingale wrote in her letter, "when we parted, we had got up so high into *Vestiges* that I could not get down again, and was obliged to go off as an angel."[10] In keeping with her ironic tone, Nightingale implied that such lofty matters sat oddly in the opulent context of "such a dinner and such plate as has seldom blessed my housekeeping eyes."[11] Angelic transport was the only escape. This again was a reference to *Vestiges,* which suggested that angels would follow humans in the evolutionary sequence. Nightingale was drawing out the increasingly dominant stereotype of the ideal wife as an "angel in the home," while implying that such lavish dinners were wasteful.[12] As she wrote in another letter, "Is all that china, linen, glass necessary to make man a Progressive animal?"[13]

The formality of such a dinner meant that nothing beyond an acquaintanceship could be formed; Nightingale's reference to "eternal friendship" is (like everything else in her letter) ironic. Talk about God's providential laws could smooth over these kinds of potentially awkward social situations. Natural theology offered a way to express shared religious sentiments while avoiding quarrels over doctrine. It has long been recognized that arguments "from Nature to Nature's God" could be used to sidestep political and religious controversy; but natural theology's mediating role was most critical in conversation. Reflective works of science had achieved their status partly as cornucopias of polite party talk—as a kind of sublime version of discussing the weather.[14] *Vestiges* could be read as the latest in a long tradition of natural-theological conversation books.

General advice manuals often recommended reading science. "Books, balls, bonnets and metaphysics" were suitable topics for talk, according to Captain Orlando Sabertash's semicomic *Art of Conversation* (1842). More seriously, the evangelical Sarah Ellis's *Daughters of England* (1842) proposed God's role in sustaining the universe as an appropriate subject for women:

> 'Science!—what have we to do with science?' exclaim half a dozen soft voices at once. Certainly not to give public lectures, nor always to attend them. . . . Neither is it necessary that you should sacrifice any portion of your feminine delicacy by

9. *V*1, 276.

10. F. Nightingale to [?P. Nightingale], [Feb. 1845], in Cook 1913, 1:37.

11. Ibid.

12. For the stereotypical aspects of this image, see Peterson 1984.

13. F. Nightingale to Madame Mohl, July 1847, in Cook 1913, 1:42.

14. The classic analysis of the mediating role of natural theology is Brooke 1991a, 192–225, although this does not discuss how readers used the works in question.

> diving too deep, or approaching too near the professor's chair. A slight knowledge of science in general is all which is here recommended, so far as it may serve to obviate some of those groundless and irrational fears, which arise out of mistaken apprehensions of the phenomena of nature and art; but, above all, to enlarge our views of the great and glorious Creator, as exhibited in the most sublime, as well as the most insignificant, works of his creation.[15]

Science offered a route to moral regeneration. A knowledge of nature made women better Christians and conversationalists, able to listen intelligently when men expatiated on the laws of providence.

Men might have other aims. Captain Sabertash (actually a pseudonym for Society habitué Major General John Mitchell) suggested astronomy as a surefire way to attract the opposite sex. The ideal setting was a warm summer's evening at a country-house party, and the first step was to escape the distracting bustle of the crowd. Alone in the garden, the couple could admire the calm sublimity of the heavens. Women were especially open to the romance of astronomy:

> The stars,—their lustre, number, incalculable distance,—the immensity of space required for their mighty orbits . . . produce strange thoughts in female hearts. Women have more feeling than we have,—their minds are more easily moved by whatever is great, glorious and sublime; and when so elevated they are more open to the impressions of *la belle passion;* which, with them, is always—in its origin, at least—of a pure and ennobling nature.[16]

A discourse on the stars—rehearsed from books, spoken as if on impulse—was thought to have direct physiological effects on the female body. "I have generally observed," Sabertash noted, "that during such astronomical lectures the pretty dears drew closer to me, and leaned more perceptibly upon my arm."[17]

Vestiges offered enticing opportunities. Take, for example, the opening paragraph ("The mind fails to form an exact notion of a portion of space so immense"), and the last sentence before the conclusion ("Thinking of all the contingencies of this world as to be in time melted into or lost in the greater system"). Or take the purple passage at the end of the astronomy chapter ("'Man pauses breathless at the contemplation of a subject so much above his finite faculties, and can only wonder and adore'"). Such passages struck just the right note of tasteful sublimity.[18]

15. Ellis 1842, 68–69.

16. Sabertash 1842, 135–36. Curtin 1987, St. George 1993, Morgan 1994, and Barnes 1995 provide helpful discussions of conduct manuals as a genre.

17. Sabertash 1842, 136.

18. *V*1, 1, 386, 26.

Fashionable readers, both women and men, scanned the reviews for such passages, and more generally for hints about what to say about the latest books. Even printing brief excerpts tended to disaggregate their contents, juxtaposing them in ways appropriate for polite discourse. Reviews broke down unilateral arguments, pointed out fine passages, and gave clues about how the flood of publications might be assessed. In this sense, periodical reviews served as highly specific literary advisors, targeted adjuncts to the literature of conduct. Representing a vast array of diverse opinions, they became crucial aids in opening up possibilities for talk. The reviews were widely valued for this purpose, not least in aristocratic and genteel circles. As one editor of a metropolitan weekly noted (with tongue firmly in cheek), the aim was "to enable people who cannot read all the books published, to talk about them as if they had read them."[19]

Didactic manuals about behavior in general, on the other hand, were dismissed as relevant only to readers whose parents might not be familiar with approved manners. In the real world of Society, the first rule was to avoid being a bore.[20] Talk was an art: quick, witty, cutting, clever. Contemporary diaries and letters show that rules were never static or fixed. Here *Vestiges* had the advantage of making an orthodox subject into something just dangerous enough to be attractive. The book could be read as elaborating the conventional belief that God sustained creation through law; but it drew on naughty new ideas about reproductive physiology and the status of the soul. This gave divine creation a topical frisson for the first time in years.

Conversations verging on the risqué required managing intimacy and bodily deportment. There was much "quizzing" about tails and monkeys. The mathematician Augustus De Morgan, who loved puzzles and conundrums, joined in conversations about the book by putting forward an alternative theory, in which new physical characteristics emerged through mental exercise. He recalled an accomplished guitar player who never practiced, but who mentally rehearsed the motions until his fingers learned their habits. Parodying the *Vestiges* style, De Morgan wondered "if this should be a minor segment of a higher law? What if, by constantly thinking of ourselves as descended from primaeval monkeys, we should—if this be true—actually *get our tails again?*" De Morgan did not attribute this kind of psychic evolution to *Vestiges,* but he did use it to search for the unknown author, who in thinking habitually about human origins would "naturally get the start of his species" and grow a tail. On hearing any man of eminence discourse knowledgeably about *Vestiges,* De Morgan would take "a curious glance at his proportions" to check for the hidden sign of

19. *Critic,* 3 May 1845, 1–2, at 2.

20. There is no good general history of boredom, although its literary manifestations are sketched in Spacks 1995.

5.4 [John Leech], "The Rising Generation," *Punch,* 27 Mar. 1847, 128.

authorship.[21] Such violations of etiquette, suitable only in an all-male setting like the Athenaeum Club, acknowledged fears raised by the prospect of bestial origins, while mocking them into insignificance.

Badly managed talk on such subjects in mixed company could be ridiculed as ungenteel and unfashionable. *Vestiges* figured large in party jokes and puns. As part of a series in *Punch* parodying "The Rising Generation," a slovenly young aristocrat declares at a private party that "Woman is decidedly—aw—an inferiaw—aw—animal" (fig. 5.4). The women he addresses are of all types, from the young musician at the piano to the formidably intellectual matron; their skeptical glances and obvious accomplishments make this particular Vestigian doctrine look crude and improbable. If anyone was "inferiaw," it was the ill-mannered male, with his low forehead and unprepossessing chin. The correlation between intelligence and brain size assumed in such a cartoon was al-

21. De Morgan 1872, 211.

most universal, although phrenology had become deeply unfashionable in some Society circles. The phrenological character of the later chapters of *Vestiges* could have become the book's most significant hostage to conversational fortune, for single-minded advocacy of what Sabertash called "this exploded old subject" had become the sure mark of a bore. However, the book was generally thought to have handled phrenology with discretion—reference being made only to "the system of mind" of Franz Joseph Gall, and not to bumps on the head or other soft targets for satire.[22] All in all, *Vestiges* offered wonderful opportunities for displaying conversational skill.

Binding Intimacy

Cosmic evolution had not been in vogue since the 1790s. Now it was. Before *Vestiges,* the origins of humanity could be discussed only after the ladies left the men to their port and cigars.[23] Now the subject was talked about in mixed company and among women. To examine the uses of *Vestiges* in more intimate settings, and in more private parts of London town houses than the dining room or salon, we can turn to Elizabeth Barrett's celebrated sickroom. The Barrett family lived on Wimpole Street, a less fashionable district suitable to the family's status as decayed gentry. For years Barrett had lain on her couch, her spaniel Flush at her feet, reading, writing poetry, and entertaining selected visitors. In December 1844 her most important new visitor was Anna Jameson, a leading writer on Christian art. Like many gentlewomen, Jameson preferred the moral uplift of sickroom visiting to parties in high Society. She had earlier noted that a visit to the invalid Harriet Martineau "will bring *me* the highest advantages that mind can impart to mind."[24]

Barrett described her conversations with Jameson in letters to established friends. Their first talk was closely tied to the press: Jameson praised Barrett's *Poems,* which were sharing the review columns with *Vestiges* in the weeklies; they then turned to the announcement of Martineau's mesmeric cure. This brought them together, for in sharing fears about mesmerism, they almost seemed to be demonstrating its sympathetic powers. Despite a striking contrast in their conversational styles—Barrett found Jameson "rather analytical & examinative than spontaneous & impulsive"—they embraced at parting, an unusual degree of physical intimacy after only an hour's conversation.[25]

22. *V*1, 322, 341.

23. See, for example, C. Lyell to S. Lyell, 19 Mar. 1837, in Lyell 1881, 2:7–9, at 8, which reports a conversation between Lyell and Lord Holland, cited in Topham 1998, 254.

24. A. Jameson to O. von Goethe, 24 July 1842, in Needler 1939, 138–39. The power of the female invalid is characterized in Winter 1995 and Bailin 1994.

25. E. Barrett to J. Martin, [26 Nov.] 1844, and Barrett to R. H. Horne, 3 Dec. 1844, in Kelley et al. 1984–98, 9: 244, 255.

When the two women next met, *Vestiges* occupied a role akin to that of a religious tract, had two evangelical women been conversing in the sickroom. Jameson tried to convince Barrett that the book was "comforting," although Barrett had found reading it a depressing experience. *Vestiges,* she lamented, was "one of the most melancholy books in the world," especially distressing in degrading the status of the soul. "I persisted," she reported, "in a 'determinate counsel' not to be a fully developped monkey if I could help it."[26] Here the awful implications of transmutation could be turned aside with a joke.

Vestiges had other associations that revealed further barriers to intimacy. The discussion became caught up with a disagreement about Lady Noel Byron, whom Barrett had detested "all my life long." Lady Byron was Jameson's best friend, and might well have recommended *Vestiges* to her; it is also likely that Barrett knew that the book was often attributed to Lady Byron and especially to her daughter, Ada Lovelace.[27] Jameson defended Lady Byron's character. The circumstances of the separation from Lord Byron were misunderstood, she claimed; and since Jameson had herself suffered from a disastrous marriage, she was in a position to know. She added that, despite a reputation to the contrary, Lady Byron had said "she knew *nothing of mathematics, nothing of science.*" Lying on her sickroom couch, Barrett could not believe that the worldly Lady Byron really shared her own taste for the poetical and spiritual. The claim that "she cares much for *my* poetry!" was met with blank incredulity. This time there seems to have been no close embrace.[28]

Even this brief report shows how ambiguities in *Vestiges* could be used to explore the possibilities of friendship between Victorian gentlewomen. In this respect, however, the book proved less effective a topic than mesmerism. Many women thought mesmerism inherently "sympathetic," so that even talking about it could awaken hidden bonds of sympathy. For the forthright, brisk Jameson, *Vestiges* was (with *Eöthen*) one of the "two most interesting books which have appeared lately." It offered spiritual comfort and supported her belief in "that progressive principle which is the vital part of our constitutional government."[29] For Barrett, *Vestiges* was a "melancholy" book, tied to a dull materialism. Others among her circle agreed.

Barrett's own feeling about creation had been expressed in "A Drama of Exile," the longest and most ambitious poem in her recently published collection. The Morning Star recalls the nebular origins of the universe in a song to the fallen Lucifer:

26. E. Barrett to J. Martin, [25] Jan. 1845, in Kelley et al. 1984–98, 10:41.

27. E. Barrett to J. Kenyon, [3 Jan. 1845], in Kelley et al. 1984–98, 10:5.

28. E. Barrett to J. Martin, [30] Jan. 1845, in Kelley et al. 1984–98, 10:41.

29. A. Jameson to O. von Goethe, 21 Feb. 1845, 25 Apr. 1848, in Needler 1939, 152, 162.

Until, the motion flinging out the motion
To a keen whirl of passion and avidity,
To a dim whirl of languor and delight,
I wound in gyrant orbits smooth and white
With that intense rapidity.
Around, around,
I wound and interwound,
While all the cyclic heavens about me spun.
Stars, planets, suns, and moons dilated broad,
Then flashed together in a single sun,
And wound, and wound in one:
And as they wound I wound,—around, around,
In a great fire I almost took for God.[30]

In returning the copy of *Vestiges* she had borrowed from Kenyon, Barrett noted that the "writer has a certain power in tying a knot—(in mating a system)—but it is not a love-knot."[31] Everything had been joined and fitted, but without feeling. In conversation, the book's failings become tied up with those of a suspected author, Lady Byron, whom Barrett saw as a calculating woman bereft of poetry. Barrett and Jameson became friends, but never achieved a "sisterly" intimacy. "First I was drawn to you," Barrett wrote the following year, "then I was and am bound to you, but I do not move into the confessional notwithstanding my own heart and yours."[32]

For a glimpse of how *Vestiges* entered into domestic settings involving both women and men, we can turn to the most famous household in the country, Buckingham Palace. Early in 1845, Prince Albert read *Vestiges* aloud to the young queen each afternoon. The twenty-five-year-old Albert had scientific interests dating back to his childhood in Coburg, where tutors had taught him chemistry, natural philosophy, and natural history. Albert hoped to bring advances in German science to the British, and *Vestiges,* which made the latest Continental embryology accessible, was sure to receive a ready audience. They

30. "A Drama of Exile," lines 854–66, in Browning 1900, 163. Barrett returned to the theme in *Aurora Leigh* (1856).

31. E. Barrett to J. Kenyon, [3 Jan. 1845], in Kelley et al. 1984–98, 10:5. The painter Benjamin Robert Haydon, another intimate of Barrett's circle, was also unfavorable toward the book, as shown by a diary entry for 26 Nov. 1845: "Sunday. Prayed sincerely & wrote for 2 hours. Read Vestiges of Creation—did not like it" (Pope 1963, 5:489).

32. E. B. Browning to A. Jameson, [4 Aug. 1846], as quoted in the entry on E. B. Browning in *DNB.* The letter is dated in Kelley et al. 1984–98, 13:225. For more on the relationship, see Foster 1988, 132–35; and for Barrett and the intellectual authority of women, David 1987.

discussed it in relation to mesmerism and phrenology, subjects of established interest at court. Certainly Victoria's maid of honor, Amelia Murray, heard enough to be able to talk about the palace readings at Lady Byron's party. *Vestiges,* she said, was "being extensively read in the highest circles" and was "generally attributed" to the tutor of the Lovelace children, the physiologist William Carpenter.[33]

Albert and Victoria often read aloud to each other, mostly novels, biographies, and books of travels. "When I read," she noted in her journal, "I sit on a sofa, in the middle of the room, with a small table before it, on which stand a lamp & candlesticks, Albert sitting in a low armchair, on the opposite side of the table with another small table in front of him, on which he usually stands his book."[34] The queen was bored by technical monographs and British Association meetings, which meant that science tended to take Albert away from her side. Reading *Vestiges* together offered an opportunity to resolve this tension. With its accessible presentation and attractive narrative, the book served much the same function in their domestic relations as did the scientific talks later given in the palace to the royal children. *Vestiges* could make science a shared interest.[35]

The same seems to have occurred in many families, in which *Vestiges* was often given by men to women. Woronzow Greig sent a copy to his mother, the scientific author Mary Somerville, and also to Ada, countess of Lovelace, which her husband thought "a most acceptable present."[36] The poet and philanthropist John Kenyon loaned his copy to Elizabeth Barrett.[37] Copies were often given as presents for special occasions such as birthdays or Christmas, the latter custom having just come into prominence. The available evidence suggests that men usually made the first move in these transactions. *Vestiges* was too controversial, too "advanced," to be offered the other way. Presented as a gift, *Vestiges* could serve to express family ties or affection. One of the country's great naval heroes, Admiral Sir Richard Waller Otway, gave

33. Carpenter 1888, 34. In his account of the *Vestiges* reception, Alexander Ireland (1884, xviii) mentioned that Prince Albert was suggested as an author, and subsequent works have treated him as a leading candidate. However, there are no contemporary references to Albert other than the report of a joke made at a Liverpool party, in RC to AI, 8 Jan. 1845, NLS Dep. 341/110/139–40.

34. Journal entry for 3 Nov. 1844, the Royal Archives, Windsor Castle.

35. Bennett 1977, 236–37.

36. Earl of Lovelace to Lady Noel Byron, [17 Nov. 1844], Bodleian Library, Dep. Lovelace Byron, 372; M. Somerville to W. Greig, 28 May 1845, in Somerville 1873, 278. *American Book Prices Current 1992* (New York: Bancroft-Parkman, 1993), 407, lists a copy of the second edition inscribed to Lady Lovelace from Greig, and from her to him in return.

37. E. Barrett to J. Kenyon, [3 Jan. 1845], in Kelley et al. 1984–98, 10:5.

a copy to his daughter, the youngest of twelve children. As she wrote on the flyleaf:[38]

Martha Stewart Otway
from my beloved father
1844

The Aristocracy of Nature

That an earnest book of science could become the subject of such fascination among the gentry, aristocracy, and even royalty is a sign of how the metropolitan elite had redefined their role as arbiters of taste. Fashion was no longer frivolous. In intellectual life, as in politics, clothing, and so many other matters, the hereditary aristocracy was expected to provide national leadership. Extravagant social display and riotous dissipation were no longer acceptable, as they had been in the prince regent's day. The aristocracy evinced a new seriousness of purpose, with attitudes shaped by the evangelical moralizing evidenced in so many other changes of the period. The presence of science and scientific men in elite Society vouchsafed the aristocracy's right to govern on behalf of the people. Heroes of knowledge, such as John Herschel or Adam Sedgwick, made an evening party something other than a political cabal or a frivolous indulgence.[39]

Many circles—such as Holland House for the Whigs, Lady Palmerston's at Cambridge House for the Tories—centered on political groupings, as they had since the eighteenth century (fig. 5.5). Three issues dominated Parliamentary debate during the 1844–45 season: the Maynooth controversy, involving state endowment for the Roman Catholic college in Ireland; the reformers' campaign to repeal the excise duties on imported corn; and the railway mania. These in turn related to the fundamental political and economic issues of the industrial era: the relations between church and state; the campaign for free trade; and the impact of technological progress. Although these debates were never far away, the soirées and parties in London's West End town houses offered rather different possibilities for discussion than those found in Parliament, not least because the presence of women was thought to "soften" party feeling. At dinners such as the one at which the Nightingales met the Ashburtons, politics could form a common set of concerns with literature, art, music, and natural science.[40]

38. *V1*, private collection. Otway is identified from Burke and Burke 1847, entry under Sir Robert Waller Otway.

39. Mandler 1990.

40. Davidoff 1973 is an invaluable account of the organization of London Society, although focused on the later nineteenth century; also important are Mandler 1990 and the literature cited

5.5 Charles Robert Leslie, "Lord and Lady Holland, Dr. Allen and William Doggett in the Library at Holland House, 1839." This conversation piece, an oil painting exhibited at the Royal Academy in 1841, shows the importance of learning within aristocratic Whig tradition. The portrait in the lower left corner is the eighteenth-century Whig essayist Joseph Addison. In 1845, after Lady Holland's death, the picture passed to the Whig statesman Viscount Howick, the third Earl Grey.

Aristocratic leadership had been central to the reforming Whig administrations of the 1830s; Whig peers engaged in intellectual pursuits and were active in the Royal Society, the British Association, and the London specialist societies. Such men were often fantastically wealthy, with all the resources of great landed estates at their disposal. Many had huge libraries of thousands or even tens of thousands of volumes. A new book like *Vestiges* would typically be taken out of its cloth casing and expensively rebound in leather, with an armorial

there on political circles in the first half of the century. Dodds 1953, 197–237, gives a good sense of the topics of conversation among the metropolitan elite in 1845, as does Hayter 1965 for the summer of the following year.

bookplate to mark it as a permanent addition to a collection gathered over many generations.

The high aristocracy also purchased spectacular natural history specimens and objects of art and set up private laboratories, museums, and observatories. George Howard, Viscount Morpeth (later the seventh earl of Carlisle) had virtually turned Castle Howard in Yorkshire into a laboratory for the study of mesmerism.[41] Edward Seymour, the eleventh duke of Somerset, had presided over several of the metropolitan scientific societies and entertained scientific men at Wimbledon Park, his country house south of London. Another Whig peer, William Parsons, the earl of Rosse, built the world's largest telescope on the grounds of his Irish estate at Birr with local labor and his wife's money (fig. 5.6).[42] The thick, crenelated walls, which in earlier centuries would have kept out Celtic barbarians or Catholic rioters, enforced stability and scientific precision. The telescope occupied a position in the landscape that might once have been occupied by a picturesque folly, but now the purposes were serious: improvement, utility, the quest for knowledge. Placed in the grounds of Rosse's country house, it was a neogothic ornament of aristocratic Whiggery.

By the mid-1840s the Whig grandees had been in opposition for several years, and they never fully recovered political power. However, their enthusiasm for science was increasingly shared by some among the more progressive elements of the Tory party, which had formed the government since the election of 1841. Prime Minister Sir Robert Peel was a liberal Tory interested in the kind of questions *Vestiges* raised. He regularly invited leading men of science to his residence in Whitehall and his country seat, Drayton Manor, to discuss natural history, physiology, and racial development. What he thought about the book is unknown, but the men of learning who strolled through his famous picture gallery certainly were talking about it.[43]

Progressive politicians like Peel thus valued men of science for their advice. Lord Francis Egerton, first earl of Ellesmere and a leading Peelite Tory, welcomed the Scottish geologist Hugh Miller's refutation of *Vestiges*. Ellesmere needed no men of science to rescue him from the book, but appreciated their expertise. "I read the Vestiges with disbelief and detestation," he told Miller, "but it has been a deep satisfaction to me to find such impressions justified and buttressed by arguments which I could not supply, and to feel assured that my own incredulity and aversion are not based merely on my own ignorance and

41. Winter 1998a, 146–53.

42. Schaffer 1998, Hoskin 1990, Chapman 1998.

43. R. Owen to E. Owen, 9 Oct. 1846, in Owen 1894, 1:288. Peel also held a dinner in honor of Buckland and Owen in the spring of 1845, at which *Vestiges* would presumably been a topic of conversation: *Times*, 18 Mar. 1845, 5. On Peel's scientific patronage more generally, see Desmond 1989, 354–56; Rupke 1994, 52–54.

5.6 Aristocratic science in action. "The Great Telescope, (Of 52 feet, 6 feet clear opening of the speculum) erected at Birr Castle in Ireland, by the Earl of Rosse, President of the Royal Society." Lithograph, c. 1850, by William Bevan after a drawing by Miss Henrietta Crompton, with figures added by Herbert Crompton Herries.

incapacity." *Vestiges* had become widely diffused by the time Ellesmere wrote, and he congratulated men of science for neutralizing "the emanations of a poisonous book." This was their proper role as engineers of the public mind: to undertake "with signal felicity and success the office of a sanitary commissioner of science."[44]

Among the Whigs, the townhouse of Lord and Lady Lovelace on Saint James's Square became an important site for discussing *Vestiges*, and the authorship was often traced to its door. Both the earl of Lovelace, William King, and Ada's mother, Lady Byron, were occasionally named, as was their tutor William Carpenter. So too was their close friend, Andrew Crosse of Fyne Court in Somerset, whose electrically created mites played a big part in the book. Suspicion that Ada Lovelace had written *Vestiges* was almost universal in some circles. She had received her copy from Mary Somerville's son while they were in the country in November 1844, and their common friend Charles Babbage recommended that she read it—that is, if she hadn't written it. As

44. Lord Ellesmere to H. Miller, 27 Nov. 1849, in Bayne 1871, 2:415.

the earl told Lady Byron, he was impressed by the moderate tone, "more humble & pious & calculated to beget that feeling in its readers, more than any sermons."[45]

As a Unitarian, Lord Lovelace embraced *Vestiges* as a blow against the Established Church. Other Whig aristocrats, most of whom belonged to that church, read the book avidly but were more equivocal. The Whig reform program had been founded on a view of nature that was both progressive and operating according to law. Now similar ideas were being used to forward a disagreeable evolutionary cosmology. The duke of Somerset was shocked that "so ambitious a title" should be chosen by an Englishman. "A Frenchman might well do that, and perhaps, after all, his universe was bounded by the Pyrenees on one side and by the British Channel on the other." As Somerset advised his son, who was cruising the Mediterranean, it was far better for men of their class to attend to the practical aspects of physical science, such as navigation and engineering.[46]

Whig aristocrats with estates in Ireland, whose power depended on Protestant ascendancy, tended to view *Vestiges* with similar suspicion. Lord Rosse was torn between genteel caution and a wish to see his giant telescope play the decisive role in national debates about developmental cosmology. The great six-foot reflector could aid in undermining *Vestiges* (especially if the author were a mere journalist, as Rosse eventually came to suspect was the case), but only if his own relationship to any refutation was discreetly managed.[47]

There was no strict doctrinal or ideological line capable of enforcement among the aristocracy, even on matters of party politics. A true gentleman had to make up his own mind. Readings of *Vestiges* varied widely, precisely because aristocratic authority rested on independent, nondemocratic, masculine judgment. Many aristocrats kept commonplace books or diaries toward this end (fig. 5.7). The entries relating to reading are not the tortured soul-searchings of middle-class evangelicals, but the judicial entries of county magistrates coming to an opinion.

Take the example of Lord Morpeth, who read *Vestiges* in his town house on Grosvenor Place in Mayfair. The book, he noted in a diary entry for 26 March 1845 just after he began, "is supposed to lean to deism," and there were rumors about mistakes. Two weeks later he made his judgment. *Vestiges* "has much that is able, startling, striking," although "it obviously strikes out often at random." He agreed that progressive development did not "conflict more with the Mo-

45. Earl of Lovelace to Lady Noel Byron [17 Nov. 1844], Bodleian Library, Dep. Lovelace Byron, 372.

46. Duke of Somerset to E. Seymour, 20 Jan. 1845, Mallock and Ramsden 1893, 268.

47. Earl of Rosse to M. Somerville, 12 June 1844, quoted in Hoskin 1990, 338; Cobbe 1894, 1:195.

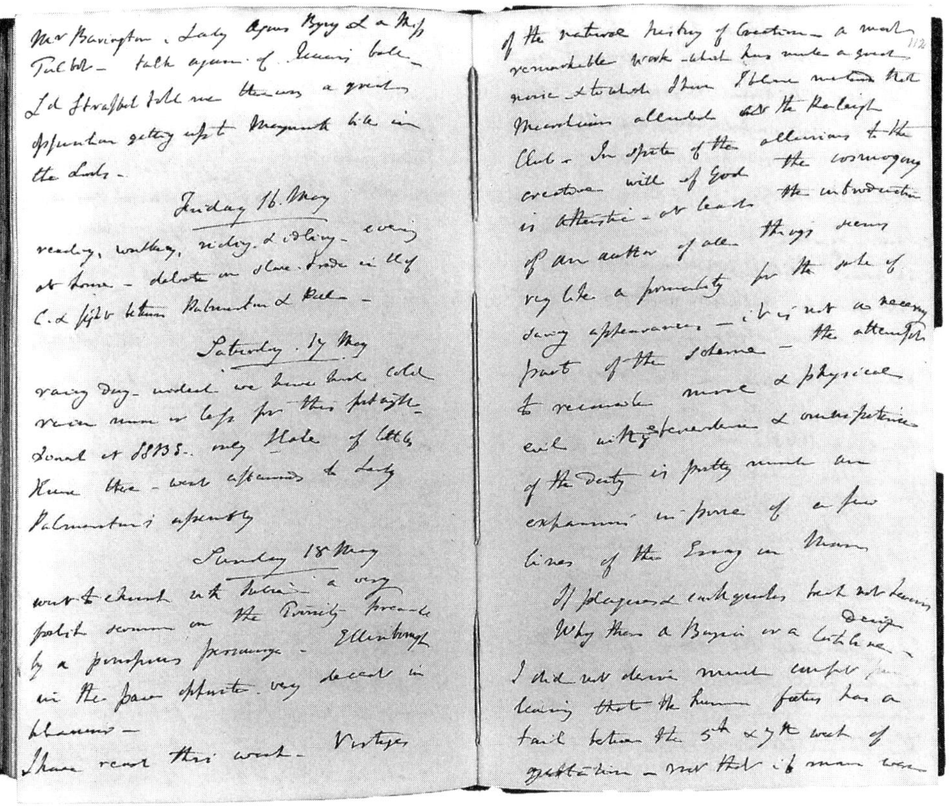

5.7 Opening from Sir John Cam Hobhouse's diary with his reading of *Vestiges.* BL Add. mss. 43747, ff. 111v–112r.

saic accounts than the received theories of modern Geology; the order assigned to the appearance of man certainly harmonises with them." Morpeth disliked the book's views on human origins ("I do not much care for the notion that we are engendered by monkeys"), but objected more strongly to its claim that all the planets were similar to our own. Morpeth would have preferred an ordered diversity of worlds; *Vestiges*—somewhat alarmingly—stressed that the earth was "a member of a democracy."[48] The negative aspects of Morpeth's reading were confirmed in the July *Edinburgh,* the main quarterly organ of the Whigs.

48. Lord Morpeth, journal entry for 7 Apr. 1845, from the Castle Howard Archives, J19/8/7, p. 2. For Morpeth, see Mandler 1984 and Hilton 1994. Mandler 1984, 99 reads the relevant passage in the opposite way, arguing that Morpeth appreciated the "vision of a riotous diversity, of other worlds unlike his own." Morpeth's diary passage is ambiguous, but *Vestiges* is not. The similarity of all planets to the earth was usually taken as a central claim of the book, one that Morpeth clearly took issue with. For the solar system as a democracy, see *V*1, 32.

The review was "long, elaborate, deep, not enlivening, but very conclusive against the book"—although he recognized that Harriet Martineau, his associate in mesmeric experiments, "would not think so."[49] In the end, Morpeth liked to believe that theological orthodoxy—such as he found in reading the Nonconformist *British Quarterly*'s review—would never lead him to condemn the book outright. Rather, he kept his diary to work out views independent of party or sect.

The former Whig minister Sir John Cam Hobhouse, the second Baron Broughton, also recorded his judgments of books in a diary. He lived on Berkeley Square in Mayfair, had been a regular guest at Lady Holland's, and as an old friend of Lord Byron knew the Lovelaces well. Hobhouse first discussed *Vestiges* in March 1845 at the Raleigh Club, a group in London founded to encourage smoking and male camaraderie. He heard gossip from scientific friends that the book "was full of blunders." At a dinner at the Lovelaces at the end of May, his host told him that Ada Lovelace was suspected to be the author.[50]

Shortly before talking to Lord Lovelace, Hobhouse had finished reading the book for himself. It might be thought that a man in his position would dismiss it out of hand: Whewell, author of the anti-Vestigian *Indications of the Creator*, had been his tutor at Cambridge, he had converted from Unitarianism to support of the state Church, and his scientific friends had cautioned him against it. But like Morpeth, Hobhouse made his own judgment:

> I have read this week—Vestiges of the Natural History of Creation—a most remarkable work—which has made a great noise—& to which I have I believe mentioned that Murchison alluded at the Raleigh Club. In spite of the allusions to the creative will of God the cosmogony is atheistic—at least the introduction of an author of all things seems very like a formality for the sake of saving appearances—it is not a necessary part of the scheme—the attempt to reconcile moral & physical evil with y[e] benevolence & omnipotence of the deity is pretty much an expansion in prose of a few lines of the Essay on Man
>
> If plagues & earthquakes test not heaven's design
> Why then a Borgia or a Catiline—

Hobhouse read with special attention to the origin of man. He took no comfort from learning that the human fetus had a tail at certain stages of gestation, nor that there was analogy between human developmental stages and the record of fossil life. He found equally alarming the suggestion that a prematurely born

49. Lord Morpeth, journal entry for 26 July 1845, from the Castle Howard Archives, J19/8/8, p. 27.

50. Entry for 17 Mar. 1845, John Cam Hobhouse diaries, BL Add. mss. 43747, f. 82. Geikie 1875, 2:21–22, describes the founding of the Raleigh Club.

man would approximate to an ape or a woman. Yet the style and tone were "good," especially at the difficult point describing the introduction of the human race. "It does not meddle with revealed religion," he concluded, "—but unless I am mistaken the leaders of revealed religion will meddle with it—."[51] Angry clerics would attack, but Hobhouse reserved the right to form his own opinion.

Visionary, daring, unfettered by prejudice: these were reasons why *Vestiges* appealed to the progressive Whig aristocracy. The book was also relevant to the very different intellectual projects of the ultraconservative faction of the Tories. In the political wilderness since their failure to defeat reform in 1832, these men pursued science as a form of esoteric metaphysics, often in connection with studies of antiquities or the Bible. Their cosmological systems might appear to us as "maverick" and "eccentric," but this is simply a consequence of their origins in an unfamiliar form of scientific practice. The Welsh lawyer Samuel Bosanquet's exposé of *Vestiges* as the whore of Babylon was the product of such a quest. Some ultra-Tory evangelicals, like the collector of monster saurians Thomas Hawkins, developed apocalyptic visions of cosmic history. He reviled any kind of naturalistic development, and had called the French naturalist Jean-Baptiste Lamarck "an assassin" who "waylays the Doctrines of Sin, and the Righteousness of God." Hawkins pointed out in a letter to the *Times* that Michael Faraday's latest discoveries suggested that the nebular "Fire-mist" of *Vestiges* must be wrong. If atoms were immaterial "spheres of power," as Faraday had announced at the Royal Society, then a theory based on the condensation of matter made no sense. Faraday's work, which became the basis for electromagnetic field theory, was of compelling interest to the transcendental metaphysicians.[52]

The ultra-Tories took special pride in their independence of judgment, and felt even less constrained than most by the findings of scientific astronomy and geology. Peel's evangelical brother-in-law, William Cockburn, the dean of York, developed a Bible-based theory of the strata in opposition not only to *Vestiges* but to all forms of modern geology. As he complained in a letter to the *Times,* the book shared with the geologists a tendency to assume "things neither proved, nor probable, nor possible."[53] The paternalist earl of Stanhope agreed that geology was mere "fiddle faddle," but from a perspective that was

51. Diary entry for 18 May 1845, BL Add. mss. 43747, f. 112; the verses are paraphrased from Alexander Pope's *Essay on Man,* epistle i, lines 155–56. Hobhouse went on to say that "[t]he author is thought to be a tutor of Lady Lovelace—a Mr Nicholls I believe." He had obviously heard the rumor about Carpenter, and perhaps had confused it with other suspicions that John Pringle Nichol was the author.

52. Hawkins 1840, 1, and Hawkins, "To the Editor of *The Times,*" 15 Jan. 1846, 3.

53. W. Cockburn, "To the Editor of *The Times,*" *Times,* 10 June 1845, 6.

far from religious orthodoxy. He adopted the Bavarian wild boy Kaspar Hauser and contemplated the philosophy of dreams and mesmeric sleep. Stanhope made notes to the physician Edward Binns's *Anatomy of Sleep* (1842), supporting the electrical nature of thought and the reasoning power of animals. When publishing these notes in the second edition of 1845, Binns pointed out how closely they matched the "beautiful and profound hypothesis" of *Vestiges*.[54]

For all their variety, the readings by Whigs and Tories consistently placed *Vestiges* within the antidemocratic politics of aristocratic authority. Who else could the text mean when it mentioned those "original, inventive, and aspiring minds" and "a nobler type of humanity"?[55] The generic instability of the text also hinted at an author of high birth. How else could the book's boldness and independence be explained, or its references to Sir Thomas Hope's esoteric work? *Vestiges* envisioned nature as a garden of the kind that surrounded the great houses in the seventeenth century: "we begin to see parterres balancing each other, trees, statues, and arbours placed symmetrically, and that the whole is an assemblage of parts mutually reflective."[56] The Divine governor watched over this ordered landscape as a wealthy gentleman might conduct his affairs—never intervening directly, working always through the agency of others to bring about improvements. Epic grandeur, philosophical detachment, and unwillingness to be constrained by orthodoxy or genre: all these could be taken to point to an aristocratic hand.

The Invisible Lion

But who was the author? Deep anonymity created something of a crisis for "literary lionism," one of the key institutions of the fashionable culture of print that united intellect and birth. The "lion" (fig. 5.8) was a writer, musician, traveler, talker, or other accomplished individual who appeared at Society functions while his or her name was in the news. An article in the *Westminster Review* attributed literary lionism to the spread of print culture and the ease of authorship: "A wise man might, at the time of the invention of printing, have foreseen the age of literary 'Lionism,' and would probably have smiled at it as a temporary extravagance."[57]

Literary lionism maintained the ideal of a select readership among the gentry and aristocracy, who could converse directly with leading authors. Aristocratic society thereby could claim to provide intellectual leadership for the

54. Binns 1845, including comments by Stanhope on *Vestiges* on 449–50.

55. *V*1, 276, 321–22.

56. Ibid., 250–51.

57. H[arriet] M[artineau], "Literary Lionism," *Westminster Review*, Apr. 1839, 32:261–81, at 262; see also Martineau 1877, 1:271–97.

5.8 George Cruikshank's "A horrible Bore in the Company," and "*The* Lion of the party!" "Social Zoology," in Gilbert Abbott À Beckett, *George Cruikshank's Table-book*, July 1845, facing p. 141.

nation, while remaining distinct from the ordinary public, whose access to literary celebrities was only through print. This dominance, however, was beginning to be challenged as authors appealed directly to audiences through the lecture circuit. Platform appearances defined a new relation between middle-class authors and "the people," which reached its apotheosis in Dickens's public readings in the following decade.[58]

The author of *Vestiges* was an invisible lion. This challenged the established relation between literary celebrity and the metropolitan elite; "Mr. *Vestiges*" had not paid homage to Society. There was no lion to be celebrated, attacked, admired—or tamed. The other anonymous sensation of the season, *Eöthen*, shows a more typical situation, in which the author's name—Alexander Kinglake—was an open secret known to anyone who attended the right parties or corresponded with someone who did. Kinglake, a barrister, had kept his name off the title page because he feared a public reputation as an author might harm his professional prospects. In Society circles, on the other hand, Kinglake became so identified with his work that friends addressed him in conversation as "Eöthen."[59] Similarly, most people in Society could easily ascertain that Harriet Martineau had written the *Westminster* essay on literary lionism. To recall so darkly hidden a secret as the *Vestiges* authorship, people had to turn back to the Waverley sensation or the political letters of Junius.

The secrecy that often surrounded aristocratic participation in the world of print made a titled author a distinct possibility. Anonymity had long been used within court culture to manipulate and maintain the lines between public and private. Authors could be acknowledged within their coteries while not suffering the taint of literary commerce. It is thus not surprising that fashionable speculation about *Vestiges* centered on names that were aristocratic and genteel. Could the author be Lord Lovelace or Lord Thurlow, known to be keen supporters of the work? Or was it Sir Charles Bunbury or his brother Edward, both sons of a famous Whig dynasty? By comparison, the names of Carpenter, Nichol, and so forth were relatively unknown and easily confused.

For over six months, the most widely canvassed name was that of Sir Richard Rawlinson Vyvyan (fig. 5.9). The case of Vyvyan demonstrates the impossibility of linking *Vestiges* in any unequivocal way with middle-class liberal political reform. Born in 1800, he was perhaps the most unbending Tory aristocrat in Parliament. He had been a leader of the Country Gentlemen faction, which opposed the Reform Act, working-class education, Roman Catholic emancipation, repeal of the Corn Laws, and all other liberal measures.[60] Vyvyan had

58. Small 1996.

59. Paston 1932, 47–48; Gaury 1972, 47–52.

60. Bradfield 1968; for further details and bibliography, see Boase and Courtney 1878–82, 2:840–41, 3:1357.

5.9 Richard Vyvyan, the leading candidate for the authorship of *Vestiges* during the 1844–45 Season.

an intense interest in science, and possessed a laboratory and "a most choice library" at his estate at Trelowarren in Cornwall, of which he made "a very scholastic use."[61] In London he had a spacious villa in Regent's Park. Vyvyan kept a "night journal" in Italian, recording his dreams and metaphysical reflections, and wrote many philosophical works, including an anonymous privately printed *Essay on Arithmo-physiology* (1825) and the first volume of a *Psychology* (1831), also anonymous, on the immortality of the animal soul. The later work had been withdrawn from circulation after being accused of materialism.

Out of Parliament for several years, Vyvyan completed in 1841 a two-volume treatise, *The Harmony of the Comprehensible World.* This was a metaphysical work of extreme transcendentalism, based on abstract mathematical principles combined with a self-developing power. A deductive scheme of progress, Vyvyan argued, could be applied to the phenomena of the natural world. The *Harmony* supported geological progress, abstract mathematics, spontaneous generation, transmutation of species, phrenology, and a physiological influence between the planets. It argued that matter needed to be seen in terms of the forces between particles and not in terms of the particles themselves.[62]

61. Caroline Fox, diary entry for 7 Jan. 1837, in Pym 1882, 1:40–41.

62. As copies of Vyvyan's books are exceptionally difficult to find, I have referred to the 1845 edition of the *Harmony* in the British Library. For the suppression of the 1831 volume, see *Notes and*

Just as Vyvyan believed that political decision making should be left in the hands of the landed aristocracy, so did he wish the *Harmony* to have a select readership. Potentially dangerous works such as those of Lamarck could be read and discussed, but only among the chosen few. Vyvyan accordingly had the *Harmony* privately printed and presented to a handful of leading men of science for their comments.[63] As a Fellow of the Royal Society, Vyvyan was well-known in scientific circles; he had begun his work after conversations with gentlemanly geologist Henry De la Beche, who visited Vyvyan at Trelowarren many times during his offical geological survey of Cornwall.[64] Vyvyan seemed a likely author for *Vestiges* among the few who knew of the earlier work, who then spread the rumor—perhaps partly to throw the scent off themselves.[65]

To modern readers, Vyvyan might appear an improbable choice for the authorship. If species transformation, phrenology, and the other doctrines in *Vestiges* were grounded in liberal politics, then an ultra-Tory is the last sort of person who would be expected to hold such views. Even to phrase the issue in this way is misleading, however, for it makes Vyvyan and the other aristocratic natural philosophers into curious eccentrics. For Vyvyan and others like him, politics was about aristocratic authority, not the acceptance or rejection of specific doctrines. It would have been part of his independence, as a hereditary peer, to choose to support a cosmic developmental law. (In fact, the main argument against Vyvyan's authorship was that *Vestiges* would have even "wilder" had he written it.) Independence, even in issues of party, was crucial to the politics of the Country Gentlemen.[66]

Vyvyan detested being identified with *Vestiges.* This is not because evolutionary doctrines ruled out all but infidels as possible authors, but rather because of the wide readership the work appeared to target. To be thrown into the public arena went against everything that Vyvyan believed about the constitution of authority. He issued all his books privately and told a leading anatomist that he would not publish "until I have the opinions of some of my practical

---

Queries, 23 Aug. 1879, 5th ser., 12:149; 25 Oct. 1879, 5th ser., 12:333–34; 1 Nov. 1879, 5th. ser., 12:357; 17 Sept. 1887, 7th ser., 4:235.

63. R. Vyvyan to R. Brown, 31 Dec. 1841, BL Add. mss. 39954, f. 368; R. Vyvyan to R. Owen, 8 Jan. 1842, NHM, Owen Correspondence, additional letters and papers received, 1982–.

64. "I make no apology for submitting this to you. Our *conversations* determined me to undertake it, and four years retirement from politics have afforded me leisure": R. Vyvyan to H. T. De la Beche, 10 Jan. 1842, NMW, Geology Dept., De la Beche papers; see also Caroline Fox, diary entry for 7 Jan. 1837, in Pym 1882, 1:40–41.

65. This attribution is also included in the publisher's list bound with the second American edition of *Vestiges,* issued in 1845; see figure 11.6.

66. Bradfield 1968, 729.

friends upon them." He saw working geologists and naturalists as underlaborers in an esoteric quest whose fundamentals were to be established by the aristocracy. A new, one-volume edition of the *Harmony* issued late in 1845—like the original, anonymous and privately printed—made the difference clear. A preface praised that "very popular publication" for its "skilful arrangement" and "language as appropriate as it is elegant." But praise of form implied criticism of approach and audience. Vyvyan's search began with first principles, "a method which may be tedious" to less discriminating readers who had been attracted to *Vestiges*.[67]

The flood of gossip put Vyvyan in a difficult position. At first he disdained publishing a denial and made the rounds of his friends, disowning the work in conversation, as did his younger brother Edward.[68] This only increased suspicions. After five months their campaign began to take effect, but the rumors would not die. In October 1845 the *Falmouth Packet,* a Cornish newspaper, claimed to have "good reason" for pegging the authorship on Vyvyan. The story was picked up, first by a competing paper in Vyvyan's constituency and then throughout the country. Forced to respond, Vyvyan asked the *Times* to issue a denial on his behalf.[69] This seems to have done the trick, although a few periodical editors continued to attribute the book to him.[70] Vyvyan still had no idea who the real author was, even after *Vestiges* had been out for over a year.[71]

The second of the suspected authors of *Vestiges* was Ada, countess of Lovelace (fig. 5.10). Twenty-nine years old and the mother of three, Lovelace was Lord Byron's only legitimate daughter and a friend of Anna Jameson and other literary women.[72] Her husband was extremely wealthy; besides their London mansion, they owned two large country houses, Ockham Park in Surrey and Ashley Combe in Somerset. For a woman of her position, anonymity and pseudonymity could allow participation in the world of research and discovery—while maintaining a distanced relation with the market.[73] In contrast, her

67. R. Vyvyan to R. Owen, 8 Jan. 1842, NHM, Owen Correspondence supplement 3, additional letters and papers received 1982–; Vyvyan 1845, x.

68. E. Barrett to J. Kenyon, [3 Jan. 1845], in Kelley et al., 1984–98, 10:5.

69. *Falmouth Packet,* 18 Oct. 1845, 8; *West Briton,* 24 Oct. 1845, 2; *Times,* 28 Nov. 1845, 4; *Times,* 3 Nov. 1845, 6 (first publication of Vyvyan's denial); *West Briton,* 7 Nov. 1845, 2; *Falmouth Packet,* 8 Nov. 1845, 8; *Examiner,* 8 Nov. 1845, 712.

70. "The Delusions of the Day," *Torch,* 10 Jan 1846, 1:21–23, at 23.

71. Vyvyan 1845, x; *Times,* 3 Nov. 1845, 6, asserts his failure to discover the real author.

72. There is a substantial literature on Lovelace; see Woolley 1999, Winter 1998b, Stein 1985, Moore 1977, Baum 1986.

73. Her authorship of the notes (signed "A.A.L.") on Babbage's Analytical Engine was mentioned in print, with Lord Lovelace's permission, for the first time in Charles Weld's history of the Royal Society; see Weld 1848, 2:386–87, and Baum 1986, 87.

friend Harriet Martineau, whose status depended on her pen, could have no such qualms —although for this Martineau paid the price of being caricatured as the type specimen of a "strong-minded woman" (fig. 5.11). Lovelace was known as a protegée of Babbage and De Morgan and associated with the electrical philosopher Crosse and her children's tutor Carpenter. All these men moved in liberal Whig or Radical circles and were suspected of writing *Vestiges* themselves. In 1843, working with Babbage, Lovelace had prepared an annotated translation of General L. F. Menabrea's technical memoir on the Analytical Engine, the mechanical computer that was featured in *Vestiges.* Lovelace was passionately interested in science, in which she had long planned an ambitious work that would transform cosmology.

The struggle to evolve a science in accord with feeling had led to intense difficulties in developing an appropriate role. Lovelace alternated between despair at her ignorance and grandiose plans for revealing cosmic secrets. "I believe myself to possess a most singular combination of qualities exactly fitted to make me pre-eminently a discoverer of the hidden realities of nature," she told her mother. She would do wonders in mathematics, become "the Deborah, the Elijah of Science" or a "Newton for the Molecular universe."[74] Lovelace became involved in mesmerism, phrenology, calculating engines, and the transmutation of species. What she proposed was nothing less than a romantic science that would reconcile her gender, her status as the wife of a Whig peer, and her love of natural philosophy. Her body itself would become a "molecular laboratory" through which she would obtain a mathematical understanding of mental phenonema. It was easy to see why Babbage and other friends wondered if she might have written *Vestiges.*

Conversations about the authorship easily moved toward the question of whether women should write on such subjects at all. Middle-class comic magazines and conduct manuals had few doubts about the answer and drew upon the stereotype of the female pedant already evident in the "Lady of Scientific Habits" (fig. 2.3). Thus the text for George Cruikshank's "Social Zoology" (fig. 5.8) noted that literary lionesses—though increasingly common—were both "exceedingly harmless" and "hardly ever subject to be pursued." Literary bores, it went on to say, were more dangerous, for there was always the

74. Stein 1985, 86, 150, quoting from letters to her mother Lady Noel Byron. The instability of the role Lovelace attempted to establish is shown by the variety of opinions about her; Babbage and De Morgan devoted considerable time to teaching her, while John Herschel was highly critical: "As regards Lady L's notes I do not find them quite so distinct. It is very difficult to say how far she does and how far she does not understand either the engine she treats of or the mathematics she explains; at least I find myself in reading her notes at a loss in the same kind of way as I feel when trying to understand any other thing which the explainer himself has not clear ideas of." J. F. W. Herschel to C. Lyell, 15 Nov. 1844, RSL, Herschel Letters, vol. 22, ff. 210–11.

5.10 Ada Lovelace, c. 1844. Tinted photograph, from a daguerreotype. The lace trimming and full skirt were typical of women's high fashion during the 1844–45 season; gentlemen typically wore black cravats, fancy silk waistcoats, and closely fitting jackets with tails.

5.11 *Below:* Female subspecies from *The Natural History of "Bores,"* drawn by the comic artist Henry George Hine. *Upper left:* the "strong-minded woman" bore, who drives men away with cigar smoke and dogmatic views on the laws of creation. *Upper right:* young man vainly trying to converse with the "silent woman bore." *Lower left:* "Miss Lucretia Lutestring," an unmarried poetic bore in her forties, angry at the railways and soaring alone into the empyrean. *Lower right:* the female literary bore, drinking from a pot of blue ink. The stereotypical female reader of *Vestiges* was the strong-minded woman, who "affects the high mathematics, or talks about Fourier and communism; Malthus and the theory of population—and reads the Reports of the Statistical Society." Reach [1847], 56–78.

possibility of being "bored to death."[75] The standard comic taxonomy, *A Natural History of Bores* (1847), featured women prominently: the "strong-minded woman bore," the pretty, empty-headed "silent woman," the poetic "woman of mind," the intellectual bluestocking, her tail dripping ink (fig. 5.11). Masculine bores were rendered as highly specific types, but the feminine classification was so inclusive that any woman could be cast as one. The only proper habitat for a woman was in the home.[76]

Such comments are often used to show that women were retreating into a separate, domestic sphere; but in many ways the reverse was true. Crude stereotyping reflected uncertainty about women's place, especially among the middle classes—not any widening gulf between public and private.[77] What we are witnessing is a reaction against the increasingly large part that women were taking in national life. In genteel circles, the rise of smaller, intimate parties enabled women to take a greater role in intellectual discussion. In progressive Whig and Radical circles, a feminine hand in *Vestiges* could be taken to signal that women were at last coming into their own. This may be why Anna Jameson was so keen on the book; it is certainly what Ada Lovelace's husband believed. The earl was flattered to find *Vestiges* attributed to his wife. This was the kind of oracular epic that might be produced by Byron's daughter, one who aspired to become "the High Priestess of God Almighty Himself."[78]

An invisible lion was stalking the metropolis. During the winter and spring of 1844–45, when interest in *Vestiges* was at its peak, the author was hunted without success. The two principal suspects in fashionable London could scarcely have been more different. One came from a well-known family of liberal Whigs, the other was a Tory of the most conservative kind; one had been brought up as a Unitarian, the other was an Anglican; one could not vote, the other was in Parliament. Given the value placed in aristocratic culture on independent judgment, there was no way that the theoretical positions in the book could be mapped onto a single set of political or religious doctrines. Potential

75. Gilbert Abbott À Beckett, "Social Zoology," *George Cruikshank's Table-book,* July 1845, 141–43.

76. Reach [1847], 56–78.

77. The work of Vickery (1993) on the eighteenth century, and that of Peterson (1984), provides a vital basis for reconsidering the older thesis of "separate spheres" in the Victorian period, which had the unintended effect of limiting women's participation either to abject domesticity or active resistance.

78. Stein 1985, 126, quoting from a letter to her husband William Lovelace; Earl Lovelace to Lady Noel Byron, [17 Nov. 1844], Bodleian Library, Dep. Lovelace Byron, 372. Winter 1998b, 233, discusses the tensions in Lovelace's project: "She could not succeed in making her brain and body into a laboratory precisely because this involved renouncing control over the very processes that produced knowledge."

authors found their faces watched for signs that the mask might be cracking. But the lion remained invisible.

The End of Conversation

The appearance of *Vestiges* on drawing room tables reveals a characteristic phase in civic culture. Conversation about books had been at the center of the great European salons at least since the mid-eighteenth century. The industrial transformations of the early nineteenth century did not destroy or marginalize this culture, but produced a new and relatively stable balance between conversation and print.[79] The vast machinery of print that developed in the 1830s shaped talk about books in a pervasive way. In the eighteenth century, review journals were few and often aimed at nothing more than suggesting the contents of a work. By 1844 periodicals had become organs of opinion, so that conversation about books meant conversation informed by readings of periodicals and newspapers.

The effects of this outpouring of print upon the oral culture of aristocratic life was much debated. Was "steam intellect" destroying polite society? A *Quarterly Review* essay on the rights of women summed up one vision of the change in December 1844, just as books like *Eöthen* and *Vestiges* were the talk of London. This essay, as fashionable readers knew, was written by "Eöthen" himself, Alexander Kinglake. In his view, print culture had superseded conversation and letter writing:

> Its effect upon society, in this respect, is analogous to that of our stupendous machinery upon individual industry. A hundred years ago the poor English matron could sit at her spinning-wheel, with the prospect of a certain, though humble reward. Time passed; men had made their iron-limbed Frankenstein—had given him steam for the breath of his life,—and soon he stood up against his makers—a terrible rival—a giant asking for work. He works well, and will earn his hire; but work he must have—more work than women could do by their ten hundred thousand fire-sides. . . . And so the functions of talking and letter-writing are usurped by the press.[80]

The press was seen as an all-encompassing monster, a steam railway whose tracks were being laid down through drawing rooms (fig. 5.12). The machines of men were thought to be destroying a culture of talk and letter writing in which women had been full participants.

The *Quarterly Review* essay lamented that the plethora of reviews reduced the possibilities for employing books in conversation. Only through what King-

79. Vincent 1981, 1989; Webb 1955.

80. [A. Kinglake], "The Rights of Women," *Quarterly Review*, Dec. 1844, 94–125, at 117.

5.12 Steam technology interrupts a party at dinner; "Very Alarming Railway Accident" from George Cruikshank's *Comic Almanack* for July 1846. Cruikshank 1870, 2:137.

lake termed the "mental ownership" of any single volume could books play their full role in conversation. Readers (especially women) who read only the reviews were eating "sugar-plums" and "sweets," "tasting the cream that is skimmed for them," a process that meant that they were "well nigh starved." Genuine "mental ownership," the *Quarterly Review*'s anonymous author said, came through careful reading, not by unthinking consumption.[81]

Disraeli's novel *Tancred* (1847)—the sequel to *Sybil*—parodied the supposed decline of conversation in the context of the *Vestiges* sensation. The eponymous hero is introduced to Lady Constance Rawleigh at a fashionable dance. She is "a distinguished beauty of two seasons," "very clever," but reportedly able "to breathe scorpions as well as brilliants and roses." Tancred, Lord Montacute, regrets her tone when they meet at a breakfast party (organized by a woman whose self-possession is among the "mechanical inventions of a high class") but remains fascinated.[82]

Shortly afterward, Tancred calls for a more extended conversation with Constance, whose mother and best friend are also present. She goes into raptures about the latest cosmic sensation, the *Revelations of Chaos:*

81. Ibid., 119.

82. Disraeli 1927, 107, 89–90, 111.

> "To judge from the title, the subject is rather obscure," said Tancred.
>
> "No longer so," said Lady Constance. "It is treated scientifically; everything is explained by geology and astronomy, and in that way. It shows you exactly how a star is formed; nothing can be so pretty! A cluster of vapour, the cream of the milky way, a sort of celestial cheese, churned into light, you must read it, 'tis charming."
>
> "Nobody ever saw a star formed," said Tancred.
>
> "Perhaps not. You must read the 'Revelations'; it is all explained. But what is most interesting, is the way in which man has been developed. You know, all is development. The principle is perpetually going on. First there was nothing, then there was something; then, I forget the next, I think there were shells, then fishes; then we came, let me see, did we come next? Never mind that; we came at last. And the next change there will be something very superior to us, something with wings. Ah! that's it; we were fishes, and I believe we shall be crows. But you must read it."
>
> "I do not believe I ever was a fish," said Tancred.
>
> "Oh! but it is all proved; you must not argue on my rapid sketch; read the book. It is impossible to contradict anything in it. You understand, it is all science; it is not like those books in which one says one thing and another the contrary, and both may be wrong. Everything is proved: by geology, you know."[83]

Constance is no "spiritual mistress," but a fool who reduces the origins of the human race to a confused catalogue of creatures and a recipe for cheese making. Even the reference to crows is a hit against *Vestiges*, which in its earliest editions had argued that humans were to the rest of the primates as crows were to other birds. Tancred, who had been on the point of falling in love, is providentially saved, and the young lady misses out on an advantageous marriage. It is a turning point in the plot, for the hero is so disgusted that he quits Britain for the Holy Land, where his search for spiritual regeneration continues. "I must get out of this city as quickly as possible," Tancred concludes, "I cannot cope with its corruption." The diarist Henry Crabb Robinson thought it "a capital satire" on *Vestiges*.[84]

The most telling feature of this encounter is that by Victorian lights it was not really a conversation at all. The *Revelations,* reported as a bald set of assertions, reveals a potential soul mate as a witless bore. Constance had "guanoed her mind by reading French novels" and talked "like a married woman," throwing off opinions on all topics "with unfaltering promptness and with the

83. Ibid., 112–13.

84. Ibid.; H. C. Robinson, diary entry for 3 Apr. 1847, in Morley 1938, 2:664.

well-arranged air of an impromptu."[85] From Disraeli's point of view, speculative science was unsuitable for women; such books, mechanical both in form and meaning, undermined social intercourse and symbolized the dull materialism that was destroying the inner life of the nation. It is telling too that Constance's real character was revealed not at public parties, but at home, the place where character was thought to be formed. Disraeli offered a romantic Tory criticism of the machine as a destroyer of civil society.

Yet rather than demolishing civility, the industrialization of print could be seen as offering new settings for conversation. The press kept talk going at a time of divisive religious and political debate. The revolution in communications and the aristocratic ideals of polite society came together in creating a common agenda of fashionable reading—works lasting, like the latest styles in dress, for a single season. Virtually all the major literary classics of the nineteenth century first came to public notice in this way. Thomas Carlyle, one of the great lions, understood the symbiosis between the salons and the steam press as much as he despised it. In Carlyle's "clothes-philosophy," *Sartor Resartus,* today's fashions became tommorow's rags, and ultimately the raw materials for the paper out of which the intellectual systems of the future were being made. These systems, "the complete Statute-Book of Nature," left no room for spirit, mystery, or a sense of wonder; they reduced nature to a great machine, grinding blindly according to law. It is not surprising that Carlyle later condemned *Vestiges* as one more sign of the nation's spiritual bankruptcy: "Dull book (quasi-atheistic), much talked of then."[86]

85. Disraeli 1927, 111. Strikingly, it is Lady Constance who offers to lend the book to Tancred, one of the few occasions I have found in which a woman takes this initiative. "I will lend it you if you like: it is one of those books one must read."

86. Quoted in Ryals et al. 1993, 19:149.

CHAPTER SIX

Science in the City

In like manner I walk through the streets of this great town . . . and I say "*A wonderful town is this* LIVERPOOL! Well, who made it? *"It came by itself!"* Indeed. Where from? Did it come by railway? Did it walk? Or how did it get here? Some men would say "*It is the production of some forces in nature*—some laws of nature."

REVEREND HUGH STOWELL (1852)

JANE WELSH CARLYLE, LONG-SUFFERING WIFE of the gloomy sage, was bemused. Used to seasonal cycles of metropolitan fashion, she was amazed to find provincial friends still discussing *Vestiges* ten months after publication. On a visit to Liverpool in August 1845, she recorded a conversation with the novelist Geraldine Jewsbury and the Unitarian Reverend James Martineau, Harriet Martineau's brother. Martineau sat down between Jewsbury and Carlyle and immediately began "to speak in the most disrespectful manner of—mechanics Institutes 'and all that sort of thing'!!" They then "got on that eternal *Vestiges of Creation*" which Martineau "termed rather happily '*animated mud.*'"

The oddest thing, from Carlyle's point of view, was that her friends wanted to talk doctrinal theology. The contrast with London was striking:

> Geraldine and M*rs* Paulet were wanting to engage him in a *doctrinal discussion* which *they* are extremely fond of—"Look at Jane" suddenly exclaimed Geraldine—"she is quizzing us in her own mind—You must know (to Martineau) we cannot get Jane to care a bit about *doctrines.*" "I should think not" said Martineau with great vivacity "M*rs* Carlyle is the most *concrete*—woman that I have seen for a long while"—"Oh said Geraldine *she* puts all *her* wisdom into what she calls *practice*—and so never gets into *scrapes*"—"Yes" said Martineau in a tone "significant of much"—to keep out of *doctrines* is the only way to keep out of *scrapes*"! was not that a creditable speech in a Unitarian?[1]

1. J. W. Carlyle to T. Carlyle, [16 Aug. 1845], in Ryals et al. 1993, 19:149.

Unitarians (who denied the doctrine of the Trinity) were noted for theological disquisition, and Jane Carlyle for sardonic skepticism; but the encounter also made an obvious contrast. No one back home debated theology at parties.

It took just six hours for books to reach Liverpool from London by train, but they were read differently when they arrived; in the case of an anonymous work, even the suspected author could change. The great port city on the Mersey is of special interest as the site of the most prolonged newspaper war about *Vestiges.* Such disputes achieved a significance beyond a single town through new means of communication. Local stories were recycled nationally; provincial editors were men of influence and reputation; the better-known titles circulated in reading rooms across the country. Industrial communication did not fix provincial debate to a metropolitan pattern, but provided new opportunities for recognizing and reinforcing differences. Thus the debate did not merely echo arguments in London; rather, it was through the Liverpool newspapers that readers throughout the country first learned that men of science condemned the book.

The Commercial City

As one critic said, the city was "Liverpool—Tory-ridden, monopoly-ridden, clique-ridden, job-ridden, priest-ridden Liverpool."[2] Liverpool was the first port in the empire and a center for shipbuilding and trade. The town grew enormously, so that well before the flood of famine Irish in 1847, it was Britain's most overcrowded city (fig. 6.1). In 1841 population density for England and Wales averaged 275 persons per square mile; in Liverpool, the figure was 138,224, even more than Manchester's 100,000. Fifty-two out of every one hundred children died before the age of five. Liverpool was the "unhealthiest town in England," "the black spot on the Mersey." With its huge laboring class, Liverpool was the scene of sectarian conflicts and riots, and the middle-class elite was riven by theological, political, and economic divides. After the 1842 election control of the council shifted from the liberal Whigs to the Tories, who achieved ascendancy by fanning the flames of anti-Catholic sentiment on the streets.[3]

Liverpool's cultural life was equally fractured. Before 1850, at least thirty-nine formal associations presented science to the public, from the Mechanics'

2. An English Mechanic, "Education in Liverpool," *Liverpool Journal,* 30 Nov. 1844, 3.

3. For these details, see Neal 1988, esp. 1–79. The best general account of Liverpool, although it focuses on a later period, is Waller 1981; see also Collins 1994. Moore 1992 provides a major reassessment of Liverpool politics in the 1840s, although his attempt to downplay sectarian conflict is not entirely convincing. The wider context of politics in the northern cities is dealt with in many works, including Fraser 1982, Garrard 1983, and the classic Briggs 1963.

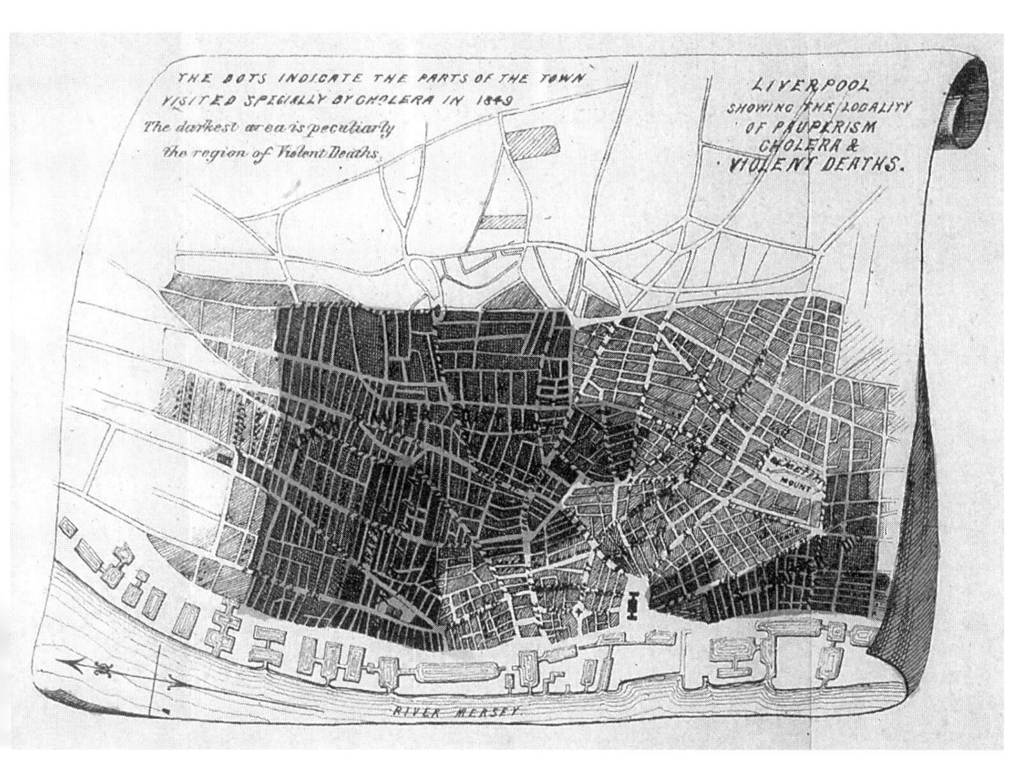

6.1 The black spot on the Mersey: map of Liverpool compiled by the pioneering statistician and antiquarian, the Reverend Abraham Hume. The two main districts subject to pauperism are shaded, and a smaller area, "peculiarly the region of Violent Deaths," is black. Small dots indicate streets that suffered the worst outbreaks of a cholera epidemic in 1849. Vignette from folding map in Hume 1858.

Institution (f. 1825) to the socialist Hall of Science (f. 1839). Visiting lecturers peddled a bewildering variety of scientific wares.[4] The Mechanics' Institution (fig. 6.2), with its array of courses in "useful knowledge" subjects like astronomy, mathematics, and surveying, had been founded on a Unitarian initiative.[5] It was countered after 1839 by the Collegiate Institution (fig. 6.3), run by the Anglican Church. The curricula at the two schools were in many ways similar, but the Collegiate taught science as part of orthodox Christian knowledge. Unitarians were specifically excluded as heretics. The contrast was built into stone: a classical facade for the Mechanics' Institution, Tudor gothic for the Collegiate. The two sites had to be kept well apart, for it was feared that

4. Inkster 1985; see also Kitteringham 1982.

5. Sellers 1969, 247; Tiffen 1835. For a good contemporary account, see "Mechanics' Institution at Liverpool," *CEJ*, 31 Dec. 1842, 396–97.

6.2 Classical (Ionic) facade of the Liverpool Mechanics' Institution, designed by Arthur Hill Holme and built in 1835.

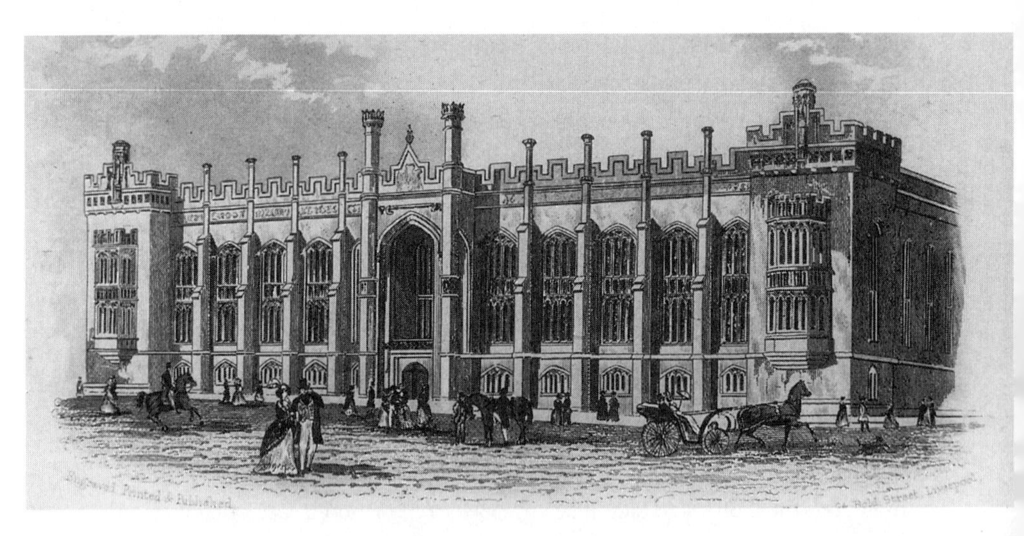

6.3 Tudor gothic facade of the Liverpool Collegiate Institution, designed by Harvey Lonsdale Elmes and completed in 1843. Elmes also designed civic buildings in a classical style, such as Saint George's Hall, but this was not deemed appropriate for an institution devoted to education on religious principles. From [Stonehouse] 1846, facing p. 164.

otherwise the boys would fight.[6] As the Reverend Hugh McNeile—brilliant preacher, evangelical Anglican, and fierce opponent of popery—proclaimed, the Collegiate's success had come through "laying aside all the new-fangled plans of liberalism" and constructing knowledge on the literal truth of the Bible, "that rock on which alone we can build either for our happiness or salvation hereafter." McNeile, a dominant figure in municipal Tory politics, was received "amidst the most deafening cheers."[7]

The liberal, Unitarian ideal of a nondenominational education, which would reduce sectarian violence and keep the masses from Mormonism, atheism, and other aberrations, was anathema to McNeile's followers. True learning was godly learning, with a specifically Anglican orientation. In their view, secular education opened the door to an unholy alliance between popery and atheism. This was a viewpoint widely shared in working- and middle-class households threatened by the shiploads of Irish Catholic immigrants arriving daily at the docks.[8]

Science had a tenuous toehold in urban intellectual combat. The problem had been clearing a neutral ground where the middle classes could discuss natural knowledge without getting embroiled in politics and religion. The principal forum, as in most cities, was the Literary and Philosophical Society, founded in 1812 and meeting behind the elegant facade of the Liverpool Royal Institution (fig. 6.4). The 135 members were a cross-section of local professionals and men of business. No women were admitted, even as guests.[9] Although the society had been started by radicals and Unitarians, from the early 1840s its council was dominated by Tory Anglicans, largely medical men and clerics, who hoped the arts and sciences would damp down sectarian strife. Meetings were held fortnightly on Monday evenings, with a preannounced paper followed by discussion from the floor. Every effort was made to distinguish the proceedings from the doctrinal heckling that characterized much of Liverpool's public discussion; as Edward Baines said, they were "*regulated conversations*" rather than "debates." The rules forbade "all discussions on the particular party-politics of the day, and the peculiar need of any sect of Christians."[10] There were good reasons for this. During the 1790s, suspicion of seditious tendencies had extinguished a literary discussion club around the local Jacobins. Fifty years later, memories of repression continued to be strong.

6. Wainwright 1960, 35.

7. "Liverpool Collegiate Institution—Delivery of Prizes," *Liverpool Journal,* 15 June 1844, 2.

8. Walker 1968; Murphy 1959 and Roderick and Stephens 1978 discuss the educational debates.

9. There is no history of the Literary and Philosophical Society, although its foundation is discussed in Inkster 1985 and Kitteringham 1982. The minute books and other records are preserved in the Local History Library of the Liverpool Central Libraries.

10. Cited in Kitteringham 1982, 336.

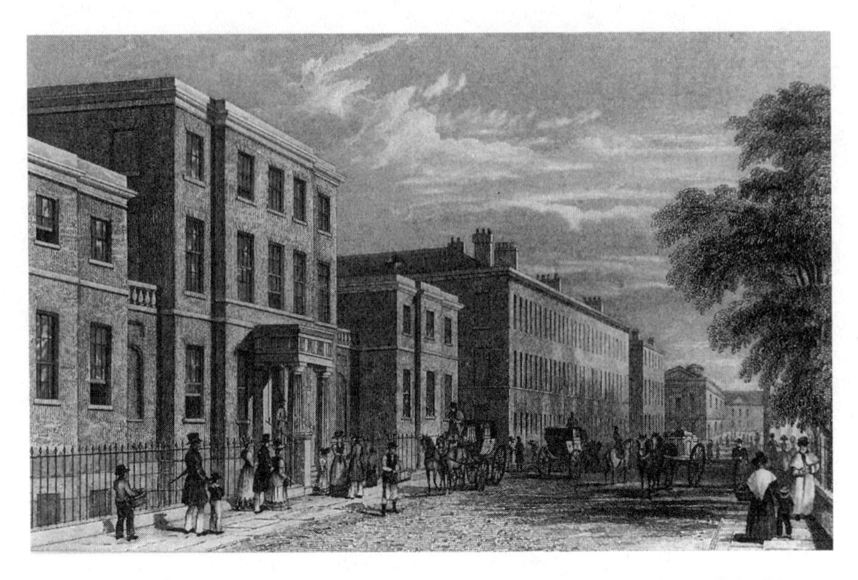

6.4 The Liverpool Royal Institution on Colquitt Street, meeting place of the Literary and Philosophical Society. Such buildings were an important source of civic pride. Engraving by F. Hay from a drawing by G. and C. Pyne, in Austin et al. 1831, facing p. 41.

Scientific books could not be discussed in this context without arousing passions. When thousands were living in cellars, when church accommodation was inadequate, when secular education was divisive, when pollution poisoned the air and water, the issues had immediate practical consequences. The physician William Duncan, treasurer of the Literary and Philosophical Society, had shown that cramped cellar dwellings had produced the unhealthiest town in England. *Vestiges* could be read as addressing these problems. Progress, it claimed, depended on physical conditions: bad air, food, and water could halt the advance of humanity and might even move it backward. A note suggested that mothers living in cellars produced an unusual proportion of defective children.[11] In the "hungry forties" such passages were read as contributions to the debate about the price of food.

Its potential relevance for urban reform created a brisk demand for *Vestiges* in Liverpool and other northern centers. Local booksellers obtained copies from Simpkin and Marshall, the usual London supplier to the provincial trade. The book was widely consulted in public reading rooms. The Mechanics' Institution library, with over eleven thousand volumes the largest of its kind in the country, had a presentation copy of the first edition. Copies would also probably have been available at the Mechanics' and Apprentice's Library, the Athenaeum, and

11. *V*1, 228–29.

other subscription libraries. The Liverpool Library, Britain's oldest circulating collection, bought two copies (something it rarely did) together with *Explanations* and almost all the rest of the pamphlet literature. This was part of a fine collection of natural history, natural philosophy, and theology. Subscribers (893 in 1850) paid a guinea a year.[12]

Conversation, as in London, was initially the main forum. By the beginning of 1845 *Vestiges* was being extensively talked about in Liverpool at dinner parties and soirées in middle-class social circles, particularly among the town's beleaguered, tight-knit circles of Unitarians, radicals, and free-trade liberals. Alexander Ireland traveled from Manchester during the first week of January to attend a gathering at which the book was much discussed. He feigned ignorance of the authorship, even when someone went so far as to attribute the book to Prince Albert.[13] This implausible suggestion has since been repeated as an important feature of the national debate, but at the time it could only be taken as "a first-rate joke." Albert would never have exposed himself in this way.[14]

Public discussion of the book in most cities remained confined to conversation. A survey of newspapers reveals only offhand references in articles devoted to related subjects, or in occasional public lectures, reviews, or letters to an editor. The weekly *Cambrian* reported a lecture at the Swansea Society for the Acquirement of Useful Knowledge; there were a few letters to the *Manchester Guardian* and other English newspapers; in Cornwall, Vyvyan's suspected involvement led to claims and counterclaims in the local press.[15] But sustained disputes in print are hard to find. Liverpool was different. On what might seem to be unpromising ground (compared, say, with Manchester) the town became the site of extended controversy.

Why? Here an understanding of the setting is critical. In metropolitan Society the main potential authors were Lovelace and Vyvyan; but in Liverpool the leading suspect was the twenty-nine-year-old William Hodgson, principal of the Mechanics' Institution (fig. 6.5). Son of an Edinburgh printer, Hodgson had undergone a religious crisis to emerge as a disciple of Combeian phrenology. He was vigorously anticlerical and committed to free trade and moderate

12. Liverpool Library 1850, 90. The other libraries are described in [Stonehouse] [1846], 167–71, and in Perkin 1987.

13. Details of Ireland's report are inferred from RC to AI, 8 Jan. 1845, NLS Dep. 341/110/139–40. The *Liverpool Journal* ("The Prospective Review," 8 Feb. 1845, 2) noted that *Vestiges* was "a strange book, which everybody now reads and talks about."

14. RC to AI, 8 Jan. 1845, NLS Dep. 341/110/139–40.

15. Hughes 1989, 414; "The Milky Way: Our Astral System," *Manchester Guardian,* 22 Jan. 1845, 8; H. H. Jones, "Motion of the Solar System," *Manchester Guardian,* 29 Jan. 1845, 8; J. G. S. [Chambers], "Motion of our Solar System," *Manchester Guardian,* 8 Feb. 1845, 5. For the reports in Cornwall, see *Falmouth Packet,* 18 Oct. 1845, 8, and 8 Nov. 1845, 8.

6.5 William Ballantyne Hodgson in later life. During the mid-1840s he was principal of the Liverpool Mechanics' Institution and a suspected *Vestiges* author. Mills 1899, 139, portrait from 1867.

reform. "To make the past the model and measure even of the present, and much more of the future, is madness, is Toryism"; he wrote, "but to obliterate the past, and build the future on any other foundation, is very midsummer madness—it is Radicalism."[16] Hodgson corresponded regularly with Combe, Chambers, Nichol, Carpenter, and Ireland, partly as a relief from the "undisputed sway" of evangelical Toryism in Liverpool. "The tyranny of the priesthood is said to be great in Scotland," he told his future wife, "but really I think it is much worse here."[17]

As head of the Mechanics' Institution, Hodgson occupied the key position in Liverpool secular education. Forty-eight teachers were on his staff. Nationally known lecturers spoke twice each week; fifteen hundred students (mostly from the lower middling orders) attended classes: there were a large library and museum. The institute was the largest in Britain outside London. "With too many," Hodgson proclaimed in an address, "popular education is a low, and vulgar, and revolutionary thing." Instead he hoped to "train up a race of young

16. W. Hodgson to Wotherspoon, 9 Nov. 1841, in Meiklejohn 1883, 20.

17. W. Hodgson to J. Cox, 28 Sept. 1841, in Meiklejohn 1883, 35–36. For Hodgson's circle, see Meiklejohn 1883 and the letters of reference for his application to Liverpool in Hodgson [1839].

men" who could go forth to educate others.[18] If authored by such a powerful figure, *Vestiges* would constitute an immediate danger to public morals.

Fictions of Development

The Reverend Abraham Hume (fig. 6.6), a thirty-year-old Anglican priest and lecturer in English Literature at the Collegiate Institution, raised the alarm. He delivered an elaborate attack on *Vestiges* at the Literary and Philosophical Society on 13 January 1845, the first meeting of the new year. Every seat was taken, with numerous visitors and intense interest throughout the city. At the next meeting two weeks later, local specialists responded at length and Hume read out a remarkable series of letters from the nation's leading men of science. Both sessions were reported in the liberal free-trade *Liverpool Journal,* the usual outlet for the society's proceedings, and newspaper correspondence continued for months.

Although unusually intense, these exchanges drew upon established forums for communication and shows how progress, development, and evolution were negotiated in a major urban center. Through the newspapers, a key forum for intellectual controversy, the local debate became nationally known. Hume publicized his attack by supplying material to the *Liverpool Journal* and distributing the result as a pamphlet to Literary and Philosophical Society members and intellectual leaders across Britain. Newspaper essays of more than ephemeral interest were often reissued in this way; from the printer's point of view it only involved breaking up columns of newspaper type into short lengths. Because Hume's pamphlet was so widely circulated, the society's recently commenced *Proceedings* included only an abstract.[19]

At the first meeting, Hume praised the Literary and Philosophical Society for affording a "common ground of philosophical inquiry" free from sectarian tensions. From his point of view, this made it all the more effective as a public platform, more so even than the pulpit. The Scriptures, Hume claimed on another occasion, had been written for a "popular" audience, and could not be used to decide "questions of a purely philosophical or scientific kind."[20] His greatest fear was that materialism would hijack the authority of science, but his paper referred only in passing to such concerns, focusing on what he saw as the

18. Hodgson 1845, 10–11.

19. The only copy of the pamphlet (Hume 1845) I have found is in the Wren Library, TC, bound with Whewell's copy of the fourth edition of *Vestiges* (22.c.84.4). For the abstracts, see "Remarks on the Theory Advanced in 'Vestiges of the Natural History of Creation,'" and "Adjourned Discussion on the Theory Advanced in 'Vestiges, &c.,'" *Report and Proceedings of the Literary and Philosophical Society of Liverpool* 1 (1845): 37–40.

20. A. Hume, "Intellectuality of the Lower Animals," *Proceedings of the Literary and Philosophical Society of Liverpool* 4 (1848): 59–73, at 73.

6.6 The Reverend Abraham Hume, antiquarian, Anglican priest, and teacher of English at the Collegiate Institute.

facts. William John Conybeare, the Collegiate's principal, agreed; he disliked *Vestiges*,[21] but felt that ultimately faith and reason would coincide. This strategy of accommodation—labeled by Conybeare a few years later as characterizing a "Broad Church"—was rejected by McNeile's evangelicals, who saw science as a paltry substitute for Scripture. Conybeare and Hume hoped to forge an alliance between metropolitan scientific authorities, local medical men, engineers, and naturalists, and the Anglican Establishment.

Hume appealed to a mathematically rigorous conception of scientific reasoning. If even one of the book's fundamental assumptions was wrong, he said, "it was like introducing a cipher among arithmetical factors" and the whole equation of progressive development would fall to the ground. Because facts were "incontrovertible and unassailable," they could resolve otherwise intractable debates.[22]

21. W. J. Conybeare to A. Sedgwick, [1851], in Clark and Hughes 1890, 2:193; A. Hume, "Remarks on the 'Vestiges of the Natural History of Creation,'" *Liverpool Journal,* 1 Feb. 1845, 5.

22. A. Hume, "Remarks on the 'Vestiges of the Natural History of Creation,'" *Liverpool Journal,* 1 Feb. 1845, 5.

The main body of the address showed that *Vestiges* could not be squared with standard texts of specialist science. The development hypothesis, Hume said, was compounded of the nebular hypothesis and geological progression; "the one is *not proved,* and . . . the other is *not true.*" He rejected the analogy between a growing child and a developing nebula, pointing out that astronomers had never observed such a sequence in progress. Nor did Hume find any evidence in the fossil record. He relied on Lyell's denial of geological progression in the *Principles of Geology* and emphasized that fossil discoveries were pushing back the point at which "advanced" forms appeared in the strata record. Fish had now been found "in one of the very lowest" beds; the plants of the coal era were not only simple monocotyledons, but also more complex forms; what seemed to be a human fossil footprint had been discovered in Scotland only a few weeks before, and although Hume was cautious about this latter fact, it pointed to the bankruptcy of progressive development.[23]

Hume restricted his attack to arguments about fossils, nebulae, and embryos, but his concerns ran deeper. The new developmental cosmology, which he privately linked to Unitarians and liberals, threatened Tory Anglican attempts to recolonize the territory of learning. He thought the *Vestiges* author lived in Liverpool, which made the *Vestiges* controversy the latest battle in a long-term struggle for authority within the local middle-class elite.[24]

Hume was ideally placed to be a standard bearer against dissolvent rationalism. Born in Ireland, he had been educated at the Royal Belfast College, the University of Glasgow (where he studied mathematics), and at Trinity College, Dublin. For four years after arriving in Liverpool in 1843 he served without stipend as curate of Saint Augustine's in Shaw Street; in 1847 he became vicar of Vauxhall, one of the most destitute parishes in the country.[25] For 10,662 parishioners there were seventy-six public houses and gin shops (and probably at least as many brothels) but only a hundred seats for worship in the Established Church. Known for his dedication and sense of humor, Hume showed an active interest in Liverpool's cultural life, joining the Literary and Philosophical Society soon after his arrival. During the next forty years he worked to improve the physical, moral, and spiritual condition of the people, developing paternalist schemes for "civilizing" the urban poor who lived among the coal wharves and warehouses along the Leeds and Liverpool Canal. Hume compiled statistics and maps to underline the appalling conditions and mobilize charitable support; these became models for later national surveys, especially the 1851 Religious Census.

23. Ibid.

24. A. Hume to R. Murchison, 28 Jan. 1845, GSL M/H66/2.

25. For biographical information, see Morley 1887; "The Late Canon Hume," *Liverpool Courier,* 22 Nov. 1884.

The attack on *Vestiges* was part of a campaign for improving public morals. Hume's first paper at the Literary and Philosophical Society, only a few weeks before, was on "The Nature and Influence of Modern Works of Fiction." Hume had defined as fiction narratives that, although not factual, "lie so completely within the limits of nature and possibility, that they might be facts at any time." He argued that with few exceptions (Scott's novels, for example), such works were "injurious," attended with "danger and disadvantage." Science, history, and biography cultivated the rational intellect, which raised man above bestial sensuality. Agreeing with those who thought British science was in decline, Hume lamented that scientific works led publishers and authors into ruin, and blamed book societies and circulating libraries for creating a nation of novel readers. "There is danger," Hume said, "that as one particular kind of diet injures the digestive organs, so the limitation to one particular kind of mental food may destroy the mind too."[26] This was the danger of *Vestiges.*

Like other "slum parsons" Hume was horrified at the spread of infidelity. Catholic priests claimed that Anglican science led to unbelief, and it seemed evident that lack of church provision led shopkeepers and mechanics toward Mormonism, Nonconformity, or outright heathenism. Hume had been shocked to find an infidel tract—an example of these "cheap and poisonous publications"—with a Liverpool bookseller on the title page. But he made this discovery in London at the British Museum. Socialism of the kind led by the philanthropist Robert Owen, often associated with irreligion, was effectively a spent force in Liverpool after 1842.[27] With its high proportion of unskilled laborers, the town lacked the artisan population that gravitated toward infidel socialism elsewhere in England. Only a few years earlier, the local branch of the Owenites had built a spacious Hall of Science on a prime site in Lord Nelson Street, costing five thousand pounds and holding fifteen hundred people. Facilities included a library, a newsroom, a school, and even a rooftop observatory.[28] Christians of all denominations deplored the raising of such a prominent piece of urban architecture on foundations of freethought and rejoiced when it failed and became a theater.[29]

The corruption of popular taste, Hume believed, had made *Vestiges* a sensation "in certain quarters." Like a novel, its attractions were seductive: the author

26. A. Hume, "On the Nature and Influence of Modern Works of Fiction," *Report and Proceedings of the Literary and Philosophical Society of Liverpool* 1 (1844): 18–25, at 18, 23, 24. The wider debate about novel reading among the working classes has been extensively discussed; see Altick 1957, Brantlinger 1998, Flint 1993, James 1974, and Rose 1995.

27. Hume 1851, 12.

28. Rose 1957, 175.

29. Rose 1957, 177.

"leaves the mind open to the wildest inferences, and to daydreams, to which the ordinary play of fancy can present no parallel." The book had an imposing appearance, with "its peculiar arrangement, its flowing style, and its adaptation to lead the reader insensibly forward." These rhetorical blandishments—"like the ornaments of a confectionery cake"—were read as signs of the anonymous author's bad faith. This was no spiritual quest for truth, but a skillfully manipulated tissue of high-sounding language and strong plotting that occupied "the debatable ground between science and fiction."[30] Narrative fantasy had nothing to do with the mathematical logic of true science.

Hume concluded by presenting the anonymous author with an unusual challenge. Labeled as science the book was dangerous and misleading; but recast as romance, with obviously fictional material added, there would be nothing to fear:

> Why should we not have the scientific novel, as well as the historical, in which one would gladly forgive the departure from fact, as he does for the pleasure of viewing an impersonation like Rob Roy, or a pageant like the tournament of Ashby? In its present state, the book is not without its uses, but it should be read with caution, and its statements, especially, should be received "with a grain of salt." But should this suggestion be adopted in the third edition, we may yet be able to speak of the book in terms of praise as unqualified, as, I am sorry to say, we must now speak of it in terms of disapproval.[31]

Vestiges had to be unmasked if science was not to be undermined as a support for Christian faith.

The Unitarian Threat

Hume's fears emerged in the context of the religious conflicts that swept Liverpool in the 1840s. The philosophy of natural law had often been associated with Nonconformists, especially the Unitarians with their stress on rational debate and doctrinal rigor. There were three Unitarian chapels in the city: Paradise Street, led by James Martineau; Renshaw Street, with John Hamilton Thom as the preacher; and Toxteth Park Chapel, overseen by the young John Robberds. Although numerically insignificant (and despised even more than other Dissenters), Unitarians had been civic leaders, often well educated and wealthy.[32]

30. A. Hume, "Remarks on the 'Vestiges of the Natural History of Creation,'" *Liverpool Journal,* 1 Feb. 1845, 5.

31. A. Hume, "Adjourned Discussion on the 'Vestiges of the Natural History of Creation,'" *Liverpool Journal,* 8 Feb. 1845, 2.

32. See the important articles by Webb (1978 and 1990); also Sellers 1969, 217–68; Beard 1846, 324–25, 333.

Their pulpit orations, evening parties, and periodicals became a focus for discussion about *Vestiges,* which was suspected of being their pet project.

Five years earlier, conflict between Anglicans and Unitarians had come to a head when the Reverend Fielding Ould, evangelical incumbent of Liverpool's Christ Church, castigated what he called "the degrading assumptions of the God-denying heresy of Unitarianism." Ould was joined by McNeile and other Anglican evangelicals, who lectured on the role of reason and common sense in interpreting the Bible. As the preface to their collected discourses put it, Unitarians "have deified their own fallible conjectures, instead of humbly acquiescing in the plain meaning of Scripture."[33] This conflict defined the national position of Unitarianism for decades and raised issues of intellectual freedom, secular education, and individual salvation. The attack on *Vestiges* opened a campaign in which battle was to be joined by leaders of learning from the Church of England.

It is telling that a Unitarian, the Reverend John Robberds, was the book's sole defender in the published Literary and Philosophical Society debate. Born in 1814, he was the second son of the Reverend John Gooch Robberds, a leading Manchester Unitarian. The young Robberds had ministered to his congregation at Toxteth Park Chapel from 1840, advocating abolition of capital punishment, free trade, and other liberal causes.[34] Robberds began his response to Hume by defending the nebular hypothesis, citing Nichol's claim that it "reached the very verge of ascertained knowledge." In any event, he did not see the nebular hypothesis as essential to the theory. It mattered little whether man and the "*privileges* of mind" had originated through electricity, or through the direct exertion of creative force; both could express the divine will. Robberds read the book as well intentioned and based on "deep reflection and extensive research."[35]

Robberds would probably have been shown the proofs for the review of *Vestiges* in the first number of a new Unitarian quarterly, the *Prospective Review,* in February 1845. The *Prospective* was published in London but edited by four prominent Unitarian ministers in the north of England, including Martineau and Thom in Liverpool. The result was well received by wealthy Unitarian families; Elizabeth Greg Rathbone mentioned what a good impression the pe-

33. *Unitarianism Confuted* 1839, xii. For an account of the debate, see C. Wicksteed, "The Liverpool Unitarian Controversy of 1839," *Theological Review* 14 (Jan. 1877): 85–106.

34. "Liverpool Anti-monopoly Association," *Liverpool Mercury,* 28 July 1842, 252; Robberds, "Capital Punishment," *Proceedings of the Literary and Philosophical Society of Liverpool,* 1847, 3:121–25. For the more general concern of Unitarians to improve conditions in Liverpool, see Packer 1984.

35. "Discussion on the 'Vestiges of the Natural History of Creation,'" *Liverpool Journal,* 15 Feb. 1845, 3.

riodical was making, and sent a copy of the article to her son, who was extending their cotton business in China in the aftermath of the Opium War. Their partners, the Worthingtons, had already sent him *Vestiges*.[36] Some newspapers attributed the review to Martineau, but the Rathbones and other Unitarians knew that its author was the young theological liberal Francis Newman.[37] Associated with the sect though not a member, Newman had advocated a "church of the future," which would avoid dogma and rest on ethical foundations.

Much behind-the-scenes work had been required to make Newman's review suitable for publication. Thom worried that passages on the relation between matter and the soul would be misunderstood, and inserted a statement to the effect that a philosophy of materialism did not preclude free will, an independent moral realm, or an immortal soul. Newman recognized that *Vestiges* was not always clear on these issues, but recommended it all the same, defending its use of the notorious experiments of Andrew Crosse:

> Suppose that a future Mr. Crosse should succeed in constructing a living dog out of inorganic matter, by a series of galvanic operations, and that this dog should display all the sagacity and affections of other dogs; this would be the most decisive imaginable proof of the identity of that substance by which brutes think, feel, and live, with electric and other forces which act on unorganised matter. Yet such an experiment would not have the most remote tendency to undo our experience and our internal perceptions that truth, justice, disinterestedness, humility, compassion, purity, are better than their opposites; it could not justly lower our reverence and admiration for the great Power who presides over the universe which we behold, or alter in any point the posture of our hearts and spirit towards him.[38]

Within a few months Newman was among the growing list of those tagged as the mysterious author.[39]

Vestiges posed other problems for Unitarian doctrine. Robberds pointed out that the book was inconsistent in distinguishing miracles from natural law. "It was unfair to assume," he said to the Literary and Philosophical Society, "that the conception of direct influence from the Creator implied that He did not act

36. E. G. Rathbone to S. G. Rathbone, 6 Feb. 1845, University of Liverpool Library, Rathbone Papers, xi.1.60.

37. *Liverpool Journal*, 8 Feb. 1845, 2; E. G. Rathbone to S. G. Rathbone, 6 Feb. 1845, Liverpool University Library, Rathbone Papers, xi.1.60.

38. [F. W. Newman], "Vestiges of the Natural History of Creation," *Prospective Review* 1 (Jan. 1845): 49–82, at 57; F. W. Newman to J. H. Thom, 12 Dec. 1844, 20 Jan. 1845, Liverpool University Library, Rathbone Papers, xiii.1.131, 132. See also [F. Newman], "Explanations," *Prospective Review* 2 (Jan. 1846): 33–44.

39. *Literary Gazette*, 12 July 1845, 455.

according to a uniform plan, and it was an assumption to suppose that the existence of laws could supersede the necessity of a constantly acting power." Such laws, Robberds concluded, "were merely expressions of the order and manner in which creative energy was observed to operate," so that direct creative interference could be part of a uniform scheme.[40] (This was precisely what *Vestiges* could be claimed to have said, but many were reading the text otherwise.)

The London-based Unitarian monthly *Christian Reformer* was less willing to be charitable. Its review early in January 1845 damned *Vestiges* for making natural law "something which is separate from the Deity himself,—some positive and tangible existence which he employs as the medium of his operations." "It may seem ungallant to the sex," the reviewer noted, "but it is fair to science, to say that the argument is womanish." The *Reformer* thought *Vestiges* "insidious," "objectionable," with "certain languid attempts to give a possible Christian character to the speculations." This was the voice of Old Dissent, arguing that thrusting law between God and Nature was not only unphilosophical, but led to atheism. The Liverpool Unitarians were more flexible, so that even the city's most infamous Owenite deist, John Finch, remained a member of Robberds's congregation.[41] They were unwilling to dismiss *Vestiges* as the product of a closet skeptic, preferring to clarify its ambiguities. Clarity was essential, for the status of miracles was fiercely debated; this was, notably, one of the issues that distinguished the Liverpool circle from traditional Unitarians who maintained doctrines of biblical inerrancy. Led by Thom and Martineau, the Liverpool theologians had begun to move away from a faith based on scriptural exegesis toward one centered in emotional commitment. The inner light combined with the traditional Unitarian emphasis on reason to give what they saw as a more consistent way of understanding the Bible.[42]

Martineau later summed up the *Vestiges* debate in the *Prospective* by stressing that miracles must be seen as part of law. The discovery of natural laws should not be seen as a "loss of God":

> Who can doubt that this feeling is at the foundation of the hostility displayed against the "Vestiges of the Natural History of Creation?" The author has no doubt committed errors in detail, and availed himself of questionable hypotheses, in order to connect the parts of his system, and complete his generalisation. But the detection of these imperfections has been sought with an eagerness not to be misunderstood; and has brought relief to the awe-struck imagination of many a reader,

40. "Discussion on the 'Vestiges,'" *Liverpool Journal,* 15 Feb. 1845, 3.

41. "Vestiges of the Natural History of Creation," *Christian Reformer; Or, Unitarian Magazine and Review,* n.s., 1 (Jan. 1845): 34–39. On Finch, see Rose 1957, 171.

42. Webb 1990 provides a clear outline of the "new school" of Unitarians.

> to whom the spreading tracks of law . . . seemed but a highway for the exile of his God. Science thus becomes burdened with a tremendous responsibility: wherever it works, it is engaged in superseding Deity: it drops, as a deadly night-shade . . . benumbing all that was divine; and as the narcotic circle widens, the awful sleep extends.[43]

For the Liverpool Unitarians, faith could not be grounded in divine intervention: "The speculative convert to miracles," Martineau told his congregation, "is the practical Atheist of nature."[44]

The Uses of Expertise

After a brief exchange following the reading of Hume's paper at the Literary and Philosophical Society, Robberds had moved to adjourn discussion. The next meeting witnessed not only Robberds's lengthy reply, but also brought a different kind of testimony to bear, as local men of learning responded to what Hume had said. With one exception, they were young and had moved to the city recently, part of the new generation that replaced the society's founders. All had published scientific papers and used science in their professional work; and several had national reputations. Trained in London and abroad, these men deployed antidevelopmental arguments with local resonances. They were predominately Tory Anglicans, so Hume and the Establishment could draw on powerful support.

Thomas Inman, who had been an outstanding medical student, was only twenty-five. He had recently been appointed house surgeon at the Liverpool Royal Infirmary and (later in life) became a prolific author on medicine, antiquities, and mythology. In his comments, Inman remained skeptical of finding pattern or law in the early history of life. Sustaining Hume's denial of progression, he sketched a hypothetical scenario for the fossil deposition that rendered geological progression an artifact of geographical distribution. It was "perfectly natural" to assume that all organisms had been created in a single center (probably in Asia) and subsequently diffused outward. The lower organisms would migrate most rapidly, depositing their remains before the higher types could catch up. Hence the "progressive" nature of the record might be only apparent. Inman did not mean this scheme to be taken literally—although it was conveniently open to a scriptural reading—but wanted to illustrate just how tentative the record of progress really was.[45]

43. [J. Martineau], "Theodore Parker's Discourse of Religion," *Prospective Review* 2 (Jan. 1846): 83–118, at 93.

44. Martineau 1839, 24.

45. "Discussion on the 'Vestiges of the Natural History of Creation,'" *Liverpool Journal,* 15 Feb. 1845, 3.

That Inman should make such a critical reading is not surprising. His father, a director of the Bank of Liverpool, had sent him to the Tory-Anglican foundation at King's College in London, where he qualified as a surgeon. He would have learned his comparative anatomy from Thomas Rymer Jones, author of a cautiously conservative textbook. Inman's concern about *Vestiges* was part of his desire to raise the status of scientific pursuits by professional men. As he later said from the Literary and Philosophical Society's presidential chair, pursuing science involved a calculated risk:

> A clergyman who is an adept in geology is too frequently considered little better than an infidel, and if he adopts chemistry he is supposed to have dealings with the devil; as was Friar Bacon in days gone by. If a lawyer shows a familiar acquaintance with the laws of physics he is at once considered to be an unsound jurist. . . . If a schoolmaster is known as a naturalist he is in danger of losing his pupils. . . . Dr. Mantell lost almost all his practice because he was known to be an ardent geologist; and a physician is supposed to be lost to his profession when he writes upon metaphysics.[46]

These concerns led Inman to denounce mesmerism and table-turning, and to prevent self-proclaimed cosmologists from trumpeting speculations through local scientific channels; this was vital if Liverpool savants were to occupy a respectable place in the national network of specialist science.[47]

The chemist R. H. Brett shared Inman's concern. He had recently obtained a doctoral degree in science on the Continent and served as lecturer on chemistry, pharmacy, and medical jurisprudence at the Liverpool Infirmary Medical School, a post that owed its existence to the impetus for providing medical men with scientific training.[48] He also lectured on science at the Collegiate Institution and was glad to put his expertise at Hume's disposal. On the issue of geological progression, Brett cited Buckland on the first appearance of fishes in the same general group of strata as the early invertebrates. Even if the succession was as *Vestiges* claimed, it seemed absurd to adopt a reproductive law of development that no one had ever witnessed in action. Brett preferred to see species as "distinct creations" associated with physical changes on the earth's surface.[49]

Brett then focused on embryology. He pointed out that Fletcher's *Rudiments of Physiology*, among the most sophisticated sources in *Vestiges*, actually opposed transmutation. Brett read the embryology of *Vestiges* as contradictory; the ap-

46. T. Inman, "Inaugural Address," *Proceedings of the Literary and Philosophical Society of Liverpool* 11 (1857): 18–30, at 19.

47. Ibid., 19, 21; Bedford 1854, esp. 22–23.

48. Stephens and Roderick 1972, 84.

49. "Discussion on the 'Vestiges of the Natural History of Creation,'" *Liverpool Journal*, 15 Feb. 1845, 3.

pearance of the human brain at various developmental stages was confused with that in the adult forms of lower organisms. Such seemingly esoteric issues had profound consequences, for if physical conditions affected the growing embryo, then the appalling squalor of urban life would lead to physical degeneration among the working classes.

The comments of several other speakers at the meeting were not reported, but it is easy to see why they intervened, for *Vestiges* would have threatened their professional interests as well. Thomas Spencer, a Tory and advocate of technical education, was a chemist and student of galvanism, known for his invention of electrotyping—a process used for reproducing high-quality illustrations. Interested in mesmerism, religion, and cosmology, he argued many years later in a pamphlet that life was the product of physical and chemical processes acting according to a divine plan. However, Spencer never published anything on these subjects during his working life, and may have vetoed the reporting of his comments.[50] Similarly, Joseph Dickinson, a physician at the Royal Infirmary, had spoken on the causes behind distribution of life, but published only on descriptive botany.[51] Such men believed that the best way to put Liverpool on the national scientific map was through practical improvements and inventories of locally accessible phenomena.

John Cunningham too may have worried that controversy might harm his professional practice. He had been engaged as an engineer and architect for over a decade, designing many imposing civic structures.[52] This had led to a passion for geology, so that Cunningham became the local expert on the New Red Sandstone quarries of Storeton in nearby Cheshire. Such sites became sources of local pride and places of pilgrimage for visiting naturalists (fig. 6.7). Because vertebrate remains were scarce, Cunningham's findings of fossil footprints of saurians and tortoises at Storeton were often cited in debates about progression: the quarries provided traces of life high in grade but low in the geological record.[53] Cunningham, who corresponded with Buckland, recognized the importance of breaking up a progressive sequence if *Vestiges* and similar works were to be confounded. Early in spring 1845, two months after commenting on Hume's talk, he discovered what he had "long looked for": *bird* footprints, the first time in Europe that warm-blooded vertebrates had been found so low in

50. "The Late Thomas Spencer, F.C.S.," *Liverpool Courier,* 23 May 1885; Spencer 1885.

51. In 1843 Dickinson gave a paper entitled "On the Laws which Regulate the Distribution of Vegetable Forms, Over the Surface of the Globe" (see *Report of the Proceedings of the Literary and Philosophical Society of Liverpool* 1 [1845]: xxvii); but this was never published.

52. "The Late Mr. John Cunningham, Architect," *Builder,* 18 Oct. 1873, 821.

53. Bowler 1976; Desmond 1989, 311; for Cunningham's work, see Tresise 1989, 1991, and G. H. Morton, "Anniversary Address by the President," *Abstract of the Proceedings of the Liverpool Geological Society* 2 (1870): 3–29, at 8–14.

6.7 Unique scientific sites served as a focus for competition between cities. Liverpool's traditional rival, Manchester, showed off the huge fossil trees at Dixon Fold on an excursion during the British Association for the Advancement of Science meeting in 1842. Among the locals are John Dalton (who actually had been too ill to attend), Lady Harriet Catherine Egerton (author and translator), and Lord Francis Egerton (MP for South Lancashire and collector of fossil fish). The figure presenting the find is probably the railway engineer John Hawkshaw. The geologists shown on the right are (moving clockwise from the top) Murchison, De la Beche, Sedgwick, and Buckland. A similar party during the Liverpool meeting a few years earlier had been taken by John Cunningham to visit Storeton. The lithograph was prepared from a design by the London artist Robert William Buss. "Visit of Members to the Fossil Trees in the Coal Measures, near Manchester, June, 1842"; frontispiece to Heywood 1843.

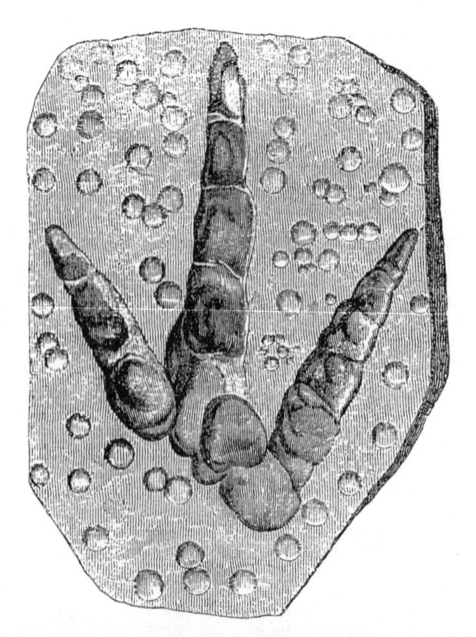

6.8 "Slab of Sandstone (new-red), showing impressions of rain-drops, and foot-print of bird" similar to those found by Cunningham at Storeton. *V*10, 62.

the sequence (fig. 6.8). Traces of birds in such ancient rocks undermined the most striking evidence for progression, and hence for transmutation. On a Saturday field excursion fifteen Literary and Philosophical Society members had traveled to Storeton, and it was while explaining his finds that Cunningham made this exciting discovery. The evidence (destroyed soon after by quarrying), was reported by Hume in the *Liverpool Journal* and by Cunningham himself in the Geological Society of London's *Quarterly Journal.*[54] As with contemporary discoveries of bird tracks in Massachusetts, this report was quickly used against *Vestiges.*

The Limits of Authority

For all the show of local talent, Hume worried that the adjourned Literary and Philosophical Society debate would produce a standoff, with one member's opinions pitted against another. To avoid this, Hume addressed a circular letter to the leading men of science in Britain.[55] With replies in hand he was confident of success. George Bidell Airy, the Astronomer Royal, was "the official authority"; the others were the most eminent specialists. The case would rest not on his own confessedly "imperfect advocacy," but on that of leading experts. Hume laid their letters on the table as the principal exhibit of his reply. Expertise should bottom the argument. "To these evidences," he claimed, "it is surely unnecessary to add anything."[56] An open forum, created for free discussion, could thus be used in an authoritarian way. The dogmatic certainty of scientific fact could close controversies and resolve disputes.

But the picture was more clouded than ever. A writer identified only as "A Mining Engineer" pointed out to *Liverpool Journal* readers that Hume's correspondents were split on geological progression; Murchison, Mantell, and James Bryce (an old Belfast friend of Hume's) defended a general picture of life's advance on earth, while Lyell undermined it, believing that all the major fossil groups extended back to the earliest strata. Buckland's letter, moreover, was not really about general issues of progress but about specific instances of degeneration—claiming, for example, that very early reptiles were of a higher type than more recent ones.[57]

54. *Liverpool Journal,* 30 May 1845; J. Cunningham, "On some Footmarks and other Impressions Observed in the New Red Sandstone of Storton, near Liverpool," *Quarterly Journal of the Geological Society of London* 2 (1846): 410; minutes for paper read by A. Hume on 25 May 1845, Liverpool Literary and Philosophical Society, Liverpool Public Library, 060 LIT 1/4.

55. See, for example, A. Hume to R. Murchison, 21 Jan. 1845, GSL M/H66/1.

56. "Adjourned Discussion on the 'Vestiges of the Natural History of Creation,'" *Liverpool Journal,* 8 Feb. 1845, 2.

57. A Mining Engineer, "Remarks on the Discussion in the Literary and Philosophical Society, on the 'Vestiges of the Natural History of Creation,'" *Liverpool Journal,* 1 Mar. 1845, 2.

More serious still were disagreements about facts. Murchison pointed out that fishes, "the lowest class" of vertebrates, had never been found in the Lower Silurian. Lyell, on the other hand, claimed to have made precisely this discovery during his recent American travels. "Presuming Mr. Lyell to be right," Mining Engineer noted, "it is rather droll that Mr. Murchison should, by implication, charge the author of the 'Vestiges' with being behind in his reading."[58] Far from closing off debate, the letters published in the *Liverpool Journal* revealed a lack of consensus on basic matters of fact. The display of expertise had backfired.

Further contradictions involved the composition of the early atmosphere. Was *Vestiges* correct in attributing the advance of living beings to a progressive diminution of atmospheric carbonic acid in relation to oxygen? Answers to this question had important consequences for understanding not only the past but the present: burning the coal-bearing strata found so abundantly in the industrial northwest had led to high levels of pollution and respiratory disease. Mantell's letter to Hume denied that ancient atmosphere had varied at all from its present proportions. Mining Engineer, however, showed that Europe's leading chemist, Justus von Liebig, had argued the opposite; and Liebig's view was "worth infinitely more than Dr. Mantell's." An editorial note in the *Liverpool Journal* attempted to resolve the discrepancy by pointing out that Liebig's disciple Lyon Playfair was lecturing on this question at the Mechanics' Institution. In his talk, Playfair argued that even the most luxuriant vegetation would have required no more carbonic acid than did the atmosphere today; this implied that the issue was effectively irrelevant.[59]

The only point on which all Hume's correspondents could agree was that *Vestiges* failed to meet scientific standards. Mantell regretted that the book's reception indicated the sorry state of popular understanding. "It is humiliating to think that a work of such flimsy pretensions should have obtained so much consideration, because it proves how little the intelligent public are aware of the real progress of science." Lyell condemned the book on the basis of reports from other geologists without even reading the copy the author had sent him.[60] Most of the other correspondents referred to the antievolutionary passages planted in their published works for use on precisely this sort of occasion. Thus Murchison pointed to the entry "Transmutation of Species" in the index to his *Silurian System* (1839), and the earl of Rosse sent a brief note referring to a paper in the *Philosophical Transactions* on his telescopic observations. Only the government geologist Henry De la Beche did not denounce "progressive development" in

58. Ibid.

59. Ibid.

60. "Adjourned Discussion on the 'Vestiges of the Natural History of Creation,'" *Liverpool Journal,* 8 Feb. 1845, 2.

any of the manifold meanings of that term. Mining Engineer agreed with this cautious approach, noting that "[g]eology wants facts, and not fancies."[61] The lesson of the debate, from this perspective, was to avoid speculation.

In London, as in some Liverpool circles, publishing these private letters was seen as bad manners and a tactical blunder. As the cocky young botanist Joseph Hooker told Charles Darwin, "Some Liverpool Parson, after reading 'Vestiges', had written to all Geologists for proofs on the contrary, & rather coolly, printed all the answers." It was, as Hooker condescendingly said, "a funny thing."[62] For geologists whose credibility was vital in campaigning against scriptural literalists, ridicule was not so amusing. Lyell's admission that he hadn't even read the book made him look like a fool. Most of Hume's eminent correspondents would never have allowed anything but the most oblique reference to *Vestiges* to appear in print under their names. By publishing private letters without permission—and worse still, by reading them at a public meeting—gentlemanly disagreements had been brought into the open.

Newspaper Savans and Newspaper Bullets

The lesson was clear: institutions like the Literary and Philosophical Society could be turned to good effect in the battle for civic supremacy, but the press could not be controlled. One indication of the ideological fissures in Liverpool was the presence of no fewer than eleven weekly newspapers. There were four liberal papers (*Albion, Chronicle, Journal, Mercury*), three conservative ones (*Mail, Standard, Times*), and several devoted solely to business affairs and advertising (*Gore's General Advertiser, Liverpool European Times, Myers's Mercantile Gazette, Williamson's Liverpool Advertiser*). There was a fortnightly paper in Welsh, and the city was the international center for Mormon publishing, including their paper, the *Millennial Star.* Circulations were often large (each issue of the *Mercury* sold over seven thousand copies in 1845), although prices of five or six pence limited purchase to the genteel and upper middle classes.[63] Most people consulted newspapers in commercial reading rooms and clubs; these were the best places to follow the controversy.

Provincial papers were much more than local news sheets; copies could be found in London, Boston, Calcutta, and cities throughout the empire. They printed parliamentary reports, international news, and extensive accounts of public meetings. Their editors were powerful figures, as the papers were central

61. A Mining Engineer, "Remarks," 2.

62. J. D. Hooker to C. Darwin, [28 Apr. 1845], in Burkhardt and Smith 1985–99, 3:184.

63. *Supplement to the Liverpool Mercury,* 17 Oct. 1845, 409; on the Liverpool newspaper trade more generally, see Perkin 1987 and 1990, and "The Provincial Press of the United Kingdom," *Reynolds's Miscellany,* 19 Dec. 1846, 106–7. Also useful are standard contemporary reference works, especially the *Newspaper Press Directory,* which started publication in 1846.

vehicles for local political activism and party propaganda. The burly Irish Catholic liberal Michael Whitty took control of *Liverpool Journal*—the main forum for the *Vestiges* debate—after serving as chief of police, where his responsibilities had included breaking up rival Catholic and Protestant street gangs.[64] Editing a major paper ensured an entrée into the corridors of power.

Within the politicized bullpit of newspaper controversy, there were few places for science, far fewer than in their modern equivalents. "Pressmen, in general," as a *Liverpool Journal* editorial admitted, were "totally unacquainted" with science. Only London offered opportunities for paid science writing. Newspapers had no equivalent of the modern science journalist or science page. Astronomy, botany, geology, and chemistry were "magic to the million," alien to readers and publishers alike.[65] The main editorial columns had something to say about science only when it impinged on politics, religion, or education. Science coverage in newspapers could signal civic aspirations, broaden readership, or fill empty columns. Some editors published reports of traveling lecturers or local societies; most issued accounts of British Association meetings, sometimes as substantial supplements. Editors also scissored and pasted paragraphs from new books into their columns as space fillers; the *Albion* quoted *Vestiges* on the time it would take a racehorse to traverse the solar system, and later extracted a passage about extraterrestrial life.[66] Such extracts gave readers something to talk about, and added to a paper's diversity and interest. Books were rarely reviewed, although the quarterlies were sometimes surveyed. The liveliest columns were devoted to correspondence, usually unsigned or pseudonymous. Religion and politics so split the different papers that even the most neutral articles and extracts took on a polemical coloring.

What codes regulated civic life when parties in dispute could not accept their opponents' good faith? Many correspondents identified violations of propriety in what Hume had done. "G. S.," in a letter with a Manchester postmark, wrote to the *Liverpool Journal* complaining that the meeting had misrepresented an "interesting" and "well-argued" book. *Vestiges* did not claim the nebular hypothesis was "veritable fact," only that it "'may be considered as *verging upon the region of ascertained truths.*'" G. S. also highlighted discrepancies in the savants' replies, condemned Hume for twisting them to his own ends, and stressed the need for "calmness and modesty."[67]

64. Neal 1988, 42. For the significance of local editors, see Fraser 1985.

65. "Science in Parliament and at Cambridge," *Liverpool Journal,* 28 June 1845, 8.

66. "Extent of the Solar System," *Liverpool Albion,* 18 Nov. 1844, 2; "Remote System of Stars," *Supplement to the [Liverpool] Albion,* 6 Jan. 1845.

67. G. S. [Chambers], "Discussion on the 'Vestiges of the Natural History of Creation,'" *Liverpool Journal,* 22 Feb. 1845, 5.

At its most fundamental, the controversy concerned the ethics of public debate. The *Liverpool Journal* prided itself on supporting free trade not only in agricultural produce but in ideas. As a liberal editor, Whitty believed in fair play, which is why he published Hume's paper and so much of the ensuing discussion on both sides. Whitty might not have printed the letter from Manchester, however, had he known that G. S. was the *Vestiges* author praising his own book under the cover of a pseudonym. This was a common practice in newspaper correspondences, in which all sorts of interested parties could participate without revealing their identities.

Although Whitty did print letters both for and against *Vestiges,* the space given to attacks suggests deliberate policy. An editorial comment praised the *Prospective*'s learning and liberality, but registered "dissent in toto" from its favorable review. Whitty also printed a letter from R. R., which echoed Hume in condemning *Vestiges* as a closet speculation that "may rank amongst books of science, as some novels rank amongst those of history":

> Read the "note conclusory" of our author. "Thus ends a book, composed in solitude, and almost without the cognizance of a single fellow-being." Yes, yes, this retiring and secrecy is the peculiar characteristic of your race; it is this peeping at nature through a skewer-hole that fills your honest heads with such monstrous one-sided ideas, and leads to speculations without end, until your abortive bantlings come forth, like bats and owls, screaming and flapping about until they scorch their wings at the lights of science or knock out their brains against its pillars; and so, after surprising, terrifying, and annoying the world, the product of your "solitude" dies an unnatural death.[68]

Whitty's *Liverpool Journal* thus offered an open forum, but one tipped against heterodoxy.[69]

The Tory press rejected the *Journal*'s liberal model for resolving disputes and took a very different view of the natural world, one in which biblical chronology and scriptural precedent reigned supreme. The *Liverpool Mail,* which supported High Church principles and protectionism in trade, reported British Association meetings in its "comic department."[70] To condemn this as bigoted obscurantism, however, would be a mistake; questions were at stake that could not be compromised. For many of the *Mail*'s readers, science was associated

68. R. R., "Had Man Ever a Tail?," *Liverpool Journal,* 14 Feb. 1845, 2.

69. A year later, the *Journal* complained that the *Prospective*'s notice of *Explanations* was "written in too friendly a spirit to satisfy the expectancy of just criticism." *Liverpool Journal,* 14 Feb. 1846, 2. See also the free-trade *Liverpool Chronicle,* 31 May 1845, 2. On liberalism and the politics of the market for newspapers, see Jones 1996, 146–55.

70. *Liverpool Mail,* 28 June 1845, 8.

with infidelity and failed to answer the great questions of salvation.[71] Insofar as the *Mail* reported science at all, it was through declamatory forms related to the pulpit. Thus the editor Robert Alexander regularly printed lectures from the Collegiate Institution, on subjects ranging from "the Tower of Babel" to "physiology and the laws of health." Several of these implicitly attacked *Vestiges.*

Anglican Tory evangelicals, in the ascendant at this time, read the *Liverpool Standard,* which doubted that science could defeat secularism and would have nothing to do with the Literary and Philosophical Society. The newspaper's priorities are illustrated by its support, during the weeks when *Vestiges* was being debated, for designing steamships on scriptural principles. These state-of-the-art vessels symbolized Britain's commercial prowess and Liverpool's status as an international gateway. Speaking at the reopening of the Collegiate Institution's lecture hall, McNeile had praised a leading Merseyside shipbuilder for constructing steam vessels on the instructions given by God for Noah's ark (fig. 6.9). "Now," he told the ecstatic audience, if "the very best directions for shipbuilding which the most improved scientific artist, in this time of science, can adopt, are those given to Noah by his Divine Instructor, we have another argument for the wisdom of that volume baffling the inventions of men,—baffling all the ingenuity of infidelity to gainsay."[72] Knowledge without humility led to socialist freethought, papist corruption, and Chartist violence. Technological progress would come not through flimsy man-made knowledge but from the divine blueprints in Genesis.

From the opposite end of the political spectrum, the Literary and Philosophical Society agenda was also rejected by Liverpool's middle-class radicals. Their paper, the *Albion,* had a national reputation as an advocate for progressive reform. The current editor, Charles Macqueen, published several letters about *Vestiges.* One correspondent parodied the conflict as a medieval tournament; Hume, as Critic, had broken his lance not only against *Vestiges,* but on the shields of the scientific authorities whose authority stood fully behind the nebular theory. The letters claimed that Hume's onslaught would reduce geologists "from the high pedestal of philosophy to be mere breakers of stone and collectors of curiosities." Mocking concerns about romance-reading, the letters were written as if "from the unpublished chronicles of Froissart." Hume, as knight, had failed to abide by the codes of scientific controversy:

> The knight is wroth and very sulky in his contest with geological development, and will not be adjudged by the gentle and courteous rules of our tournament.

71. "Literature and Science: Saving Wisdom," *Double Supplement to the Liverpool Mail,* 18 Oct. 1845, n.p.

72. *Liverpool Journal,* 18 Jan. 1845, 5; for controversy with the *Standard* on this issue, see "The Rev. H. M'Neile on Shipbuilding," *Liverpool Mercury,* 2 Feb. 1845, 58.

6.9 The animals entering the steam ark. From Grandville 1844, facing p. 285. BL, 1458.k.10.

> He scorns that intellectual liberality which is ready to adopt whatever is rendered highly probable, and brusquely asserts, that "*it may be true a priori*"; "*but I am not called upon to admit it till he has shown it, undoubtedly shown it to be so. . . .*"
>
> He built up a man of straw and called him the "Unknown author of the Vestiges" and then wrote to certain savants for paper bullets from their books with which he pelted the said man of straw. The paper bullets came to his order and were fashioned from his questions.[73]

Hume, in short, had used the society to avoid "fair play." To make his case, the *Albion* correspondent eschewed staid sobriety and drew on ribaldry, innuendo, parody, caricature, verging on the rhetoric of the unstamped press.

The letters were signed "T.," but regular *Albion* readers would have recognized them as the work of John Taylor, a Liverpool cotton broker, astronomer, and poet.[74] Taylor advocated civil and religious liberty, collected books on the French Revolution, and condemned those who blew "the ecclesiastical trumpet" in scientific disputes. The struggle to establish the sciences of progress was, as were other causes supported in the *Albion,* part of "the hard battle of Reform."[75] Taylor served notice of his disdain for scientific elitism by publishing in newspapers. He poked fun at the Literary and Philosophical Society and targeted squibs at pompous gentlemen who aimed to control science. Throughout the early months of 1845, he was embroiled in bitter controversy over the local observatory with a leading London astronomer, Richard Sheepshanks. Taylor wanted Liverpool to have an independent observatory equipped to make original contributions. He bitterly resented those who would limit authority to those who had "an M.D.^{999}, an F.R.S.^{999}, an F.R.A.S.L.^{999}" appended to their names.[76] *Vestiges* provided another pin to prick the bubble of orthodoxy.

Taylor was a maverick who rebelled against the increasingly sober tone of public debate. As one Literary and Philosophical Society member—probably Hume himself—suggested in the *Liverpool Journal,* "His next visit in cap and bells will probably be more in character."[77] Sheepshanks, although scarcely an unbiased source, expressed a widely held view when he claimed that "Mr. Taylor's return of scientific and literary attainments must be nil." The London *Athenaeum* agreed, complaining that men of science should have to combat "the

73. John Taylor [T., pseud.], "Scientific Quixotism, or, the Critic *v.* Herschel and Laplace," *Liverpool Albion,* 17 Feb. 1845, 2; also "The Second Tournament: From the Unpublished Chronicles of Froissart: The Author of 'The Vestiges' *v.* the Critic and Others," *Liverpool Albion,* 24 Feb. 1845, 2.

74. Smith 1868, 2:180–82; "The Late Mr. John Taylor," *Liverpool Mercury,* 14 Dec. 1857, 8; [Shaw] [1869], 35.

75. Smith 1868, 2:181.

76. From a letter dated 1 April 1845 in the *Liverpool Mercury,* quoted in Sheepshanks 1845, 19.

77. A Member of the Literary and Philosophical Society, "Vestiges of the Natural History of Creation," *Liverpool Journal,* 8 Mar. 1845, 2.

trumpery aspersions of newspaper *savans.*"[78] On the other hand, Taylor had powerful allies, including Sheepshank's nemesis Sir James South.

It was, perhaps, because T.'s mock tournament was seen to be inadequate that a much more elaborate defense of *Vestiges* began to be published in the *Albion* from March 1845. These anonymous letters displayed a thorough knowledge of science, with support for the nebular hypothesis, the progressive history of life, the unity of type, the transmutation of species, and spontaneous generation. Through four installments (the length of a short book) they claimed that Hume and other critics were using factual gaffes and printing errors to obscure familiar truths:

> And what are the views advocated by the author of the *Vestiges of Creation.* Are they new, singular or recondite? Are they not, on the contrary, clear, explicit, and familiar? Is not the general popularity of the book due to the graceful and flowing manner in which old truths are told and old conclusions moulded in harmonious combinations? Is the existence of *law* in nature's operations a new or hidden doctrine of the schools? Has it not rather become a common phrase, a household word?[79]

These letters condemned Hume's "language of draconic severity" and his publication of letters by authorities who deserved "respectful deference" but not "unconditional submission." The same view was taken a few months later in another of the liberal newspapers, the *Liverpool Mercury,* whose anonymous correspondent regretted the "ill-tempered" tone of the "schoolmen," from "our own minor reverend at the Collegiate" to Adam Sedgwick in the *Edinburgh.* "In polemics or criticism," the article noted, "nothing can be more unfair than to raise the hue and cry of materialism."[80]

William Hodgson—the reputed author of *Vestiges*—agreed with this assessment, and could well have written it. Nothing short of a revolution in the conduct of reading practices could resolve the differences that were tearing apart his adopted city. Disagreeing with extreme Protestants like McNeile, and those at the Collegiate concerned about the spread of dangerous books, Hodgson advocated "the smallest possible restrictions" on reading, even by children. No book ought to be forbidden; the alternative was "a war of literary and literal extermination" in which each party read only works supporting their own opinions. Debate fostered individual judgment, the pursuit of "that ambiguous and indefinite thing which we call Truth." Toward this end, Hodgson founded a Mental Improvement Society, which aimed to go beyond self-development by

78. *Athenaeum,* 11 Oct. 1845, 985–96.

79. "The Vestiges of the Natural History of Creation. Reviewed in Letters to the Rev. Dr. Hume," *Liverpool Albion,* 24 Mar. 1845, 2; 7 Apr. 1845, supplement, 1; 21 Apr. 1845, 2; 5 May 1845, 2.

80. "The Edinburgh Review, and the Vestiges of the Natural History of Creation," *Liverpool Mercury,* 17 Oct. 1845, 3d supplement, 1.

emphasizing mutual instruction. He supported women's education, adding a class for female teachers and a day school for girls to the Mechanics' Institution. Wives and daughters were invited to join the Institution itself—again in contrast with the Collegiate. "'Try all things,'" he told a friend; only through open discussion could knowledge progress.[81]

Urban Developments

Jane Welsh Carlyle was not alone in being struck by contrasts between literary life in Liverpool and London. Victorian cities were defined as much by intellectual culture as by geographical boundaries and census results. The Literary and Philosophical Society discussed a range of subjects that would have been anathema at London's learned societies; newspapers and their editors played far more central roles in managing debate; theology could be discussed at evening parties. Maintaining a forum isolated from politics and religion was almost impossible. Conflicts between Tory and liberal, Anglican and Dissent, Catholic and Protestant demanded choices about what kind of knowledge about nature was appropriate to a commercial city.

In Liverpool, the proper conduct of reading became part of the problem of defining the middle class. Debating *Vestiges* contributed to fracture lines *within* the middle class; they were intra- rather than interclass, with a crucially important religious dimension. As in many northern cities, conflict across classes was very little in evidence.[82] For working people themselves—some 70 percent of the total population—the issues were very basic, involving food, housing, and fundamental political rights. At the same time, the cotton brokers and professional men who settled down to read *Vestiges* or the *Liverpool Journal* in the comfortable chairs at the Liverpool Athenaeum could not help but do so in light of the crisis facing the city. Passages on human degeneracy were hard to read as abstract philosophy when tens of thousands lived in cellars; the chemical composition of the atmosphere had political implications when industrial pollution was debated at city council meetings; the relations between Scripture and science gained urgency when steamships were designed to the measurements of Noah's ark. In an aggressively commercial town, debates about natural law easily shifted into issues of political economy and the laws of trade.

Anyone who lived in Liverpool during its rampant expansion was acutely aware of the significance of urban geography. Several years after the events in this chapter, McNeile and other Anglican preachers rented the former Owenite Hall of Science on Lord Nelson Street, transforming what had once been a

81. W. Hodgson to Wotherspoon, 9 Nov. 1841, in Meiklejohn 1883, 20–22.

82. Seed 1982 discusses this issue in the context of Manchester, where the Unitarians had far more power than in Tory-dominated Liverpool.

center for infidel socialism into a place for proclaiming the divine word. In talks printed locally and sold for three halfpence, they reveled in their victory over the forces of darkness. The Reverend Hugh Stowell of Manchester, one of the most acclaimed pulpit orators in Britain, took the opportunity to attack the doctrine of progressive development that had been proclaimed in the same place only a few years before. He ridiculed the idea that the self-sufficient laws of nature could produce a "great town" like Liverpool. Did one law result in streets and squares, another in docks, another in warehouses?

> A man is either a liar or a fool who says that this great commercial city came of itself—that all England came of itself—that all Europe came of itself—that the system which we call *the world* came of itself—that all the worlds, which dazzle with their brightness and astonish with their beauty, and by their courses guide man upon this earth—that all these things *came of themselves!*

The city's extraordinary development told against the doctrine of natural law. Or as Stowell went on to say, the universe could no more emerge by chance than a sermon or a tract could print and bind itself. "We know God is there in the making of things," he recognized, "just as we know a book is *made* by man."[83]

83. Stowell 1852, 5.

CHAPTER SEVEN

Church in Danger

I think you could smash him and I wish you would. . . .
Already some consider this book as the signal of the
Revolt against the Church.

GEORGE W. FEATHERSTONHAUGH
to the Reverend Adam Sedgwick, 16 November 1844

LIVERPOOL'S EXPLOSIVE GROWTH was not the only sign that the Church of England's pastoral ideals had become anachronistic. The Establishment had traditionally looked toward the country rather than the city, with national solidarity maintained by the network of parish clergy. But faced with Dissent, irreligion, and indifference, the state Church increasingly appeared to be one sect—"Anglicanism"—among many.[1] Clerical authority in science was implicated in the balance between sacred and secular power that became central to dealing with *Vestiges.* For all its theistic gloss, this was not a work emanating from the bastions of Establishment learning, nor was it likely to be by anyone with orthodox Christian beliefs. For the Church to retain any claim as cultural arbiter of the English people, it had to confront this reading of the divine book of nature.

In *The Way of All Flesh* (1903), Samuel Butler satirized the dilemmas that had faced young Church of England clerics who were unprepared for *Vestiges.* At Cambridge, the novel's hero Ernest Pontifex comes under the influence of the "Sims," the evangelical followers of the Reverend Charles Simeon. Ernest learns nothing about evolution and not much about theology. Recently ordained and aiming to transform the nation's spiritual life, he settles in squalid London digs as a first step to bringing the poor to Christ. His first shock comes when a tinker he attempts to convert from freethinking knows the New Testament better than he does. Later the same day, Ernest wanders to the British Museum read-

1. For the challenges confronting ordinary clergymen in the Church of England, see Knight 1995; Machin 1977 surveys relations with the political establishment. Hilton 1988 and Turner 1993 review the intellectual debates.

ing room, and spends the afternoon reading *Vestiges*. This is his second shock, for his mind is in an "embryo" state in which all new developments are so sudden it seems scarcely possible to survive them; after a few more shocks he is in prison for indecent assault and has abandoned Christianity.[2]

As the freethinking tinker had said, "'[Y]ou Oxford and Cambridge gentlemen think you have examined everything. I have examined very little myself except the bottoms of old kettles and saucepans, but if you will answer me a few questions, I will tell you whether or no you have examined much more than I have.'"[3] How did the Church of England respond to the threat represented by *Vestiges?* In some ways, we know more about this subject than almost any other discussed in this book. The reactions of William Whewell and Adam Sedgwick have been extensively analyzed, for in England they became identified as the leading champions against the book. Their readings seem to have an evident importance that the Reverend Hume's, say, lacks; they represent religious orthodoxy, "liberal Anglicanism," or the imaginary town of Oxbridge. Alternatively, they are seen as unique, idiosyncratic individuals. But why these particular individuals should have been placed in positions of authority—and the fragility of their status—has rarely been discussed.

Clerical Magistrates of Nature

From the 1790s onward, the natural sciences became an increasingly significant part of the Establishment's response to unbelief. A theology of nature, taught at the ancient universities, was one way of maintaining the authority of the state Church. Science teaching at Cambridge and Oxford was not intended to institute a modern, professional education, but rather to educate Christian gentlemen. Half of the students became clergymen, and most of the college fellows were in holy orders. Geology, chemistry, and botany were taught as optional lectures, not examinable until after midcentury. A decentralized structure militated against the expansion of science, for formal education was carried out within the colleges by individual tutors; science provision was largely the responsibility of a small number of professors and readers appointed by the university—a far less powerful body.

By the mid-1840s advocates of scientific education despaired of Oxford, where studies were devoted almost entirely to classics and theology. Attention focused on the Tractarian (so called because of their publication of *Tracts for the Times*) revival led by John Henry Newman, the elder brother of Francis Newman. Matters had been brought to a head in January 1845 by the attempt of the Board of the Heads of Houses to strip W. G. Ward of his Balliol fellowship for

2. Butler 1993, 214–15, 248, 272–74.

3. Ibid., 272–73.

claiming that no Anglican doctrines were inconsistent with Roman Catholicism. This attack alarmed not just Newman's followers, but liberals too. As one of them noted, "If the Heads of Houses may sit in judgment on Ward's book to day, they may try Buckland for his geology to-morrow."[4] Buckland viewed his efforts to introduce science as a complete failure. His coarse humor and blue bag of fossils seemed out of place in the asceticism that permeated Oxford life; arguments for the divine design of hyena dung had little attraction for those who saw the earth as the scene of spiritual probation.[5]

At Cambridge (fig. 7.1), the centrality of mathematics in the examination system had led to a greater commitment to the nondegree science subjects and gave apologetics based on natural theology more currency. A broad-based consensus in the colleges supported the idea that museums, lectures, libraries, laboratories, and other facilities could be arrayed against threats to the social and moral fabric. But the mid-1840s were difficult times. Lavish funding for new science facilities had brought the university close to bankruptcy, so that proposals for further development fell on deaf ears. A postgraduate theological examination was instituted, which although voluntary, was usually regarded as a prerequisite for ordination. Students who hoped to take holy orders turned their attention to preparing for this, and attendance at the extracurricular science lectures plummeted. Instead of being taught geology, many now prepared for examinations in scriptural chronology, where they were expected to affirm the reality of the Flood, and the Creation of the world in 4004 B.C. As the hostile *Times* said in 1846, "[T]here are no guests at the table, the invitations are refused, and the hated meats are spoiled."[6]

Even during the slump of academic science in the 1840s, the ancient universities could display men who combined exemplary piety with European repu-

4. F. D. Maurice to J. Hare, 15 Jan. 1845, in Maurice 1884, 1:398–400. For the Tractarian movement, see Chadwick 1971, 167–231, and the recent survey in Nockles 1997; on academic roles generally in Oxford, Engel 1983.

5. Rupke 1997, 561–62; 1983, 267–74.

6. "The University Professorships," *Times,* 9 May 1846, quoted in Becher 1986, 80, with further discussion of the problems of the 1840s and a survey of the early period. Other important studies of Cambridge include Rothblatt 1968, Geison 1978, Garland 1980, and Fyfe 1997.

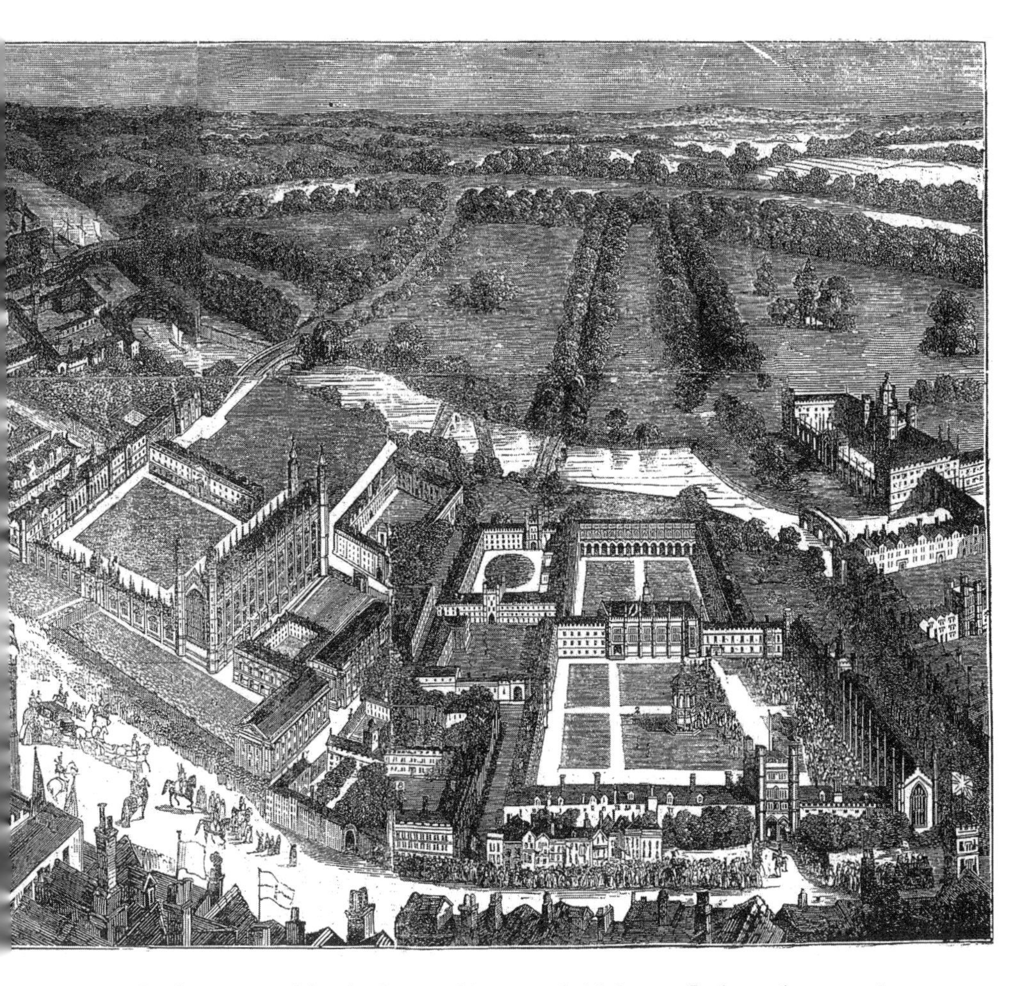

7.1 Bird's eye view of Cambridge, marking a royal visit in 1843. It shows the procession along Trumpington Street, past King's College and Senate House, to waiting crowds at Trinity College. Although the colleges retained their medieval aspect, many of the buildings were new, and the railway line to London opened less than two years later. *ILN,* 4 Nov. 1843, 296.

tations. Most were beneficed clergy. Their combination of learning, seriousness, and clerical status was celebrated in the newspapers; as shown by portraits in the *Illustrated London News,* men like Whewell (fig. 7.2) were middle-class culture heroes.

The clerical professoriat, although marginal within the larger structure of Anglican learning, had had a major impact in shaping the vocational ideals, pastoral roles, and publication patterns of the new sciences. What the clerical

7.2 William Whewell praying in Great Saint Mary's, the University Church in Cambridge. Seating in "the Golgotha" was reserved for Masters of the Colleges. *ILN*, 28 Oct. 1843, 284.

magistrate was to the English law, the clerical man of science could become to the laws of nature. Endowed with the moral authority of their status, men like Buckland, Sedgwick, and Whewell were expected to speak to large theological and philosophical issues. Buckland's lectures, for example, warned students about "the Fallacies of the Doctrine of *Development* maintained by the anonymous author of the 'Vestiges of Creation'"; and Cambridge students (including a future archbishop of Canterbury) recalled that *Vestiges* "supplied a perpetual thread of criticism" in Sedgwick's geology course.[7] Within limits, it was part of their job to be dogmatic—to transcend the conventions that held back their metropolitan counterparts among men of science. Their prominence was not simply an outcome of their qualities as individuals, but a manifestation of the way in which the sciences were handled within the Anglican hierarchy.

7. Rupke 1983, 179; Clark and Hughes 1890, 2:500; see also 2:350–51.

As the *Quarterly* and *Edinburgh* began to sound out possible reviewers for *Vestiges,* they looked to the universities, and especially to Cambridge and to Cambridge-trained men. The *Edinburgh* had often criticized the universities, but now wanted to make amends, not least because its leading scientific contributor, the Scottish Presbyterian David Brewster, had begun to write for the new *North British Review.* The editor Macvey Napier even asked Whewell, Brewster's old nemesis and a frequent target of the *Edinburgh*'s venom in the past, to review the book. "The work in question is more than likely to influence the younger class of inquirers," Napier warned Whewell, "and it therefore becomes . . . a duty to show them how far it is calculated to prove a safe guide, or in what respects it is apt to mislead, if it has any tendencies to false doctrine, or unwarranted modes of philosophizing."[8]

Whewell said no to the *Edinburgh,* but had encouragement from other quarters. William Smyth of Peterhouse, Cambridge's gouty Regius Professor of Modern History, encouraged him to attack *Vestiges* as a product of "those washed & unwashed Radicals" who had kept the country in a state of alarm since the Reform Bill.[9] The liberal Tory diplomat and geologist George W. Featherstonhaugh warned his Cambridge friends that a reply was needed right away. "Perhaps it may be thought best to say little about it," he told Sedgwick a month after publication of the book; "if not, it is to men like yourself and Whewell that we must look for the antidote."[10] The book was conquering the philosophical shallows of the metropolis.

Whewell responded first. His *Indications of a Creator,* published in mid-February 1845, was perfectly calculated for readers in the West End salons. A hurriedly prepared anthology drawn from his previous works in the philosophy and history of science, *Indications* was dedicated to Smyth and targeted "general readers" who could not be expected to be familiar with five fat tomes. Queried by Herschel for "giving indulgence to the craving for *quintessences* & brief exposes,"[11] Whewell replied that *Indications* had a very specific target. It was not, he stressed, "a wanton selection of elegant extracts, but a compulsory selection of theological extracts."[12] The best way to understand the aim of *Indications* is to look at its physical form. Whewell specified the character of the production to the university publisher, John W. Parker, for he wanted to match

8. M. Napier to W. Whewell, 8 Feb. 1845, TC Add. mss. a.210[10].

9. W. Smyth to W. Whewell, 29 Jan. [1845], TC Add. mss. a.212[III].

10. G. W. Featherstonhaugh to A. Sedgwick, 16 Nov. 1844, CUL Add. mss. 7652.I.E.105. Featherstonhaugh's interest in *Vestiges* is documented in Berkeley and Berkeley 1988, 272–73.

11. J. F. W. Herschel to W. Whewell, 10 Mar. 1845, RSL Herschel Letters, vol. 22, f. 225.

12. W. Whewell to J. F. W. Herschel, 12 Mar. 1845, in Todhunter 1876, 2:325–26.

Vestiges in page size and type. He had copies specially bound up in white and gold, and wondered if additional copies of the second edition should be in this binding. Length, size, paper, and typography marked *Indications* as a work "daintily dressed for dainty people."[13]

Publishing in this format was the most damning criticism of the pretensions of *Vestiges* that Whewell could make. From the perspective of Cambridge, with its all-male colleges and strict moral regimen, the book could be read as displaying the worst features of metropolitan Society: superficial, speculative, and dominated by women. Books needed to be studied, not skimmed for polite conversation. To avoid giving unwonted dignity to a "bold, unscrupulous and false" work, Whewell refused all requests for a review, and *Indications* in its first edition never named its target.[14] He never bothered to speculate on who might have written *Vestiges.* Behind the genteel silences, however, was scarcely concealed fury at the metropolitan cosmological chatter. *Indications* was the earliest separately published antidote, begun after just two editions of the offending work. Far from being reluctant, as is often assumed, Whewell acted quickly and for maximum impact.

At this point Whewell thought a comprehensive "doctrine by doctrine" review would be a waste of time. From the perspective of the Master's Lodge at Trinity, real science could only be communicated through a disciplined readership. As he told a friend:

> I believe that the promulgators of long pondered truths ought to be prepared to wait a while for the gratitude of the world, for they cannot mix themselves with popular and periodical literature, or with London coteries, in such a way as to find a set of ready made admirers when they publish. But this is not to be regretted, for truths of any broad philosophical kind do not admit of transmission through admirers so made. I have been much amused, in this point of view, with the success of the *Vestiges of Creation.* No really philosophical book could have had such success: and the very unphilosophical character of the thing made it excessively hard for a philosophical man to answer it, and still more to get a hearing if he did.[15]

Truth had no power for those without properly prepared minds, who read books in the wrong way. Superficial skimmers craved answers that "real" philosophy and "real" science could not provide.

13. Ibid.; for the binding, see W. Whewell to J. W. Parker, 10 Dec. 1845, CUL Add. mss. 4251(B)1483.

14. Whewell 1845b; W. Whewell to F. Myers, 16 Mar. 1845, in Stair-Douglas 1882, 317–19, at 319.

15. W. Whewell to R. Jones, 18 July 1845, in Todhunter 1876, 2:326–27. For Whewell's priorities more generally, see Yeo 1993 and the essays in Fisch and Schaffer 1991.

"It may be said," Whewell wrote in the second edition (where, stung by *Explanations,* he attacked the book by name), "if the hypothesis of creation put forward in the *Vestiges* is not the true one, what then is? To this question, men of real science do not venture to return an answer." Take away the fine words and appealing narrative, and what was the book?: *"a System of Order in which life grows out of dead matter, the higher out of the lower animals, and man out of brutes,"* a phrase Whewell used three times in his text.[16] Attempts to demonstrate the origins of the universe, of life on earth, or of the human race were not illegitimate in themselves, but were bound to fail.[17] Here Whewell repeated comments he had rehearsed in correspondence with his close friend and relation the Reverend Frederic Myers, curate of Saint John's Keswick in the Lake District. Myers had ventured to praise *Vestiges* for its range and ambition, if not its actual performance (and he continued to be impressed, as evidenced by his sermons in Cambridge during the following year). Speculative systems, Whewell replied bluntly, always attract readers who "have no power or habit of judging scientific truth."[18]

The difficulties were bound up with the position of natural science at the ancient universities. The Church had acknowledged the sciences as forms of expertise, as well as creating the possibility of a vocation as "metascientific" commentator. But the elementary, extracurricular nature of science teaching meant that only a handful of clerics could speak to the variety of subjects in *Vestiges.* In demonstrating that the book was superficial, a critic would have to be superficial in return, violating specialist boundaries and drawing facts from subjects in which he had no firsthand experience. In the Liverpool debate, the Reverend Hume had tried to get around the problem by asking orthodox specialists to comment on their areas of expertise; but this had only exposed inconsistencies and disagreement.

Anglican theologians tended to excuse themselves from replying to *Vestiges* on the grounds of their lack of knowledge. After the archbishop of York and the bishop of London appointed the Reverend Frederick Denison Maurice to give the Boyle lectures for 1846, established in honor of the chemist and theologian Robert Boyle as a forum for defending Christianity against atheists and deists, *Vestiges* was suggested as a suitable subject. But Maurice, who was professor of English literature and history at King's College in London, feared his "gross ignorance of physics, in every branch of it," would make him a poor critic.[19] This

16. Whewell 1846, 12, 19, 30.

17. Whewell 1846, 9, in reply to *E*1, 127.

18. F. Myers to W. Whewell, Mar. 1845, and W. Whewell to F. Myers, 16 Mar. 1845, in Stair-Douglas 1882, 316–19; Myers 1852, 13–16.

19. F. D. Maurice to S. Clark, [Aug. 1845], in Maurice 1884, 2:416.

sidestepping of science by academic theologians would never have occurred a few decades earlier. A similar situation occurred with the Reverend Thomas Worsley, master of Downing College at Cambridge, who served as the university's Christian advocate from 1844 to 1850. Although holders of this post had often defended the established Church from science-based infidelity, Worsley said nothing in public about *Vestiges* or any other "prevalent errors," confining his attention to analogical studies of the New and Old Testaments.[20] Men with credentials in astronomy, geology, and other scientific subjects were now expected to lead the orthodox counterattack.

This is further illustrated by the small number of separately published replies. Pamphlets remained the usual way of conducting theological controversy, but aside from Whewell's *Indications* and Hume's privately issued lecture, there were only two short works against *Vestiges* written during the debate's crucial early months, neither by the established clergy. One was the heavily footnoted version of a lecture delivered to the Literary and Scientific Institution in Frome, Somerset, by the Anabaptist preacher John Sheppard. This circulated primarily among Dissenters, although it was also read in evangelical circles in the Establishment. The other was that of Samuel Bosanquet, who had taken his degree at Oxford in mathematics and classics.[21] Bosanquet's outburst appeared a week after Whewell's *Indications,* and was sometimes paired with it in reviews. Among several favorable notices the High Church monthly *Christian Remembrancer* praised it as a "seasonable and searching antidote."[22]

The *Remembrancer* and most Establishment periodicals, however, did not support Bosanquet's heady combination of paternal despotism, Pentecostalism, and premillenarianism. Nor did they share his overt hostility to the natural sciences. The Reverend John Oliver Willyams Haweis of Norwood in Surrey had welcomed Bosanquet's entry into the debate, but was disappointed by the result, which showed little respect for science and less knowledge than *Vestiges.* Haweis spoke out in a letter to the High Church *British Magazine,* blaming both the *Vestiges* sensation and the poverty of Bosanquet's response on the neglect of natural theological apologetics in education. Graduates could leave their colleges without understanding the relation between God's word and works. Such men "come to think some lying legend of a saint a more profitable subject of contemplation than the architecture of the heavens, and leave professors of natural philosophy to lecture to empty benches." Haweis recognized that it might be "indecorous" for the master of Trinity to attack the nameless *Vestiges*

20. Worsley 1845–49.

21. Hilton 1988, 96–97.

22. *Christian Remembrancer* 9 (June 1845): 612.

outright, but hoped that someone would do so. Bosanquet's tract was worse than nothing.[23]

The paucity of Anglican replies contrasts with the outpouring on disestablishment, ecclesiastical architecture, and Newman's conversion. The most violent controversy had arisen in January 1845, when Parliament voted a permanent grant to the Roman Catholic college at Maynooth in Ireland. A government supposedly devoted to the Protestant faith was agreeing to fund the training of Catholic priests on a long-term basis. Such issues were central to Church doctrine and to Church-state relations; they had to be tackled in every parish in England, and every parish priest potentially had something to say about them. The threat of infidel materialism was real enough, but involved issues that Myers identified as "unecclesiastical theology."[24] Most churchmen agreed that the book's claim to be based on science demanded an expert informed answer. An early review in the short-lived monthly *Parker's London Magazine* pleaded with "the Christian philosophers, of whom the British Association can boast so many," to expose the errors and fallacies before *Vestiges* did any more damage.[25] By the spring of 1845 the situation was desperate. No sooner had the first edition of *Indications* reached the booksellers than the third edition of *Vestiges* sold out on publication. The bishop of London, the Reverend Charles James Blomfield, had read *Indications* "with great pleasure" and thought it "no doubt will do much good," but remained alarmed about the continuing sensation in the metropolis.[26] In April *Vestiges* went into its fourth edition without any substantial refutation from the Establishment. The nonreligious press continued to be favorable, and the major quarterlies had not found an appropriate champion. Something more forceful than Whewell's little book would be needed if the clerical professoriat's authority was not to be damaged.

The Reviewer as Reader

The favored candidate for this task was the Reverend Adam Sedgwick, Woodwardian Professor of Geology, vice-master of Trinity College, and canon of Norwich Cathedral (fig. 7.3). His activities as a critic are important because he became the book's most consistent and controversial opponent. Sedgwick's reading practices, examined closely, reveal how he used books in general and *Vestiges* in particular. This should help in understanding a characteristic form of

23. J. O. W. H[aweis], "Vestiges of Creation and its Answers," *British Magazine* 27 (1 May 1845): 522, 525, 526. Haweis is the only person in the Clergy List for 1845 with these initials, and his other writings suggest his interest in these subjects.

24. F. Myers to W. Whewell, Mar. 1845, in Stair-Douglas 1882, 316–17.

25. *Parker's London Magazine,* Feb. 1845, 95–104, at 104.

26. C. J. Blomfield to W. Whewell, 22 Mar. 1845, TC Add. mss. c. 87[79].

7.3 Adam Sedgwick, in his academic gown on the far right, shows students the wonders of the Geological Museum in Cambridge, newly opened in 1842. Large vertebrate fossils were especially important for debates about the Flood and the relations of humans to the most recent creation. Lithograph by George Scharf.

intensive academic reading. Sedgwick's famous review in the *Edinburgh* is often taken as typical of contemporary opinion, but is in fact highly idiosyncratic, embodying its author's unique status at Cambridge, in science, and within the Church.

Sedgwick, who turned sixty in 1845, was the most popular man in Trinity, distinguished for his conversation, learning, and hypochondria. He had a high reputation as a pioneering investigator of the oldest fossil-bearing strata. Appointed to the Woodwardian Chair in 1818, he had built up the university's geological collections and its reputation for original research. A liberal Whig, he supported the SDUK, the mechanics' institutes, and other movements toward popular education and the diffusion of knowledge. His religious views became increasingly evangelical, although he had little patience with missionary societies, prayer meetings, and Simeon's more ardent followers.[27]

Sedgwick's services as a champion of science and faith were in heavy demand. A few weeks before *Vestiges* was published, he had achieved national celebrity for replying to the Reverend William Cockburn at the York meeting of the British Association, when the latter had assailed modern geology as antiscriptural.

27. Clark and Hughes 1890, 1:269, 2:588.

Early in the century, Cockburn had served as Christian advocate at Cambridge, where he had attacked nebular astronomy as a revolutionary import.[28] As dean of York, Cockburn occupied a high position in the ecclesiastical hierarchy and his views had considerable clerical support. At the York meeting, the entire chapter house at the cathedral refused to sit down to dinner with Sedgwick. The editors of the *Times*, the conservative evangelical *Record*, and key provincial papers also registered their opposition. George Hudson, England's uncrowned railway king, expressed a widely held view when he reportedly said that "[i]f it comes to a vote, I shall vote for the Dean and Moses."[29] The York confrontation was a key moment in the battle over the relations between Scripture and science, and Sedgwick's courage was hailed across the full spectrum of the liberal press, including the *Spectator* and *Chambers's*, the freethinking *Movement*, and Hugh Miller's evangelical *Witness*.[30] Sedgwick's many friends now looked to him to redress the balance. The dazzling combination of fact and wit that had defeated bigotry from within could defend against infidelity from without. "We trust," *Parker's London Magazine* said, "the same powerful arm will be raised against the monster of cosmogony whenever and wherever it shall appear."[31]

At first Sedgwick declined to respond. He does not appear to have been sent a presentation copy, nor did he read *Vestiges* carefully for several months. But the subject kept recurring in conversation and in letters. In November Featherstonhaugh told him of the sensation in London, noting that Vyvyan was the probable author, and encouraging him to write a review.[32] The dean of Ely, the Reverend George Peacock, had already mentioned that the authorship was

28. Cockburn 1804, which also argued that a "much-lengthened duration of the world" would "materially affect the credibility of our religion."

29. Shipley 1913, 43; "British Association for the Advancement of Science," *Times*, 30 Sept. 1844, 6; "Geological Controversy," *Magazine of Science and School of Arts* 7 (1846): 287–89; "British Association for the Advancement of Science," *Record*, 3 Oct. 1844, 3, and esp. the editorial in the *Record*, 10 Oct. 1844, 4. Similar views are recorded in several of the York papers. The dean of York issued his lecture and some of the subsequent correspondence in Cockburn 1845. The confrontation is discussed in Morrell and Thackray 1981, 243–45.

30. "Our Weekly Gossip," *Athenaeum*, 5 Oct. 1844, 897; "The Scientific Meeting at York, *CEJ*, 23 Nov. 1844, 321–24; "Geology and the Bible," *Movement*, 23 Oct. 1844, 394–95 (with excerpts from the *Globe* and *Spectator*); W[illiam] C[hilton], "Theory of Regular Gradation," *Movement*, 6 Nov. 1844, 413–14; P. G., "Geology and the Bible," *Movement*, 11 Dec. 1844, 451–53; [H. Miller], "The Dean of York's Theory," *Witness*, 9 Oct. 1844; [H. Miller], "The *Record* and Mr Sedgwick," *Witness*, 16 Oct. 1844. For Prime Minister Robert Peel's doubts about his brother-in-law's conclusions, see the letter printed in W. Cockburn, "Memoir of the Late Sir Robert Peel," *New Monthly Magazine* 91 (Jan. 1851): 1–13, at 10–12.

31. *Parker's London Magazine*, Feb. 1845, 95–104, at 95.

32. G. W. Featherstonhaugh to A. Sedgwick, 16 Nov. 1844, CUL Add. mss. 7652.I.E.105.

attributed to Carpenter, and Sedgwick (knowing the earlier fracas about the *Principles of Physiology*) found him "a far more likely man than the high-flying Baronet." Whoever had written it, *Vestiges* undermined everything Sedgwick stood for. "It would be idle for me to talk of answering a Book I have never seen," Sedgwick wrote, "Tho' I hate as mischievous, & abhor as false, the . . . theory of which you have set out an outline."[33]

By the end of January Sedgwick had "seen" *Vestiges,* but told Lockhart and Napier that he was too busy to write a review, pleading lectures, correspondence, and the anticipated onset of spring gout.[34] He studied the book closely for the first time in March after his return to Cambridge, and would have heard more talk about it at several large conversaziones in London. Pressed by numerous friends, he began to regret his refusal as an act of moral cowardice.

At a Sunday breakfast in the cathedral town of Ely, Sedgwick expressed his contempt for the book to leading East Anglian clerics, including the Reverend Joseph Romilly, a senior fellow at Trinity. Romilly recorded the conversation in his diary for 6 April:

> Sedgwick discussed "Vestiges of Creation," against which work he & all scientific men are indignant. The authorship is still uncertain: Sir Ri. Vivian, Lady Lovelace, & some Scotchman or other, are talked of:—Sedgwick thinks the internal evid[ce] in favor of an authoress (in spite of the indelicacy of details about gestation) from the hasty jumping to conclusions. The book is rank materialism. Sedgwick exposed Crosse's Galvanic productions:—the *acarus* he thought he had produced was found to be full of eggs & was derided by the French philos[rs], & the *seed* was found to be a carrot seed (he having made his experiments in a garden pot).—Sedgwick has unfortunately declined the invitations of the Editors of the E[dinburgh] & Q[uarterly] to answer this book which has run thro 3 editions in a few months.[35]

Four days later Sedgwick hinted to Napier that he was reconsidering, in a vivid style that showed what he would do in a review. As it happened, Napier also had in hand an unsolicited article from David Ansted, who taught geology in London. The young man, who had helped out in the Woodwardian Museum, was competent but dull. Napier snatched up Sedgwick's offer.[36]

Why had Sedgwick changed his mind? He had had every reason to object to *Vestiges* from the beginning, but determined to reply only after he read it

33. A. Sedgwick to G. W. Featherstonhaugh, 24 Nov. 1844, CUL Add. mss. 7652.III.A.5.

34. A. Sedgwick to M. Napier, 27 Jan. 1845, BL Add. mss. 34625, ff. 34–35; also printed in Napier 1879, 489–90.

35. CUL Add. mss. 6823, pp. 22–23.

36. A. Sedgwick to M. Napier, 10 Apr. [1845], BL Add. mss. 34625, ff. 98–102, also in Napier 1879, 490–92; A. Sedgwick to M. Napier, 24 Apr. [1845], BL Add. mss. 34625, ff. 152–54; M. Napier to A. Sedgwick, [Apr. 1845], in Clark and Hughes 1890, 2:87.

closely and pinpointed its source: a woman, "a right true blue," probably Ada Lovelace. As he told Lyell:

> I cannot but think the work is from a woman's pen, it is so well dressed, and so graceful in its externals. I do not think the "beast man" could have done this part so well. Again, the reading, though extensive, is very shallow; and the author perpetually shoots ahead of his [*sic*] facts, and leaps to a conclusion, as if the toilsome way up the hill of Truth were to be passed over with the light skip of an opera-dancer. This mistake was woman's from the first. She longed for the fruit of the tree of knowledge, and she must pluck it, right or wrong. In all that belongs to tact and feeling I would trust her before a thousand breeches-wearing monkeys; but petticoats are not fitted for the steps of a ladder. And 'tis only by ladder-steps we are allowed to climb to the high platforms of natural truth. Hence most women have by nature a distaste for the dull realities of physical truth, and above all for the labour-pains by which they are produced. When they step beyond their own glorious province . . . they mar their nature (of course there are some exceptions), and work mischief, or at best manufacture compounds of inconsistency.

No woman could completely transcend the limitations of her sphere: even Harriet Martineau, "mesmeric dreamer" and "economist in petticoats," could not escape.[37] Sedgwick was unusual only in expressing the commonly held doctrine of complementary natures at such length. Men were logical, rigorous, and concrete; women commanded sentiment, feeling, and imagination. For Sedgwick, the distinction between the sexes was, ideally, absolute.

Sedgwick came to his conviction about the author solely through internal evidence. Her character could be judged by the character of the text; as he had explained to his niece a few years earlier, good reading practice involved connecting the lives of writers with their works, as part of an intensive process of study that connected texts to history, biography, and circumstance.[38] This was a common technique in the evangelical literature of self-help. Because books spoke so eloquently of their authors, *Vestiges* could be read as embodying the fatal essence of femininity as applied to science.

Through Sedgwick's rhetoric, the book *became* a woman. With its "charm of manner & good dressing,"[39] *Vestiges* was the temptress Eve, the whore of Babylon, or a common prostitute, disguised in the modest garb of philosophy. Two decades before, Sedgwick had prowled the Cambridge streets as university proctor, in one month alone arresting fifteen women for prostitution. *Vestiges* deserved the same treatment. His review would tear off the pretty clothes;

37. A. Sedgwick to C. Lyell, 9 Apr. 1845, in Clark and Hughes 1890, 2:83–85.

38. A. Sedgwick to F. Hicks, 30 Dec. 1840, in Clark and Hughes 1890, 2:7–8.

39. A. Sedgwick to J. F. W. Herschel, 11 Apr. 1845, RSL Herschel Letters, vol. 15, f. 425.

he planned to "strip off the outer covering and show its inner deformity and foulness."[40]

Sedgwick made every effort to exemplify the patient, manly, Christian reader that the *Vestiges* author so clearly was not. His comments about the book discriminate carefully among different ways of reading. Unlike those he criticized in London coteries, he was no bluffer: firsthand knowledge of a book was distinguished from hearsay, gossip, or reviews. Even after spending some time with the first edition in January, he still spoke merely of having "seen" the book, although he already had enough knowledge to "abhor" it and sketch the outline of what a review would look like. Sedgwick distinguished this kind of casual perusal, which (like most readings by his contemporaries) left few or no traces, from actual "study."[41]

The intensive nature of studied reading is apparent in Sedgwick's copy of the third edition. This involved a line-by-line engagement with the author's evidence and argument (fig. 7.4). Sedgwick used a simple system of pencil annotations, usually a plus sign against passages he agreed with, a minus sign against those he disliked. Further comments in the margins were signaled by a double *x* next to the offending passage. Most of the time he was concerned to note the more telling mistakes. Below one passage in which the author claimed priority for his embryological views, Sedgwick wrote "*xx* Lamarck did know the fact of development." Commenting on the claim in *Vestiges* that creation was going on only in the "obscure" corners of life, he noted "*xx* Why only in the obscurer? We ought rather find nature still busy at her *last work* viz. rubbing off monkeys' tails & making men of them. But we do not."[42]

Sedgwick handled books roughly. He took possession of them by marking his name and address in the front matter, scribbled on any available space in the margins, and folded down pages to keep his place or mark significant passages. Marked pages were listed, often with further angry comments, at the front and rear (fig. 7.5). Books were for use, not for collecting or aesthetic pleasure: unlike the wealthy or the aspiring, he rarely removed advertisements or had cloth-cased volumes rebound in leather. His library was large (several thousand volumes in the mid-1840s) but with few older works or rarities. However, the range of titles was remarkable, with strengths in divinity, philology, natural philosophy, and especially in geology. He subscribed to both the *Edinburgh* and the *Quarterly,* as well as the *Philosophical Transactions* and other scientific

40. A. Sedgwick to C. Lyell, 9 Apr. 1845, in Clark and Hughes 1890, 2:83; Desmond and Moore 1991, 54.

41. A. Sedgwick to M. Napier, 27 Jan. 1845, BL Add. mss. 34625, ff. 34–35; also printed in Napier 1879, 489–90.

42. *Vestiges,* 3d ed., copy in TC Library, Adv.d.8.4, pp. 180, 239, passim.

EARLY HISTORY OF MANKIND. 295

mental phenomenon, language, as the communication of ideas, was no new gift of the Creator to man; and in speech itself, when we judge of it as a natural fact, we see only a result of some of those superior endowments of which so many others have fallen to our lot through the medium of a superior organization.

The first and most obvious natural endowment concerned in speech is that peculiar organization of the larynx, trachea, and mouth, which enables us to produce the various sounds required. Man started at first with this organization ready for use, a constitution of the atmosphere adapted for the sounds which that organization was calculated to produce, and, lastly, but not leastly, as will afterwards be more particularly shewn, a mental power within, prompting to, and giving directions for, the expression of ideas. Such an arrangement of mutually adapted things was as likely to produce sounds as an Eolian harp placed in a draught is to produce tones. It was unavoidable that human beings so organized, and in such a relation to external nature, should utter sounds, and also come to attach to these conventional meanings, thus forming the elements of spoken language. The great difficulty which has been felt was to account

xx But why do not monkeys talk?

7.4 "But why do not monkeys talk?" A page from Sedgwick's copy of the third edition of *Vestiges,* annotated as part of close study in preparation for reviewing. TC, Adv.d.8.4.

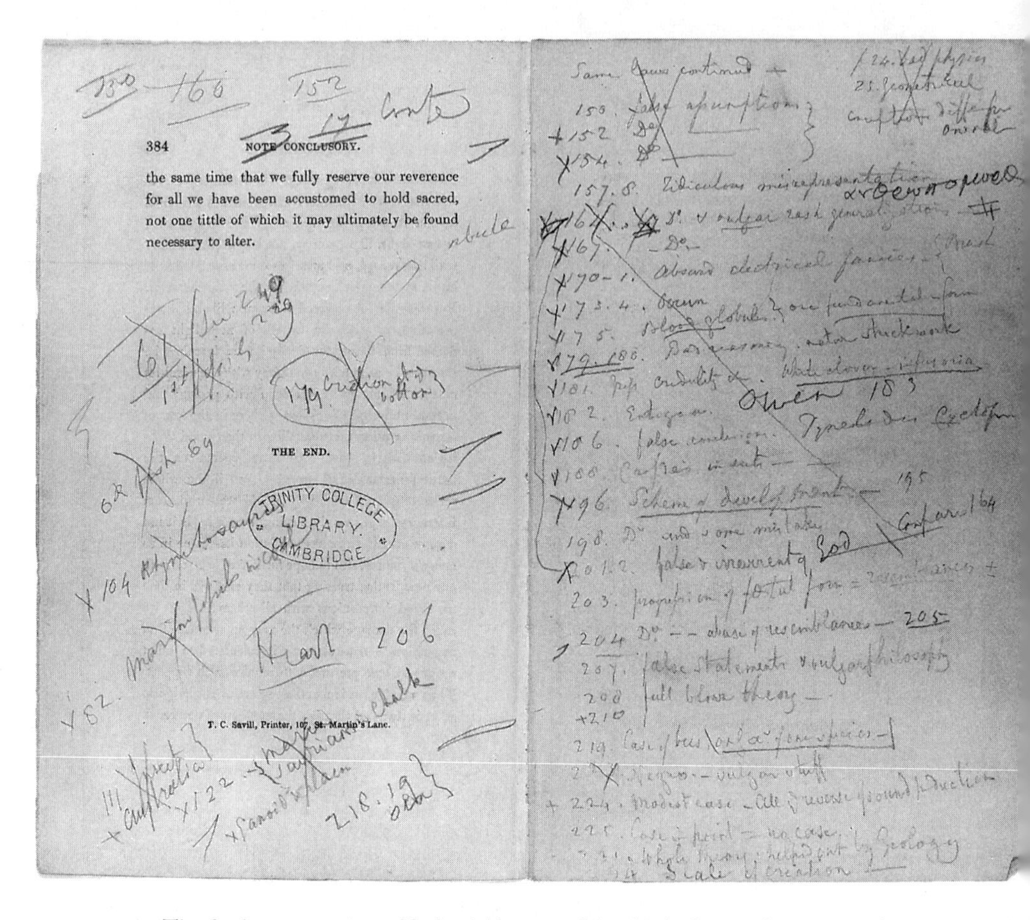
384 NOTE CONCLUSORY.

the same time that we fully reserve our reverence for all we have been accustomed to hold sacred, not one tittle of which it may ultimately be found necessary to alter.

THE END.

T. C. Savill, Printer, 107, St. Martin's Lane.

7.5 The final page opening of Sedgwick's copy of the third edition of *Vestiges,* with marked pages and further comments. The library stamp was added after Trinity College purchased the book at the auction of Sedgwick's books. TC, Adv.d.8.4, pp. 384–85.

journals. Stacks of unsorted books and papers were crammed into every available corner of his rooms. If Sedgwick could not find something in this unruly chaos, he had easy access to the libraries of his college, the university, and the Cambridge Philosophical Society, as well as the collections of friends. He annotated *Explanations,* for example, in a copy owned by Whewell.[43]

The technique of close study had its origins in the traditions of divinity and classical scholarship pursued in the great European universities since their foundation. Their application to the sciences was well established too, so that similar systems of markings are found in books annotated by many contemporaries.

43. *Explanations,* 1st ed., copy in TC Library, 22c.84.5. For Sedgwick's library, see Swan 1873.

Annotations intervened in the printed text, softening it up for criticism. In this way, the writing of a learned commentary (or, in this case, a review essay) began in the very process of reading. By annotating extensively and often in complete sentences, Sedgwick rehearsed the rhetoric of his review; he also did this in conversations and letters. This gave opportunities to try out strategies for writing. His friend Romilly also helped by listening to drafts and checking proofs. Sedgwick encouraged Napier "to use your editorial shears without mercy" and (unlike *Edinburgh* regulars like Carlyle and Macaulay) had no qualms about changes to his prose.[44]

Sedgwick needed all the help he could get: he had never written a review before, and gout made sitting irksome. He wrote at his desk in his rooms at Trinity in the early morning, the only time free of distractions and rheumatic pain. The back numbers of the *Edinburgh* on his shelves were a helpful guide on how to proceed, and a daunting reminder of the great literary traditions of the Whigs. He longed for the manly vigor of John Playfair in dealing with astronomy and natural philosophy, and the light touch of Sydney Smith, whom he had often met at Holland House.[45] In its early years the *Edinburgh* had been noted for paradox, sarcasm, and brilliant rhetoric—a style that certainly shaped Sedgwick's own. But applying the lessons to a work he hated so passionately was not easy. "Unmitigated contempt, scorn, and ridicule are the weapons to be used," he noted as he began writing, and not surprisingly the result was often heavy-handed.[46] Napier, initially delighted at the prospect of a new star in the Whig literary firmament, was faced with one of the great editorial nightmares of his career.

The worst problem was length. All quarterly editors tried to keep contributions punchy and short, in order to compete with the burgeoning monthlies; and despite monsters such as Macaulay's hundred pages on Francis Bacon, the average length of articles in the *Edinburgh* had actually declined under Napier. When asked about how much space might be needed, Sedgwick's answer was disconcertingly vague: "*it must be long,* as far as I can see my way." There were also difficulties with structure, as Sedgwick (who was disorganized at the best of times) never had the whole text under his eye at once. Napier normally had entire articles in hand before sending them to the printer, but as the *Edinburgh* was so late in noticing *Vestiges,* he had decided to have the manuscript set into type as it arrived, usually in batches that filled a sheet or two of sixteen printed

44. J. Romilly, journal entries for 3 May, 13 May, 27 May, 5 June, 6 June, and 11 June 1845, CUL Add. mss. 6823, pp. 30–36. See also A. Sedgwick to M. Napier, 17 Apr. 1845, BL Add. mss. 34625, ff. 113–19; also in Napier 1879, 492–94, at 493.

45. A. Sedgwick to M. Napier, 17 Apr. 1845, BL Add. mss. 34625, ff. 113–19, also in Napier 1879, 492–94; A. Sedgwick to M. Napier, 9 May [1845], BL Add. mss. 34625, ff. 188–89.

46. A. Sedgwick to M. Napier, 17 Apr. 1845, BL Add. mss. 34625, ff. 113–19, also in Napier 1879, 493.

pages. This ensured that the review would appear on time, but also meant that (as Sedgwick told his niece) that one part was being printed, "while the other part was slowly uncoiling from my brain in Cambridge."[47] He was projecting eighty pages in mid-May, but the manuscript kept coming, so that Napier called a stop. When the article appeared in July 1845 it was eighty-five pages, among the longest the *Edinburgh* had ever published. Napier offered £110 in payment—more than the *Vestiges* author received for the first two editions—which Sedgwick wanted to return.[48] He never wrote for the quarterlies again.

The Fall of Babylon

In Sedgwick's rooms, framed in gilt, was a fine proof mezzotint by John Martin of the fall of Babylon, a great city that had been corrupted by intellectual ambition and lascivious indulgence (fig. 7.6).[49] The moral was obvious. But while writing, Sedgwick soon decided that he had been wrong to suspect a metropolitan "blue" of authoring *Vestiges.* Rumors about Combe and Mantell were circulating at Cambridge dinner parties, and most of his London friends suspected a man, probably Richard Vyvyan.[50] Sedgwick wrote his article with the Tory aristocrat in mind, although gossip was rife about other possibilities at the British Association meeting at Cambridge in June. Hedging bets and wary of offending so powerful (and male) an opponent as Vyvyan, the review admitted that the anonymous author was "accomplished, and, in a certain sense, well-informed," as well as credulous, superficial, skeptical, imaginative, childlike, crazed, hypothetical, and fantastical. *Vestiges* might be by a man, but it was not manly. The article also reaffirmed that an author's character could be dissected from the character of his book. Long passages about initial suspicion of a female author point up Sedgwick's continuing conviction of the feminine qualities of the argument. Napier, writing from Edinburgh, had agreed that these had ensured "its popularity with *learned* women, and half-read and shallow men."[51]

Sedgwick was most worried about susceptible young women readers in the metropolis, particularly in Whig literary circles. "You have no conception what mischief the book has done & is doing among our London blues," Sedgwick told him, "& God willing I will strive to abate the evil."[52] He wanted to warn

47. A. Sedgwick to M. Napier, 17 Apr. 1845, BL Add. mss. 34625, ff. 113–19; A. Sedgwick to I. Sedgwick, 3 July 1845, in Clark and Hughes 1890, 2:87.

48. A. Sedgwick to M. Napier, 21 July 1845, BL Add. mss. 34625, ff. 288–89.

49. Swan 1873, 42; Wees 1986, 36–37. Martin's painting of this subject had been exhibited in 1819, and a mezzotint based loosely upon it was first issued in 1831.

50. Romilly, diary entry for 14 Apr. 1845, CUL Add. mss. 6823, p. 25.

51. M. Napier to A. Sedgwick, [Apr. 1845], in Clark and Hughes 1890, 2:87; [A. Sedgwick], "Natural History of Creation," *Edinburgh Review* 82 (July 1845): 1–85, at 3–5.

52. A. Sedgwick to M. Napier, 4 May 1845, BL Add. mss. 34625, ff. 175–76.

7.6 "The Fall of Babylon," mezzotint by the apocalyptic artist John Martin, 1831. Assisted by a divine storm, the forces of Cyrus the Great defeat the Chaldeans, who have been corrupted by female temptation and ambitions to rival God with their huge towers. British Museum, Department of Prints and Drawings, Mm 10-6.

Leonard Horner's five daughters, to whom he had already addressed the postscript of a letter about the book.[53] Among the "Horneritas," Mary was the wife of Charles Lyell, and Frances had married the botanist Sir Charles Bunbury. Susan, Joanna, and Katharine had keen interests in botany, travel, and history, and later published translations and original works of their own. With their interests in German literature and science, they exemplified the kind of intelligent female readers liable to corruption.

For Sedgwick, the book's popularity also threatened to provoke a conservative counterreaction, intellectual famine following the poison of progressive development. The *Edinburgh* had to expose *Vestiges* lest Cockburn and other Tory diehards within the Church use "this efflorescence of folly" as an occasion for "a starving treatment," with taxes on knowledge and limits on literacy.[54] As an evangelical Whig, Sedgwick believed that religious revival should go hand in hand with the diffusion of knowledge from credible sources. *Vestiges* threatened

53. A. Sedgwick to M. Napier, 24 Apr. [1845], BL Add. mss. 34625, ff. 152–54.
54. A. Sedgwick to M. Napier, 31 May [1845], BL Add. mss. 34625, ff. 239–41.

this union and had to be destroyed. These fears were not unfounded, for the prominence given to Sedgwick's clash with Cockburn showed that geology's triumph was by no means assured. The *Times* during the late spring and summer of 1845 featured a lively debate in its correspondence columns about *Vestiges,* in which the scriptural validity of geology was a central issue. It began as a contest between Cockburn and the Whig peer Lord Thurlow, but soon drew in a half-dozen other pseudonymous correspondents. "Anti-megatherium" (who shared the suspicion that a woman had written *Vestiges*) raged against the clerical strata hunters and complained that Whewell's "well-meant" *Indications* failed to "contradict the geological phantasies."[55]

In Sedgwick's view, original science should not be debated in the newspapers. Nor should it emerge from drawing rooms and dinner parties, the lectern, or the pulpit, but only from the field, the laboratory, and male clubs like the Geological Society. Those qualities that made woman admirable—her trusting faith and love, for example—made her unable to participate in the making of science. In their right place, women were welcome: Sedgwick encouraged ladies to attend his geology lectures and addressed them in his review. He loved the metropolitan soirées and at Cambridge enjoyed mixed conversation at dinner and tea parties, often in his rooms at Trinity. But actual scientific discovery was different, for it involved an "enormous and continued labour" for which women were unsuited.[56] When women tried to do science, or were associated in any way with its production, the result was disastrous—what Sedgwick called privately a "deformed progeny of unnatural conclusions." Such violent rhetoric must have had something to do with his realization, during the same month he read *Vestiges,* that he would never marry. His museum would have to serve in place of wife and children.[57] Sedgwick could not rage so forthrightly against the book in print as he did in letters, but the implication was clear that true science was a masculine birth. As he told Napier, the reviewer should stamp with "an iron heel upon the head of the filthy abortion, and put an end to its crawlings."[58]

For this approach to work, the vanquisher of *Vestiges* had to present himself as a leading man of science. Sedgwick's anonymous article spoke with a voice that combined the magisterial editorial "we" of the *Edinburgh* with a personal authority gained from years of work in science:

55. Anti-megatherium, "To the Editor of the Times," *Times,* 27 June 1845; Anti-megatherium, "Canons Recupero, Buckland, and Sedgwick," *Times,* 4 July 1845.

56. [Sedgwick], "Natural History of Creation," 4.

57. A. Sedgwick to I. Sedgwick, 1 Mar. 1845, in Clark and Hughes 1890, 2:80; also 2:350.

58. A. Sedgwick to M. Napier, 10 Apr. 1845, BL Add. mss. 34625, ff. 98–102, also in Napier 1879, 492.

> But we know, by long experience, that the ascent up the hill of science is rugged and thorny, and ill-fitted for the drapery of a petticoat; and ways must be passed over which are toilsome to the body, and sometimes loathsome to the senses. And every one who has ventured on these ways, has learned a lesson of humility from his own repeated failures. He has learned to appreciate the enormous and continued labour by which every new position has been won. . . . No man living, who has not partaken of this kind of labour, or, to say the very least, who has not thoroughly mastered the knowledge put before his senses by the labours of other men, has any right to toss out his fantastical crudities before the public, and give himself the airs of a legislator over the material world.[59]

Although all reviews in the *Edinburgh* were anonymous, Sedgwick made every attempt to make his authorship known, and newspaper reports widely credited it to him. By placing himself in the position of "a legislator over the material world," he created a voice that was masculine and Christian, that spoke on behalf of an imagined united scientific community. The networks of Establishment knowledge had not been constructed to attack such a wide-ranging synthesis, but they could be used toward that end. Even in areas close to his own knowledge Sedgwick consulted a range of leading specialists.

Making science speak with a single voice required ironing out discrepancies and disagreements. Consider progressionist geology, the issue on which the Reverend Hume's campaign had foundered. Like Buckland, Sedgwick had built his reputation advocating progress in the earth's history; and also like Buckland, he switched his emphasis after reading *Vestiges.* Progress, the review said, was as much apparent as real: the record of life was full of breaks, gaps, and signs of degeneration.

At the time of his review, Sedgwick was embroiled in a controversy about the oldest rocks with fossils, which were crucial both to his own specialist research and to the claims in *Vestiges.* His friend and former collaborator Murchison grouped all the older rocks in a "Silurian" system, while Sedgwick put only the uppermost in the Silurian and held the rest in a tentative "Protozoic." This Protozoic had no distinctive fauna, but the complete Silurian fauna gradually emerged during it. Thus just before *Vestiges* appeared, Sedgwick had built a progressive advance from invertebrates to vertebrates right into the heart of a classification. Sedgwick's system became even more fraught with problems after 1846, when he equated Murchison's entire "Lower Silurian" with what he henceforth called the "Cambrian system." This gave his strata a defining fauna, but it was a fauna lacking vertebrates. Prima facie, Murchison's classification was better suited to attacks on Vestigian progression, for the enlarged Silurian

59. [Sedgwick], "Natural History of Creation," 4.

included vertebrate fishes and all the major invertebrate classes in a single group that extended to the base of the fossil-bearing geological column. The "highest" form of life could then be said to appear in the oldest geological group.[60]

Sedgwick thus faced a tension between his strategies for defeating the Silurian on the one hand and *Vestiges* on the other. For the most part, he maintained a sharp boundary between the two disputes, debating Murchison behind the closed doors of the Geological Society, in private correspondence, or in the controlled circumstances of the British Association. In print he drew a moral from his failed quest to find a distinctive, lowly fauna: "We have spent years of active life among these ancient strata—" he noted, "looking for (and we might say longing for) some arrangement of the fossils which might fall in with our preconceived notions of a natural ascending scale." He had looked in vain, and had been forced "to bow to nature."[61] Equanimity in defeat showed that science could be a model of manly philosophic calm, above passions and personal interest.

The qualities used to give science its special character, however, made it vulnerable to a wide-ranging cosmological synthesis. By "running from one thing to another without any system," *Vestiges* had broken one of the fundamental rules of reading practice Sedgwick had outlined to his niece.[62] No one could claim to be an authority on an entire field of science, let alone all of them. The ways in which the author had succumbed to this temptation were evident in the tiniest of mistakes. For example, *Vestiges* gave the proportion of the equatorial to the polar diameter of the earth as 230 to 229, rather than the ratio of 300 to 299 used in contemporary observatories. Sedgwick knew nothing of "the high flying modern analysis," but remembered from his student days that the *Vestiges* figure was an old theoretical derivation by Newton. Slips of this kind pointed to "science gleaned at a lady's boarding-school."[63]

Unlike Whewell, Sedgwick was optimistic about retraining the minds of ordinary readers. "Let the people have food," he told Napier, "but let it be wholesome."[64] His *Edinburgh* article was precisely the serious, fact-packed, "doctrine by doctrine" reply that Whewell had initially thought so ill-advised. The body of the review followed the structure of *Vestiges,* matching its reading of the "book of nature" point for point.[65] Tracking the argument through astronomy,

60. Secord 1986a, 117–18, 139–43.

61. [Sedgwick], "Natural History of Creation," 31.

62. A. Sedgwick to F. Hicks, 30 Dec. 1840, in Clark and Hughes 1890, 2:7–8.

63. [Sedgwick], "Natural History of Creation," 27; A. Sedgwick to M. Napier, 9 May [1845], BL Add. mss. 34625, ff. 188–89.

64. A. Sedgwick to M. Napier, 31 May [1845], BL Add. mss. 34625, ff. 239–41.

65. [Sedgwick], "Natural History of Creation," 62.

progressive geology, spontaneous generation, and embryology, the review showed that the latest evidence undermined continuous transitions of all kinds in nature. Fire-mists do not become nebulae; nebulae do not become planets; matter does not become life; invertebrates do not give rise to vertebrates; reptiles do not become mammals: there are no gradations or ambiguities.

A great chasm in nature separated man from beast. The crucial point was that *Vestiges* had "*annulled all distinction between physical and moral.*"[66] Animal instinct had not "developed" into human reason; the chattering of monkeys had not "progressed" to become language; physical reality was not moral reality. Unlike physical truths, moral truths could not be read directly through the senses. The moral lessons of nature's book could only be read analogically, through the exercise of reason in great men:

> Truth is always delightful to an uncorrupted mind; and it is most delightful when it reaches us in the form of some great abstraction, which links together the material and moral parts of nature—which does not annul the difference between material and moral,—but proves that moral truth is the intellectual and ennobled form of material truth, first apprehended by sense. And believing that all nature, both material and moral, has been framed and supported by one creative mind, we cannot believe that one truth can ever be at conflict with another.[67]

Human understanding, because divinely inspired, could "ennoble" physical as moral truth. Thus the mathematical regularity of the cells of bee hives, evidence in *Vestiges* for the material continuity of instinct and reason, could be recognized instead as proof of divine action.[68] When the senses were prepared, all nature could speak of God. Sedgwick believed in suspensions of natural law, and thought the Irish famine of 1845–46 was God's judgment on human sin. God also acted directly in creating new species of animals and plants. But the foundation for his faith was not in divine intervention, but in the scriptural assurance that humanity—and hence human understanding—was made in God's image.[69]

By using natural law to explain the soul, *Vestiges* threatened the fine balance between faith and science that Sedgwick had maintained since his childhood in the Yorkshire Dales. In the village of Dent he had watched Tom Paine burnt in effigy and as an undergraduate found the "immortal works" of Robert Hall, the

66. Ibid., 3.

67. Ibid., 56.

68. Ibid., 14.

69. Ibid., 3, 79. In contrast, Gillispie 1951, 164–70, identified the chief source of Sedgwick's objections as his belief in miracles, rather than his belief in man's special status.

celebrated Baptist preacher in Cambridge, intensely inspiring.[70] Hall's responses to the French Revolution revealed the secularizing dangers of William Paley's stress on utility as a measure of perfection. Affinities with Dissent, cemented by his family background and experiences at Cambridge, led Sedgwick to adopt an evangelical religion of the heart, emphasizing personal salvation, Christ's atonement, and emphatic pulpit rhetoric. As the Cambridge Tory M.P. Henry Goulburn said, Sedgwick was "intellectually powerful and so very highly cultivated," and yet "exceedingly narrow in his religious opinions."[71] He found reviewing more like writing a sermon than composing a scientific paper.[72]

The *Edinburgh* offered an opportunity to sound a warning. Readers of *Vestiges,* without realizing what was at stake, were on the verge of abandoning their hopes of immortality. The book "comes before them with a bright, polished, and many-coloured surface, and the serpent coils of a false philosophy, and asks them to stretch out their hands and pluck forbidden fruit." "Our glorious maidens and matrons" could be saved only if *Vestiges* were unmasked as debased, fallen, feminized, and unwholesome.[73] Sedgwick's article was welcomed by those journals that had already opposed the book, such as the *Literary Gazette,* which saw "the scourging and irrefragable review" as a vindication of what they had told readers six months before.[74] More tellingly, the review also found friends in quarters in the Church where the sciences, and especially geology, had been viewed with suspicion. These ranged from the evangelical *Christian Observer* to the Anglo-Catholic *Christian Remembrancer,* which praised Sedgwick's "masterly essay" for placing the dangers of the development hypothesis "beyond all further controversy."[75]

The review's crude vehemence, however, did not go down well in polite drawing rooms. Whewell recognized that it would shoot wide of its fashionable target. "How do you like Sedgwick's Article in the Edinburgh?" Whewell asked a friend. "To me the material appears excellent, but the workmanship bad, and I doubt if it will do its work."[76] The sexual rhetoric, the stress on "disgusting

70. On Paine: Clark and Hughes 1890, 1:31; on Hall: A. Sedgwick to H. C. G. Moule, 16 Dec. 1868, in Clark and Hughes 1890, 1:80–81.

71. Clark and Hughes 1890, 2:588.

72. E.g. A. Sedgwick to M. Napier, 17 Apr. 1845, BL Add. mss. 34625, ff. 113–19, at f. 115, which speaks of giving "a sermon on our conceptions of a *first cause*" (this passage is not in the published version of this letter). Sedgwick later had to reassure Napier that he wrote from "no narrow or clerical feeling"; see [28 May 1845], BL Add. mss. ff. 233–35.

73. [Sedgwick], "Natural History of Creation," 3.

74. *Literary Gazette,* 12 July 1845, 455.

75. *Christian Remembrancer,* June 1845, 612; "Vestiges of Creation," *Christian Observer,* Sept. 1851, 599–610, at 602.

76. W. Whewell to R. Jones, 18 July 1845, in Todhunter 1876, 2:326–27.

views" of physiology, the fears of moral disintegration were associated with more extreme evangelical Anglican and Dissenting press. This sounded too much like Bosanquet. It bore out the fears of Mary Somerville, living by this time in Italy, that the book would "offend in some quarters," and Sir John Cam Hobhouse's prediction that it would attract orthodox ire. The thoroughness of the *Edinburgh* critique may have convinced Whig aristocrats in a way that merely "orthodox" reviews had not, but many of them thought the article "not enlivening." Henry Holland lamented its "lengthy inefficiency." Lockhart agreed that the article was heavy reading—this was not what would be wanted for the *Quarterly*. Like most Tories, he suspected that Sedgwick was protecting himself now that the tendency of geology had been exposed: "The truth is all the *savants* are sore at the vestige man because they are likely to be in the same boat as him."[77]

Outside the Establishment, liberal theologians who had been cheered by the slaying of Cockburn were appalled to find Sedgwick cozying up to their opponents. Cynics condemned his outburst as a calculated attempt by a place-seeking parson, envious of Cockburn's status or of Buckland's elevation to the deanery of Westminster. John Forbes, a close associate of Churchill and editor of the *British and Foreign Medical Review*, thought that this was the only way to explain Sedgwick's "self-lashed furor and rabidness."[78] More commonly, the more extreme liberal press suspected that the violence meted out to *Vestiges* was a quid pro quo for the attack on Cockburn; in the words of the *Liverpool Mercury*, "a mere anonymous bookmaker might well be sacrificed to evidence the orthodoxy of a Cambridge divine." As the *Prospective* noted, some specialists hoped to secure "immunity to their own speculations, by a cheap display of eloquent zeal against all who dare to go beyond their measure."[79]

"The True Ichthus" and the Body of the Church

As read in the courts of Trinity, *Vestiges* reeked of the charnel house and dissecting room; its destruction required a knowledge of anatomy and developmental physiology. Because Cambridge and Oxford had no medical faculties until much later in the century, it has been assumed that they were both weak in these subjects. In fact, sophisticated resources were ready to hand.

77. M. Somerville to W. Greig, 28 May 1845, in M. Somerville 1873, 277–8; Hobhouse diary entry for 18 May 1845, BL Add. mss. 43747, f. 112; Lord Morpeth, Journal, 26 July 1845, from the Castle Howard Archives, J19/8/8, p. 27; H. Holland to W. Whewell, 26 Jan. 1846, TC Add. mss. a.206.[107] J. G. Lockhart to R. Murchison, 20 July 1845, BL Add. mss. 46127, ff. 104–5.

78. J. Forbes to RC, 17 Jan. 1846, NLS Dep. 341/94, f. 68.

79. [F. Newman], "Explanations. A Sequel to the Vestiges of the Natural History of Creation," *Prospective Review* 2 (Jan. 1846): 33–44, at 37; "The Edinburgh Review, and the Vestiges of the Natural History of Creation," *Supplement to the Liverpool Mercury*, 17 Oct. 1845, 1.

7.7 Lecturing on comparative anatomy at Oxford c. 1840. The emphasis in this heavily symbolic painting is on the connections of comparative anatomy to book learning. Books and fossils crowd the picture, with human skulls in a case at the left. A bust of John Hunter, faced by the skeletons of birds, surmounts the classical doorframe. The lecturer is probably John Kidd, Regius Professor of Medicine from 1822 and keeper of the Radcliffe Library (shown through the window) from 1834. Kidd gave elementary lectures on anatomy and physiology to a declining number of students in the decades before his death in 1851.

Anatomy, like other sciences, was taught at Oxford and Cambridge as a philosophical subject (fig. 7.7); although human dissections were occasionally carried out, would-be physicians always completed their practical studies at one of the hospital schools in London or Edinburgh. Since 1817 the professor of anatomy at Cambridge had been the Reverend Dr. William "Bone" Clark. Clark opposed plans for setting up a metropolitan-style medical school, an idea supported by several medical men in the university.[80] Clark saw no place at Cambridge for "such endless multiplicity of minute teaching" nor did his counterparts at Oxford. University training, as Clark said, "cultivates the taste and matures the reason, and fits men to secure for themselves the mastery in every subsequent professional pursuit."[81]

80. Relations between anatomy and medicine at Cambridge are discussed in Weatherall 2000.

81. As stated in his evidence to the University Commissioners (United Kingdom, Parliament 1852, "Evidence," 110); Acland 1848, 12–17.

Anatomy did not have to be old-fashioned to be genteel. True, the Saxon naturalist Carl Gustav Carus found Oxford anatomy in about as lively a state as the shriveled remains of the dodo in the Ashmolean Museum. The anatomy theater was "quite in the antique style," and he thought the fine natural history books of the Radcliffe Library would be better moved to Dresden, where someone might read them. The genial and disheartened Regius Professor of Medicine, John Kidd, continued to lecture to small groups of students on comparative anatomy and Paley's argument from design.[82] Henry Wentworth Acland condemned Oxford for ignoring the broad principles that would have prepared him for medical study at Edinburgh. "I was no more instructed in the ordinary rudiments that constitute the basis of a Professional mind, than a cow" he told his father in March 1845.[83] But with Acland's appointment as Lee's Reader in Anatomy at Christ's Church in that autumn things began to change. A man of birth, breeding, and influential friends, he revitalized the lectures, assembled a physiological series modeled on the Hunterian collection of the Royal College of Surgeons, and worked behind the scenes to gain support.[84] Even he nearly despaired of success.

The situation at Cambridge initially seemed more promising.[85] Clark was a lively and sociable man who had transformed the university's small anatomical museum. He had bought extensively at auctions and added embryological models made by his wife Mary Willis Clark, preparing so many dissections that he developed a premature stoop. In common with Trinity men such as Whewell, Clark drew on German romantic practices to combat unbelief and materialism among Cambridge students. He had learned medicine at Saint Bartholomew's in London, and like his teachers argued that organic matter was not the cause of life, but only its manifestation. In 1831 his first paper assumed a uniform succession of developmental stages advancing "to a still increasing degree of perfection." But recapitulation did not imply transmutation, for monsters exhibiting an "excess of development" (such as the partial Siamese twins the paper described) never represented a higher class.[86] In a review paper for the British Association, Clark became one of the first in Britain to adopt Karl Ernst von Baer's view that vertebrate and invertebrate embryos differed in

82. Carus 1846, 186–88. As Robb-Smith 1997 shows, Kidd had been an active holder of the chair in the 1820s.

83. H. W. Acland to T. Acland, 23 Mar. 1845, Bodleian Library, Ms. Acland d.30. ff. 101b–104, at f. 103.

84. Atlay 1903, 130–41; Fox 1997; Robb-Smith 1997.

85. Carus 1846, 152–53.

86. W. Clark, "A Case of Human Monstrosity, with a commentary," *Transactions of the Cambridge Philosophical Society* 4 (1833): 219–55, read 16 May 1831. For details of Clark's life, see C. Watson, "A Cambridge Professor of the Lost Generation," *Macmillan's Magazine*, Jan. 1870, 267–72; and Shipley 1913.

their very earliest stages. He argued, as always, against transmutation: all living species were "originally created" as we see them today. An anonymously published survey of research on the ovum demonstrated Clark's command of the main German-language authors.[87]

Cambridge was thus well stocked with anatomical ammunition against *Vestiges*. Whewell had already consulted Clark for his earlier books, and recycled the relevant sections in *Indications*.[88] Sedgwick too looked across Trinity high table to his old friend for help, as he knew little about theories of generation. It was almost certainly at his request that Clark read a paper against *Vestiges* at the Cambridge Philosophical Society on the evening of 28 April 1845, for Sedgwick chaired the meeting and quoted largely from the talk in his *Edinburgh* critique. The society, which had both a library and lecture hall, met in purpose-built rooms near Trinity College (fig. 7.8).[89] Clark emphasized that the great Continental authorities rejected linear recapitulation, although some of their disciples were less careful, making claims "now feebly re-echoed in this country." The danger, as Clark pointed out, was not materialism but transcendental idealism in the style of the German *Naturphilosophen*. But higher types did not pass through developmental phases permanent in lower ones; changes in conditions had no effect on the process; there was no "law of development" common to all animals. Such claims seemed to justify a reading of the *Vestiges* embryology as an "absurdity" based on a "fallacy."[90]

But just how antiquated or crude was the embryology in *Vestiges?* The answer depended on what readers demanded of science. From many points of view the book was remarkably up-to-date. Sympathetic readers could point to passages that unequivocally rejected simple embryonic recapitulation; these passages had an impeccable pedigree, through Carpenter's *Principles*, to von Baer. They were, however, not cited in the first four editions, which meant that the author could be accused of failing to consult any recent literature. Moreover, the summary of von Baer's law in *Vestiges* followed a long account of the recapitulatory view, cited from Fletcher and Lord. This strategy might have advantages in making the latest findings comprehensible in a racy, popular work, but for the intensive reading encouraged at Cambridge it could be accused of

87. W. Clark, "Report on Animal Physiology," *1834 British Association Report* (London: John Murray, 1835), 95–142, at 114, 115; [W. Clark], "Von Baer, Valentin, Wagner, Coste, Eschricht, &c. on the Early Development of the Ovum," *British and Foreign Medical Review*, Jan. 1840, 1–35.

88. W. Whewell 1845b, 26.

89. "Professor Clark offered remarks on the theory of development as that theory is propounded in a work entitled Vestiges of the Natural History of Creation." Entry for 28 Apr. 1845, CPS, minute book of the general meetings, vol. 2, CPS archives. Clark's paper was never published, but what he said can be reconstructed from [Sedgwick], "Natural History of Creation."

90. [Sedgwick], "Natural History of Creation," 79, 85.

7.8 Lecture room of the Cambridge Philosophical Society, where William Clark testified against the embryological conclusions of *Vestiges* in April 1845. *ILN,* 28 June 1845, 405.

being muddled. Clark read *Vestiges* with the rigor he thought should be applied to any scientific text.

More generally, knowledge about human anatomy had been strictly controlled at Cambridge. While Carpenter wrote accessible textbooks, manuals, and primers, Clark published little and built up a museum collection for would-be medics. Access was contested from the very beginning. Soon after opening in 1832, the museum had been stormed by poor townspeople who wished to recover the body of a workhouse pauper destined for dissection. Mistaking wax models for human remains, they inflicted "a vast amount of damage" on the collections. Iron bars were installed afterward to protect the dissecting rooms from the fury of "low and turbulent blackguards."[91] There were more subtle boundaries as well. Once within the museum, ordinary visitors were only allowed to see the comparative specimens, not those illustrating human pathology or sex. Ladies could see skeletons and stuffed animals, but dissecting rooms were male preserves. Similarly, although Clark had stressed the antievolutionary potential of von Baer's embryology since the 1830s, this was primarily

91. Watson, "Cambridge Professor," 270; Shipley 1913, 4; Richardson 1987, 263, 264–65.

within the cloistered walls of the university, and not through publications. Women, weak creatures of feeling, might be tempted Eve-like to taste the forbidden fruits of skeptical materialism, to "soil their fingers with the dirty knife of the anatomist." These were subjects (as Sedgwick's review said) that "could not so much as be named without raising a blush upon a modest cheek," fit only for professional medical texts.[92]

Vestiges had forced these esoteric questions of embryology onto the public stage. It did this not because of mistakes and journalistic simplifications, but because the text was unusually well-informed about the latest findings from the Continent. Most readers learned of the new views for the first time through the controversy about *Vestiges.* Distinctions of which nonspecialists like Sedgwick had been unaware—between von Baer and his predecessors, for example—became important matters of controversy. To be an effective source in replying, Clark had to be accorded an authoritative status, although he was little known outside the university. In quoting from his talk, Sedgwick subdued its polemical anti-Vestigian origins, footnoting it simply as a memoir on "foetal development."

Other sources of anatomical expertise also gained prominence. By the mid-1840s Richard Owen had emerged as the leader of British comparative anatomy and a key exponent of a natural theology that looked for design in archetypal plans rather than in special adaptations. The Cambridge men profoundly admired Owen's work, helped to get him a government pension, and hoped that he might take up a university chair.[93] Owen's mind was seen to be preeminently teleological, perfectly adapted to seeing nature's underlying relations. He was, in consequence, among those approached by the *Edinburgh* and *Quarterly* as a potential reviewer. But he had turned them down, ostensibly because this "would give 'Vestiges' an importance calculated to add greatly to its mischief."[94]

In a strategic alliance with his Cambridge patrons, Owen's greatest generalization—the archetype of the vertebrate skeleton—became imbued with the virtues of Christian Platonism (fig. 7.9). The archetype was a blueprint of creation in the divine mind. For Sedgwick, Whewell, and other Cambridge men

92. [Sedgwick], "Natural History of Creation," 3. Sedgwick at first asked female readers to skip the pages on the "foetal question." By the time he had finished, he told Napier that "any man or woman must be *nice* in Dean *Swifts' [sic] sense*" to object to what had been written, so the "mock modest passage" was revised (A. Sedgwick to M. Napier, 17 June 1845, BL Add. mss. 34625, ff. 265–66). "We wish with all our hearts we could pass this subject over," he told his readers in the published version, "for it is fit only for professional books, and it requires illustrations which we cannot give here" ([Sedgwick], "Natural History of Creation," 74). Clark, by publishing almost nothing and most of that in specialized places, implicitly maintained the same boundary.

93. Corsi 1988, 267–68; Desmond 1989, 353–58; Richards 1987, 153–67.

94. R. Owen to W. Whewell, 14 Feb. 1844 [1845], in Brooke 1977, 142.

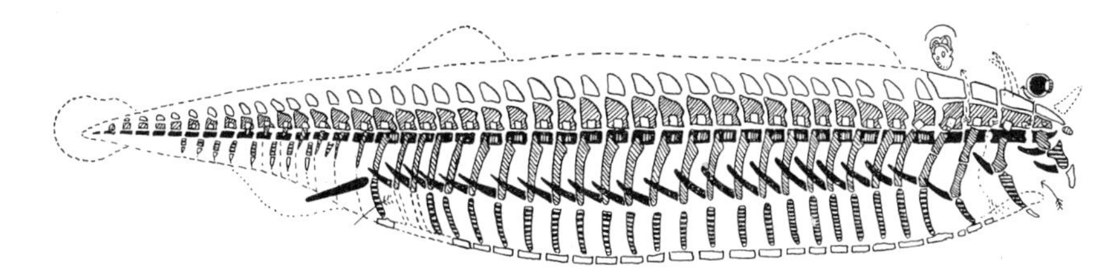

7.9 Richard Owen's archetype of the vertebrate skeleton. From the traditions of academic theological scholarship this could be seen as "the true ichthus," Christ's presence expressed analogically in nature. From redrawn version in Russell 1916, 105.

of science, the archetype freed the "foetal question" from any taint of materialism and the moral corruption associated with practical dissection. For the Reverend Thomas Worsley, the archetype could be read symbolically, as the fish symbol of the early Church. The analogy of nature could enrich parallels between the Old and New Testaments. The human body could be understood scientifically, but not explained through material causation. Whewell, one of the electors who had appointed Worsley as Christian advocate, welcomed Owen's work as among the supreme conclusions of inductive science. As Worsley had suggested in a letter to Owen, "the best way of overthrowing the Dragon-worshippers with their idols is to set before them the true ichthus."[95]

Reconstructing Science and Religion

Vestiges afforded very different opportunities at Oxford, not least because science had less importance there than at Cambridge. The Tractarians had a reputation for opposing science, as famously illustrated when Mary Buckland half-jokingly wondered if her husband would be burnt at the stake. Certainly the Tractarians believed that the sciences had little to do with the Christian idea of a university. John Henry Newman's defection to Roman Catholicism in October 1845 proved a catastrophe for his followers, but opened the way for fresh developments among those who stayed within the Anglican fold.

A number of young Oxford men were swinging toward liberal divinity or outright unbelief, and for many *Vestiges* was their first exposure to a cosmology based on natural law. James Anthony Froude, a young fellow at Exeter College, lamented that the Tractarian belief in the authority of church institutions provided insufficient defenses against secular doubts. Reading *Vestiges* led him to reject miracles in nature, at the same time he was abandoning freedom of the will and divine inspiration of Scripture. Once he published a fictionalized version of his fall from grace in the notorious *Nemesis of Faith* (a book publicly

95. R. Owen to H. W. Acland, 28 Dec. 1848, Bodleian Library, Ms. Acland d. 64, ff. 135–36, quoted in Rupke 1994, 203.

burned in Exeter College hall), he lost his college fellowship too. In disgrace, Froude stayed with the Reverend Charles Kingsley at Ilfracombe in Devon, where they discussed *Vestiges* and other books as they walked along the moors.[96]

The young poet Arthur Hugh Clough, also associated with the Oxford Tractarians, was deeply affected by *Vestiges,* which he read in Liverpool in the Easter vacation of 1845. Clough built his experience of the book into his poems of religious doubt, most notably in verses that became famous in later editions of his poems as "The New Sinai":

> And as of old from Sinai's top,
> God said that God is One,
> By Science strict so speaks He now
> To tell us, There is None!
> Earth goes by chemic forces; Heaven's
> A Méchanique Céleste!
> And heart and mind of human kind
> A watch-work as the rest![97]

Clough expressed what many young dons and students felt after the shattering effect of Newman's apostasy.

It became vital to come to terms with the sciences of progress and the dangers of skepticism. Leading this movement was Henry Acland, who pursued reforms with great consequences for relations between science and the established Church. Acland announced his program in "The Bodily Nature of Man," his inaugural address as Lee's Reader in October 1845. Basing his ideas on his education at the University of Edinburgh, Owen's work, and *Vestiges,* Acland argued that knowledge of the human body and its place in nature was essential to all students.

> No cosmical law was altered or modified by the introduction of man into the world. You would expect, or not be surprised that in many of the corporeal habitudes of the New Being, there would be a great similarity with the old. Some characteristics of vegetables and more of animals are accordingly found united in him. As in other organized beings, these are developed in him, in such forms as are most agreeable to his nature, adapted to his wants, and calculated for the end and means of his existence. He superadded the spiritual gifts which ["separated"

96. Dunn 1961, 1:147–48.

97. "The theory that the human race is a distaillation [*sic*] of the monkey-kind has recently been revived with great eclat in a work called Vestiges of the Creation, ascribed by some people to Lady Lovelace = 'Ada sole daughter.'" A. H. Clough to J. P. Gell, 2 Apr. [1845], in Mulhauser 1957, 1:112. The verses were first published in February 1849; Lowry, Norrington, and Mulhauser 1951, 17.

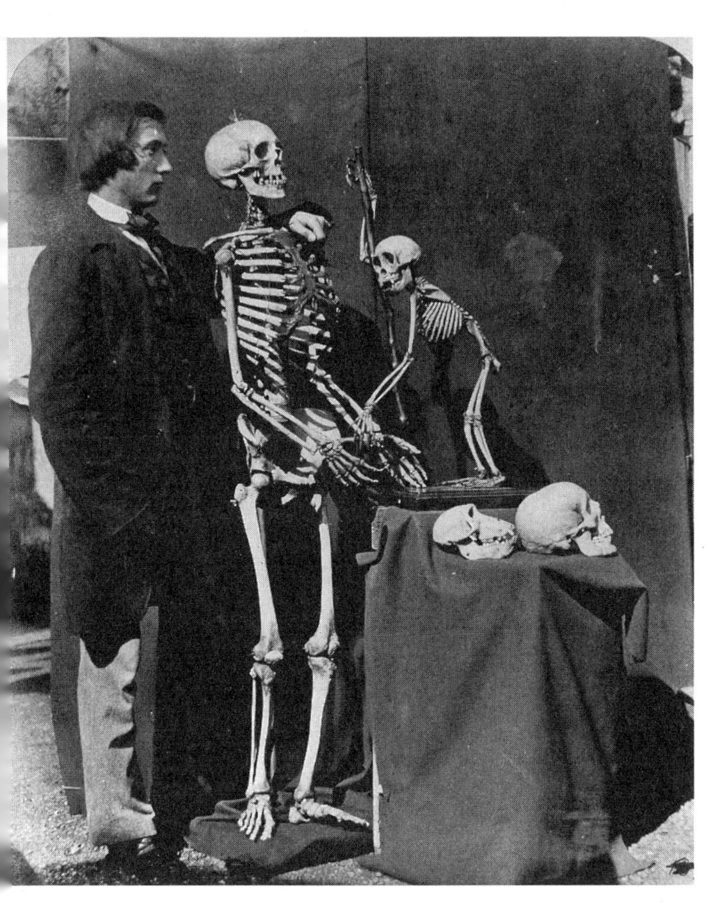

7.10 Henry Acland's success in establishing scientific education at Oxford drew on traditions of comparative anatomy and museum-building. In this photograph taken by Charles Dodgson (Lewis Carroll) in June 1857, Acland's student Reginald Southey is shown with the skeletons of an ape and a human.

> *deleted*] sever him from the world beneath, and bind him to higher nature, and, in common with them, to his Maker.[98]

Acland stressed that training need not be superficial or popular to be elementary, for science was "facts connected, illuminated, interpreted, so as to become the intelligible embodied expression to His creatures of the will of God." He condemned as "absurd" theories that ignored man's spiritual nature and brought lower races out of higher ones, but encouraged close comparative study of animals and humans (fig. 7.10). In generalizations like Owen's archetype, anatomy incorporated the highest science within the spiritual values of Protestant Christianity. Owen's vision provided a foundation on which to rebuild developmental anatomy, to show that science need not end in skepticism and

98. H. W. Acland, "The Bodily Nature of Man—An Inaugural Lecture Delivered in the Anatomy school Oxford Oct. 23. 1845," Bodleian Library, Ms. Acland e. 1, ff. i, 14–15.

arrogance. In the Oxford setting, the archetype was compatible with many aspects of *Vestiges*.[99]

Acland's tact was vital. His inaugural address helped convince influential men like the Reverend Doctor Edward Pusey—canon of Christ Church and Regius Professor of Hebrew—to maintain an open mind on developments in science. The Reverend Charles Marriott, another leader of the Tractarians after Newman's defection, attended Acland's entire course. Still other High Churchmen, such as the Reverend Doctor William Sewell of Exeter College, remained deeply concerned about skeptical tendencies. His *Christian Vestiges of Creation*—written in 1852 and published nine years later—pointed to parallels between God's role as the "invisible fountain-head" of the Church and the natural world.[100] At best, Acland could hope that such men would not block his reforms outright.

The most important advocates of the new strategy were a younger generation of Oxford Tractarians, including Richard William Church, James Mozley, and Frederick Rogers, who founded the weekly *Guardian* to bring the movement onto a national stage. Church surveyed the controversy in an anonymous review of *Explanations*, which he wrote while attending Acland's Owen- and *Vestiges*-inspired anatomical lectures and practical evening classes in histology and embryology. The review announced an important shift in the way the Tractarians would deal with science. It argued that while the *Vestiges* author and his critics had come to a draw on the factual battleground, the latter had shown themselves "not quite prepared for the contest." What counted were the higher realms of theology and philosophy, for too long despised as mere Scholasticism and "mystical superstition."[101] *Vestiges* represented "the revenge of moral and metaphysical science" on geology and astronomy; a knowledge of nature, whatever else it could do, was not a necessary pointer to salvation.[102]

From the *Guardian*'s perspective, to ground faith in miraculous exceptions to natural order—as it read the *Edinburgh* reviewer as advocating—was to court disaster. What if transmutation should prove true? What if Crosse's experiments *should* produce insects? Heaps of facts, eighty-five pages of them, could not obscure the direction in which research seemed to lead:

> The *Vestiges* warns us, if proof were required, of the vanity of those boasts which great men used to make, that science naturally led on to religion. (It may lead beyond the experiment and the generalization, to vast theories,—visions and histories

99. Acland 1848, 12, 16.

100. "All life and all organization around us springs, and is developed from an invisible fountain-head; as the Church ultimately issues out from Him, whom the eye does not behold." Sewell 1861, 96–97.

101. [Church], review of *Explanations, Guardian,* 18 Mar. 1846, 141–42, at 142.

102. Ibid.

> for the imagination, realities of order and law for the reason—to a *substitute* for religion.) In a world of widening and self-sustained order, an Epicurean atheism is not so difficult; something deeper than the facts of natural science is required to undercut its premises. It is the metaphysician—the abstract thinker—who is wanting in the field.[103]

As the *Guardian* saw it, *Vestiges* demonstrated the poverty of the schemes developed by Buckland, Sedgwick, and their allies, which relied on design and miraculous creation as testimony to God's existence. The "starveling argument" of Paley no longer sufficed to keep science safe. A more dramatic move was needed, one that accepted the possibility that the direction charted in *Vestiges* and similar works might be true. The Church had to be ready to accept the sufficiency of natural laws in explaining the natural world. "It is not special pleading," the review concluded, "or timid indecisive fighting about details, which will meet the march of that science which openly threatens to be infidel, because no one will help it to be Christian."[104]

The local target for the *Guardian*'s remarks was the Reverend Baden Powell, Oxford's most radical defender of natural theology. Powell had been appointed to the Savillian Professorship of Geometry in 1827 with ultraconservative High Church support. But during the 1830s he made a dramatic volte-face, arguing in *The Connexion of Natural and Divine Truth* (1838) that contemplation of the natural order provided the only rational route to Christian belief. Miracles were open to scientific explanation and could not provide a basis for faith.[105] Powell did not speak out about *Vestiges* until the 1850s, but the High Church men knew and despised his efforts to forge a strong connection between natural law and the truth of Christianity. The *Guardian* used *Vestiges* to break this link. The physical and the moral had to be separated by "impassable limits." Religion had to be grounded in faith, not reason; science, freed of its foundational role, could then move forward unimpeded.

Prominent Oxford men followed the *Guardian*'s lead. The Reverend Mark Pattison, Oxford's most successful tutor, included *Vestiges* along with Herschel's *Preliminary Discourse* and Humboldt's *Cosmos* as one of the leading works of the literature of science.[106] Pattison, Pusey, and Marriott continued to believe in

103. Ibid., 141–42. From this point of view, the appointment of Samuel Wilberforce as bishop of Oxford was a setback, both because of the form of natural theology he espoused and because it led to conflict with all forms of evolution, as at the 1860 British Association meeting.

104. [Church], *Guardian*, 18 Mar. 1846, 141–42; for comments on the review, see Brooke 1991b, 224–25, and Yule 1976, 266–68.

105. Powell's views are subtly analyzed in Corsi 1988.

106. Tollemanche 1885, 3; Pattison, however, distinguished such surveys from "science itself," which involved coming into chapel "with the mud of Shotover and the railway cuttings sticking to your knees."

the superiority of classics, philosophy, and mathematics as forms of intellectual discipline, but agreed that the natural sciences could have a place within a Christian university. When plans for a university museum came to a vote in convocation in 1854, their support proved critical. *Vestiges* might prove true, but this would not affect the foundations of faith. "Keep in view," the *Guardian* had said, "the great principle that belief in God does not depend upon the natural—that nature is not the real basis of religion, and we can safely afford full and free scope to science."[107]

Perhaps the most significant reader from an Oxonian High Church perspective was William Ewart Gladstone, a cabinet minister in Peel's second administration and a key figure in the emergence of liberalism. Raised as a devout Anglican evangelical in Liverpool, Gladstone had joined the High Church party while a student in the early 1830s. He was a voracious reader, spending a fortune on his library and keeping careful records of everything he read throughout his long life (fig. 7.11). During the summer of 1845, concerned to develop a new basis for individual conscience during the growing crisis in Church-state relations after Maynooth, Gladstone began systematic study of the works of the eighteenth-century bishop Joseph Butler. Two years later he read *Vestiges* in six days and wrote about it for his planned volume.[108]

This reading appears to have shaped attitudes Gladstone would hold for the next half century. The book, from his perspective, was skeptical but useful in clearing Christianity of needless encumbrances. This position was close to that being developed at Oxford and in the *Guardian:* Gladstone was an old friend of the politician Sir Thomas Acland, and knew about the reforms Acland's son was attempting. *Vestiges,* on Gladstone's reading, could be reconciled with the literal truth of Genesis and the possibility of pre-Adamic men. The "dust of the earth" out of which Adam had been created might have passed through many forms. The book showed the successive stages whereby God revealed through secondary causes "the map of his own counsel." As Gladstone later wrote, "Indeed, I must say that the doctrine of Evolution, if it be true, enhances in my judgment the proper idea of the greatness of God, for it makes every stage of creation a legible prophecy of all those which are to follow it."[109] Creation by natural law was compatible with Christian doctrine and individual responsibility.

The Varieties of Belief

Two decades after publication, Gladstone could recall no book that had made such a stir. Among students and the younger college fellows, *Vestiges* continued

107. Fox 1997, 649, 653–54; Atlay 1903, 143, 415–18; [Church], *Guardian,* 18 Mar. 1846, 141–42.

108. Gladstone, diary entries for 12–17 July 1847, 6 June 1880, in Foot and Matthew 1968–94, 3:634–35, 9:535.

109. Gladstone to S. Jevons, 10 May 1874, in Lathbury 1910, 2:101; discussed in Hilton 1988, 344.

7.11 A close reader in action: William Ewart Gladstone in old age in his private library. Gladstone continued to be concerned with *Vestiges* to the end of his life; he referred to it in debating Huxley in the 1890s and had reread Thomas Monck Mason's refutation in the previous decade. Portrait of 1895 by John M'Lure Hamilton in Lathbury 1910, 2: frontispiece.

to be widely read well into the 1850s. A group of young university men read it together while relaxing in the Lake District during the long vacation in 1853. As one of them, the promising Cambridge mathematician James Clerk Maxwell, told a friend, the book "excited thought & talk." But *Vestiges* evidently did not deserve serious study. He expressed his view of such works in an essay on "idiotic imps" at Cambridge's exclusive debating society, the Apostles. The worst of what he termed the "dark sciences" were spiritualism and table turning. "The most orthodox system of metaphysics," Maxwell explained, "may be transformed into a dark science by its phraseology being popularised, while its principles are lost sight of."[110]

Maxwell's reaction was characteristic of those who had welcomed science as a path to faith. As the historian John Brooke has said, the book sold the pass. This was true primarily at Cambridge, where design, progress, and adaptation had been used to support Christian truth; now, these same arguments seemed to lead to the denial of miracles and the unique status of humanity.[111]

110. J. Clerk Maxwell to R. B. Litchfield, 23 Aug. 1853, and "Idiotic Imps," in Harman 1990–95, 1:223–26; Campbell and Garnett 1882, 166–92. On Gladstone, see Morley 1903, 2:165.

111. Brooke 1979, 50.

Among the strict evangelicals who agreed with Dean Cockburn, *Vestiges* confirmed that the book of nature could be read only in the light of Scripture. And among conservative traditionalists, it could be taken to show that Paley offered little more than a longer path than Paine to the dreary wastes of deism. Clerical men of science like Sedgwick and Whewell found themselves caught in the middle. Their moderate evangelical belief in a law-bound universe, subject to occasional miracles, seemed increasingly inconsistent. As Featherstonhaugh had warned, "If some step is not taken the odium theologicum against the lovers and cultivators of natural science will wax great."[112]

In Oxford, on the other hand, *Vestiges* had facilitated new and more constructive relations between science and faith, a shift that attracted national attention. The *Guardian* editors were elated to receive a letter of approval from Owen, who was chafing between Sedgwick's empiricist evangelicalism and Powell's icy rationalism.[113] Acland, too, was "very vigorous" about the turn among the Oxford theologians.[114] *Vestiges* remained an awkward book that could be turned to local advantage. But what happened when the readers were closer to those anticipated in the text? To answer that question, the next chapter returns to Edinburgh.

112. G. W. Featherstonhaugh to A. Sedgwick, 16 Nov. 1844, CUL Add. mss. 7652.I.E.105.

113. J. B. Mozley to A. Mozley, 4 Apr. 1846, in Mozley 1885, 176–77; for Owen and Sedgwick, see Richards 1987, 166–67; for Owen and Powell, see Desmond 1982, 44–48.

114. J. B. Mozley to A. Mozley, 4 Apr. 1846, in Mozley 1885, 175–76.

CHAPTER EIGHT

The Holy War

For months the name of the book was in every mouth, and one would be accosted by facetious friends, "Well, son of a cabbage, whither art thou progressing?"

JAMES BERTRAM, *Some Memories of Books Authors and Events* (1893)

THE PHENOMENAL CAREER OF *Vestiges* in Edinburgh is recalled by James Glass Bertram, a twenty-year-old clerk who had just finished his apprenticeship. Bertram first heard of the book from an itinerant preacher, "who went about denouncing it as something which would in time sap the influence of the Holy Scriptures." Speculation about the authorship in printing houses was rife, as were jokes in shops, taverns, and inns. Bertram remembered a "merry party" at the Rainbow Tavern, where the after-dinner speaker began by addressing his listeners as "'descendants of Apes.'" Other memoirists recalled the sensation too. Edinburgh was a small place in which learning and literature remained key industries, so that acquaintances could joke about new books while passing on the street.[1]

Edinburgh (fig. 8.1) was still reeling from the greatest event in Scotland's nineteenth-century history. This was the Disruption of 1843, when the evangelical party had broken away from the Established Church of Scotland to form the Free Church.[2] The Disruption climaxed an old split between the evangelical and moderate parties within the national Church and permanently altered the role of religion in political life. It is often claimed to have weakened the churches against the forces of secular liberalism—but what those forces were is often unclear. The reaction to *Vestiges* shows how specific threats were dealt with at this crucial juncture, as reading was contested with unprecedented intensity. It becomes possible to see how questions of practical politics were

1. Bertram 1893, 132–35.

2. Surveys of the literature on the Disruption include Withrington 1993, Brown 1997, and Brown and Fry 1993. General accounts of Scottish church history include Drummond and Bulloch 1975 and Brown 1987.

8.1 Edinburgh from the Calton Hill in 1848. On the left are Arthur's Seat and Salisbury Crags. The castle dominates the skyline of the Old Town; the main thoroughfare of Princes Street marks the division with the orderly streets of the New Town on the far right. From Steuart [1848], a folding panorama sold to tourists.

questions of perception: most fundamentally of all, how controversy turned on the politics of reading. What processes—physical, psychological, spiritual—were thought to make meaning from the letters printed on a page?

Circles of Reform

Vestiges targeted a national audience but—like all books—was written from a local perspective. In 1844 Edinburgh could be reached only by carriage or by sea, and hence was relatively isolated; the first railway connection to the south opened only three years later. The city, with a population of some 164,000 in the 1841 census, was divided into the Old Town with its famous castle and narrow wynds, and the New Town with its spacious public squares and classical architecture. Edinburgh's diverse, overlapping, and fractious readerships had been incorporated into *Vestiges*. Its specific understanding of liberalism had been forged among the phrenologists who lived in the New Town; its projected images of radicalism had been shaped by the experience of the city's commercial thoroughfares; and its implicit anticlericalism was a reaction to evangelical Presbyterians. The book was hailed in phrenological circles but damned by the "saints" as a trashy nightmare.

Interest was most intense in the circles around George, Cecilia, and Andrew Combe. By the early 1840s, their campaign to enthrone phrenology as the key science of progress was flagging: innovation seemed at a standstill, the number of societies was declining, and even on Andrew Combe's estimation the *Phrenological Journal* printed "second or third rate gossipy speculations" (it ceased publication in 1847).[3] Ostracized from the higher echelons of Edinburgh soci-

3. Gibbon 1878, 2:274; A. Combe to G. Combe, 29 Aug. 1845, NLS mss. 7274, ff. 91–93; H. G. Atkinson to G. Combe, 25 May 1846, NLS mss. 7278, ff. 3–4; Cooter 1984, 88–90.

ety, the Combes and their friends felt themselves a beleaguered minority, an island of advanced liberalism in an evangelical sea.[4] *Vestiges,* a brilliantly written work by an avowed phrenologist, was "a great hit" in their long struggle. Many in Combe's circle received presentation copies, and others bought theirs from Maclachlan, Stewart, and Co., publishers of phrenological works and a center for liberal conversation.

Combe noted with satisfaction that the sensation had created a ferment among the genteel Whigs around the *Edinburgh Review.* Even the aging Lord Francis Jeffrey, who had disdained contact with the phrenologists for years, praised the book as a work of "great talent, altho' full of wild theories."[5] Like Napier, he recognized the "clerical" tone in Sedgwick's review, and condemned it as "that monster paper . . . from which so much was expected."[6] What could be accepted as orthodoxy in England came across in post-Disruption Scotland as sectarian obscurantism. By placing phrenology alongside prestigious sciences such as astronomy, the book gave new life to the battle against what phrenologists and the older Whigs alike condemned as "pure priestly prejudice."[7] In Edinburgh, it seemed a gift to the Combeites from an unknown benefactor whom Tait—himself no friend to the system—acknowledged as "a philosophical phrenologist."[8]

4. See A. Combe, "Remarks on G. C.'s letter to von Struve," 21 Oct. 1845, NLS mss. 7274, ff. 97–99; G. Combe to A. Combe, 9 Nov. 1845, NLS mss. 7390, ff. 207–13; and H. C. Watson to G. Combe, 25 Aug. [1846], NLS mss. 7282, ff. 132–33.

5. As reported in G. Combe to J. Clark, 9 Nov. 1844, NLS mss. 7388, ff. 792–95.

6. F. Jeffrey to M. Napier, 8 Oct. 1845, in Napier 1879, 506.

7. G. Combe, diary entry for 14 Aug. 1845, NLS mss. 7425, f. 21r.

8. *Tait's Edinburgh Magazine,* Dec. 1844, 800–801, at 800. Shapin 1975 discusses the social divisions within the Edinburgh phrenological debates.

Combe received one of the earliest free copies. Within a few days he had skimmed it and read the first third with care; at this point he had no idea who the author was. On a Saturday walk in mid-October 1844, he told his friend Robert Chambers to read it, expressing delight that the author of such a theory should share their belief in phrenology's truth. Combe also mentioned that he planned to write a congratulatory letter to the author through the publisher. This he did two weeks later, after reading the whole book carefully. His feeling on turning the pages was a mingled sense of "pleasure and instruction"; it combined "all the sublimity of a grand poem, and the sober earnestness & perspicuity of a rigidly philosophical induction." His letter compared *Vestiges* to "a new sun" in the scientific firmament, which "will probably collect around it innumerable facts, until at length it shall develop itself into a Theory as perfect as a planetary system."[9] It was with some disappointment, then, that Combe read the *Edinburgh*'s critique the following August. He was staying in the German spa town of Bad Homburg, and spent a rainy few days reading a copy borrowed from a prominent Scottish Whig, who told him that Sedgwick was rumored to be the reviewer. Combe copied long extracts into his diary. The article, he noted, betrayed ignorance and prejudice, "a mind of the second or third order"—strong in the limited field of geology, but flawed in his own areas of logic and mental physiology. It led him, however, to be more circumspect about praising the book's use of the term "law" and its evidence for the origin of life from matter.[10]

Given the qualified nature of his support, Combe could not understand why the evangelicals continued to label him an advocate of the development hypothesis—perhaps even as the *Vestiges* author. His nephew Robert Cox explained that the general tenor of the *Constitution of Man* made these suspicions understandable. "No doubt," he wrote in a letter, "you have never meant to adopt the 'Vestiges' theory of the development of new species of animals in geologic times; but taking the present system of things only into view, you have taught that the world contains within itself the elements of improvement; and though you allow and believe in the possibility of supernatural influences, and admit and *seem* to believe in their actual agency, still you give them entirely the go-by in what you teach."[11]

The Edinburgh phrenologists did their best to spread the word further afield, commending *Vestiges* in conversation and correspondence. "His hypothesis is not *proved*," Combe told a Continental phrenologist, "but it is very ably

9. G. Combe, "To the Author of 'Vestiges of the Natural history of Creation,'" 30 Oct. 1844, NLS mss. 7388, ff. 780–81.

10. G. Combe, diary entries for 10–16 Aug. 1845, NLS mss. 7425, ff. 19v–23v.

11. R. Cox to G. Combe, 26 Apr. 1848, NLS mss. 7291, ff. 87–90.

supported."[12] "The work displays great scientific learning"; he explained to the educational reformer Lucretia Mott, in what was probably the first news of the book to arrive in the United States, "it is clearly and calmly written, and if printed cheaply, would be another battery erected against superstition."[13] The *Phrenological Journal*'s January number, edited by Cox, carried one of the earliest reviews in the quarterlies. This praised the book for bringing the evolution of animals and plants to the forefront of scientific attention, and for its advocacy of law. Combe, who wrote it anonymously, also added a brief defense to subsequent editions of the *Constitution*.[14]

The phrenologists also pushed *Vestiges* on visitors to their immediate circle, including the flamboyant London-born actress Fanny Kemble, who sometimes stayed with her cousin Cecilia Combe when performing in the Edinburgh theaters. Kemble complained that phrenology entered so thoroughly into all aspects of life in the Combe household as to limit friendships with outsiders; ordinary conversation could not be followed without an understanding of relevant terms. George Combe could only attribute her skepticism to the weakness of her organ of Causality,[15] and as part of his efforts to train her mind, suggested that she should read *Vestiges*.

Judging from the scanty evidence, women in phrenological Edinburgh may have had strong views about the book, but they kept them largely to themselves. Anne Chambers had obvious reasons for silence, but her friends said little about the book either. None seems to have been sent a presentation copy. The popular author Camilla Toulmin, asked at a dinner party for her opinion, could only remember that "I was not capable of forming one."[16] Cecilia Combe longed to write a children's book of conversations on natural law, but her husband discouraged the project. Even so independent a woman as Kemble, who criticized Cecilia for being George's "echo" on phrenological matters,[17] discounted her own capacity to deal with the facts of *Vestiges* ("I am quite too ignorant to dispute"), while couching criticism of its logic in the conventional

12. G. Combe to E. Hirshfield, 16 Dec. 1844, NLS mss. 7390, ff. 4–6; also G. Combe to R. R. Noel, [Dec. 1844], NLS mss. 7390, ff. 7–9; G. Combe to R. R. Noel, 18 June 1845, NLS mss. 7390, ff. 141–43; G. Combe to C. E. Cotterell, NLS mss. 7390, ff. 24–25; G. Combe to G. D. Berney, 31 Dec. 1844, NLS mss. 7390, ff. 22–23.

13. G. Combe to L. Mott, 20 Oct. 1844, in Gibbon 1878, 2:187–88.

14. [G. Combe], "Vestiges of the Natural History of Creation," *Phrenological Journal, and Magazine of Moral Science* 18 (Jan. 1845): 69–79, at 78; Combe 1847a, 3–4. Other friends such as the chemist Samuel Brown, not directly in Combe's circle, were less favorable. See RC to AI, [late Feb. 1845], NLS Dep. 341/110/86–87.

15. Kemble 1878, 1:245–47; F. Kemble to H. St. Leger, 26 June 1840, in Kemble 1882, 2:27.

16. Toulmin 1893, 86.

17. F. Kemble to H. St. Leger, 30 Sept. 1839, in Kemble 1882, 1:265–68.

language of sewing: "many of the conclusions in particular instances appear to me to be tacked or basted (to speak womanly) together loosely and clumsily." (The same metaphor, as we have seen, cropped up in Barrett's reaction.) Kemble claimed to have no theory of her own, preferring religious hopes and a strong sense of duty. She thought, however, that women who broke out of conventions were not necessarily to be admired. It was all too easy to become lost in the "dark and unfathomable abyss" of cosmology. "I thank God," Kemble wrote to her closest confidant, "I have not the mental strength, *and infirmity* to seek to grapple with this impossible subject." Too much independence could thus be portrayed as feminine weakness. The novelist Catherine Crowe, with her interest in ghosts, demons, and spiritualism, was often suspected of possessing "a most preposterous organ of wonder."[18] She became enthusiastic about *Vestiges* when, in the week of publication, her neighbor Robert Chambers read her a chapter aloud.[19]

The Edinburgh phrenologists debated how best to use the book in an increasingly hostile religious environment. George Maclaren, geologist and editor of the liberal *Scotsman,* recommended the work for its "*suggestive* tendencies" but sounded a more cautious note than Combe had done. The *Scotsman* was the most widely circulated paper north of the border, opposed to church establishments and favoring religious Dissent, free trade, and educational reform. So as not to lead readers to a work they might find objectionable, the review warned that it placed human origins under the reign of law. This, the review said, was the part of the book where mistakes in science most showed, however "curious" its arguments. With sales of the *Scotsman* ten times that of the *Phrenological Journal*'s pitiful three hundred, Maclaren was aware of the religious sensibilities of his readership. Yet the review stressed that the author was "no surly or discontented infidel," but "of strong mind, inquisitive, and well cultivated."[20]

Others in Combe's circle wished the book had been bolder. Sir George S. MacKenzie, mineralogist, phrenologist, and aristocrat, took it as a starting point for his anonymous pamphlet of 1846, *Vestiges of Error in Religious Doctrine.* This continued themes he had developed in previous assaults on the "priestcraft" of all organized churches. Through "jesuitical chicanery" and "pharisaical externals," they followed neither the example of Jesus nor "the pure religion of nature" that he had taught.[21] MacKenzie had planned to discuss the authorship and explicitly to defend Chambers, but Combe dissuaded him from

18. F. Kemble to H. St. Leger, 14 Nov. [1847], 19 Nov. [1847], in Kemble 1882, 3:242–47; on Crowe's Wonder, see Kemble 1882, 2:83.

19. RC to AI, 15 Oct. 1844, NLS Dep. 341/111/19–20.

20. [G. Maclaren], "Literature," *Scotsman,* 16 Nov. 1844, 4.

21. [MacKenzie] 1846. His probable authorship can be inferred from letters cited below.

this course. He also feared that MacKenzie might have trouble finding a respectable publisher. "The *style* of your vestiges is so vehement & uncompromising," Combe warned him privately, "and the errors attacked are so sacred in the eyes of the vast majority of the community, that I should fear even Churchill would shrink from bringing them out."[22] Assaults on orthodoxy smacked of the freethinking gutter press. Combe proved to be right. While MacKenzie's pamphlet fell dead from an obscure London publisher, its celebrated instigator continued to sell thousands of copies.

For the phrenological propagandists, the book played its most important role in the political arena, proclaiming that liberal reform demanded a natural order that worked through law. Thus Combe's review in the *Phrenological Journal* stressed its support for improvements in sanitation, parks, and public baths. Ultimately an improved understanding of causation would produce a reformation in literature, religion, and society. The management of criminals could be transformed by reading *Vestiges*.

> We deliberately allow all the natural conditions calculated to produce and multiply these unfortunate beings to flourish around us, and then wonder at the increase of crime and destitution; nay, we go farther, we wreak our vengeance on offenders against the law for becoming what our treatment has tended to make them; for what can be the effect of short imprisonments of young delinquents in common jails, but to ripen the beginners into mature and dangerous criminals?[23]

Combe proposed a "moral" regime of self-rehabilitation as an alternative, so that prisoners would start in solitary confinement and enforced idleness, until "the mental depression of *ennui*" forced a change in habits. They would gradually be given liberty, with punishment for the slightest breach of discipline. Terms of imprisonment would be indefinite, with power of release dependent on the decision of government commissioners. Such a scheme, the *Phrenological Journal* noted, was wonderfully fortified by *Vestiges*.

Exercising the Mind

Reading was central to the phrenological reformers, for the printed word was among the most effective ways for a small party to contribute to local and national debates. In phrenological terms, the act of reading literally changed minds and was therefore central to education. Books were intellectual gymnasia, where readers could exercise their brains just as athletes exercised their

22. G. Combe to G. S. MacKenzie, 16 Oct. 1846, NLS mss. 7390, ff. 203; also 20 Nov. [1846], NLS mss. 7390, f. 539.

23. [G. Combe], "Vestiges," 78–79; also Gibbon 1878, 2:155–57, and de Giustino 1975, 145–62.

8.2 George Combe holding a book. Calotype by the Edinburgh photographers David Octavius Hill and Robert Adamson.

muscles. As a theory of mind and brain—the only complete theory, according to Combe and his friends—phrenology explained how this happened, how books gained meaning, and what their effects on brains would be (fig. 8.2).[24]

Reading, for the phrenologists, began after the impressions of light were sent to the brain. The eye itself could not form ideas, but only received, modified, and transmitted physical sensations. Unlike philosophers who argued that sight had to be learned, Combe attributed the rudimentary sight found in children or the formerly blind solely to the imperfect state of their organs of vision. The mutual adaptation between a fully functioning eye and the impressions of light from the outside world could be taken as one of the many signs that nature worked according to regular laws.[25]

At its simplest level, recognizing the characters on a page brought the phrenological organ of Form into action. Individuals in whom Form was large learned to read with great ease, even in unfamiliar languages or when the book was at

24. I am indebted to Johns (1996; 1998a, 380–443) for the approach taken in the following section.

25. Combe 1843, 2:21–27.

an odd angle or upside down. Making meaning out of these letters involved the interplay of all the faculties. In dealing with the natural phenomena that were the subject of *Vestiges*, the crucial phrenological organs were Individuality, Eventuality, Comparison, and Causality, each of which had to be brought into play. Rely only on the first three, and "the universe might appear as nothing more than an assemblage of objects undergoing changes; but Causality adds a perception of the adaptation of means to produce ends." The *Phrenological Journal* took pride in the fact that *Vestiges* reinforced this vision of mental action.[26]

Phrenologists distinguished between two kinds of reading. The first gratified the feelings and involved novels, poems, and works of pious sentimentality. Reading such works was akin to dram drinking, an artificial stimulus to those mental faculties that already predominated in the brain, raising them into "fervid and irregular action." In contrast, the second type of reading involved the philosophical study of human nature and the duties of life—works such as the *Constitution of Man* or *Vestiges*. Such reading cultivated the intellect, stimulating the weaker faculties and subduing those that were too active already. As Combe explained in an early letter of advice: "difficulties are met with, which the intellect has to struggle with to overcome, but this struggle gives it strength; the subjects are often least pleasing at the first, but become interesting in proportion as they are understood. The mind in such studies strikes out multitudes of ideas of its own, and feels itself increasing at once in its stores of acquired ideas, and in its capacity to combine them and originate others."[27]

Combe directed these remarks at Fanny Kemble, whom he thought suffered peculiar disadvantages as a reader because she was a woman. Phrenological studies showed that the brains of women weighed, on average, four ounces less than those of men.[28] Fortunately, Kemble had "a favourable combination of faculties," allowing her to act and think in a superior way by instinct alone. But exercising the intellectual faculties through reading about the laws of nature would enable her to advance still further. Causality might be deficient, but the right kind of reading—especially in an enticing narrative form—would bring it into healthy action. This would convince her of phrenology's validity and inform her portrayal of classic Shakespearean roles with a scientific understanding of the human mind. Combe could recommend *Vestiges* both because of what it said and because the act of reading would provide appropriate mental exercise, even if the reader resisted its message. From the perspective of religion, no reader with a balanced range of faculties could fail to admire the author's "reverential spirit."

26. Ibid., 2:39; [Combe], "Explanations," *Phrenological Journal* 19 (Apr. 1846): 159–75, at 161.
27. G. Combe to F. Kemble, 23 July 1830, in Gibbon 1878, 1:230–31.
28. Combe 1843, 1:114.

Kemble based her own reflections on a different and less rigid sense of the physiological experience of reading. She expressed her reaction in a letter to her confidant, Harriet St. Leger, as a kind of horrified fascination, the combination of conflicting emotions for which she was celebrated in polite society: "Its conclusions are utterly revolting to me,—nevertheless, they may be true." The facts in *Vestiges* were of "a thousand times more interest than the best of novels," but Kemble was disgusted by its reliance on material causation and its conclusions about the origins of humanity.[29] The emphasis on the future progress of the race somewhat reconciled her to the book, for it resonated with her own hopes for human perfectibility. At the same time, Kemble felt these views were not a consequence of the rest of the author's argument, which seemed poorly constructed and illogical.

While valuing Combe's advice, Kemble was as little convinced by his phrenological theory as by *Vestiges.* In this she was not alone. The marginal position of the phrenologists within Edinburgh meant that their model of reading was widely ridiculed. Most commentators shared Combe's admiration for Scottish common sense philosophy,[30] but interpreted it in an antiphrenological way, arguing that "organs" such as Wit, Causality, and Veneration were silly, hypothetical, and at least potentially irreligious. Like Combe, they had to come to terms with the fact that the ideal reader ought also to be a Christian. Combe, however, thought this issue should be tackled only when the mind was properly exercised and aware of the importance of the natural laws. Only at that point would the reader be ready to receive higher forms of moral and religious instruction, and this task he left to the clergy. Combe's schemes for national secular education, in which the Bible was to be read only after a thorough program of mental exercise, developed from his phrenological ideas on reading.

For evangelicals, in contrast, the Bible was transcendently the most important of all books, from the earliest stages of literacy and missionary education to the most advanced scientific and scholarly inquiries. What, then, did they think happened in the process of reading? How did the testimony of any book, and particularly the Bible, carry conviction (fig. 8.3)? The best way to approach these questions is through the writings of the influential Free Church leader, the Reverend Thomas Chalmers, who commented extensively on problems of perception, the philosophy of mind, and the interpretation of testimony in relation to the Bible.[31]

29. F. Kemble to H. St. Leger, 14 Nov. [1847], in Kemble 1882, 3:242.

30. Combe called Thomas Brown "the sun of metaphysics," but relied on an interpretation of his work rejected by those who did not accept the principles of phrenology; see Gibbon 1878, 1:166–68.

31. On Chalmers's theology, see esp. Hilton 1988, Topham 1999.

8.3 Hugh Miller reading, c. 1854. NLS 10774, no. 80.

Evangelical discussions of reading begin with light. In Genesis light appears as the first created of all things, made by God to be perfectly adapted to the needs of man. The light of nature, as Chalmers wrote, came upon mankind slowly, just as the spiritual light conferred by the Bible came as a series of gradual revelations. "These truths . . ." he wrote, "may be regarded as so many objects on which visibility has been conferred by so many successive communications of light from heaven."[32] The truth of revelation, like the printed letters on the page, was made manifest through the agency of God's provision of light. The light needed to interpret the Word was very bright and near, but the spiritual blindness of humanity kept mankind in darkness:

> Let the organ of discernment be only set right; and the thing to be discerned will then appear in its native brightness, and just in the very features and complexion which it has worn from the beginning, and in which it has offered itself to the view of all whose eyes have been opened by the Spirit of God, to behold the wondrous things contained in the book of God's law.

32. Chalmers 1835–42, 22:12.

It was as though, Chalmers continued, a "little tegument" blocked the channel of communication between the reader and the visible truth. Cut this by "a medicating process upon your own faculties," and belief would follow as a matter of course.[33] The effects of reading were to be judged, not in terms of the achievement of a harmonious balance of the faculties, but through conviction of scriptural truth.

Drawing on those interpretations of common sense philosophy that Combe rejected, the Scottish evangelicals argued that the human mind was so framed that a moral character or disposition could be traced in a rudimentary form from a very early age, even before the "embryo intellect" could read.[34] Conscience, operating in the human mind as a kind of moral governor, meant that all people were in a position to receive the Bible's message. The felt supremacy of conscience led believers, through a single inference, to a conviction of the existence of God. This "master faculty" of conscience was a book of divine testimony planted in the mind, what Chalmers liked to call the "portable evidences" of Christianity.[35] Conscience was an observable fact just as much "as the regulator in a watch is a thing of observation."[36] All people—both men and women—possessed this "master faculty" of conscience; it was only waiting to be brought into effective action, either by preaching or (what was far preferable) the actual reading of the Bible.

The reading of all books other than the Bible was undertaken to minister toward salvation. Learning in itself was nothing: "The sacred page may wear as hieroglyphical an aspect to the lettered as to the unlettered."[37] But once this was understood, books could aid the Holy Spirit in its work of clarifying perception. The reading of scientific works, like the close investigation of nature itself, was encouraged when it contributed to this end, "to move away that film from the spiritual eye."[38] Chalmers stressed that nature and revelation spoke of their common author, just as the study of mental philosophy could strengthen the internal argument for Christianity's truth. "There is naught either in true poetry or in true philosophy," he wrote, "that is adverse to revelation."[39]

Science was thus not a substitute for Scripture, but a means of inculcating the humility that could lead to approaching God's Word in the correct frame of mind. The laws of association of ideas—what the common sense philoso-

33. Ibid., 22:27, 28.

34. Chalmers 1847–49, 7:69. The ambiguities involved in drawing on common sense philosophy for an evangelical philosophy of mind are discussed in Rice 1971.

35. Chalmers 1835–42, 4:171.

36. Ibid., 1:311.

37. Ibid., 9:15.

38. Ibid., 22:25.

39. Chalmers 1847–49, 9:xxxvi–xxxvii.

pher Thomas Brown had termed the "law of suggestion"—could reinforce not just connections between individual ideas, but between tendencies of thought. Baconian induction provided a template for spiritual understanding:

> Give me the truly inductive spirit to which modern science stands indebted both for the solidity of her foundation and for the wondrous elevation of her superstructure, and this, when transferred to the study of things sacred . . . would infallibly lead, in the investigation, first of the credentials, and then of the contents of revelation, to the firmer establishment of a Bible Christianity in the mind of every inquirer.[40]

Sound theology, like sound philosophy, had to be based on facts such as those in what Chalmers recognized as the "reliable testimony" of Scripture.

How then did Chalmers and the other Scottish evangelicals believe that we should approach the work of reading Creation? The answer to this question can be answered by looking at their commentaries on the book of Genesis. The danger was always that biblical phrases would be repeated without thought for their meaning. Rote learning could breed inattention as the verses gained "mechanical currency" in the reader's mind. "My God," Chalmers prayed in his Sabbath Scripture readings on Genesis, "let the idea of Thee as my Creator come to me not in word only but in power. . . . And O may Thy Spirit, who caused this world of beauty and order to emerge from a chaos, operate with like effect on my dark and turbid and ruined soul." Only through Christ, "'the light that shineth in the heart'" could salvation come.[41]

The Reverend Robert Candlish of Free Saint George's in Edinburgh made a similar point in his celebrated Genesis commentary of 1843. Do not read the Bible's first verses, he warned, thinking about astronomy or geology, about resolving difficulties, nor about the laws of nature. These could come later. Start with the simple, extraordinary fact that God was speaking to you, directly in the pages of his book:

> Thus, in a spiritual view, and for spiritual purposes, the truth concerning God as the Creator must be received, not as a discovery of our own reason, following a train of thought, but as a direct communication from a real person, even from the living and present God. This is not a merely artificial distinction. It is practically most important. Consider this subject of creation in the light simply of an argument of natural philosophy, and all is vague and dim abstraction. It may be close and cogent as a demonstration in mathematics; but it is as cold and unreal; or if there be emotion at all, it is but the emotion of a fine taste, and a sensibility for the grand or the lovely in nature. But consider the momentous fact in the light of a

40. Ibid., 7:250.
41. Ibid., 5:1–2.

> direct message from the Creator himself to you. Regard him as standing near to you, and telling you, himself, personally, all that he did on that wonderful week. Are you not differently impressed and affected?[42]

The Bible's opening words could be believed, Candlish said, as "the direct assertion of a credible witness" and taken in with the humility of a child. Once fully realized, this idea would enhance the reader's understanding of the great message of salvation told in the rest of the Bible. If we start from nature, Candlish said in his famous sermon "Paul Preaching at Athens," we approach God as a dim and vague abstraction. If we begin with Christ, and then go into nature, "we see the God of judgement blazing everywhere."[43]

By this standard the God of *Vestiges* was distant, dry, and cold. Evangelicals in Scotland were no more fooled by references to a deity than were their counterparts in England. The *North British Review* did not ask the author to renounce his infidel opinions, only to stop parading atheism under a Christian banner:

> We know the danger, and feel the cruelty of rashly judging a neighbour's heart; but if that heart is too liberal of its issues, we may judge of the fountain by its stream. If it is seen beating through the skin, we may at least count its throbs. The author whom we pity and rebuke, has given full vent to his inmost thoughts on the exciting subjects of the origin, the condition, and the destiny of his species. He confesses that they are hostile to "existing beliefs, both philosophical and religious." He allows that the "collision" may be "vexatious," and he knows that they are utterly irreconcilable with those cherished truths which are the safeguard of states, and the best securities for domestic and individual peace—the only truths which restrain in temptation—console in sorrow, and smooth the rugged passage to the grave.[44]

The quarterly in which these comments appeared, which had been founded with the support of the Free Church, echoed the *Edinburgh*'s fears that the book was continuing to make its way among readers in London's drawing rooms and boudoirs. The review acknowledged that *Vestiges* did not advocate atheism, but stressed that unprepared readers were being led in that direction.[45]

As in other cities, young men from the lower middle classes and women were at greatest risk. Both lacked the masculine strength of mind to detect fallacious reasoning. The evangelical publisher on Princes Street, Johnstone and Hunter, summed up the problem in its prospectus for a series of cheap books, to be called the Christian Athenaeum, which would provide the right kind of

42. Candlish 1843, 1:13.

43. Candlish 1859, 129–58, at 155–56.

44. [D. Brewster], "Explanations," *North British Review*, Feb. 1846, 487–504, at 503.

45. [D. Brewster], "Vestiges of the Natural History of Creation," *North British Review*, Aug. 1845, 470–515, at 503.

reading material. This noted that the shorter working hours had given "young men in shops, warehouses, offices, and other places of business" more time for reading; similarly, ladies no longer wasted their time in "minute and profitless needlework"—a work of the hands only—but were applying their "intellectual and moral powers."[46]

Evangelical publishers and authors alike would have been shocked to learn that Combe had recommended *Vestiges* to Fanny Kemble. Both sexes had the faculty of conscience, but women were characterized by a "yielding submission" and "soft and gentle temperament" that could easily be led astray; Kemble, as an actress, was already in special danger. For a Christian nation, the effects of education on families could be disastrous. As the *North British* said: "It would augur ill for the rising generation, if the mothers of England were infected with the errors of Phrenology: it would augur worse were they tainted with Materialism."[47]

The fundamental issue was divine grace and the experience of conversion. Because the *North British* was not overtly sectarian, its review appealed "to Reason and not to Faith,"[48] aiming to convince readers on factual grounds similar to those Sedgwick had used in the *Edinburgh*. The *North British* went on to quote Scripture, but unlike the *Edinburgh* it had no apology to make for doing this:

> If it has been revealed to man that the Almighty made him out of the dust of the earth, and breathed into his nostrils the breath of life, it is in vain to tell a Christian that man was originally a speck of albumen, and passed through the stages of monads and monkeys, before he attained his present intellectual pre-eminence. If it be a received truth that the Creator has repeatedly interposed in the government of the universe, and displayed his immediate agency in miraculous interpositions, it is an insult to any reader to tell him that that being slumbers on his throne, and rules under a "primal arrangement in his counsels," and "by a code of laws of unbending operation."[49]

The advocacy of natural law was unobjectionable, for evangelical Presbyterians recognized that readers could be drawn toward God by seeing his hand in the ordinary course of nature. It was just that in *Vestiges* there seemed nothing more, no indication that "any law was thundered from Sinai, or preached from Mount Tabor."[50]

46. *A Catalogue of Works Published by Johnstone and Hunter Edinburgh* (Edinburgh: Johnstone and Hunter, 1851), bound with CUL 8100.d.783, including "Prospectus" for the Christian Athenaeum.

47. [Brewster], "Vestiges," 503.

48. Ibid., 484.

49. Ibid., 474.

50. Ibid., 503.

Spiritual Sanitation

"We seem," one of the Scottish evangelical monthlies warned, "to be standing on the verge of a vast volcano, ready to explode, and overwhelm us with terrible destruction."[51] The proper conduct of reading was central to avoiding spiritual catastrophe. To see what this involved, take the most militant and best-organized body of evangelicals in Britain—the Free Church of Scotland—and how it responded to *Vestiges.* David Octavius Hill's painting of the climactic moment in the founding of the Free Church, "Signing of the Deed of Demission," illustrates the importance that evangelical Presbyterians attributed to the Word (fig. 8.4). The Deed, through which a third of the clergy in Established Church of Scotland gave up all claims on and positions in that church, is only the most prominent document in the huge range of print and manuscript material crowding the picture. Chalmers preaches from the Bible, the table is strewn with papers and letters, and many of the men are shown reading or with books in their hands. The geologist and editor Hugh Miller, a shepherd's plaid over his shoulder, writes for the Free Church newspaper the *Witness,* copies of which lie in a pile before him. David Welsh, to the left of Chalmers, holds the protest he had read at the general assembly of the Church of Scotland; within a year, he would be the *North British*'s first editor.

The *Vestiges* sensation offered a perfectly timed opportunity to display the Free Church's credentials as defender of the faith scientific. Rather than constructing a natural theology based on reason, without reference to revelation, the Free Church united the two in what historians have called a "theology of nature."[52] This broader project, typical of many evangelical churches, encouraged commentary from those whose expertise was theological rather than scientific. Although discussion of the Free Church response to *Vestiges* has always been limited to a handful of scientific authors, the range of those who spoke out was much wider. Many believed that the book offered an opportunity to cement the alliance between Calvinist religion and modern science. The classicist John Stuart Blackie discovered this the hard way at a philosophical dinner in Edinburgh in the summer of 1845. A self-confessed heretic who followed Combe and Francis Newman in rejecting orthodox religion, Blackie found himself alone in defending *Vestiges* against the men of the Free Church.[53]

51. "Periodicals for the People," *Lowe's Edinburgh Magazine,* n.s., 1 (Jan. 1847): 192–200, at 200.

52. Topham 1999, 144–45, with reference to earlier work by John Brooke. On the Free Church and science, see Baxter 1985, Baxter 1993, Livingstone 1987, Smith 1998, Topham 1999. Much of the literature focuses on individuals especially active in scientific practice, rather than attitudes within the Free Church more generally.

53. J. S. Blackie to E. Blackie, 11 Aug. 1845, in Walker 1909, 118.

8.4 "Signing the Deed of Demission." Engraving after part of David Octavius Hill's oil painting. Brown 1884, frontispiece.

From the New College, built like a fortress on Castle Mound, the Free Church armed ministers for combat against theories of development falsely based on science. The New College assembled a remarkable teaching staff, all of whom lectured, spoke, or wrote against *Vestiges* and related works in the 1840s and the 1850s. The Reverend John "Rabbi" Duncan, the professor of Hebrew and a former missionary in Jerusalem, told students in conversation that the scientific evidence was insufficient to support so momentous a conclusion as the emanation of spirit out of matter.[54] Chalmers, the college's principal and professor of divinity, never mentioned *Vestiges* in his writings, but implicitly targeted the work in the *North British*.[55] The Reverend James Buchanan, a distinguished preacher, was more forthright. In 1845 Buchanan was appointed professor of apologetic theology at the New College and began lecturing against *Vestiges*. His two-volume *Faith in God and Modern Atheism Compared* (1855) was based on these early lectures and published by his son, who

54. Knight 1907, 146–47.

55. [T. Chalmers], "Morrell's Modern Philosophy," *North British Review* 6 (Feb. 1847): 271–331, at 315.

took the unusual step of issuing five of the chapters as shilling pamphlets, including one on "development." Buchanan acknowledged that criticizing a work on scientific grounds was not his "proper province," but he summarized the issues as they affected the case for belief in God. Development was inherently incapable of proof and had been undermined by recent discoveries; and even if shown to be probable, the theory need not lead to atheism.[56]

In establishing the New College, the Free Church debated how far its limited resources should be applied to support instruction in the sciences. The issue came to a head at the church's assembly in September 1845 at Inverness in the Scottish Highlands. Some ministers, led by Candlish, gave priority to setting up a larger national network of Free Church colleges; others, including Chalmers and Miller, argued that the focus should be on a single center in Edinburgh, which could offer advanced theological teaching and expertise in areas, including science, that would not otherwise be taught. In the end, Candlish's position prevailed, but a single chair in natural science was created for the New College. In this case, as at the ancient universities in England, *Vestiges* helped make the case that science needed a place in Christian education, to "silence the infidel prattle."[57]

The distinguished naturalist appointed to the post, the Reverend John Fleming, had written widely on zoology and geology. At the start of his speech acknowledging his appointment, Fleming contrasted the positive attitude of the Free Church of Scotland with the Anglo-Catholic High Church party in England, which was "crushing science." The materialism that infected *Vestiges,* he argued, showed that these sciences needed to be cultivated in a truly Protestant context. Not wanting to be contaminated by association, Fleming claimed never to have read the book. The extracts he had seen, however, demonstrated to him that it was "full of the grossest materialism" and written by a deluded visionary. Some thought metaphysics alone was enough to combat such heresies and that scarce funds should not be wasted on the sciences. Fleming disagreed. If the natural science chair was properly funded, he said, every Free Church cleric "would be better qualified to be an expounder of the Word of God—better qualified to hold intercourse with his parishioners—and better qualified during his walks to hold intercourse with his maker."[58]

Sermons could be used to instill a sense of the proper relations between science and faith to Free Church believers, most of whom attended services at

56. Buchanan 1855, 1:420–510, esp. 428. For the shilling pamphlets, see the publisher's prospectus, Bodleian Library, John Johnson Collection, prospectuses box 17, item 1003.

57. Baxter 1993, 108; "Meeting of the Free Church Assembly," *Witness,* 30 Aug. 1845, 3; [Brewster], "Vestiges," 505; and *Scottish Guardian,* 23 Sept. 1845, 1.

58. *Witness,* 30 Aug. 1845, 3.

least once and usually twice or more every Sunday. During the rest of the week they could read an extensive range of periodicals.[59] The most authoritative, the *North British*, was strictly speaking not a Free Church organ; instead it claimed to offer a perspective more thoroughly imbued with Protestant Christianity than the existing quarterlies.[60] It targeted a wealthy readership with essays that attempted to be substantial and authoritative. Most of its numerous articles on the *Vestiges* controversy were by David Brewster, celebrated for his optical researches and prolific journalism. "No writer of the present age unites a higher degree of literary ability to exact science," noted the *Witness;* "no writer of our own country unites them in a degree equally high."[61] Sales had been good (about two thousand per issue, one-fifth of the *Edinburgh*'s ten thousand), but confined largely to Scotland, and Brewster's articles were a major factor in giving it a presence south of the border too. His demolition of *Vestiges* was especially widely read in America, where publishers reprinted it with most editions from late 1845 onward.

The *North British* insisted, more explicitly than the *Edinburgh* ever could, on a deity who occasionally subverts nature's laws through special providence. On a human scale, such interventions were rare, but on a geological one they occurred often—every time a new species was created, for example. In his last essay for the quarterly, Chalmers stressed the need for miracles:

> It is when new systems emerge from the wreck of old ones, and from the ruins of a former catastrophe there is built up another modern habitation, and peopled with new races both of animals and vegetables—it is then that we demand the interposal of a God. Whence did those new genera and species come into being? Nature gives no reply to this question; and, though ransacked throughout all her magazines, the secret of these actual and present, and altogether new organisms, is nowhere to be found.[62]

Brewster's anonymous slashing of *Vestiges* in the *North British* also supported this notion of miracles. It even advocated a literal notion of "a mighty deluge" and looked to the day when submarine deposits "may yet give up their dead, and exhibit to some inquiring pilgrim the history of his race written on stone." The review did not claim that fossilized archbishops and kings had been found, but

59. North 1989 provides a virtually complete and exceptionally useful guide to surviving nineteenth-century Scottish periodicals.

60. Shattock 1982.

61. [H. Miller], "The General Assembly," *Witness*, 27 May 1841; also in Miller 1861a, 306. On Brewster, see the essays in Morrison-Low and Christie 1984, esp. by Baxter and Brock.

62. [Chalmers], "Morrell's Modern Philosophy," 315. Chalmers's views on miracles are discussed in Hilton 1988, 84–85, 108–14.

expected they would be.[63] It is hard to imagine Sedgwick hoping to come upon such specimens on one of his field trips.

Most periodicals associated with the Free Church did not attack the book directly, preferring to present science without reference to the pestilent airs of infidelity.[64] The nearest approach to coverage comparable to the *North British* was the monthly *Lowe's Edinburgh Magazine* (f. 1846), sponsored by the Free Church leaders to give a promising young writer from Dundee, James McCosh, an editorial post in Edinburgh. McCosh's first book, *The Method of the Divine Government* (1850) offered a sophisticated justification for the Free Church position in natural theology, based on efforts to bring the common sense tradition into line with Continental philosophy and the writings of Immanuel Kant. Unlike the more extreme English evangelicals, who often rejected these developments, the men of the Free Church brought the latest philosophy to bear on Christian theology. McCosh's attack on *Vestiges* continued the campaign that he had begun as a periodical editor. Like most Free Church journals, *Lowe's* allied itself with a populist, liberal Whig political line and condemned those who would foster class antagonism.[65]

What of the *Witness* (fig. 8.5), the semiweekly newspaper associated with the Free Church? In a bold move, Candlish had hired as editor the virtually unknown Hugh Miller, who had worked as a stonemason and then as a bank clerk before becoming the "sledge-hammer" for the evangelical party. A contributor recalled the tiny *Witness* office on publishing nights: Miller toasting cheese on the fire shovel, boiling huge pots of coffee, and putting the final touches on his editorials.[66] He had begun to establish a reputation in geology with his articles, especially after these were collected into a book, the *Old Red Sandstone* (1841). Miller had opposed Lamarckian evolution in that work, but only wrote briefly against *Vestiges* in the *Witness*, calling it "one of the most insidious pieces of practical atheism that has appeared in Britain during the present century."[67]

63. [Brewster], "Vestiges," 515.

64. For examples of brief comments or reviews in the Scottish religious press, see *Edinburgh Christian Magazine* 10 (Feb. 1859): 350–52 (review of book by P. J. Gloag, with comparison of *Vestiges* and socialism); *Presbyterian Review and Religious Journal* 19 (Oct. 1846): 539 (review of Mason 1845); "Creation or Development," *Palladium* 1 (1850): 348–61 (review of Miller 1849 and Anderson 1850); *Presbyterian Review and Religious Journal* 21 (July 1848): 361–62 (review of Oken, *Elements of Physio-Philosophy* [1847]); and *Torch* 1 (10 Jan. 1846): 30–32.

65. "Periodicals for the People," *Lowe's Edinburgh Magazine* 1 (Jan. 1847): 192–200, at 197; McCosh 1850. McCosh's theological career is surveyed in Hoeveler 1981 and Livingstone 1992.

66. Rainy and Mackenzie 1871, 134–38. On Miller and his contributions to the *Witness*, see Shortland 1996b, 287–369, although the attributions in the bibliography need to be used with caution. For Scottish newspapers more generally, see Cowan 1946, esp. 271, which lists a few further reviews of *Vestiges* in the Glasgow newspapers.

67. [H. Miller], "The Physical Science Chair," *Witness*, 17 Sept. 1845, 2–3.

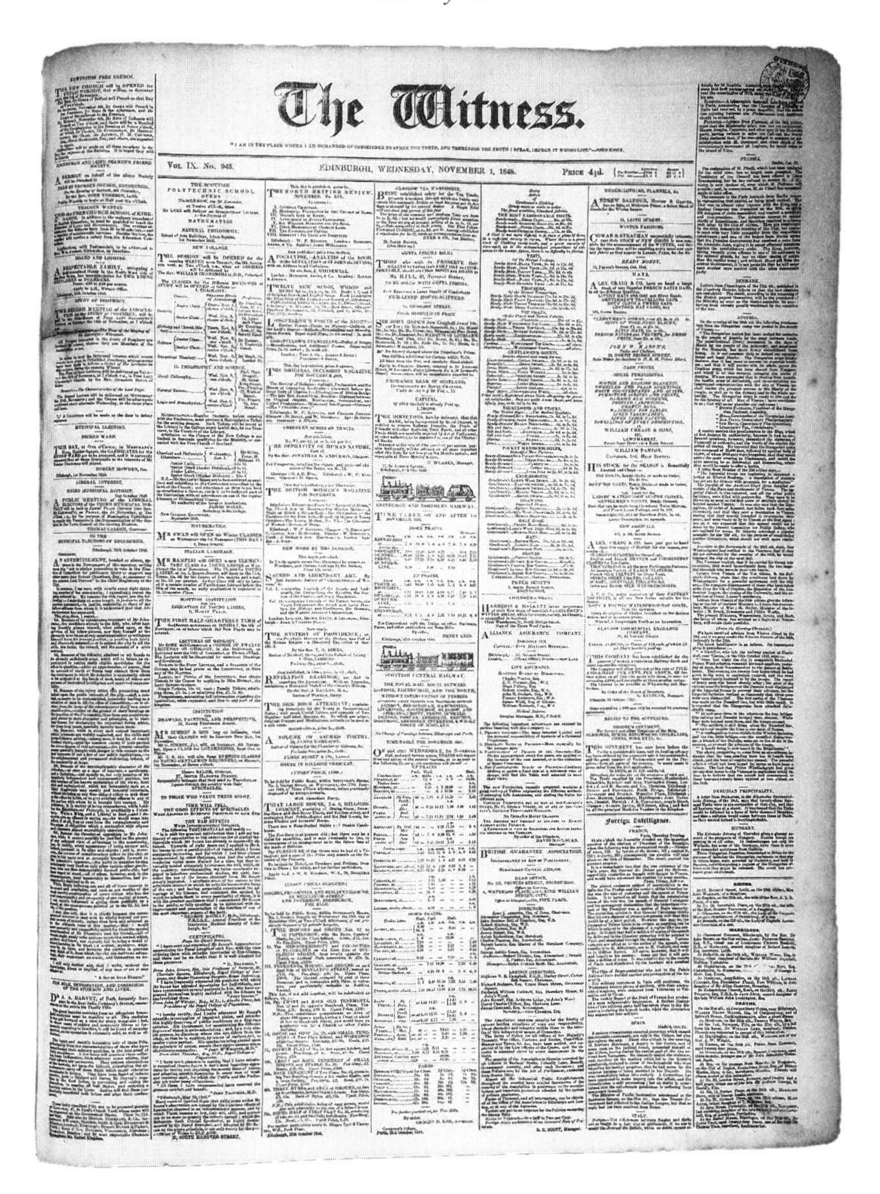

The Witness.

Vol. IX. No. 945. EDINBURGH, WEDNESDAY, NOVEMBER 1, 1848. Price 4½d.

8.5 *The Witness,* 1 Nov. 1848, 1. This issue included the letter that forced Chambers to withdraw from the election for lord provost.

Miller waited to launch a book-length attack until the cheap editions of *Vestiges* began to sell. His *Foot-prints of the Creator* (1849) attempted to reclaim lost readers through an engaging discussion of a giant and very early fossil fish. Unlike most of his writings, *Footprints* was not serialized in the *Witness,* presumably because parts were necessarily "prolix" and "repulsive."[68] However,

Miller used all of his literary skill to make esoteric details about fish fossils accessible to readers seduced by the cheap edition of *Vestiges*.[69] *Footprints* was printed (on the *Witness* presses) in the same small size as the work it attacked, and sold five thousand copies in the next four years. *Footprints* was praised in all the evangelical literature in Scotland and England as the most convincing refutation available. It went out of print in 1853 after its evidence about the first appearance of vertebrate fossils proved faulty.

The central issue was God's constant watchfulness over the creation and his ability to act in any way. Miracles were central to the Calvinist tradition, in which natural theology had absolute limits determined by revelation.[70] Interpretations of the Flood and other biblical events might change in the light of science, and the evolutionary development of the body might even be true; but one miracle—the creation of the human soul—remained inviolate. While writing *Footprints,* Miller had read a newspaper account of a lecture by the geologist David Page that made the development hypothesis indifferent to Christianity's truth.[71] Miller was outraged. If Adam merely took the first human step in an ongoing upward march, then there was no need for that "second Adam," Jesus Christ, to restore mankind from an abject state of sin. Readers were told that the inexorable upward progress of the development hypothesis denied the need for the miracle of Christ's atonement on the cross, the central tenet of Calvinist theology.

The history of mankind—and of nature—was not progressive, but a story of degradation. The opening vignette of *Footprints* expressed this ironically: a putto hammers a rock to find a grinning phrenological skull (fig. 8.6). Miller thought the gestation of the human brain well worth investigating, despite its discovery by transcendentalist infidels in Germany and France, and pursued their idea that man might represent the summation of all creation. He denied, however, any parallels with the history of life: "it is a fact of *foetal* development, and of that only."[72] (As the later editions of *Vestiges* pointed out, the *Old Red Sandstone* had expressed a very different view.) Miller underlined the absence of progress by describing a bizarre armored fish, the *Asterolepis,* at Stromness on the shores of the Loch of Stennis in the Orkney Isles. With its immense size and complex cranial buckler (fig. 8.7), the *Asterolepis* could be used to suggest that the onward narrative of geological progress in *Vestiges* was mistaken. Above all, the last days of human history would not be an advance on the present, but would represent the millennial transformation, the final triumph of Christ over

68. Miller 1849, vii. Brooke 1996 and Henry 1996 discuss Miller's theology in *Footprints.*

69. Miller 1849, vii.

70. This point is stressed in Henry 1996.

71. Miller 1849, 17–18. Miller referred politely only to "a gentleman of Edinburgh"; that Page was the lecturer is clear from the *Fifeshire Journal,* 21 Sept. 1848, 4.

72. Miller 1849, 291; Richards 1990, 138–39.

WHEN engaged in prosecuting the self-imposed task of examining in detail the various fossiliferous deposits of Scotland, in the hope of ultimately acquainting myself with them all, I extended my exploratory ramble, about two years ago, into the Mainland of Orkney, and resided for some time in the vicinity of Stromness.

8.6 Vignette from the opening page of *Foot-prints of the Creator.* Note the autobiographical opening to Hugh Miller's text, stressing science as a form of self-culture. Miller 1849, 1.

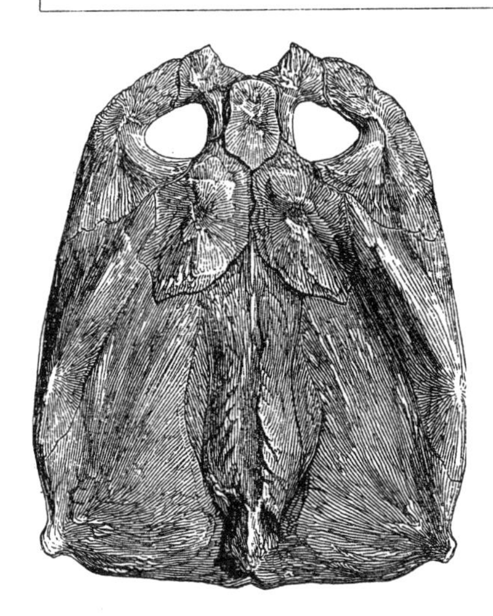

8.7 Inner surface of the cranial buckler of *Asterolepis,* showing the interior traces of the advanced brain of this Old Red Sandstone fish. Miller 1849, 75.

sin and death. The possibility of salvation set man apart from the mere animal vitality of the brutes. If transmutation were true, there was no reason we should not be "by nature *atheists,*" like foxes, dogs, or wolves. The failure to recognize this was the great desolating blindness of *Vestiges.*

Against this hellish nightmare, the *Witness* posited the Free Church as the basis for a godly commonwealth that would encompass all the Scottish people. The *Witness* shared Chalmers's hope that the Disruption would mark the rebirth of a national church based in local charity and communal solidarity. Spiritual renewal—starting with the Bible—would end crime, poverty, and destitution.[73] The problem involved the political economy of reading, just as ridding Edinburgh of cholera was a problem in the political economy of

73. Brown 1982, esp. 282–379; for the *Witness,* see Macleod 1996, 200.

sanitation. As the *Witness* said in a revealing comparison, "the lower levels of society" had sunk into a "miasmatic marsh," out of which "poor law assessments, fierce revolutionary outbreaks, plagues, and pestilence, threaten to arise and envelope in indiscriminate ruin the classes above."[74] Books like *Vestiges* drew the people down into this moral swamp. Its readers exchanged a place in God's kingdom, Miller wrote, for "a horrid life of wriggling impurities, originated in the putrefactive mucus."[75] The Free Church's responses to *Vestiges* were works of spiritual sanitation, attempts at "draining and purifying the bog."[76]

Evangelical Unions

The Free Church prided itself on leading the onslaught, but opposition to *Vestiges* was widespread in the fissiparous world of Scottish religious life. A plethora of Presbyterian denominations, notably the United Secession Church, were left over from withdrawals from the Establishment prior to the Disruption. There were also Dissenting sects (the Scottish branches of the Congregationalists, Methodists, and Baptists) that favored getting rid of any form of Presbyterian Establishment altogether. Antidotes to *Vestiges* from these diverse groups ranged from the briefest of penny tracts to elaborate treatises like the Reverend Graham Mitchell's *Young Man's Guide Against Infidelity* (1848). This fat manual, intended for young travelers and others imperiled by a "shipwreck of faith," pitted science, scriptural natural history, and biblical miracles against the skeptics. An entire chapter offered a rebuttal to *Vestiges.*[77]

The "Voluntaries," as the Dissenters were known, stressed conversion as a personal decision for Christ, without imposition from a state-supported church.[78] Independence was especially prized by Congregationalists, who rejected all forms of formal organization beyond the individual church. Like their English counterparts, the Scottish Congregationalists were among the most active opponents of *Vestiges.* Their monthly magazine, for example, had a substantial review, surveying the evidence against the book, arguing against the separation of natural law from divine will.[79]

In Edinburgh, Congregationalist opposition to *Vestiges* was spearheaded by a layman. George Wilson was Regius Professor of Technology in the univer-

74. [H. Miller], "The People their Own Best Portrait Painters," *Witness,* 5 Dec. 1849, 2.

75. Miller 1849, 313.

76. [H. Miller], "The People their Own Best Portrait Painters."

77. Mitchell 1848, 32–52.

78. Brown 1997, 690–91. A similar position was taken by those within the Free Church who disagreed with Chalmers's vision of a godly commonwealth and welcomed the Disruption as marking the end of church establishments in Scotland.

79. J. M——m, "Vestiges of the Natural History of Creation," *Scottish Congregational Magazine,* n.s., 5 (Sept. 1845): 409–17.

RELIGIO CHEMICI.

ESSAYS

BY GEORGE WILSON, F.R.S.E.

Late Regius Professor of Technology in the University of Edinburgh.

All things were made by Him; and without Him was not anything made that was made. In Him was life; and the life was the Light of Men.—JOHN i. 3, 4.

London and Cambridge:

MACMILLAN AND CO.

1862.

8.8 Science in the light of the Word: Noel Paton's title page engraving to George Wilson's posthumous *Religio Chemici: Essays* (1862).

sity, a historian of science, and a chemist who followed Christ's "own bleeding footsteps to the rest that remaineth for the children of God" (fig. 8.8).[80] He wrote anonymously against the work in the Dissenting *British Quarterly* and may have written the long reviews of the book and its sequel that appeared there.[81] In the summer of 1846, Wilson gave four sixpenny lectures against *Vestiges* to the Edinburgh Young Men's Society. As he told his mentor, "I have

80. Wilson 1860, 459.

81. [G. Wilson], "Chemistry and Natural Theology," *British Quarterly Review* 7 (Feb. 1848): 204–38; see also "Vestiges of the Natural History of Creation," *British Quarterly Review* 1 (May 1845): 490–513; "The Law of Development in Nature," *British Quarterly Review* 3 (Feb. 1846):

too much wrought only at science and literature, hoping thereby to secure a position which would enable me to serve Christ effectually. But many things warn me that my life will be a short one, and that what I can do, must be done swiftly." The young organizers praised Wilson's exposure of "the scientific errors, false reasonings, and infidel tendencies" of the work. Six publishers offered to issue the lectures, but Wilson never completed the revisions.[82]

By the start of the next decade, there was no shortage of book-length antidotes in Scotland. Following in the track of Miller's *Footprints,* clergymen from different denominations issued works that combated transmutation. The Reverend David King, minister of Glasgow's United Secession Church, wrote *Principles of Geology* (1850), which reconciled the sciences with revealed and natural religion. King's small volume went through four editions and probably sold as many copies as *Footprints* during this period.[83] The *Principles* was published by Johnstone and Hunter in the Christian Athenaeum series, which also included a work of botanical theology that condemned progressive development.[84] The author, John H. Balfour, was not a member of the Free Church but maintained close ties with it, and his election to the chair of medicine and botany in the University of Edinburgh was seen as a major evangelical victory.

At least a few who remained in the Established Church of Scotland also wrote against *Vestiges. The Course of Creation* (1850) was a handsome octavo by the Reverend John Anderson of Newburgh, which summed up his previous work on the Old Red Sandstone fossil fish. Published by Longman in London, Anderson's book surveyed the geology of Scotland, England, France, and Switzerland, and then turned to theoretical problems, attacking transmutation and estimates that the earth was millions of years old.[85] There were articles, too, in the Establishment press: one monthly with a circulation of several thousand

178–90. For the *British Quarterly,* see Houghton et al. 1966–89, 4:114–25. The two main reviews are attributed (p. 128) to "John David" on the basis of a letter from John Brown to his cousin John Taylor Brown, printed in Brown and Forrest 1907, 61. As John David is otherwise unknown, there may be a mistranscription involved; the articles are very much in Wilson's style and unusual in stressing chemistry. The *British Quarterly* was a new periodical that gave a voice to Dissent among the chorus of quarterlies. Wilson's brief anti-Vestigian comments were reissued in the Traveller's Library, a cheap series sold at railway bookstalls (Wilson 1852).

82. Wilson 1860, 327; "Dr Wilson's Lectures on the 'Vestiges'" *Witness,* 11 July 1846, 2. The advertisement shows that they were delivered on 23, 26, and 30 June and 3 July 1846: see *Witness,* 20 June 1846, 1.

83. King 1850, 1853. King was the physicist William Thomson's brother-in-law; see Smith and Wise 1989, 127.

84. Balfour 1851, vii-viii, 190–92; Fleming 1851. For Balfour's religious affiliation, see "The Late Emeritus Professor Balfour," *Scotsman,* 12 Feb. 1884, 3.

85. Anderson 1850, esp. 325–55, 388–407. The reviewer in the monthly *Edinburgh Christian Magazine* 3 (1851): 75–82, preferred Anderson's work to Miller's, at least for general readers.

had a two-part notice of *Footprints* as part of its program of Sabbath reading.[86] The aim was to protect readers from a common enemy; but the division between the Established Church and the Free Church was too deep and too recent to be easily forgotten. At one level the disagreement was about the relation between expertise and faith. Miller's *Witness* contrasted "*Moderate* science personified in Dr Anderson of Newburgh—a dabbler in geology, who found a fish in the Old Red Sandstone, and described it as a beetle" with "science, not *Moderate,* on the other side, represented by Sir David Brewster."[87] Anderson was willing to forget these hard words by praising *Footprints;* but he also tried to set the record straight, outlining his own version of events surrounding the discovery of the fish fossils. More fundamentally, Anderson argued that natural theology provided proofs of God's existence and a short time span for earth history. Free Church men of science, in contrast, hardly ever made such claims, believing instead in a theology of nature grounded in revelation.[88]

The moderatism associated with the Establishment was not deism, but in the bitter battles of post-Disruption Scotland it could be portrayed as having tendencies that way. *Vestiges* seemed a reversion to the rationalism of the eighteenth century—insofar as it was religious at all. If this was all there was to faith, as Miller put it, then Christianity would be "an idle and unsightly excrescence on a code of morals that would be perfect were it away." Miller attributed such a view to a few members of the moderate party in the previous century,[89] but by implication he spoke of at least some of their modern representatives—those ministers who had not signed the Deed of Demission. Moderate policy, which gave authority over ecclesiastical appointments to secular authorities, could not on this view be squared with evangelical doctrine. *Vestiges* could be used to illustrate the final issue of the spiritual "deadness" attributed to those who had remained within the Establishment.

Evangelicals in Edinburgh did what they could to work together. This was part of a much larger movement toward evangelical union, forced especially by the outrage they all felt toward the parliamentary grant for the Roman Catholic college at Maynooth. Contacts became stronger between the Voluntaries and

86. B. B., "Footprints of the Creator," *Edinburgh Christian Magazine* 2 (May 1850): 56–61; (June 1850): 80–85.

87. [H. Miller], "The Disruption," *Witness,* 20 May 1843, quoted in Brooke 1996, 173; also in Miller 1861a, 476.

88. Anderson 1850, 60–68, 83–89, 335. The *Free Church Magazine* condemned the YMCA lecture at Exeter Hall by John Cumming, evangelical leader of the Establishment-affiliated Presbyterian Church in London, as "the specious glitter of fluent declamation," which tripped over itself in attacking the blunders of the development hypothesis. "Lectures to Young Men," *Free Church Magazine* 8 (July 1851): 208–12, at 210.

89. Miller 1849, 17.

the Free Church in any event, particularly after Chalmers's death. The Free Church press was effusive in its praise of King's *Principles,* seeing the work as proof that at least one minister in the United Secession Church "has not suffered himself to fall behind his age." The *Witness* also approved Wilson's "highly interesting and able lectures" against *Vestiges.*[90] Conversely, the periodicals of other evangelical denominations gave glowing reviews to Buchanan's *Faith in God* and Miller's *Footprints.*[91]

Weapons from all sides could be used in the war against unbelief. Of all the Church of England clerics who dealt with science, the evangelical Sedgwick had the most affinities with the Free Church perspective and was often called into service as an ally, a connection that he welcomed. The minimal reference to revealed religion in his review was never seen as a failure of personal faith, for it was understood that references to the saving power of Christ would not have been allowed by the editor.[92] The *Edinburgh* was notorious among Scottish evangelicals for having "crucified Jesus"[93] in its early years, so that Sedgwick's notice could be taken to reflect an editorial change of heart.

There were signs of a national alliance between evangelical Protestantism, specialist science, and opposition to developmental theories of progress. Evangelical union became an attractive option, motivated by the Maynooth controversy, the weakening effects of the Disruption—and by the threat of infidel uses of science. There were local initiatives along the pattern of the national Evangelical Alliance, which met for the first time in Liverpool in 1846. The most revealing (though short-lived) was the Scottish Association for Opposing Prevalent Errors, founded by a broad-based coalition that met for prayer early on Monday mornings.[94] This informal group had sponsored two thousand copies of the Reverend C. J. Kennedy's *Nature and Revelation Harmonious* (1846). Kennedy, a Dissenter of the United Secession Church, was a respected teacher who lectured on natural philosophy in Paisley. He thought phrenology might be true, but argued against extending natural law to the origin of living beings.[95]

90. [H. Miller], "Dr King's Principles of Geology," *Witness,* 1 May 1850, and the extracts bound at the back of the copy of the second edition of King's *Principles* at CUL 8340.d.250; [H. Miller], "Dr Wilson's Lectures on the 'Vestiges,'" *Witness,* 11 July 1846, 2.

91. See the extracts from fourteen periodical reviews in the publisher's prospectus, Bodleian Library, John Johnson Collection, prospectuses box 17, item 1003.

92. Gillispie 1951, 149–80, makes the classic contrast between Sedgwick and Miller. A. Sedgwick to H. Miller, 3 Sept. 1849, in Clark and Hughes 1890, 2:159–62.

93. *Fife Herald,* 18 Oct. 1849, 794.

94. Cairns 1860, 265–67.

95. See T. Murray to G. Combe, 12 Jan. 1847, NLS mss. 7287, ff. 44–45, for biographical details about Kennedy. Combe 1847b is a reply to Kennedy.

The threat of *Vestiges* within the constellation of heresies was raised at the association's first public meeting in 1847. Assembling on the neutral ground of a hotel, the audience was treated to three hours of speeches against popery, Tractarianism, antispiritualism, socialism, Combe's *Constitution,* and *Vestiges;* all those "holding Evangelical sentiments" could join, although only a few Establishment men appear to have done so. Fear of common enemies brought together men who had not been on speaking terms for years. The Reverend William Cunningham was one of the so-called "wild men" of the Free Church, professor of Church history in Edinburgh New College and a fiery polemicist. He was on the organizing committee, while the Reverend Dr. John Brown, the United Secession Church's professor of exegetical theology, was vice president and chairman. The two men had just been reconciled through the Evangelical Alliance.[96]

"Infidelity is of the spirit of the present," Brown declaimed in his keynote speech, "and of the spirit of the future. It is coming in stronger and stronger."[97] The Reverend Andrew Thomson, also of the United Secession Church, addressed the menace of infidel materialism, including *Vestiges.* He was worried about the superficial attractions that materialism had for "youth in our workshops and untutored minds in our manufacturing districts." Outright infidelity might have had its day, but the same ideas seemed to be returning in new disguises. "Infidelity," Thomson lamented, ". . . like the spider, was hanging out its webs on many a beautiful tree of knowledge, and but all too often, alas! it succeeded in ensnaring the unwary passer by." The peril could be overcome by public debates, expertly written tracts, and above all the learning of men like Brewster, Buckland, and Sedgwick. Only those "mightiest minds of our age" could "repel scientific cavil with scientific reply" and meet the challenge.[98]

The difficulty of pan-evangelical union is evident in the Scottish Association's fate. In May 1848, in the wake of European revolutions and the national Chartist meetings, the association began a statistical survey of infidelity; but this came to nothing, and the group faded away.[99] As with the national Evangelical Alliance, finances never matched ambitions; the association's only fund was a hundred-pound donation from a wealthy layman. The Voluntaries were not very powerful in Scotland, and with hundreds of churches to build, the Free Church was in no position to lead a campaign against all the errors of the day. Those, like Miller, who were skeptical about any alliance that might weaken

96. Scottish Association for Opposing Prevalent Errors 1847. On Cunningham, see Rainy and Mackenzie 1871, 255–56.

97. Scottish Association for Opposing Prevalent Errors 1847, 4.

98. Ibid., 11–12.

99. W. Thomson to R. Paul, 10 May 1848, NLS. mss. 5140, ff. 61–62.

the Free Church claim as the national Establishment tended to stay away. The union against secular liberalism was unsystematic, underfunded, and undermined by sectarian strife.

The Politics of Authorship

Revealing the secrets of a title page was bad manners. It was acceptable to make a passing guess, but attempts to root out someone who did not wish to be discovered were viewed with distaste. Anonymity was a private interest not to be violated unless in exceptional circumstances. These taboos acted with particular force once *Vestiges* began to be condemned in terms that made it clear why the author might want to remain unknown. Thus the *Edinburgh, North British,* and *British Quarterly* would have nothing to do with naming individuals in print, however much their reviewers hated the book.

Many Scottish evangelicals spoke of the author as a "spirit" or a "shade," preferring to think of the volume before them rather than an individual author. Miller made no guesses about the authorship, neither in *Footprints* nor in any of his known correspondence. There were, however, hints; *Footprints* identified *Vestiges* with the "hapless school" of Scottish phrenology, and characterized the author as "a practised and tasteful writer," whose work was "charged with glittering but vague resemblances" best suited to "some light literary game of story-telling or character sketching." Miller respected Chambers, who had helped him at the beginning of his career, too much to try to smear his name. Others within the evangelical party agreed. "Evil surmising," Kennedy told Combe, "is not a practice in which I indulge."[100]

Yet some felt the author needed to be unmasked. Respect for a private interest could not be allowed to override the interest of the public at large. In the case of *Vestiges,* anonymity seemed to be misleading readers and allowing an unknown materialist, perhaps responsible for educating children, to infect society from within. "It may often be as useful to find out the author of an unsound book," one of them said, "as to find the true cause of a disease."[101] There were several elaborate attempts to identify the agent of infection, using techniques of textual comparison honed in biblical scholarship and the Junius controversy. In 1846 the *American Review* compared passages from *Vestiges* with the writings of Isaac Taylor, the author of the *Physical Theory of another Life* (1838). This attribution, like most American discussions, had little impact outside the United States. Serious efforts to nail the author began in the months after the people's edition gave *Vestiges* a wider audience. One visitor to Chambers's house claimed to have seen a suspiciously comprehensive run of editions on the library

100. Miller 1849, 256–57, 263; C. Kennedy to G. Combe, 1 Feb. 1847, NLS mss. 7286 ff. 25–26.
101. *Macphail's Edinburgh Ecclesiastical Journal and Literary Review* 4 (Nov. 1847): 251–66, at 251.

shelves, and on that basis attributed the authorship to "the younger of two literary brothers in Edinburgh, the cojoint proprietors and editors of a well known popular weekly journal." This story (which could be doubted) was first printed in the *Manchester Courier* for 3 November 1847.[102]

Two days earlier, a much more damaging attempt to unmask Chambers appeared in *Macphail's Edinburgh Ecclesiastical Journal,* a monthly supporting the Established Church of Scotland. The author of this unsigned attack had had a vision of the book in a hideous dream:

> It is anything but a bringer of sweet sleep—we have found it rather productive of an opposite effect—of night-mare. We dreamt the other night . . . that we saw Babbage's Machine, mounted stride-leg on an Acarus Crossii, and galloping away from the Author of the Vestiges, who had the charge of them both, and who was vehemently pursuing, in a *solitary fluttering garment.* The scene unaccountably changed; the Author of the Vestiges had turned into Babbage's Machine, and was devouring Acarus Crossii—Old Red Sandstone—and every thing that came in the way!!![103]

Who was the half-naked figure chasing the contents of the book? The article paralleled passages from writings acknowledged by Chambers with those from *Vestiges* (fig. 8.9). Four volumes of previously anonymous essays had just been published under his name, which made identifying his style considerably easier. Rather cannily, the *Macphail's* essay never directly accused Chambers, as this could have led to a libel suit. Instead, readers were asked to draw their own conclusion about "Vestiges of the Author of 'the Vestiges of Creation.'"[104]

The simultaneous attributions in the *Manchester Courier* and *Macphail's* offered a prime opportunity for those who wished to destroy everything Chambers stood for. Before this, his name had been guessed as one among many, and had rarely appeared in print; but the *Courier* went so far as to claim that the authorship was "no longer a secret." The parallel passages in *Macphail's* were not much more convincing than those in the *American Review,* but they established an occasion for a public denial. The enemies of secular development had Chambers cornered. He could never acknowledge *Vestiges,* for this would precipitate a full-scale boycott of the publishing business and undermine his family's position in Edinburgh society. Yet an outright lie (if proved) would have the same effect.

After the *Macphail's* attribution the Scottish religious press declared open

102. *Manchester Courier,* 3 Nov. 1847, 701.

103. "Vestiges of the Author of 'the Vestiges of Creation,'" *Macphail's Edinburgh Ecclesiastical Journal and Literary Review* 4 (Nov. 1847): 265.

104. Ibid., 147–57, with additional note in 4 (Dec. 1847): 362–63.

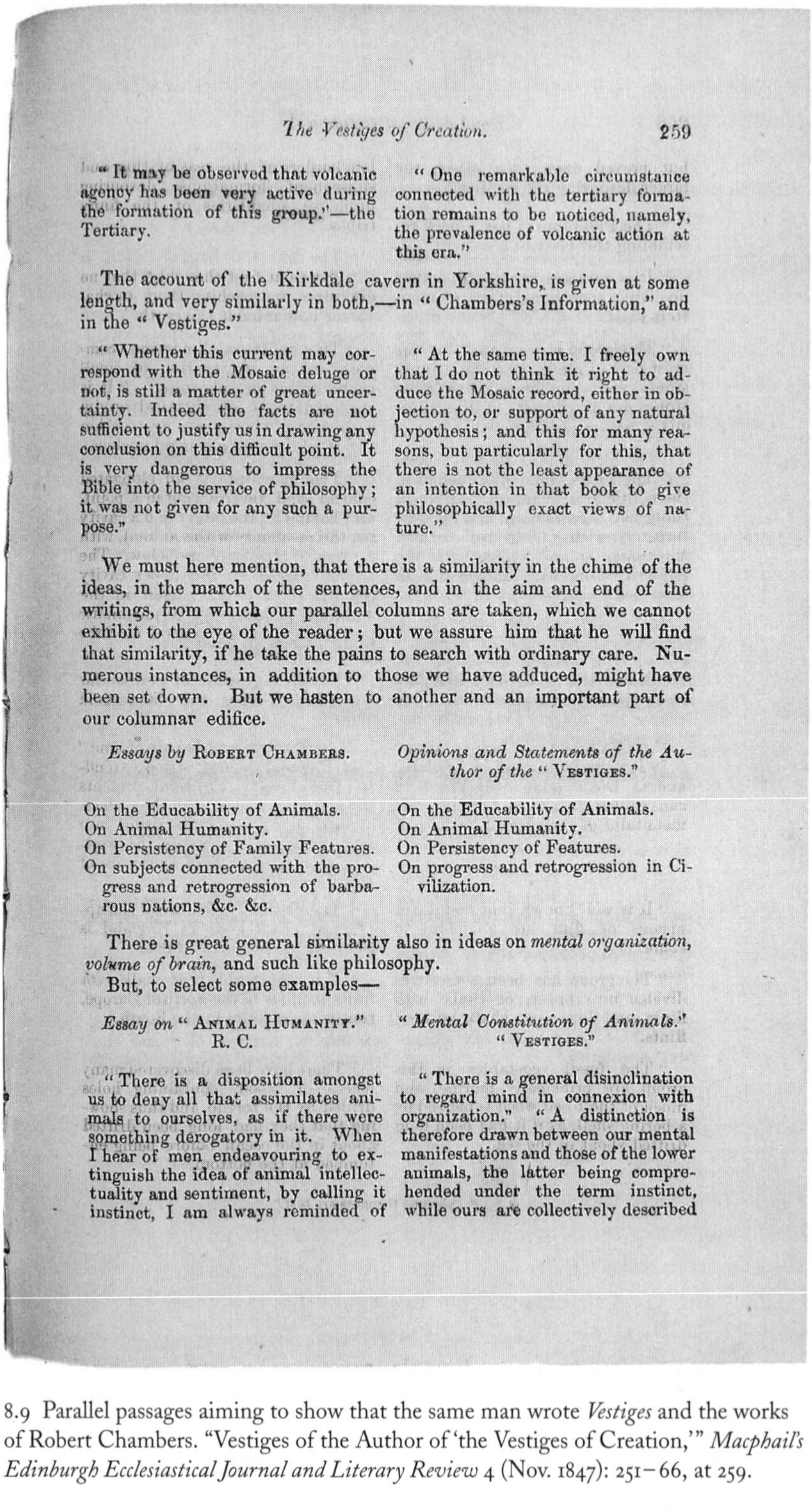

The Vestiges of Creation. 259

"It may be observed that volcanic agency has been very active during the formation of this group."—the Tertiary.	"One remarkable circumstance connected with the tertiary formation remains to be noticed, namely, the prevalence of volcanic action at this era."

The account of the Kirkdale cavern in Yorkshire, is given at some length, and very similarly in both,—in "Chambers's Information," and in the "Vestiges."

"Whether this current may correspond with the Mosaic deluge or not, is still a matter of great uncertainty. Indeed the facts are not sufficient to justify us in drawing any conclusion on this difficult point. It is very dangerous to impress the Bible into the service of philosophy; it was not given for any such a purpose."	"At the same time. I freely own that I do not think it right to adduce the Mosaic record, either in objection to, or support of any natural hypothesis; and this for many reasons, but particularly for this, that there is not the least appearance of an intention in that book to give philosophically exact views of nature."

We must here mention, that there is a similarity in the chime of the ideas, in the march of the sentences, and in the aim and end of the writings, from which our parallel columns are taken, which we cannot exhibit to the eye of the reader; but we assure him that he will find that similarity, if he take the pains to search with ordinary care. Numerous instances, in addition to those we have adduced, might have been set down. But we hasten to another and an important part of our columnar edifice.

Essays by ROBERT CHAMBERS.	*Opinions and Statements of the Author of the* "VESTIGES."
On the Educability of Animals.	On the Educability of Animals.
On Animal Humanity.	On Animal Humanity.
On Persistency of Family Features.	On Persistency of Features.
On subjects connected with the progress and retrogression of barbarous nations, &c. &c.	On progress and retrogression in Civilization.

There is great general similarity also in ideas on *mental organization, volume of brain,* and such like philosophy.

But, to select some examples—

Essay on "ANIMAL HUMANITY." R. C.	"*Mental Constitution of Animals.*" "VESTIGES."
"There is a disposition amongst us to deny all that assimilates animals to ourselves, as if there were something derogatory in it. When I hear of men endeavouring to extinguish the idea of animal intellectuality and sentiment, by calling it instinct, I am always reminded of	"There is a general disinclination to regard mind in connexion with organization." "A distinction is therefore drawn between our mental manifestations and those of the lower animals, the latter being comprehended under the term instinct, while ours are collectively described

8.9 Parallel passages aiming to show that the same man wrote *Vestiges* and the works of Robert Chambers. "Vestiges of the Author of 'the Vestiges of Creation,'" *Macphail's Edinburgh Ecclesiastical Journal and Literary Review* 4 (Nov. 1847): 251–66, at 259.

season. They recognized that this might be seen as unfair, but believed that the means justified the end of saving souls for Christ. One of the most vindictive attacks came from the liberal *Fife Herald* of St. Andrews, which was edited by a former minister of the Dissenting United Presbyterian Church. The attacks by scientific men had not been enough; *Vestiges* had sunk only after Chambers was identified as its author:

> *He* stands up against all science, philosophy, and religion! *He* writes a new and better book of Genesis! We are entitled now to lift up the mask which the author wore, and, verily, under the visor and the mail of Goliath, it is Tom Thumb. A fifth-rate *litterateur,* and a fifth-rate amateur in science. . . . Mr Robert Chambers *versus* the Bible, universal knowledge, and all men of talent and genius! The name was enough, and the book did not survive its baptism. . . . That man should spring from the loins of a monkey was scarcely more distasteful than that a theory of creation should be propounded by Mr Chambers.[105]

With the appearance of the article in *Macphail's,* some evangelicals began again to talk of boycotting W. & R. Chambers, but this was unsuccessful, largely because exposing a secret that would damage a man's personal affairs was so widely thought to be unacceptable.[106] Moreover, uncertainty continued to hover over the authorship so long as no one unequivocally claimed it. Statements that *Vestiges* was "universally" attributed to Chambers were, invariably, hostile attempts to establish that very fact and thereby destroy him. Denials, just as categorically, were efforts to defend him, even by those (like Brewster) who remained convinced that the book was so evil that Chambers could not possibly have written it.

The conflict between the saints and the sciences of progress was not only a clash of principles, but of urban party politics. The evangelicals were still in the ascendant in Edinburgh in the late 1840s, however much their attempts at union had failed. At the 1847 parliamentary election, they successfully unseated the historian Thomas Babington Macaulay, the candidate of Lord Jeffrey's Whigs, and they had also done well in city elections. The key issue had been Macaulay's support for Maynooth. Their electoral aims were not narrowly clerical though; they centered on finding liberal candidates of good morals who would support extensions of the franchise and access to educational and hospital facilities by Dissenters.

The political implications of *Vestiges* came to a head early in November 1848, when Chambers allowed his name to go forward as a candidate for lord provost of Edinburgh, the city's highest civic honor. The only other name in the field

105. Review of H. Miller, *Foot-prints of the Creator, Fife Herald,* 18 Oct. 1849, 749.
106. "The Monthlies," *Scotsman,* 6 Nov. 1847, 2.

was William Johnston, cofounder of another of Edinburgh's publishing houses. Johnston was a Dissenter and a liberal Whig but had worked against Macaulay in the parliamentary election, and so the latter's supporters were eager to find an alternative. They approached Sir Adam Hay, a High Church Tory aristocrat, as Johnston promised to stand down in the event that he came forward; but Hay was not registered to vote in Edinburgh, and so the compromise collapsed. The Macaulay Whigs settled upon Chambers only a week before the election. He was obviously flattered, both by the source of the request and the status of the office. Johnston had already gathered most of the votes, so that Chambers faced a grave disadvantage from the start; even those liberal evangelicals whose support would normally be essential for victory had not signed his requisition papers.[107]

Combe was angry not to have been consulted and thought his friend's election would be "a kind of moral miracle." In the shops Chambers was openly accused of being a Socinian and a Unitarian.[108] Worst of all, a long letter in the *Witness* exposed "grave objections" to his candidacy. The anonymous correspondent accused Chambers of being "*the mere tool of a factious and disappointed Clique*"—the Macaulay Whigs, who were hoping to capitalize on Chambers's "literary prestige and popularity." But the most serious accusations were "of a *personal* kind." The letter had pointed to "insolent attacks" on the seventeenth-century Covenanters in Chambers's *History of the Rebellions.* There could be no objection to his Episcopalianism, but the "intolerant, unbridled, and unmannerly" anti-Presbyterian bias in his writings was inexcusable, an insult to the national faith.[109]

These factors alone were enough to disqualify Chambers from political office. The letter in the *Witness,* however, went on to refer to his suspected authorship, and cited the reviews in the *North British* and *Edinburgh:*

> Every one knows that Mr Chambers studiously excludes all religious subjects and references from his periodicals; and that notwithstanding the vast multitude of papers of all kinds which he has written or published, it would be difficult to gather from any one of them that a God exists, or that a way of salvation from sin has been revealed. He is, indeed, the great representative of our non-religious periodical press. This is of itself seriously criminal. But he is charged with even worse than this. He is charged with writing and sending forth to the world a work which, not

107. Accounts in the Edinburgh newspapers include: "The Municipal Elections," *Witness,* 25 Oct. 1848, 2; *Scotsman,* 28 Oct. 1848, 2–3; "The Civic Chair," *Witness,* 1 Nov. 1848, 2; *Scotsman,* 1 Nov. 1848, 2; *Scotsman,* 4 Nov. 1848, 2, 3; "Municipal Elections," *Witness,* 4 Nov. 1848, 3; "Municipal Elections," *Witness,* 8 Nov. 1848, 2. See also G. Combe to C. Crowe, 1 Nov. 1848, NLS mss. 7391, ff. 570–71.

108. G. Combe to C. Crowe, 1 Nov. 1848, NLS mss. 7391, ff. 570–71.

109. "Advertisement. An Elector's Objections to Mr Robert Chambers," *Witness,* 1 Nov. 1848, 3.

> to speak of its false and superficial science, "expels the Almighty from the Universe," and renders "the revelation of His will an incredible superstition,"—(*North British Review*); which "tells us that our Bible is a fable, when it teaches us that man was made in the image of God" . . . (*Edin. Review*, No. clxv. page 3.) I do not say that Mr Chambers *is* the author of this revolting production; but he is *alleged* to be the author.

The letter noted that Chambers had never publicly denied the authorship. No Christian elector could lend his support to anyone under "so grave a suspicion."[110]

Two days later the newspapers reported that Chambers had withdrawn. The Whigs were left in disarray, and Johnston won an easy victory. The evangelicals were jubilant; as the *Witness* commented, supporters of the Presbyterian Establishment were being replaced "by men of a decidedly improved stamp." The *Witness,* though, expressed a certain sympathy with Chambers, remarking that his mistake had been to allow his name to be put forward in the first place.[111] Although the *Witness* could pour out the most vitriolic satire, Miller distanced the paper from personal abuse of Chambers. He allowed the anonymous letter to appear only as a paid advertisement, thereby avoiding editorial entanglements and a potential libel suit. He completely disagreed with the view that secular periodicals or secular education were "seriously criminal."

All of the nonevangelical Edinburgh liberals—Combe, Whig leaders such as Melville, and the *Scotsman*—were furious about Chambers's abortive candidacy. The sitting lord provost, the Whig publisher Adam Black, had advised Chambers to fight it out, but William Chambers was alarmed about potential damage to the business.[112] With the firm's schoolbooks and journal being linked with *Vestiges,* a boycott might succeed. The worst aspect of the affair was, as Combe said, "shewing to the Evangelicals the extent of their own power."[113] He lamented that the antiphrenological alliance, forged in the 1820s between Jeffrey's circle and the evangelicals, was now bearing its bitter fruit. As he later noted in his diary, "An ultra liberal party in politics has sprung up in favour of a large extension of the suffrage, triennial Parliaments, & vote by ballot, but cemented together by the most narrow fanatical spirit in religion."[114] Some friends thought that if Chambers had "*bearded* the Evangelicals" and fought the contest, he would have unmasked some of the loudest advocates of Church disestablishment as traitors to religious liberty. Maclaren's *Scotsman* used the

110. Ibid.
111. "Municipal Elections," *Witness,* 8 Nov. 1848, 2.
112. G. Combe to M. B. Sampson, 3 Nov. 1848, NLS mss. 7391, ff. 577–78.
113. G. Combe to M. B. Sampson, 3 Nov. 1848, NLS mss. 7391, ff. 577–78.
114. G. Combe, diary entry for 19 May 1852, NLS mss. 7428, f. 22r.

occasion to deny outright Chambers's authorship of *Vestiges*—"a book which they never read and he never wrote"[115]—and saw the election as marking a decline in the masculine energy of Edinburgh's reformers. "We Modern Athenians are not what we once were," it lamented. "Meekness best becomes our recent history. Manliness is on the decline—much strength has gone out of us."[116]

Evangelical domination of Scottish life peaked in the 1840s. The Disruption, rather than creating a new national church, ultimately marked a weakening division among the faithful. The secular reformers began to find common ground. Connections became closer, too, with some of the professors at the University of Edinburgh. Evolutionary cosmologies such as *Vestiges,* by appealing alike to Combe's circle and to readers exhausted by sectarian warfare, proved in the long run to be one way of moving the basis of reform beyond phrenology, toward sciences with a broader appeal. This shift strengthened bonds between ambitious commercial men like the Chambers brothers with the Macaulay Whigs and the *Edinburgh* clique; William Chambers was elected lord provost in 1865. The old alliance between evangelical religion and liberal politics began to break down: new readerships helped to transform the city's caste- and clique-ridden society.

115. *Scotsman,* 4 Nov. 1848, 2; J. McClelland to G. Combe, 12 Nov. 1848, NLS mss. 7295 ff. 40–44.

116. *Scotsman,* 4 Nov. 1848, 2. See also the report of the affair in the *Reasoner* 5 (22 Nov. 1848): 414—"Scotland is the great teacher of infidelity. Nowhere is religion presented in so revolting a form."

PART THREE

Spiritual Journeys

CHAPTER NINE

Sinners and Saints

Beware of bad books.

[JOHN TODD], *Self-Improvement: Chiefly Addressed to the Young* (London: Religious Tract Society, [1848])

READING IS USUALLY SEEN as an individual act. The classic images of the reader show a solitary, abstracted figure; through withdrawal and isolation, reading becomes a way of defining who we are and what our relations to others might be. Thinking of reading in this way has a long history, growing out the emergence of silent reading in the twelfth century and its spread among the laity in the late Middle Ages. The Bible, the first book to be printed, became the principal means for defining what books were for, as reading was conceived within an intense spiritual engagement, pivoted on salvation. When the Victorian antivivisection campaigner Frances Power Cobbe recalled how she "pinned her faith" on *Vestiges* at the age of eighteen, she drew upon a language of conversion with deep roots in the religious culture in which she had been raised.[1]

So far in this book I have avoided approaching reading in a highly individualized way. This part confronts the issue directly, by exploring the role of reading in the making of self-identity. The industrial revolution in communication posited a new relation between an emerging mass society and a privatized notion of the individual. To show what this might involve, subsequent chapters examine the experience of single readers—one in a northern factory town, the other the anonymous author. Neither is "typical" or "normal"—categories that are themselves part of the Victorian statistical vision of the social body; but both reveal how reading could create new forms of personal identity.

The issue of identity arose in its starkest form when readers were defined as individuals faced with the choice between salvation and damnation. My focus

1. Cobbe 1894, 1:194–95. The identity of the solitary reader is discussed in Darnton 1990, 154–87; Flint 1993; Manguel 1996; Radway 1991; and Radway 1997, among many other works. For its origins, see Saenger 1982; Taylor 1996; and Chartier 1994, 17–18.

in this chapter is on a committed cadre of libertarian freethinkers and their opponents, who campaigned to convert working- and lower-middle-class readers to Christianity. *Vestiges* offers a route into the seamy underground of urban religion, for it appeared in a crucial decade for freethought and evangelical revival. It was in the 1840s that both movements looked to the sciences for support, and the relation between reading and unbelief was fiercely debated.

In the early Victorian age, the individual was defined primarily through faith, so that later secular psychologies cannot simply be projected backward upon it. Industrialization did not drive out religion, but brought questions of spirituality to the center of national concern. In a common metaphor, print was spiritual food, which constituted each reader in relation to the body of believers. At stake was the character of Britain as a Christian nation. Freethinkers denied that nourishment could be obtained from religious dogma, but they agreed about the centrality of reading for defining beliefs and reconstructing the social body. The starkness of the options was used to define readers as individuals confronted with a choice. Every human life was a statistic, a case, and a unique spiritual journey.

Blue Books of the Soul

In the fifty years since the French Revolution, the reading habits of mechanics, artisans, and shop workers had become a major concern. Cities were Satan's stronghold, where the printed poison of infidelity penetrated most easily.[2] Or so it seemed to the missionaries who worked to save the souls of the poor among fetid alleys and packed tenements. Rural clergymen rarely found evidence of unbelief, but their urban counterparts claimed that whole streets refused to open their doors. Conditions were so bad and class feelings ran so high that missionaries paid to distribute tracts were usually working men. "It is very affecting," wrote one missionary, "to contemplate the undeniable extent to which *professed* infidelity prevails among the working orders."[3]

A fascination with statistics accompanied this focus on the individual soul. The only official national religious census was in 1851, when the numbers attending divine service were counted and analyzed, just as in the famous parliamentary "blue books" on epidemics, illiteracy, and factory labor. The evangelical religious literature went further. Converts had to be counted; there could be a statistics of belief because the spiritual state of each individual mattered so profoundly. Atheists and deists were singled out for attention because they had consciously rejected the atoning power of Christ's sacrifice, the heart of

2. Halttunen 1982 offers a stimulating account of the analogous situation in American cities.

3. Vanderkiste 1852, 115. Women, although active in philanthropic support for the London City Mission, did not serve as paid missionaries until later: Poovey 1995, 43–52; Lewis 1986, 221.

evangelical doctrine. Infidelity was particularly disturbing because it appealed to shop assistants, clerks, drapers, and the uppermost strata of the skilled working classes, the same groups from whom the missionaries were drawn.

The case books of the Church Missionary Society (f. 1799), the London City Mission (f. 1835), and similar groups in other cities provide the best evidence of how individual working people negotiated the relations between religion and science. Although biased propaganda, the reports can be read for traces of resistance and open exchange. They show that if *Vestiges* made evolutionary cosmology a topic in polite drawing rooms, it had been canvassed for decades in garrets, lodging houses, pubs, and cellars.

Take, for example, the case of a tailor living in the Saint Giles rookery in London. This was only a few streets south of the British Museum, but one of the poorest districts in London. An evangelical missionary finally gained entrance to the room by slipping in after the dog, before the door could be slammed in his face. The tailor, "more civil" than expected, handed the intruder a chair and they began to talk. The missionary found him "a very shrewd, intelligent man, who had read much on scientific subjects." He also "held Deistical principles" and opposed Christianity, "which he said was got up to keep the people in subjection to the clergy." (The missionary claimed to have "reasoned him out of that point.") The tailor had "a large parcel of books and papers, chiefly Infidel and Socialist publications," and raised problems in the biblical story of the Tower of Babel. "The writer of that portion," he was reported as saying, "must have been very ignorant of astronomy, because if he had known anything of that science he might have known that the builders never could succeed in reaching heaven with their building." The missionary replied that Moses was steeped in Egyptian wisdom, but God had not intended to teach astronomy in the Bible, but to display "the pride and sinfulness of fallen man" for future generations. The missionary reported some success. The tailor asked for a tract about the conversion of an infidel blacksmith to give to his friends, and on a subsequent visit his wife appeared to be reading an old Bible.[4]

Or take the case of a dying unbeliever in Manchester. The young man denied God even after one of his missionary visitors suggested that he believed in other things, such as the truths of physics, which could not be seen.

> We then asked, "How did you come here?" "By my parents," said he. "And how did they come?" "By theirs." "But how did the first parents come?" "By some other race of animals—such as baboons or monkeys." "And how did they come?" "By some other race of animals." We replied, "You said just now, you believe nothing

4. *London City Mission Magazine* 10 (May 1845): 52–53.

that you did not see; now, please to tell how you know the things you now assert?" He was agitated and confused.

Knowledge, and specifically scientific knowledge, could be presented as the pivot on which conversion turned. The missionary pointed to the inconsistency of accepting "such absurdities" while refusing "the most important of all truths," the saving power of Jesus Christ. Although the young man said that he had "companions" who would be able to answer, he seemed "thoughtful," and the missionaries suspected that only the presence of nine or ten women throughout the entire two-hour encounter prevented him from admitting the error of his ways. This "little gleam of hope" brightened after subsequent interviews. "He died in a happy state of mind," the report concluded, "and I trust he obtained mercy of the Lord. To him be all the praise."[5]

To ask a question in terms used by the missionaries, how "typical" were such "cases"? Missionary statistics tend to support other evidence that there were few outright unbelievers. Thus in 1845 the Manchester and Salford Town Mission reported only 12 reclaimed infidels, compared with 43 reformed drunkards, 311 converts, 322 who had been induced to attend public worship, and 2,657 who were "giving heed" to the Word.[6] Thousands did attend atheist lectures and bought their tracts, but freethought never became a mass movement like, say, Chartism.[7] A small number of active freethinkers gave form to the bourgeois nightmare of an angry mob of working-class infidels.

At one level, the war between belief and unbelief was thrust before public attention on a daily basis. The intersecting moral geographies of London make this clear. The British Museum was only a short walk from the freethinkers' John Street Institution; the evangelicals' Exeter Hall was a few hundred yards up the Strand from Somerset House, the meeting place of the main scientific societies. In the mid-1840s the wealthier classes had barely begun their flight to the suburbs: class and income divisions were structured by street rather than by distance from the center. Artisans, merchants, professionals, and aristocrats all lived crammed together in close proximity, often within a few yards of one another. It was hard to walk down certain streets without hearing exhortations against infidelity, or to travel on the train without having tracts pressed into your hand. Although certain areas could be avoided, especially at night, the

5. *Fourth Annual Report of the Manchester and Salford Town Mission, with Extracts from the Journals of the Missionaries* (Manchester: printed by William Simpson, 1841), 22–23.

6. *The Eighth Annual Report of the Manchester and Salford Town Mission, with Extracts from the Journals of the Missionaries* (Manchester: printed by Joseph Gillett, 1845), 7.

7. Despite the impression sometimes given in the best analysis of infidel science, Desmond 1987, 81–82. Royle 1974, the standard work on early Victorian freethought, tends to focus on institutional developments.

freethought debate was part of the visual and aural experience of what one author called "Babylon the Great."[8]

Any image of plebeian radicalism as atheistic, evolutionary, and anti-Christian is based in large part on fears whipped up by the orthodox press. Those who traveled among the poor knew better. They recognized that many Chartists and other political radicals were Christians who trusted the Bible but condemned the Established Church and state as priestcraft and kingcraft.[9] As the 1851 Religious Census demonstrated, indifference rather than hostility was the problem: those who had spent six days at hard labor were reluctant to spend the seventh in a pew. Drunkenness, domestic violence, and popery were much more common obstacles to conversion than abstract arguments about God's existence. Yet if deists were uncommon and atheists exceptionally rare, their theatrical tactics and large audiences were also of grave concern. Professed unbelievers attracted publicity out of all proportion to their numbers.

Through the early 1840s unbelievers usually encountered materialist views of natural law through the philosophical classics of the late Enlightenment: Baron d'Holbach's *System of Nature* (1770), Tom Paine's *Age of Reason* (1794–1807), Constantin-François Volney's *Ruins of Empires* (1791), and Elihu Palmer's *Principles of Nature* (1801). When these works had first appeared, they were expensive and discussed only in genteel drawing rooms (Paine, sold cheaply almost from the outset, was an exception). Most readers thought ideas about men with tails, such as Erasmus Darwin's evolutionary speculations, were at worst merely funny or wrongheaded. But made available to the people in street pamphlets, part works, and unstamped newspapers, the same ideas became threatening in the extreme.[10] A caricature from the peak year of counterrevolutionary paranoia, 1819, shows the attempted rape of Britannia by Radical Reform (fig. 9.1). His cloak hides immorality, blasphemy, slavery, starvation, robbery, and murder; a fire-breathing demon with hair of snakes holds aloft Paine's *Age of Reason*. Britannia, her virtue protected by royal armor, rests on the rock of religion and wields the sword of law, while the British lion (with "loyalty" on its collar) comes to the rescue. Defense of the faith is joined to defense of traditional roles for women as symbols of the nation.[11]

8. [Mudie] 1825. For moral geographies, see Driver 1988; on London street life, see Porter 1994, 185–305, with a helpful bibliography on 400–405; Dyos and Wolff 1973; and Winter 1993, 176–80.

9. Yeo 1981.

10. The literature on radical publishing before 1832 is extensive (Hollis 1970; Wiener 1969; Gilmartin 1997; Klancher 1987, 98–134; McCalman 1988, 1992; M. Wood 1994), although there is relatively little on the 1840s and 1850s: see James 1976, 153–71; Neuberg 1977; and the general literature on Chartism and Owenism. For the debate concerning revolutionary sedition and knowledge about nature, see Inkster 1979, 1981; Weindling 1980; and esp. Porter 1978b.

11. Taylor 1983, 1–18. The royal moto, in Old French, translates as "God and my right."

9.1 George Cruikshank, "DEATH OR LIBERTY.!"

Twenty-five years after this cartoon appeared, a committed ultraradical network continued to distribute antitheological writings along with reform literature, Chartist tracts, and (sometimes) pornography. Those who read such publications felt a strong sense of tradition with enlightened skepticism.[12] The persistence of this older literature, in part, simply reflected the time it took for newer titles to become available at prices that working people could afford.

Vestiges in its early editions was by this standard wildly expensive. No one below the highest reaches of the professional classes could contemplate purchasing a book that cost 7s. 6d. Working-class readers usually learned about it indirectly—through lectures, incidental references, and reviews in odd numbers of periodicals. Some pub libraries had collections of books, bought with communal funds; but many artisans with interests in science preferred to see their money spent on reference works that gave access to elite scientific culture.[13]

12. Hollis 1970, chs. 6 and 7; Johnson 1979; McCalman 1988; Royle 1974, 9–58.

13. A. Secord 1994a, 278–79, 297.

9.2 A working man searches a street bookstall in London. Stands of this kind were an important source of inexpensive reading for clerks and artisans, although it is unlikely that many copies of a fashionable book like *Vestiges* would have turned up there until later in the century. Most stalls also sold a selection of prints and engravings, several of which are hung above the books. "The Bible on the Book-stall," in [Sargent] [1859], 40.

Another way to get access was to wait for *Vestiges* to appear in the secondhand market (fig. 9.2). This was a good source for Bibles, literary classics, and older works, but not for fashionable sensations—although Charles Bradlaugh, a future leader of freethought, did find a used copy of the third edition of *Vestiges,* which he annotated extensively.[14] Other self-taught deists or atheists, such as the Coventry ribbon weaver Joseph Gutteridge, did not see the book until years later. Although a keen reader and known to local middle-class freethinkers,

14. Bonner 1891, 95. I have been unable to locate Bradlaugh's copy, which was sold at auction after his death.

THE VESTIGES of the NATURAL HISTORY of CREATION, without abridgment, corrected by the author, and all the notes. Price 2s. 6d. Also an abridgment, price 4d. G. Combe's new work on the 'Relation between Religion and Science' price 6d. Mackay's 'Voices,' price 1s. each. E. Burritt's 'Sparks from the Anvil,' price 6d. Madame D'Arusmont's Popular Lectures, price 3s. Feargus O'Connor on Small Farms, price 2s. 6d. Pitman's Manual of Phonography, price 1s. Orders for these and all other books and periodicals received and punctually attended to by Edward Truelove, Bookseller and News-vender, adjoining the Literary Institution, John-street. The Sunday newspapers, magazines, and weekly periodicals supplied as soon as published. Bookbinding performed in a superior style, on moderate terms.

9.3 An unauthorized advertisement for the people's edition of *Vestiges,* announcing that the book could be purchased in the shop of the radical bookseller Edward Truelove on John Street. It is likely that this advertisement, like many of those printed in the radical press, originated as a large placard on the front of Truelove's shop. Note the range of works that appear in the same company, from Feargus O'Connor's tract on the Chartist land plan to the phrenologist George Combe's pamphlet on religion and science. *Reasoner* 3 (7 July 1847): "The Utilitarian Record," 64; Truelove's placards are mentioned in *London City Mission Magazine* 20 (Nov. 1855): 264.

Gutteridge often faced insurmountable obstacles in getting specific titles.[15] In genteel society, books were fashion accessories with a conversational shelf life of a single season. Within plebeian circles they became known over years or even decades.

Readers in the lower reaches of the middle class obtained books more easily. *Vestiges* was widely read in mechanics' institute libraries.[16] The people's editions at 2s. 6d. reached a market previously almost untouched. Edward Truelove, who ran the bookshop next to the freethinkers' main headquarters on John Street just off Tottenham Court Road, gave the cheap reissue top billing in a list of new works in the *Reasoner* (fig. 9.3). His advertisement, an unwelcome addition to Churchill's own campaign, emphasized that the text was uncut and had all the notes. Truelove also sold the *Expository Outline* from the *Atlas,* at a price that would enable four readers to club together to obtain the gist of the debate for a penny each.[17]

15. Gutteridge 1893, 95. For more on working-class difficulties in obtaining books, see Vincent 1981; 1989, 61, 210–13. For ways around these through cooperation, see A. Secord 1994a, 277–78.

16. See chapter 6.

17. *Reasoner* 3 (7 July 1847): "The Utilitarian Record," 64.

Even with such measures, however, artisans rarely had direct contact with *Vestiges.* Most readers of the people's editions would have been in the middle classes. As Hugh Miller warned, "intelligent mechanics, and a class of young men engaged in the subordinate departments of trade and the law" were in special danger. As we have already seen, Miller's *Footprints* (1849) was one of many antidotes targeting this audience. When young men imbibed development doctrines, Miller wrote, they lost faith in Christian salvation, the core of humanity's "moral constitution," and became "turbulent subjects and bad men."[18] The social fabric crumbled. The worst, as he noted on another occasion, was "the coarse, vulgar, vigorous infidel of the people . . . sunk to the low level of the fiercer Atheists of the first French Revolution. . . ."[19] Fears about the tendency of *Vestiges* had a particular cogency when the walls of Edinburgh and other cities were plastered with atheist handbills.

Even in the comfortable libraries of the London gentlemen's clubs, the threat could not be ignored, especially after the socialist leader Robert Owen was attacked in the House of Lords for opposing Christianity (fig. 9.4). Most of his followers backed off, but a handful of extremists stepped up the offensive in the brazen penny weekly *Oracle of Reason,* claimed as "the only exclusively ATHEISTICAL print that has appeared in any age or country."[20] A sequence of blasphemy trials gave the cause tremendous publicity, briefly pushing the *Oracle*'s circulation to over six thousand at the end of 1841. A Bristol jury sent the rabble-rousing first editor, Charles Southwell, to jail for his article on "The Jew Book," an ugly attempt to exploit popular anti-Semitism to mock the Bible.[21] The second editor, the young tinsmith George Jacob Holyoake, was sent to jail in 1842, as was the third, Thomas Paterson. Paterson had enlivened the windows of the blasphemy depot on Holywell Street with huge cartoons satirizing the Old Testament, and advertised "atheism for the million." Newspapers claimed that twenty thousand people a day thronged to visit this "den of infamy."[22] Theatrical tactics like these depended on traditions of ribald

18. Miller 1849, ix.

19. [H. Miller], "The Atheist Patterson," *Witness,* 11 Nov. 1843, 2.

20. [Maltus Questell Ryall], "Preface," *Oracle of Reason* 1 (1842): ii–vii, at ii; Desmond 1987, 85–91. Authorship of articles in the *Oracle* is identified from Holyoake's marked copy at the library of the Bishopsgate Institute in London.

21. [Charles Southwell], "The 'Jew Book,'" *Oracle of Reason* 1 (27 Nov. 1841): 25. It is no coincidence that Shylock—usually played in this period as the stereotype of the greedy Jew—was the favorite Shakespearean role of Southwell, who had acting ambitions.

22. "Bruce-law and the Holywell Street Shop," *Oracle of Reason* 2 (21 Jan. 1843): 35–40, with quotations from the papers; also McCabe 1908, 1:92–93. For the campaign see Royle 1974, 72–86, and Desmond 1987, 85–88, although neither stresses sufficiently the national attention the movement received.

9.4 "Protestantism versus socialism, or the revival of the good old times." Bishops Phillpotts of Exeter and Blomfield of London drag Robert Owen to the stake, while the Mother Church and the devil look on with satisfaction. The archbishop of Canterbury holds aloft the Bible, "thy only hope of salvation." "Mother Church's fancy man," the Whig Reformer Henry Brougham (holding a "broom") blows on the fire—while looking forward to feasting on the "fat carcasses" of a clerical stew at a later date. *Penny Satirist,* 7 Mar. 1840, 1.

scurrility that had enlivened radical debate since the Wilkes affair in the late eighteenth century. Periodicals from the *Morning Chronicle* to the *Quarterly* reported the atheists' antics, so that any reader who cared about the sciences could not sidestep the issue.[23]

Street Science at the Crossroads

Despite the alarm about working-class reading habits, when *Vestiges* appeared the freethought campaign was in big trouble. The canon of skeptical

23. As the home secretary, Lord Normanby, said, "When the right rev. Prelate gave himself credit for having given a check to the Socialists, it must be admitted that he had also given them very much importance": quoted in Royle 1974, 67. On radical theatricality, see McCalman 1988, 1992; Wood 1994; and esp. Hadley 1995.

9.5 George Jacob Holyoake in 1847. McCabe 1908, 1: opposite p. 65.

works was showing its age. The wider Owenite socialist movement was being overwhelmed by the financial ruin of a communitarian experiment at Queenwood in rural Hampshire. Most of the halls of science, built as harbingers of a socialist millennium, had gone bankrupt, and some, as in Liverpool, now featured improving talks by evangelicals. The Rotunda, where freethinkers had preached in south London since the 1820s, had been taken over by Chartists who avoided irreligion as a divisive plague. The *Oracle* had ceased publication in 1843 after petty editorial squabbles, and the mainstream Owenite *New Moral World* was taken over in 1845 by its evangelical opponents.[24]

Infidels were moving away from the earthy, outrageous traditions of Enlightenment libertarianism toward a more earnest, rationalist form of persuasion that relied on geology, astronomy, and other modern sciences. They encouraged open debate on platforms and in periodicals, even welcoming hostile clerics to contribute. After his release from prison, George Jacob Holyoake (fig. 9.5), led the way in adopting these tactics, which gained ground through

24. [Ryall], "Preface," ii. Harrison 1969, 222–23, the standard history of Owenism, discusses the failures of halls of science.

weeklies like the *Movement* (1843–45) and the *Reasoner* (1846–61).[25] With a thin voice and the appearance of a seedy foreign émigré, he made finely drawn distinctions in his lectures, which had limited appeal to his working-class audiences. He was not a great orator, and wielded most of his influence behind the scenes as a publisher and editor. Labels were important in defining hoped-for alliances: "Atheism," "Infidelity," and "Socialism" became terms used by hard-liners or by evangelical opponents; Holyoake and the moderates tended toward "Freethought," "Theological Utilitarianism," "Naturalism," and "Secularism" as ways of tempting the middle classes into the fold.

Vestiges proved exceptionally welcome in extending the constituency for freethought. This is why Holyoake gave it such early and enthusiastic attention in his lectures in March 1845, one of which was devoted to "the origin of man as set forth in that extraordinary work just published, entitled *Vestiges of the Natural History of Creation.*" Delivered in the freethinkers' hall of science on Commercial Place in City Road and provocatively scheduled on a Sunday, the lecture cost only twopence and was followed by discussion. Within a year Holyoake was planning a work on *Vestiges,* although this was never completed.[26] The connection of a convicted blasphemer with the book may have been shocking, but his aim was rational persuasion. The accessible narrative was perfectly suited to gaining middle-class support.

Vestiges also provided a point of common interest with journalistic and literary circles. The artisan atheists had first learned of the book, as Tennyson had, through the *Examiner.* A notice based on that review appeared in Holyoake's *Movement* in January 1845. The author was William Chilton, a compositor who wrote zoological and geological articles for the infidel press, often while setting them into type.[27] He was well read in science, a refined and scholarly man, and the only "absolute atheist" Holyoake ever knew. Chilton immediately saw that *Vestiges* could be used to revive freethought. Tellingly, he had to base his review on the *Examiner;* a chance to read the full text came only after one of the campaign's wealthy patrons loaned a copy.[28]

So far as Chilton could tell at second hand, *Vestiges* argued "that the deity originally gave certain properties to matter, and that all natural productions were the result of those properties." Such a view, he argued, was a "transition state" to be followed by pantheism and atheism.

25. McCabe 1908, 1:62–152; Royle 1974; and Holyoake's 1892 memoirs.

26. W. Chilton to G. J. Holyoake, 1 Feb. 1846, Co-operative Union, Manchester, Holyoake Correspondence no. 155; also printed in Royle 1976, 141–42.

27. W. Chilton, "Letter from Mr. Chilton," *Reasoner* 3 (3 Nov. 1847): 607–10, at 608; for his life more generally, see Desmond 1987, 85–86.

28. W. C[hilton], "Vestiges," *Movement* 2 (8 Jan. 1845): 9–12, at 12. See also W. Chilton, "Christian Philosophers.—Dr. Dick and the Rev. A. Burdett," *Reasoner* 3 (14 July 1847): 373–80, at 376.

> There is nothing new in all this; all that the author states has been known to physiologists for many years—but the conclusions which he has drawn from these facts are new to the world at large, and will startle many a pedant from his slumbers, and awaken many a youthful mind to a sense of the bigotry and folly attempted to be crammed into the minds of the rising generation. The editor of the *Examiner* in allusion to the foregoing, says, "of these *truths* of physiology, strange as they may seem, *there is no doubt.*" The strangeness appears to me to consist in the possibility of men doubting *the truths* of physiology, geology or any other branch of science.[29]

Vestiges, Chilton thought, appeared to do little more than add a godly gloss to articles in the *Oracle of Reason* on "Theory of Regular Gradation," which had conferred a bestial origin upon humanity.

Chilton knew all too well that attacking religion with up-to-date science could pose difficulties. He had taken over the anonymous "Regular Gradation" series when Charles Southwell went to prison, and the series had rapidly changed course.[30] Southwell, a piano finisher who had ambitions as an actor, had a literary style as theatrical as his extraordinary platform performances. He had scavenged in the low-life writings of Parisian radical materialists, ready-made for the onslaught on "Priestianity." The series had opened with a half-clothed ax-wielding "Fossil Man," a racist fantasy lifted from writings of the hack naturalist Pierre Boitard (fig. 9.6).[31] In contrast, Chilton had poached from reflective works on the sciences, including the Bridgewater Treatises and Chambers's Information for the People. He printed recondite details, even classifications and species lists, taken from Cuvier, Grant, and other authorities, believing that readers needed to know the evidence on which conclusions were based. The result had been a disaster. Just before the series was cut off, Chilton had blamed himself:

> With all sincerity, then, I offer the most ample apology to the readers of the *Oracle* for the uninteresting and unpopular manner in which I have treated the important and really interesting question, when properly handled, of animal gradation. The fault was in the operator and not in the subject. I was induced to enter upon the investigation from the very favourable reception which the articles of my friend Southwell had met with—forgetful that therein lay the reason which should

29. W. C[hilton], "'Vestiges of the Natural History of Creation.' Theory of Regular Gradation," *Movement* 2 (8 Jan. 1845): 8–12, at 10–11.

30. Chilton underlines the significance of the change in "Theory of Regular Gradation," *Oracle of Reason* 1 (19 Feb. 1842): 77–78, and 2 (24 June 1843): 219–21. The series has been identified as his alone (Desmond 1987, 99; Corsi 1988, 256–57), although Chilton made it clear that the earliest articles were by Southwell.

31. [Southwell], "Theory of Regular Gradation," *Oracle of Reason* 6 1 (Nov. 1841): 5.

THE ORACLE OF REASON.

cause—but the error of supernaturalists lies here, *they* PRESUME *the universe to be an effect, and argue as though their presumption amounted to proof.* Spinoza, Vanini, Bacon, Locke, Voltaire, in short all who have written for or against the existence of a god, agree that something is uncaused, and therefore eternal. Spinoza establishes clearly that something exists, and from its existence argues its eternal existence. When Dr. Clarke tells us, in this fourth reply to Leibnitz, that the Epicurean chance is not *a choice of will*, but *a blind necessity of fate*, he wrote at random; for though chance can never be a choice of will, the word meaning as above noted, a perceived effect of an unknown cause, it is foolish as unjust to talk of the Epicurean notion of chance as signifying "a blind necessity of fate"—all Epicureans knowing, that necessity is neither *blind* nor *seeing*. The *word* necessity is expressive of the general truth, that matter does now, ever did, and ever must act *definitely* and *uniformly*.

(Fossil Man)

THEORY OF REGULAR GRADATION.

I.

I will show you how the earth has been peopled; how organic formation after organic formation has taken place, passing gradually from simple to compound bodies, and covering itself, as at this hour we find it, with plants and animals. In following matter through all its changes, noting its metamorphoses, from the most simple organisation to the most complex, we shall find, without doubt, that point where man, brutish and savage, as at first he must have been, took rank among the creatures of the universe.—Free Translation from "L'HOMME FOSSILE" (Fossil Man) of Boitard.

NOTHING is more fatal to a common-sense view of things, than the very prevalent habit of considering consequences before investigating and determining principles. It is not only illogical but absurd, and highly prejudicial to the cause of truth—for it fills timid minds with alarm, lest by a resolute and persevering search into the properties of matter and the nature of opinions, they may light upon results disastrous to their prejudices and perhaps fatal to their hopes of "singing hallelujah above the clouds." The rightly-balanced, well-ordered mind has the strongest assurance—an assurance drawn from the great fountain of human experience and the general analogies of things, that neither truth nor virtue are empty names, but real, substantial blessings, whatever clerical obstructives may preach to the contrary. Take the question, the regular gradation of the human species, and note how philosophers have occupied themselves in the solution of it—and if you do not say that merry-andrews would have gone a better way to work, it will be odd indeed. For like those creatures that, by a certain sort of instinct, to avoid pursuit, darken the medium through which they pass, so have our clerical swimmers in the sea of knowledge a happy knack of leaving a long tail of obscurity behind. Instead of carefully collecting facts, and opening their mental exchequer to receive something solid in the way of experience, they in general rest contented in most shameful ignorance, rather than their pride should be mortified by any discoveries in science, hostile to their cherished opinions. Solicitous, for their own advantage, to maintain what they call the honour of the human species, and its superiority over the brutes, all that could flatter and soothe the delightful idea is that alone which has been said and collected. Any attempt to establish a relationship, however remote, between man and the inferior animals, has always been scouted as impious, an insult to the creator, in whose image they tell us we are made, and little short of blasphemy against the holy ghost. Hundreds of sermons have been preached against the unlucky Bulliver, who insisted that in his day there was a Kentish family all tailed. Anathema after anathema was heaped upon Lochner, who, in his "Miscellanea Curiosa," relates, with great gravity and minuteness, the case of a boy with a monstrous tail. As to Dr. Ferriar, who considered that the *os coccygis* must sometimes have an accidental elongation—and Dr. Grindant, who published many cases tending to give strength to the opinion

5

9.6 "Fossil Man." Copied in this closely printed page of the *Oracle of Reason* from an otherwise unknown work, *L'homme fossile,* by the radical French naturalist Pierre Boitard. The low cranial angle and dark skin color linked this representation to contemporary illustrations of black Africans. The freethinkers also exploited racist views in attacking the Bible as the "Jew Book." [Southwell], "Theory of Regular Gradation," *Oracle of Reason* 1 (6 Nov. 1841): 5. BL, PP.632.

> check, rather than incite, me to the attempt. . . . This course in other hands might have been fraught with beneficial results, but in my case I fear it has failed.[32]

The problem, however, was more fundamental. Chilton's sources had been carefully crafted to obscure any links with religion or politics, so that co-opting Grant's fancy theories of philosophical anatomy had produced little more than dull catalogues of species. And however admirable the attempt to democratize esoteric knowledge, the result had not been what *Oracle* readers wanted. As Chilton had noted on an earlier occasion, they might find "the *technical* terms" hard to understand, but that "is a difficulty inseparable from the subject."[33] He had brought the *Oracle*'s science into the nineteenth century, but had failed to solve the problem of how the enemy's own weapons were to be turned against it.[34] The political authority of science was grounded not in doctrines of matter and natural law as Chilton had expected but in expert knowledge vouchsafed by an ideology of genius and divine inspiration.

Vestiges offered a way out. With its populist emphasis on progress, the book could vindicate freethought upon a scientistic basis. Here was a work that summed up the latest science without drowning in details. Chilton celebrated the fact that regular gradation, disparaged as his peculiar "hobby," was hailed by the *Examiner*, "a popular and erudite journal" of the "orthodox party." He drew a lesson for those within the radical movement who had opposed his tactics: atheists should rely upon physics rather than metaphysics when combating theologians. "A man may evade a syllogism but he cannot be blind to a fact—he may cut a metaphysical puzzle, but science will cut him." Chilton (who had been cut by science pretty badly himself) quoted extensively from the *Examiner* and promised a further account once he had read the book.[35]

32. W. Chilton, "Theory of Regular Gradation," *Oracle of Reason* 2 (24 June 1843): 219–21, at 220.

33. W. Chilton, "Theory of Regular Gradation," *Oracle of Reason* 1 (15 Oct. 1842): 356–57. "The Mosaic Account of the Creation and Fall of Man," in Henry Hetherington's *The Free-thinker's Information for the People* 1, no. 1–2 (1842): 1–16, took a different tack, relying on Lyell's *Principles* but focusing solely on the classic freethought issues of the biblical Flood and the age of the earth. On the origin of animals and plants, it was no more forthright than Lyell himself, noting (7), "That there is a power or energy in nature, by which new species are brought into being, appears clear, but the nature of that power is as yet unknown to man."

34. The utility of science in the atheist mission has been emphasized by Adrian Desmond (1987, 77–79), who has argued that contemporary anatomy, astronomy, and geology provided effective support for materialist views. The situation, however, proves to be closer to that described by Roger Cooter (1984, 201–55), who shows that enlisting specialist science under a radical banner proved problematic and ultimately compromising.

35. *Movement* 1 (11 Dec. 1844): 456, for receipt of the review, printed as W[illiam] C[hilton], "'Vestiges of the Natural History of Creation.' Theory of Regular Gradation," *Oracle of Reason* 2 (8 Jan. 1845): 9–12.

Some atheist radicals remained suspicious of *Vestiges,* fearing co-option by liberal political economy and a bland theism. One correspondent of the *Movement,* signed "Materialist," highlighted what the work said on these issues. What was to be done with the forthright references to "a First Cause to which all others are secondary and ministrative, a primitive almighty will, of which these laws are merely the mandates"?[36] Yet the problem was not insoluble. References to the "Eternal One" and a "First Cause" might appear unequivocal, but were regularly dismissed by all sides in the freethought controversies as meaningless conventions, tacked on to placate the saints. *Vestiges* could therefore easily be read as advocating practical atheism. Still, association with the book posed dangers of co-option. At a public debate, a minister confronted Holyoake with the fact that virtually all previous infidels accepted the divine origin of nature's laws, and *Vestiges* was one of his prime exhibits.[37]

Freethinkers had long experience of dealing with resistances of this kind in the useful knowledge works of Charles Knight and W. & R. Chambers. The latter, though suspected of partiality to freethought, were teasingly ambiguous. Holyoake had dedicated his prison pamphlet, *Paley Refuted in his Own Words,* to the firm with heavy irony—rebuking them for issuing Paley's *Natural Theology* in a 1s. 6d. People's Edition.[38] When Chilton told Holyoake of rumors that Robert Chambers had authored *Vestiges,* he noted that other writings confirmed him as a potential ally and "an advocate of the general principle" of regular gradation.[39] Such associations with the liberal middle classes, it was argued, had to be grasped if the infidel campaign was to make headway.

The significance of *Vestiges* in moderating ultraradicalism is best illustrated by the case of Emma Martin, the most celebrated propagandist for freethought and feminism. Martin, who came from a lower middle-class evangelical Baptist background, had started her public career as an atheist baiter on public platforms, distributing tracts and Bibles. In 1839 she switched sides after months of careful reading convinced her "of the evils of our social system," especially "the degraded condition of woman."[40] Martin believed that her role was to advocate

36. Materialist, "To Correspondents," *Movement* 2 (22 Jan. 1845): 32, citing *V*1, 26.

37. H. Townley and G. J. Holyoake 1852, 14; [Chilton], "'Materialism' and the Author of the 'Vestiges,'" *Reasoner* 1 (3 June 1846): 7–8.

38. Holyoake [1847], v. For the publication date of the first edition, McCabe 1908, 1:95–96; see also G. J. Holyoake, "To William and Robert Chambers, Esqrs., Editors of the 'Edinburgh Journal,'" *Reasoner* 4 (16 Feb. 1847): 168.

39. W. Chilton to G. J. Holyoake, 1 Feb. 1846, Co-operative Union, Manchester, Holyoake Correspondence no. 155; Royle 1976, 141–42.

40. Martin [1844a], 5. This pamphlet can be dated from the review by Holyoake in *Movement* 1 (4 May 1844): 166–67. Taylor 1983, 130–57, and Cowie and Royle 1982 give good accounts of Martin's life; see also Holyoake 1852.

The Reasoner
& LONDON TRIBUNE:
A WEEKLY SECULAR NEWSPAPER.

Among the few things of which we can pronounce ourselves certain, is the obligation of inquirers after truth to communicate what they obtain.—Harriet Martineau.

No. 6.] SUNDAY, MAY 6, 1855. [Price 2d

MRS. EMMA MARTIN.

INTELLIGENT WOMEN OR MEDIOCRE MEN.

In the touching biography of 'Currer Bell,' the reader saw the progress making in the firmness, self-command, intellectual strength, and social influence of women. They *can* conquer in that domain of disinterested and unnoticed purpose, where manly genius has been held to be supreme. The writer of that biography, not less than the subject, is an illustrious instance of the same welcome truth. In America, the *Una* and other papers, and a noble throng of women, are teaching the new world what we need, that more women in England should engage themselves in enforcing. The subject of this paper was an instance of the kind of intention, devotion, and good sense, which we trust to see more frequent.

Mrs. Emma Martin, so well known to the 'constant readers' of this Journal as an advocate of Freethought, was born at Bristol, in 1812, being 39 years old at the period of her death, in 1851. The engraving of her subjoined is made from the only sketch which exists, executed in water-colours, at the time of her first coming to London. The original sketch is in the possession of Mrs. Captain Grenfell, to whom we have been indebted for its use. The engraving does not perfectly recall Mrs. Martin as her last friends knew her, but it is the best portrait that can now be produced of one whose expression of feature and thought will long be held in remembrance.

good, is of all virtues, as women are now educated and situated, the most rarely to be found amongst them; they have seldom even, what in men is often a partial substitute for public spirit, a sense of personal honour connected with any public duty. Many a man, whom no money or personal flattery would have bought, has bartered his political opinions for a title or invitations for his wife; and a still greater number are made mere hunters after the puerile vanities of society, because their wives value them. As for opinions, in Catholic countries, the wife's influence is another name for that of the priest: he gives her, in the hopes and emotions connected with a future life, a consolation for the sufferings and disappointments which are her ordinary lot in this. Elsewhere, her weight is thrown into the scale either of the most commonplace, or of the most outwardly prosperous opinions—either those by which censure will be escaped, or by which worldly advancement is likeliest to be procured. In England, the wife's influence is usually on the illiberal and anti-popular side: this is generally the gaining side for personal interest and vanity; and what to her is the democracy or liberalism in which she has no part—which leaves her the Pariah it found her? The man himself, when he marries, usually declines into Conservatism; begins to sympathise with the holders of power, more than with its victims, and thinks it his part to be on the side of authority. As to mental progress, except those

9.7 Emma Martin. *Reasoner* 19 (6 May 1855): 41. BL, PP.636.

a higher morality: one not grounded in bloodthirsty superstition, a ruthless system of economic exploitation, and the subjugation of women. She began drawing crowds of several thousand to lectures and platform debates with former allies.

Martin aimed not to offend. Holyoake, a longtime friend, remarked that her "tone of mellowed soberness" contrasted with the shocking paradoxes of most infidels.[41] A posthumously published portrait reinforced this reassuring image (fig. 9.7). Her left hand raised in the traditional gesture of intelligence, the handwritten pages signaled her status as an author. At the same time, Martin's

41. "Emma Martin's Serials," *Movement* 1 (6 July 1844): 239.

demure bearing marked her as a model of respectable femininity. Woman, mother, and former evangelical, Martin was thought to be uniquely effective in reaching Christian audiences.

Yet her message about the development of man from matter was just as uncompromising. In her halfpenny tract *First Conversation on the Being of a God* (1844), two characters discuss natural theology: Theist presents the standard arguments against materialism, while Querist demolishes them. Theist argues that matter cannot have always possessed the same powers, for geology shows that races of animals have come into being at comparatively recent dates. Since the organic world is not eternal, this must "require the intervention of an intelligent cause":

> QUER. But suppose we discover, that the progression of developement in genera is matched by the progression in the developement of individual being; for instance, every human being has passed through a variety of stages, each more advanced and complex than the previous one; the embryo is in its earliest stage but an animalcule, it afterwards assumes the more elaborate character of the fish, and eventually becomes man. We know this is the case with the individual, may it not have been the case with the species also?
>
> THEIST. I do not understand you. Do you then imagine that man is but an improvement upon the lower animals?—A new product of Nature's increasing power?
>
> QUER. I see no reason to look further for a cause for his existence, than for that of the meanest insect.[42]

Four thousand copies of this eight-page tract were on the streets before *Vestiges* went to press.[43] Martin turned the genre of religious catechisms she had known in her youth to the advocacy of transmutation. Her dialogue re-created dozens of encounters between atheist lecturers and Christian ministers across the country. On the platform, she had a reputation for biting rhetoric and witty repartee. "Mistake me not, then," she wrote on another occasion, "it is not Christian differences with which I war, but the *system* itself;—not translations or commentaries, but the BOOK. Who will confront me?"[44]

Early in June 1846 placards announced that Martin would be giving a course of lectures on Tuesday evenings, devoted to *Vestiges* and *Explanations*. She spoke at the Literary and Scientific Institution on John Street, which had been the center of London freethought since 1840, when the building had been taken over

42. Martin [1844b], 5–6.

43. For the sales of this tract, often misdated to 1850, see *Movement* 1 (31 Aug. 1844): 314.

44. Quoted in Taylor 1983, 145. For these encounters, Royle 1974, 202–13; Taylor 1983, 140–41; Place 1983; Townley and Holyoake 1852.

by the Owenites. As usual, the charge was twopence, and twice that for sitting in the gallery.[45] The contents of the lectures are not known, but revival of the debate after *Explanations* made the subject timely. The development hypothesis offered an attractive way to draw listeners toward a stronger materialism.

Martin had long had an interest in physiology, and at the time she lectured on *Vestiges*, she was retraining for a new career as a midwife and lecturer on female health. In the spring of 1845, concern for her daughters, public persecution (she had been physically assaulted on several occasions), and sheer exhaustion had forced her to abandon the provincial lecturing circuit. Her aim remained political: to take medical power out of the hands of men, to bring knowledge about their bodies to women.[46] The same issue of the *Reasoner* that announced Martin's *Vestiges* lectures printed the syllabus of her lectures on the physiology of women. She spoke on "the development of fetal life," "abortion: its causes and prevention," and "gestation, symptoms, and hygiene of pregnancy." Such subjects were closely related to the model of fetal development in *Vestiges*. But for Martin, human monstrosities and birth defects were not traces of a higher creative law, but evidence for "regular gradation" and the absence of divine design. Most crucially, only by abandoning superstition could women manage their own affairs in accordance with nature.[47]

Martin spoke calmly, but advocated ideas almost universally condemned as outrageous and inflammatory. Materialist evolution would undermine the foundations of society. A naturalistic view of gestation—of the kind found in *Vestiges*—appeared to erode the family and other domestic social institutions. Middle-class commentators exploited anxieties based on the socialists' advocacy of the abolition of marriage. In figure 9.8, taken from a contemporary comic novel, a socialist lecturer holds aloft a treatise denouncing marriage, while the audience indicates its approval in an orgy of flirtation; two men court the woman on the left, one quietly removing her wedding ring. Martin had been viciously attacked for having left her first husband. Clergymen accused her of leading a life of profligate abandon, lusting after publicity and money.[48] Although Eve had been the first skeptic, there was, it was said, something unnatural about an atheist in petticoats. The appeal that the *Oracle*'s series on "regular gradation" was said to have had for women must have been particularly

45. *Reasoner* 1 (3 June 1846): 15; 1 (17 June 1846): 48. For recollections of a young listener to lectures at John Street (although not on this occasion), see Tinsley 1900, 1:43–44.

46. Taylor 1983, 146–49; [E. Martin], "A Review and a Prospect," *Reasoner* 4 (23 Feb. 1848): 177–79. See also her tract, *A Miniature Treatise on some of the Most Common Female Complaints* [1848?], mentioned in Cowie and Royle 1982, 190.

47. *Reasoner* 1 (17 June 1846): 48.

48. Taylor 1983, 183–216; Mackie [1845].

9.8 A satire on the view of marriage taken by Robert Owen's socialist followers. "The Anti-legal Marriage Association," illustration by Thomas Onwhyn from Cockton 1844a, facing p. 330.

alarming.[49] By this time, evangelical polemics often identified materialism as a feminized doctrine, a Continental fashion making an impression on women and weak-minded men. Owenite socialism, with its stress on sexual equality, was largely responsible for the change.

Hugh Miller expressed shock that hundreds of "respectable church-going people" were prepared to pay twopence to hear Martin lecture on infidel socialism. A *Witness* editorial claimed that Martin was followed by a troupe of young lads with small heads, receding foreheads, and smirking faces, suggesting that "the door of the monkey-ward in the Zoological Gardens had by some

49. [W. Chilton], "Theory of Regular Gradation," *Oracle of Reason* 1 (19 Feb. 1842): 77–78; Desmond 1987, 109.

mistake been left open, and that the inmates, escaping out into the open world, had attached themselves to Mrs Martin, and become Socialists." She told her audience that the Bible had been used to justify poverty, but the *Witness* accused her of caring only for money. The former chapel in which this "senseless and abandoned woman" spoke had been rented from a coffin maker, who had thereby turned the hall "into a sort of wholesale coffin, into which to screw dead souls."[50]

Many working-class political leaders were as eager as Miller was to distance themselves from the atheists. The president of the London Communist Propaganda Society had called the *Oracle* "a worthless production, a lilliputean printed disgrace, a pigmean illiterate dishonour to the cause."[51] Most of the remaining Owenites, while critical of organized Christianity, did not deny the existence of a deity. The erstwhile hammerer of the godly, Richard Carlile, was now advocating what he called "Sacred Socialism."[52] Even those without faith avoided offending religious sensibilities, for key leaders of Chartism were Christians, especially in the manufacturing districts. To run a great political movement aground on theological controversy would have been pointless. (Anticlericalism, where politics and religion overlapped directly, was a different matter.) With thousands out of work and starving, debating the origin of the human race or the implications of geology seemed a useless distraction. *Vestiges* was not reviewed in the Chartist press.

Working-class leaders who did combine religion with radicalism thought *Vestiges* stressed the least convincing aspects of the case against orthodoxy. Transmutation was seen as silly. The Chartist preacher, phrenologist, publisher, and deist Joseph Barker of the northern industrial city of Leeds ridiculed the book in his penny weekly *The People,* whose twenty thousand copies were read almost entirely by working men and women:

> The natural history of man is wrapped in great obscurity. Whatever supposition we adopt on the subject, we must, as it appears to me, admit that the origin of the human race must have been miraculous. Suppose we admit the theory contended for by the author of the work entitled "*Vestiges of the Natural History of Creation,*" namely, that mankind sprang from an inferior order of animals, such as the monkey or the orang-outang; the question would arise, How were those new-born or new-made creatures reared? . . . Besides;—how should it happen, that monkey

50. [H. Miller], "The Socialists at Church," *Witness,* 15 Jan. 1845, 2.

51. Quoted in Royle 1974, 75. Denouncing the *Oracle* as "pigmean" simultaneously mocked its diminutive intellectual stature and associated it with the discredited ultraradical Spenceans, whose most famous tract had been Thomas Spence's *Pig's Meat* (1793). For continuities within radicalism, see McCalman 1988.

52. Yeo 1981; the religious views of Owen and his followers are analyzed in Harrison 1969. For Carlile, see Wiener 1983.

> parents should produce offspring so widely different from themselves,—so vastly superior to themselves in so many respects, yet so greatly inferior to themselves in so many respects. . . . The theory of the author of the *Vestiges* is incredible. It appears to me about as incredible as the story contained in the Bible, that man was made out of the dust of the ground, and that woman was made out of one of his ribs while he slept.[53]

Barker, who changed creeds with dizzying frequency (from Wesleyan Methodism to Unitarianism to deism and back again to Methodism), claimed at the time of this editorial to stand apart from the war between infidelity and orthodoxy. He stressed the power of divine law, but thought *Vestiges* failed to make transmutation convincing.

The freethought leaders' theatrical tactics need to be seen in perspective. Most working people refused to link irreligion with campaigns for the vote or better living conditions. In terms of explicit commitment, Christianity—especially Methodism (and Roman Catholicism among Irish immigrants)—undoubtedly had deeper roots in plebeian culture than infidel atheism. For every self-educated artisan who rejected the Bible, thousands read it aloud to their families; for every Martin or Holyoake, dozens of evangelical preachers roamed the streets (fig. 9.9). Yet however few in number, the relentless activity of the agitators made it easy to read *Vestiges* as a trap for the unwary.

Faith in the City

Vestiges was only one of the dangers that was thought to threaten working- and middle-class readers. Combe's *Constitution of Man* (1828) and David Friedrich Strauss's *Life of Jesus* (1846) were at least as bad, and in cheap editions too. If not counteracted, reading would erode Christendom from within—which is of course what freethinkers hoped would happen. Opposition to *Vestiges* quickly became part of an evangelical crusade—although to speak of a crusade is somewhat misleading, for evangelicals were no more united than their free-thinking opponents. The evangelical revival had its origins in Methodism, which began in the 1730s and broke away from the Church of England in the 1790s. Other Dissenting sects, including the Congregationalists (or Independents) and Baptists, became strongly evangelical. Many within the Established Church made common cause with Dissenters to counter the threats of Tractarianism and infidelity.[54]

53. [J. Barker], "The Natural History of Man," *The People: Their Rights and Liberties, their Duties and their Interests* 3 (May 1850): 3.

54. Contemporary difficulties in defining "evangelical" are dealt with in Hilton 1988, 7–35, and in Bebbington 1989, 1–19. Ward 1973, Parsons 1988, Hempton 1996, Hoppen 1998, 427–71, Gilbert 1976, and Chadwick 1971 give general accounts of the large literature on Victorian religion, much

9.9 "A Street Preacher and his Audience." Heroic images of urban missionaries drew on a long tradition of Christian art. The erect, masculine figure of the preacher dominates the picture through his mastery of the book: God's Word guides both his speech and his gestures. Responses range from the scowling disbelief of the figure in the upper left corner (stereotypically portrayed as an Irishman) to the dumb wonder of the mother on the right. The seated women illustrate the progress of conversion, culminating in the figure whose arms, like those of Mary Magdalene in pictures from the New Testament, are raised in supplication. Gough 1882, facing p. 78.

The anti-*Vestiges* campaign was rooted in institutions founded in response to the French Revolution. The Religious Tract Society (RTS) had started in 1799 to counter blasphemy and irreligion in the pauper presses. The British and Foreign Bible Society was founded five years later to disseminate the Scriptures. Unlike the older Society for Promoting Christian Knowledge (SPCK), both groups were broad-based coalitions with a high proportion of Methodists, Baptists, Congregationalists, and other Dissenters. The London City Mission, founded in 1835 to evangelize the metropolis, had been even more dominated by Dissent, although by the mid-1840s the combined threats of popery and atheism had widened its basis of support to include many Anglicans.[55]

of which has until recently been focused on individual denominations and doctrinal controversies. Hempton 1996, 179–84, has a good selected bibliography. For a closely argued case that Methodist attitudes toward science were transformed by the *Vestiges* controversy, see Clement 1996, chapter 3.

55. Lewis 1986 gives an excellent analysis of the London City Mission and its role in encouraging evangelical unity. For the other groups, see Jones 1850; Howsam 1991; Clarke 1959; and Bebbington 1989, 75–150. Topham 1992, 423–29, discusses evangelical interests in popular education, and Rosman 1984 stresses their intellectual interests earlier in the century.

These organizations had evolved into efficient steam-driven machines for spreading the divine word, using the latest technologies of paper making, printing, rail distribution, and statistical reporting to achieve their aims.[56] The sheer number of publications distributed was amazing: some 573,050 tracts were handed out by the London City Mission in 1844–45 alone, with nearly twice that number three years later.[57] The famous RTS "Hawker's" or Second Series Tracts aimed to supplant the bawdy folk ballads and chapbook literature sold by itinerants. First Series Tracts targeted swearing, drunkenness, and blasphemy, as well as popery and infidelity. Titles often had cumulative print runs in the hundreds of thousands. *The Negro Servant,* contrasting the blessed destiny of a poor but pious black woman with the hellish fate of an infidel, sold 360,621 copies between 1814 and 1850.[58]

The religious organizations had responded slowly to the transformation of science publishing in the 1820s, at first confining their efforts to anti-infidel tracts and theological treatises. Alarmed by the secular tone of the useful knowledge literature, the SPCK then began issuing works on a wider range of subjects with a "decided bias" toward revelation, and over the next few years John W. Parker published about dozen small octavos on their behalf.

Fresh initiatives were needed. Too many tracts (like *The Negro Servant*) had rural settings, not urban or industrial ones. The missionary to the John Street District, site of "the head-quarters of Infidel organization for the United Kingdom," warned that the London City Mission now faced more serious threats than the familiar ones of enlightened philosophy. Just as poisons leavened the bread of the working class, he said, so too was atheism now "stealthily mixed" with much that was good. Holyoake and his associates served up secular knowledge, especially science ("too much neglected by the religious part of the community") in such a way as to bring God's word into disrepute. Of all the forms of infidelity, this was "the most difficult to grapple with."[59] Moreover, useful knowledge publishers—while no friends to atheism—had shown that the audience for cheap print was becoming more sophisticated and unlikely to be convinced by the homely lessons of a tract. Knight, for example, had been issuing weekly shilling volumes from June 1844, each with the content of a three-hundred-page octavo.[60]

56. Howsam shows this convincingly in her exemplary study of the British and Foreign Bible Society (1991).

57. *London City Mission Magazine* 14 (Feb. 1849): 23–30, at 25.

58. Jones 1850, appendix 5, 3; [Richmond] [1814]. On tracts, see Neuberg 1977, 249–64; Altick 1957, 99–108; and Topham 1992, 420–21.

59. *London City Mission Magazine* 20 (Nov. 1855): 263–66. See also the comments in [C. Williams] [1852], 188–89, quoting from the *Edinburgh Review.*

60. Knight 1864, 312–21.

To meet the challenges, the Religious Tract Society opened a magnificent new depository on Paternoster Row and commenced a series of monthly volumes "on secular subjects, treated in an evangelical manner." These little books (usually 192 pages, and priced at sixpence in paper and tenpence in boards) were to be original in matter, popular in style, and scriptural in principle. The series was aimed at families, schools, libraries, and public reading rooms.[61] By the early 1850s, several titles implicitly targeted *Vestiges*.[62] The two parts of the Scottish "Christian Philosopher" Thomas Dick's *Solar System* sold 30,510 and 26,890 copies respectively in just five years.[63] Commercial publishers seized the opportunity too. The temperance advocate John Cassell, who became a major mid-Victorian educational publisher, issued a library that included books on ethnography and astronomy that explicitly discussed the implications of *Vestiges* and ended by quoting Scripture.[64] The need for such publications was widely felt. As *Christian Observer* warned, the *Vestiges* sensation meant that skeptical arguments were reaching "ordinary readers" in "a purely popular form."[65]

Young men were in special danger. *The Problem of Life*, published by the RTS in 1846, asked readers to picture a young man, "his principles sapped by scepticism," sitting down with his Bible:

> He has imbibed a smattering of geology, and is in the outset perplexed by his inability to reconcile his favourite theory on that subject with the Mosaic account of the creation. He has glanced through some philosophical treatises on the origin of society, and is reluctant to abandon the brilliant speculations which have been hazarded on that point for the simple, unpretending account which Scripture contains of the origin of the human race. If he be a physiologist, he startles at the reasons assigned by Scripture for the introduction of death into the system of nature;

61. Jones 1850, 147–48, and advertising material included with each monthly volume. I am much indebted to Aileen Fyfe, who is currently researching the publishing activities of the Religious Tract Society in this period.

62. These included [W. Martin], *Comparisons of Structure in Animals* ([1848]); [W. Newnham], *Man, in his Physical, Intellectual, Social, and Moral Relations* ([1847]); and [W. L. Notcutt], *The Geography of Plants* ([1850]). The Reverend Thomas Rawson Birks, second wrangler at Cambridge in 1834, wrote several works for the RTS including a *Modern Astronomy* (1850), although this was not part of the monthly volume series. Birks's argument in this book on the plurality of worlds was later taken up by Whewell, his former tutor at Cambridge. See Birks 1850; on the plurality issue, see Crowe 1986, 296–97.

63. Jones 1850, appendix 5, 5.

64. F. Williams 1852, 111–23, on astronomy; Kennedy 1851, 2:112–16 on ethnography. For the library, see the advertising bound into the back of copies of Kennedy 1851. Curwen 1873, 267–78, 324–32, discusses Cassell.

65. *Christian Observer*, n.s., no. 165 (Sept. 1851): 599–610, at 601. The most comprehensive guide to infidelity, including extensive discussions of *Vestiges* (97–110), was Pearson 1853.

> if a linguist, he is equally shocked by the account of the origin of the diversity of languages at the tower of Babel. He is a believer, perhaps, in the favourite theory of the progressive advancement of the world to a state of complete happiness by the diffusion of education and the liberal arts, and is disagreeably disturbed by finding a great deal in that sacred volume to contradict this supposition.[66]

He steps out into the crowded streets of the city; the simple truths of Christ have little chance against the "voluptuous blandishments" presented on every side by commerce, luxury, and "sinful pleasures." "The age of gross and vulgar infidelity has perhaps passed away"; but the author knew that "scepticism is continually reproducing itself in new and alluring forms."[67] The city itself was portrayed as a female temptress. As one preacher told his young congregation, quoting Solomon, "'Her house is the way to hell, going down to the chambers of death.'"[68] Babylon the Great could be symbolized by the spectacle of a woman like Emma Martin lecturing on *Vestiges,* the book clothed in scarlet.

The best counsel was to avoid dangerous books, and to read all works with constant attention to their spiritual effect. Thus none of the RTS monthly volumes specifically mentioned *Vestiges,* nor did any of the evangelical magazines produced for a working-class audience, such as the *Working Man's Friend* or *Christian's Penny Magazine.* Another RTS book, *Self-Improvement,* gave advice guaranteed to raise suspicions in anyone who opened *Vestiges:*

> *How shall you begin to read a book?* Always look into your dish and taste it, before you begin to eat. As you sit down, examine the title-page; see who wrote the book; where he lives; do you know anything of the author? Where, and by whom published? Do you know anything of the general character of the books published by this publisher? Recollect what you have heard about this book.

The author advocated reading sentence by sentence, asking at the close of each "'Do I understand that? Is it true, important, or to the point? Is there anything valuable there which I ought to retain?'"[69] Answers to these questions could be marked by a system of marginal annotations (fig. 9.10). This catechismal system effectively advocated reading every book as closely as the Scriptures, to guarantee that readers would not be seduced by "unmanly" arguments.

66. [Miller] [1850], 48–49.

67. Ibid., v, 51–52.

68. Reverend Thomas Dale, "Young Men Warned Against Unsound Principles" (sermon preached at Saint Bride's Church, Fleet Street, 11 Jan. 1845), in *The Pulpit* 49 (15 Jan. 1846): 29–36, at 34; on the urban temptress, see Halttunen 1982.

69. [Todd] [1848], 74–75. This book was adapted for a British audience from an American work. The original concerns addressed by the work are discussed in Halttunen 1982, 1–6.

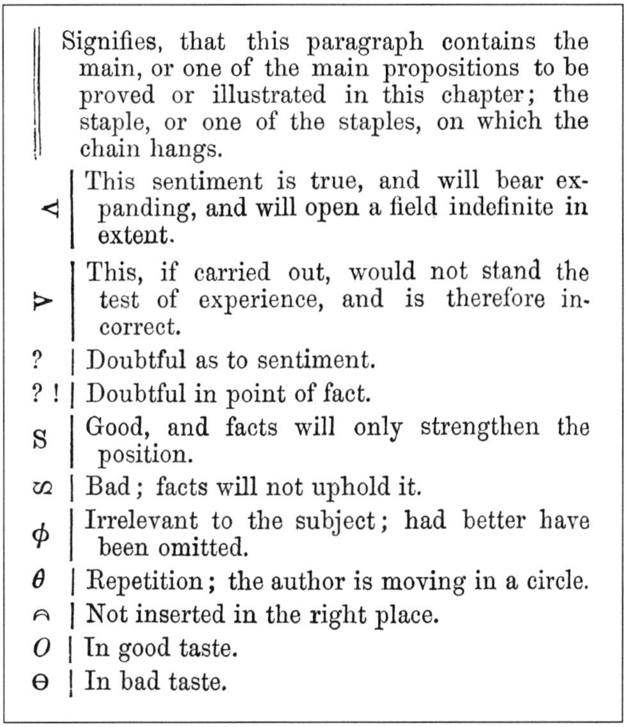

9.10 Symbols to be used by young Christian readers in annotating books. Contemporary evidence suggests that systems of annotation of this kind may have been taken up by many readers. From [Todd] [1848], 75–76.

‖	Signifies, that this paragraph contains the main, or one of the main propositions to be proved or illustrated in this chapter; the staple, or one of the staples, on which the chain hangs.
⊲	This sentiment is true, and will bear expanding, and will open a field indefinite in extent.
⊳	This, if carried out, would not stand the test of experience, and is therefore incorrect.
?	Doubtful as to sentiment.
? !	Doubtful in point of fact.
S	Good, and facts will only strengthen the position.
ꙅ	Bad; facts will not uphold it.
ϕ	Irrelevant to the subject; had better have been omitted.
θ	Repetition; the author is moving in a circle.
⌒	Not inserted in the right place.
O	In good taste.
Θ	In bad taste.

To help individuals in their journey to Christ, a common form for tracts was the conversion narrative. A reclaimed Owenite socialist wrote one such account, using the opportunity to accuse his former partner on the lecturing circuit, Thomas Simmons Mackintosh, of having written *Vestiges*. Mackintosh, a deist rather than an atheist, was a plausible candidate for the authorship. He had lectured on radical science for years and was well known in working-class circles through the cheap part work reissue of his anti-Newtonian *Electrical Theory of the Universe* (1841), a book typeset by Chilton. Man, on this view, was an organized machine.[70] Evangelicals of all parties knew that socialist infidels had long advocated regular gradation, a doctrine that they considered *Vestiges* to have dressed up in polite language.

The RTS was only one of many organizations hoping to rescue young men from the snares of dissolute reading. The Young Men's Christian Association

70. Mackintosh [1841]; Chilton mentions typesetting the cheap reissue in *Reasoner* 3 (3 Nov. 1847): 608. Mackintosh's activities are reported in the *New Moral World*, the *Mechanic's Magazine*, and the freethinking press. For a brief notice of his death, see G. J. Holyoake, "The Death of T. Simmons Macintosh," *Reasoner* 19 (14 Aug. 1850): 220; for his electrical theories, Morus 1998, 135–39. The accusation of authoring *Vestiges* is made in Maguire 1858, 68–69.

9.11 An idealized view of Exeter Hall on the Strand, the center of evangelical life in London from the early 1830s. Exeter Hall became the site for the lectures organized by the YMCA against *Vestiges* and other forms of "infidel philosophy." The Great Hall, shown here during a choral sing-along, could hold four thousand, and a meeting could be held at the same time in a smaller room below. *ILN*, 18 May 1844, 317.

(YMCA) had been founded in June 1844, a few months before *Vestiges* was published. Like the RTS initiatives, it appealed to shop assistants and clerks, those whose manly vigor was thought to be most under threat. It grew from meetings of a twenty-three-year-old draper and his friends, including several working for the London City Mission, and although dominated by Congregationalists it rapidly gained support from the main denominations. Within five years branches were established in major cities, and then around the world.[71]

A year after its foundation the YMCA began an annual series of free public lectures in which the war against *Vestiges* reached a climax. A triumph, the lectures were soon being held at Exeter Hall on the Strand, the center of metropolitan evangelical life. Preaching, the only acceptable form of evangelical theater, here addressed the threats the city posed to faith. Over the next two decades the great pulpit orators thundered the message of salvation to vast audiences (fig. 9.11). When Hugh Miller spoke there he needed an experienced

71. Binfield 1973; there is, however, no good general account of the YMCA.

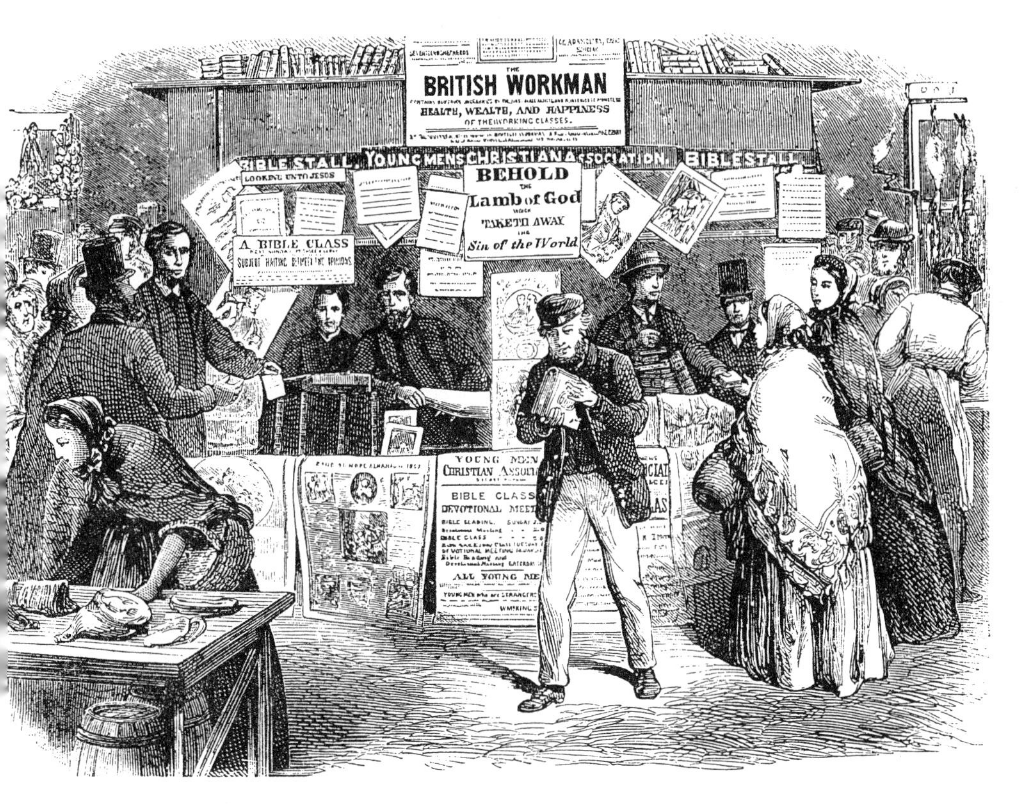

9.12 A YMCA bookstall at Leeds, showing the rich range of books, posters, and tracts that the association had published by the 1860s. In the center of the picture, a young workingman eagerly clutches the Bible he has just obtained from the stall. *British Workman,* June 1869, 168.

preacher's assistance to be heard by a crowd estimated at five thousand.[72] Published by the RTS as tracts and sold at YMCA bookstalls (fig. 9.12), the lectures reached an even larger readership, selling ten thousand copies on average and even more in the United States.

Vestiges was singled out for special treatment after it appeared in a cheap format.[73] In 1847, six months after the people's edition, the Congregationalist nat-

72. Bayne 1871, 2:440. For a contemporary history, see *Lectures Delivered Before the Young Men's Christian Association 1845–46* (London: James Nisbet, 1864), 1:ix–lxv. Other lecture series were commenced on a similar basis, notably by the Working Men's Educational Union, formed in 1852. The lectures, which attracted an attendance of 120,000 in their first year, were soon being held twice daily from premises on King William Street in the City of London. Featuring topics such as the solar system and human physiology, these illustrated talks were less overtly moralizing than those at the YMCA. Lewis 1986, 227–29.

73. A footnote in the published version of first lecture, given by the Reverend John Stoughton in the winter of 1845, had slammed the "ridiculous dreams" of Lord Monboddo and Erasmus

uralist Edwin Lankester lectured on "the natural history of creation." This was the first talk by a layman and the first ever to be published. Lankester, who had already been paid to revise and review *Vestiges* (negatively, in the *Athenaeum*), avoiding mentioning his target, referring only to "a popular treatise on the history of creation." Evidence of progress, he showed, was indeed everywhere in the fossil and living worlds—but development was impossible. His huge lecture diagrams (fig. 9.13) illustrated the "immense superiority" of the anatomical structure of the human fetus over the animals. He noted that the head bones had been separated to illustrate the vertebral theory of the skull as developed on the Continent by Lorenz Oken. Although Oken's theories were sometimes denounced in England as transcendental atheism, Lankester (following Richard Owen, from whose recent work the picture was drawn) was keen to recruit the best Continental science to a Christian cause.[74]

In Lankester's Nonconformist vision, the evidence for physical progress was paralleled by advances in the realm of mind and spirit, which would culminate by fulfilling prophecies of Christ's thousand-year reign on earth. "The time will come when God shall dwell in every heart," Lankester told his audience, "when the highest object of the mission of the Lord Jesus Christ shall be accomplished, and God shall be 'all, and in all.'" Then there would again be a "paradise upon earth, and perfect happiness shall reign." From 1850 Lankester held the part-time chair of natural sciences at New College in suburban Hampstead, where he taught similar lessons to aspiring Congregational ministers.[75]

The Reverend Hugh Stowell of Manchester delivered a blazing assault on "modern infidel philosophy" at Exeter Hall in December 1848, a few months before his triumph in Liverpool's former hall of science. Stowell had to assume that his audience of office clerks and draper's assistants might be familiar with the "wild hypothesis" of *Vestiges,* though they might not know that Continental writers had had "similar day-dreams." The popularity of such books reflected badly on the judgment of modern readers. Nothing was to be feared from science properly conducted: scientific men had assured Stowell that *Vestiges* was

---

Darwin, noting that their revival in *Vestiges* had fallen against the onslaught of the *Edinburgh Review.* Stoughton, a learned Congregationalist historian, believed that the development doctrine had been "overthrown on *scientific* grounds—the proper grounds on which it should be treated." Stoughton 1864, 25. Other relevant lectures besides those mentioned below include Archer 1848 and 1864, James 1849, and Miller 1854.

74. Lankester had first learned of these theories while attending Grant's course at University College London in 1836–37; Desmond 1989, 84.

75. English 1991, 50–53; Lankester 1848, 32. In a later, anonymous review of Lorenz Oken's *Elements of Physiophilosophy* in the *Athenaeum,* Lankester was at pains to distinguish the transcendentalist views of the German school from what he saw as their mistaken application in *Vestiges,* which "confounded a law with an idea"; *Athenaeum,* 2 Oct. 1847, 1021–22, at 1022.

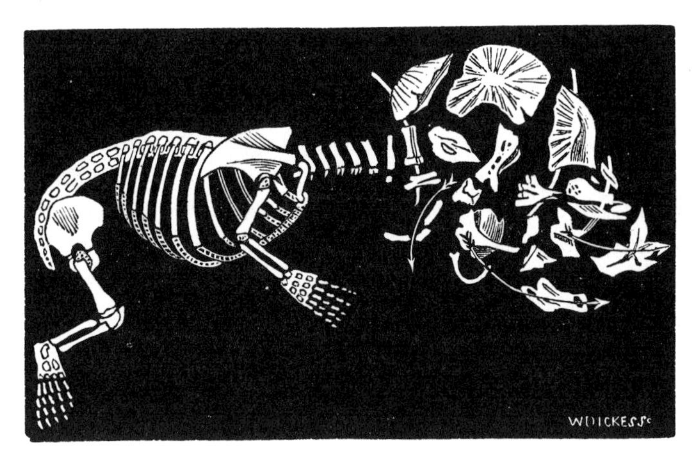

9.13 This picture of the skeletal structure of a human fetus was one of the giant diagrams used by the Congregationalist naturalist Edwin Lankester to illustrate his talk at Exeter Hall. Lankester interpreted the vertebrate archetype in man as a sign of the coming millennial reign of Christ. Lankester 1848, 30.

shallow and absurd. The real problem was the book's "profaneness" when compared with the majestic grandeur of the scriptural account of creation. However, Stowell had little patience with those he met in polite London Society who timidly regretted the harm the book might do. "But for all the pitiful patchwork theories of creation manufactured by ingenious philosophists," Stowell proclaimed, "they are as puerile and contemptible as they are irreverent and ungodly." Infidelity had dashed against the rock of faith for thousands of years. The Bible could never be "damaged." Stowell's associate, the Reverend Hugh McNeile of Liverpool, similarly emphasized that difficulties in reconciling science and religion were a moral test; the power of explanations based on natural law was no excuse for "practical atheism."[76]

The Reverend John Cumming made the link between *Vestiges* and atheism most explicit. Cumming, active in the YMCA and leader of London's Scottish Presbyterians, and opponent of the Free Church, was a staunch premillenarian, believing that biblical prophecy was being fulfilled even as he spoke. "Study all the sciences," he told his audience, "but oh! study them as they cluster round the cross; study them in the light of Him that hung upon that cross." Cumming raged against *Vestiges,* as Bosanquet did, in the conviction that the end of the world was at hand.[77] The First Vial mentioned in the Book of Revelation had

76. Stowell 1849, 164–65, 170–71, 179; M'Neile 1850, 260.

77. Cumming 1851, 226, 241; see also Cumming 1853a, 73–75; Cumming 1853b, 17–19. Cumming's equation between *Vestiges* and atheism was challenged in W[illiam] D[avidson] 1851.

been emptied as the "noisome sore" of the French Revolution in 1789. *Vestiges* had appeared at the pouring of the Sixth Vial: with the Turkish dynasty nearly over, the Jews would be returning to Palestine, while three "unclean spirits like frogs" proceeded from their respective sources, as also foretold in Revelation. The first frog was infidelity, which crept from the dragon's mouth. Cumming identified the philosopher David Hume as "a great arch-frog" whose "croaking" had "led the other unclean spirits of his day into the marshes in which he himself wallowed." During the nineteenth century, the icy rationalism of Enlightenment infidelity had been decked "in meretricious finery" and gilded with "sensuous charms" to tempt the unwary.[78] This had been done by *Vestiges,* a frog vomited from the mouth of Satan.

The anti-infidel crusade brought together a wide range of parties and sects, which on most issues were deeply divided. What "evangelical" meant was hugely controversial, but these differences could be set aside in combating skeptics. For Stowell, McNeile, or Cumming, it was a no-holds-barred fight. As the moderate Anglican *Christian Observer* admitted, *Vestiges* ought to be met with "a few sentences of vigorous invective" and a clenched fist. The book was "pigology," "undisguised materialism," "infidel and even atheistic in its tendency."[79] Similarly, the weekly *Nonconformist* newspaper, edited by the Congregationalist Edward Miall, condemned the "infidel" *Vestiges* as "that most erroneous and pernicious work."[80] On this all evangelicals could agree, despite the chasm of controversy that separated them on key issues such as the disestablishment of the Church of England. Urban missionary work against science-based infidelity (like the even more vigorous war against popery) provided a practical groundwork for the difficult project of pan-evangelical union. As already seen in Edinburgh, a shared reading of books like *Vestiges* as "infidel" paradoxically helped to define what it meant to be "evangelical."

Reading the Crisis of Faith

We are familiar with famous narratives of the so-called Victorian crisis of faith.[81] But the tracts and lectures discussed in this chapter were targeted not at

78. Cumming 1848, 405–7. For the importance of premillenarian imagery, see Harrison 1979; Hilton 1988, 95–97, 365–71; and James 1976, 49–61. For the specific dangers of the mid-1840s, Bickersteth 1846, 18–19, 32, 80–81, 195, 334.

79. "Vestiges of Creation," *Christian Observer,* n.s., no. 165 (Sept. 1851): 599–610, at 606, 601; and, for "pigology," "Claims of the Animal Creation," *Christian Observer,* n.s., no. 95 (Nov. 1845): 670–91, at 671. Yule 1976, 262–65, gives a detailed reading of *Christian Observer*'s 1851 review, stressing its more moderate aspects.

80. *Nonconformist,* 9 July 1845, 490; *Nonconformist,* 13 Aug. 1845, 569.

81. As James Moore (1990, 166) writes, "Judging from the titles of literature dealing with the conventional 'crisis of faith,' one would think that Victorian intellectuals lost their faith as the rest

gentlemen's clubs or literary salons, but at public houses, open-air debates, and twopenny halls. Men and women who lived through the Industrial Revolution experienced the changes as an intensely personal, spiritual journey. Formulating what it meant to be an individual involved exploring possibilities that would define working- and middle-class, rural and urban, Anglican and Nonconformist, masculine and feminine, believing and unbelieving. Family, faith, work associates, and local community pulled in different ways and were not always reconciled.

Three final cases show what could be involved.

The youngest daughter of the vicar of a rural Cumberland parish had been struggling with her father over religion. He abused Unitarianism, deism, and Dissent, and insisted that she read a three-volume answer to the Enlightenment rationalists. "When 'Vestiges of Creation' came out," she remembered, "our fight was serious. Giving, as it did the first idea of cosmic continuity, and the consequent destruction of the bit by bit creation of Genesis, it was a priceless treasure to me, to him a deadly and diabolical sin." At the same time, she had fallen in love with a married woman, eight or ten years her senior. The older woman was a "transcendentalist" and "pantheist." From her new perspective, *Vestiges* appeared "one of the advance guards in the forces of knowledge, as they stand arrayed against those of ignorance."

The second case is that of a Chartist and former shoemaker who had served two years in Stafford jail for sedition and conspiracy. In May 1845 he traveled to London to find a publisher for the verse epic he had written in prison. He had lost his evangelical Wesleyan Methodist faith. His poem dealt with Chartist politics and pondered the roles that priests and men of science could play in ending working-class slavery and injustice:

> I say not that there is no God: but that
> *I know not.* Dost *thou* know, or dost thou guess?—
> Why should I ask thee, priest? Darkness hath sat
> With Light on Nature,—Woe with Happiness,—
> Since human worms crawled from their languageless
> Imperfect embryons, and by signs essayed
> To picture their first thoughts. 'Tis but excess
> Of folly to attempt the great charade
> To solve; and yet the irking wish must be obeyed!———

He soon learned that similar ideas about development were in *Vestiges,* and found freethinking circles and the publishing world buzzing with rumors about

of us lose umbrellas." For a good starting point, although concentrating on the beliefs of well-known Victorian intellectuals, see Helmstadter and Lightman 1990.

the authorship. The Chartist recognized that whoever had written *Vestiges* was not an atheist. His prison poem, which drew on a remarkable range of reading, was grounded in the cosmologies of Lucretius and the Enlightenment materialists. From such a perspective, *Vestiges* held little new or startling. However, it did create common ground between radical publishers, middle-class journalists, and the freethinking circles in which he began to move.

The last case involves a boy exposed to freethought in 1837 at the age of fourteen. The son of a down-on-his-luck father who had retreated to a remote district in Wales to save money, he attended the local Anglican church twice every Sunday to hear sermons on hellfire and eternal punishment. He had gone to London to stay for a few months in Hampstead Road with his brother, who was apprenticed to a builder. He recalled evenings in one of the halls of science:

> Here we sometimes heard lectures on Owen's doctrines, or on the principles of secularism or agnosticism, as it is now called; at other times we read papers or books, or played draughts, dominoes, or bagatelle, and coffee was also supplied to any who wished for it. It was here that I first made acquaintance with Owen's writings, and especially with the wonderful and beneficent work he had carried on for many years at New Lanark. I also received my first knowledge of the arguments of sceptics, and read among other books Paine's "Age of Reason."

He claimed to have been impressed by Combe's phrenology, a cheap reprint of the anatomist William Lawrence's *Lectures on Man,* and a tract by Robert Dale Owen that damned the doctrine of eternal punishment. Back in Wales, the young man turned to a career in land surveying and became active in the Neath Mechanics' Institution and other organizations. It is here, in his early twenties, that he seems to have learned of *Vestiges* through debates reported in the local newspapers. As he told a friend in December 1845:

> I have rather a more favorable opinion of the "Vestiges" than you appear to have.
>
> I do not consider it as a hasty generalization, but rather as an ingenious hypothesis strongly supported by some striking facts and analogies but which remains to be proved by more facts & the additional light which future researches may throw upon the subject. It at all events furnishes a subject for every observer of nature to turn his attention to; every fact he observes must make either for or against it, and it thus furnishes both an incitement to the collection of facts & an object to which to apply them when collected.

Just over two years later, the author of this letter was on his way up the Amazon. Convinced that the evolutionary theory of *Vestiges* needed to be tested, Alfred Russel Wallace abandoned surveying, became a commercial collector, and traveled for most of the next decade. Like other impressionable young men, he had lost his faith and his fear of Hell. He had a commitment to socialism, self-help, and phrenology. But it was partly through *Vestiges* that he had found his vocation as a naturalist and also his problem, formulating "the law which has regulated the introduction of new species."[82]

The other two also became leaders of intellectual life. Thomas Cooper, the Chartist, had achieved notoriety through his arrest and imprisonment, but after his release his future was far from certain. He had nothing more to do with political agitation, but devoted his energies to working-class education and attacks on religion. He became a leading performer at the John Street Institution and other freethought venues, speaking on literary and historical topics until, in 1856, he regained his faith and became a traveling lecturer on the Christian evidences. In this new capacity, Cooper opposed evolution and spoke against *Vestiges*—a book that at one time had seemed tame.[83]

Eliza Lynn, the vicar's daughter, moved at the age of twenty-three to rooms near the British Museum. She associated with other bohemian authors and later married the Chartist wood engraver William James Linton, although they soon separated. Writing orientalist fiction and miscellaneous reviews, Eliza Lynn Linton became a celebrated novelist and opponent of women's rights. Many years later, she recorded her experiences in a remarkable piece of literary cross-dressing, *The Autobiography of Christopher Kirkland.* Linton outlined her science-based creed for the *National Magazine* in an essay on "the unities of nature," which hailed the developmental philosophy of *Vestiges* as pointing to "the divinity of the law of progress."[84]

These cases are important not because of any subsequent fame. In the 1840s all were of humble status and unsure future. Eliza Lynn and Alfred Wallace were in their teens or early twenties, like so many targeted in the war over freethought. Cooper had entered into a new career as an author. All had to

82. This is the title of Wallace's famous paper in *Annals and Magazine of Natural History,* 2d ser., 16 (Feb. 1855): 184–96. On the early exposure to Owenite socialism, Wallace 1905, 1:87 (although the John Street Institute is mentioned, this cannot be correct, as the Owenites occupied it only from 1840). For the letter supporting *Vestiges,* A. R. Wallace to H. W. Bates, 28 Dec. 1845, in McKinney 1969, 372. Besides Wallace's 1905 autobiography, the best general accounts of Wallace's early life are Moore 1997, 300–307; Durant 1979; and Hughes 1989. The impact of *Vestiges* is stressed in McKinney 1972. I have not cited these sources (nor those in the next two notes) in their usual place, as the cases would then tend to be read in light of later celebrity.

83. The sources used for Cooper are his poem, *Purgatory of Suicides* (1845, 191); his account of his arrival in London (1872, 259–63); and comments on *Vestiges* in Cooper 1878, 10–12.

84. E. L[ynn] L[inton], "The Unities of Nature," *National Magazine,* 28 Jan. 1859, 52–57, at 57. On *Vestiges* as "one of the advanced guards," [Linton] 1885, 1:235–36, 3:75. On her later life in London, Linton 1899, 16–39. Richards 1989a 271–75, gives a fascinating account of the dilemmas Linton faced in pursuing her interests in science.

define who they were going to be, not only in terms of this world but in terms of the life to come. From an evangelical point of view, only Cooper eventually avoided becoming part of the statistics of unbelief. Yet their references to *Vestiges* show, if a reminder is necessary, that books do not influence people in any determinate way. Books have no single meaning, no coercive "impact."

This is a familiar point, and it may seem that all individual readings of a work can do is point toward difference and heterogeneity.[85] Seen from the perspective of local reading practices, however, the cases display a remarkable degree of convergence. Each of them, like so many others discussed in this book, began from the premise that the central choice readers faced was between belief and unbelief. Evangelicals (and evangelizing freethinkers) agreed that *Vestiges,* with its seductive narrative, authoritative tone, and religious gloss, was potentially a powerful poison in the hands of unbelievers. Hence the link, implausible in other contexts, between atheism and the development hypothesis. Underlying the controversies was a consensus about what reading was for.

Reading offered important ways of defining self. For the freethinkers, the individual was a "social atom," whose significance and political power rested in group action.[86] On the other side, the autobiographies, biographies, funeral sermons, diaries, portraits, and missionary case histories that are so characteristic of the Victorian period testify to the predominant notion of life as a spiritual journey. Embodied in languages of gender, class, geography, and faith, all such narratives are simultaneously individual and social.[87] It would be a mistake, then, to see such accounts as linguistic constructions floating free from material circumstance and relatively stabilized relations of power. Defining the choice between belief and unbelief as a property belonging to an *individual* was itself bound up with larger economic and social changes. Individual identity was constructed from a range of possibilities, which can be recovered through close attention to the practices of everyday life.

The new technologies of communication offered the possibility of repairing broken bonds of community. When Wallace devoted a chapter of *My Life* to his experiences in working-class London, when Cooper recalled his experiences after leaving prison, or when Lynn Linton wrote her life, readers were invited to use past struggles to make sense of their own. Such works tied anecdotes of reading into narratives of individual experience and national spiritual progress.

During the religious revivals of the nineteenth century, the quest for iden-

85. This is the main use of actual reading experiences made in the generally admirable Flint 1993, 187, one of the pioneering works in this field.

86. Desmond 1987, 86–88; Harrison 1969.

87. Joyce 1994.

tity was increasingly shaped by the industrial culture of print. In many ways this change affected readers like Cooper and Wallace more profoundly than it did classically educated aristocrats or men and women from the professional classes; for the wealthy traditionally had access to books. Hence the growing importance of printed works in narratives of faith lost and gained among working people.[88] Each page of a deathbed conversion, a missionary's case book, or a skeptic's confession had a material basis in the economy of publishing, printing, and authorship. Individual readers of such accounts made meaning, but they did so through the practical conventions of the communities in which they moved. It may be that few men and women "lost their faith" because of reading about the earth's age or the origin of the human race or through the cataclysmic effect of any single text.[89] Yet the importance of the printed word in defining communities, symbolized by the role of debates about scriptural interpretation in defining Britain as a Christian nation, meant that books were often singled out in narratives of spiritual development.[90] In an urban and industrializing society, reading could make sense of the chaotic and emotional process of defining belief.

Tens of thousands of working men and women read *Vestiges*. But what did they experience in turning the pages of the book? How were books used in the mundane affairs of everyday life, and in making small decisions over days, weeks, and months that added up to big changes in belief and vocation? Answering these questions requires looking in depth at individuals: the next chapter analyzes a reader's experience with the same intensity that is usually applied to the process of writing.

88. Vincent 1981; 1989, 171–80.

89. This is the oft-cited result of Budd 1977, which examines Victorian autobiographies for explicit statements of reasons for loss of belief. This approach is based on a concept of opinion formation akin to that found in the older literature on the mass media, which examined the effects (say) of television violence in isolation from its settings and actual use.

90. Henderson 1989 discusses this issue in relation to some well-known autobiographies. For working-class narratives and reading, see Vincent 1981, Webb 1955, and several of the essays in Shortland 1996b, especially Shortland's own discussion of Hugh Miller's self-fashioning.

CHAPTER TEN

Self-Development

I must, however, remind my reader that the "I" who speaks in this book is not the author himself, but it is his earnest wish that the reader should himself assume this character, and that he should . . . hold converse with himself, deliberate, draw conclusions and form resolutions, like his representative in the book . . . and build up within himself that mode of thought the mere picture of which is laid before him in the work.

JOHANN GOTTLIEB FICHTE, *The Vocation of Man* (first English translation, 1848)

MOST READINGS LEAVE little or no trace—an ownership signature or a few pencil marks. Only certain types of reading, such as academic study and reviewing, produce more substantial records. Recovering the voices of readers who are not literary producers and who do not have established reputations is more difficult. Even so, the nineteenth-century material—thanks to increased readerships, the popularity of diary keeping, and the evangelical revival—is more extensive than anything available before or since. It is surprising that there are no analyses of nineteenth-century readers comparable to those available for earlier periods, such as John Brewer's study of the educated Englishwoman Anna Margaretta Larpent and Robert Darnton's account of the Rousseau-reading French merchant Jean Ranson.[1] The difficulty is not to find detailed records of reading, it is locating them for a specified title.

For *Vestiges* there is an exceptional source in the journals of the eighteen-year-old Thomas Archer Hirst of Halifax in West Yorkshire.[2] Born in 1830, Hirst

1. Brewer 1996; R. Darnton, "Readers Respond to Rousseau," in Darnton 1984, 209–49. For analogous examples, see Sherman 1994 (on John Dee); Sharpe 2000 (on Sir William Drake); Jardine and Grafton 1990 (on Gabriel Harvey). For a review, see Colclough 1998.

2. Brock and MacLeod 1980b gives the best account of Hirst's life; see also Gardner and Wilson 1993. Brock and MacLeod 1980a is a microfiche edition, with index, of the typescript

was the youngest son of Thomas Hirst, a merchant who bought wool from producers, graded it, and sold it to manufacturers. Hirst's father was a liberal Congregationalist and supporter of the Whig Lord Morpeth in the 1835 elections; his mother Hannah came from a wealthy Anglican family. The elder Hirst had died in a drinking accident in 1842, and although there was some money (through his mother) the three boys needed to find work suitable to their middle-class status.

Halifax was a bustling place with a population of twenty thousand, mostly manual laborers engaged in the textile, engineering, and machine tool industries. The steep and narrow streets lent "a touch of antiquity," but worsted mills and mill workers' dwellings circled the center. Working conditions were bad in the mills; sanitation throughout the city was appalling. As one reporter complained, Halifax was "a marvel of dirt."[3] The railway mania was at its height, so when Hirst left school in 1845 at fifteen he was apprenticed for five years to a civil engineer, Richard Carter, to train as a surveyor. This was thought to offer a good route toward a secure future as a professional man. Probably following the example of Carter's chief surveyor, the Irish Orangeman John Tyndall, Hirst began recording details of his reading in novels, periodicals, poetry, theology, and especially science. Because Hirst was so young, his journal is explicit about activities that most readers took for granted.

Hirst read *Vestiges* when he was defining his personal faith, his sense of himself, and his vocation in life. Tyndall described him as "our junior apprentice, a youth upwards of 6 feet high, and about 16 years of age—an immense development of brain which is in true keeping with his extraordinary powers of thinking."[4] Hirst explored the book in the same way as many other young working- and middle-class autodidacts in cities: through discussions with friends at home, in taverns, at work, and at the Mechanics' Institution and his local Mutual Improvement Society. He studied it closely for nearly a month, copying out long passages from it in his journal. He read reviews and Miller's *Footprints,* noting his own reactions and those of his friends. In its combination of diary, commonplace book, and conversational record, the journal reflects Hirst's immersion in contemporary periodicals, social problem fiction, and the writings of Thomas Carlyle. Hirst was precisely the kind of reader addressed by the YMCA, the useful knowledge publishers, the infidel mission, and by *Vestiges.*

---

version of Hirst's entire journal. The original of this transcription is in the archives of the Royal Institution.

3. Ginswick 1983, 1:170. For miscellaneous details about Halifax I have relied on the local newspapers and the principal contemporary street directory, Walker 1845.

4. Brock and MacLeod 1980a, 7 June 1846, f. 24 n. 38, quoting from Tyndall's *Journal,* vol. 1, f. 129.

Alternatively, Hirst could be (and usually has been) taken as a type of reader who later assumed a position on the national stage. In the 1860s and 1870s, he became known as a mathematician and for his membership of the X-Club, the inner circle of science "pure and free, untrammelled by religious dogmas."[5] Like others of his generation, Hirst emerged from a spiritual crisis to become one of the leaders of a new faith in science. Many came from backgrounds in provincial Dissent, notably the evolutionary philosopher Herbert Spencer; Tyndall shared Hirst's training as a surveyor, as did Alfred Russel Wallace. There is, however, a crushing familiarity about this way of telling the story. Only in retrospect can Hirst's case be seen as typical of the leaders of mid-Victorian science; and this view gives little idea of the struggles that were involved. Each experience of reading becomes more generally revealing the more locally it can be situated. Hirst read *Vestiges* not as a future X-Club member, but as a participant in Halifax's lively local culture of self-improvement.

Reading Vestiges

On the evening of 5 January 1848, the American transcendentalist Ralph Waldo Emerson lectured on "Napoleon as a man of action" at the Halifax Mechanics' Institution. This was part of a national tour arranged by Alexander Ireland, which included lectures on a wide variety of subjects (fig. 10.1). Although Emerson read carelessly and spoke with a nasal twang, his address was rapturously received.[6] Afterward a group of young men gathered in lodgings for discussion. Hirst, at this time seventeen, listened as the self-assured Tyndall debated Dr. William Paley, a medical man and grandson of the author of *Natural Theology*. Both Tyndall (who had moved to Queenswood in Hampshire to teach applied science) and Paley were visiting Halifax.

The subject of the disagreement was *Vestiges*. Paley thought the book's argument weakly supported, especially in relation to the nebular hypothesis. Hirst suspected that Paley was "a clever fellow, well read, but too much inclined to go along with the stream."[7] Tyndall certainly had a more positive view of *Vestiges*, for he had made a thorough "Digest" of the book in his own journal twelve months before. "There is in reality no true great and small grand and familiar in nature," Tyndall had paraphrased the text, "Such only appear when we thrust ourselves in as a point from which to start in judging. Let us pass if possible beyond immediate impressions and see all in relation to Cause and we shall chastenedly admit that the whole is alike worshipful."[8] Years later, Tyndall

5. Quoted in Barton 1990, 57.
6. "Mr. Emerson's Lecture on Napoleon," *Halifax Guardian*, 8 Jan. 1848, 7.
7. Brock and MacLeod 1980a, 5 Jan. 1848, f. 174.
8. J. Tyndall, *Journal 3*, 22 July 1847, f. 237, Royal Institution Archives.

MECHANICS' INSTITUTION.

The Directors of the above Institution respectfully announce to the Members and the Public, that

RALPH WALDO EMERSON, ESQ.,

WILL DELIVER HIS LECTURE ON

DOMESTIC LIFE,

In the ODD FELLOWS' HALL, St. James' Road, Halifax,

On MONDAY, FEBRUARY 7th, 1848.

LECTURE TO COMMENCE AT EIGHT O'CLOCK.

"The great Lessons of a practical kind which Emerson teaches, or tries to teach, are FAITH, HOPE, CHARITY, and SELF-RELIANCE."—*Tait's Magazine, January* 1848.

RESERVED SEATS, 1s. SALOON, 6d. GALLERY, 3d. Members admitted to the Reserved Seats and Saloon at half price, with the privilege of introducing two Ladies on the same terms; Gallery Free to Members only.

TICKETS to be had of Messrs. Whitley & Booth, Hartley, Birtwhistle, & M'Arthur, Stationers; and Mr. Burrows, Printer, Waterhouse Street, also at the Mechanics' Institution.

☞ The Directors are also wishful to draw the attention of the Public to the undermentioned Classes, which are in active operation under the superintendence of talented Teachers:—

THE DRAWING CLASS MEETS ON MONDAY EVENINGS.
THE GERMAN DO. ON TUESDAY DO.
THE FRENCH DO. ON FRIDAY DO.

The Adult School for READING, WRITING, ARITHMETIC, GEOGRAPHY, ENGLISH GRAMMAR, and COMPOSITION on Monday, Wednesday, and Saturday Evenings. The School for Junior Members on Tuesday, Thursday, and Friday Evenings. LATIN, MATHEMATICAL, and DISCUSSION Classes are in course of formation. Further particulars may be obtained of the Librarian at the Institution on the Evenings of Monday, Wednesday, and Saturday.

N. BURROWS, PRINTER, WATERHOUSE STREET, HALIFAX.

10.1 A placard advertising one of the series of Ralph Waldo Emerson's lectures at the Halifax Mechanics' Institution early in 1848. West Yorkshire Archives, Calderdale Central Library, Halifax, HMI:2.

continued to argue that *Vestiges* had made a strong case against direct divine interference.[9]

Vestiges served as a talking point for Halifax's young men in pubs, on walks, in their rooms, and at work. Soon after its appearance, it had been fiercely opposed in the conservative *Halifax Guardian,* the only locally published newspaper:

> If the writer is prepared to admit that these "vestiges" as narrated by geologists prove that creation was *not* effected in six days, and that death was *not* a new feature brought into the world by man's fall, he might as well admit the more consistent view of the work reviewed, and declare that all nature is progressive, that rolling stones begat lobsters; leaping lobsters, frogs; wind-inflated frogs, oxen; and that man is but a more perfect kind of monkey, beginning with the Carib and ending (at present) with the Caucasian, but capable of still further development, perhaps, into the angel![10]

Admit the findings of geology, the review argued, and the whole godless scheme followed as a matter of course. Opinion in the town on such issues ranged from

9. C. Lyell, journal entry for 23 Dec. [1858], in L. Wilson 1970, 198.
10. "Literary Notices," *Halifax Guardian,* 12 Apr. 1845, 6.

the God-denying materialism of Owenite infidels to those who anticipated the imminent reign of Christ on earth.

The first mention of *Vestiges* by Hirst in his journal is in the record of a conversation with his close friend Francis Booth in July 1847, six months before Emerson came to town. Booth, whose mother lived in a cottage on the edge of the moors, had begun as an errand boy in a printer's office but had risen to a responsible position in a local wool mill. Their talk brought up the issue of divine mystery, which they agreed was essential to reverence for the Creator. However, Booth was more skeptical in suggesting that scientific progress undercut this mystery, and hence our reverence. In reply, Hirst argued that science could only enhance devotion as it discovered general laws, perhaps even "one great natural law." Booth responded by pointing up the relevance of geology, and especially of *Vestiges,* to this debate.[11]

The enthusiasm of his friends spurred Hirst to read for himself. He borrowed the sixth edition or (more probably) its cheap reprint[12] on 1 August 1848 from the library of the local Mutual Improvement Society. His decision to read it grew out of his earlier conversations. For nearly a month the library copy accompanied Hirst wherever he went, both as a physical object and as an intellectual interlocutor. The small size of the book was convenient, but Hirst said nothing about this nor about any other aspect of the work's physical form. For him, unlike some other contemporary readers, the book was the text. He read at home, after chapel, in the evenings, in public reading rooms, and in the Mutual Improvement Society library. Sometimes he read for hours; sometimes in moments snatched from other activities. But wherever he was and whatever the time, "reading *Vestiges*"—a recurrent phrase—was an identifiable activity, to be taken up where left off. Most of Hirst's reading was done silently and by himself; and through the writing in his journal, it became part of the record of his life.

The pages of summary and reflection in Hirst's journal are the best evidence for his deep immersion in an approach to reading that he would have learned through the kind of study recommended in the learned traditions of Congregational Dissent.[13] At a basic level, Hirst was concerned to report—as neutrally

11. Brock and MacLeod 1980a, 13 July 1847, ff. 89–90; for Booth, see J[ohn] T[yndall], "Memoranda Concerning Dr. Hirst," *Proceedings of the Royal Society of London* 1893, 52:xiv–xviii.

12. Hirst does not identify the edition, but this can determined from passages quoted in the diary; compare Brock and MacLeod 1980a, f. 271 with *V6*, 238, and *V7*, 137–38. *Vestiges* is listed in the *Alphabetical and Classified Catalogue of the Library of the Halifax Mechanics' Institution and Mutual Improvement Society* (Halifax: printed by N. Burrows, 1851), 36. The earliest surviving catalogue, this was compiled after the amalgamation of the two institutions.

13. This tradition of reading has been more fully discussed in the American context; see Davidson 1988, 69–79, and Jackson 1999, 161–62.

as possible—the plain meaning of what *Vestiges* said. Many entries in the journal are long chapter summaries and quotations, written out like the documents he copied for his master, with his own views clearly distinguished. He made few notes on the geological chapters, being already familiar with Mantell, Lyell, and the "beautiful digest of the science" in Chambers's Information for the People.[14] After a week's relatively slow reading, he noted that the author "begins to announce his hypothesis, and that in a masterly manner," and from this point onward the book is abstracted at great length. By the thirteenth of August, a Sunday, he had reached the chapter on "the history of mankind, their peculiar forms, languages and colour." Here again, "all facts [were] in accordance with his theory" of a common origin.[15] The following day, he read the chapter on the "purpose of the animated creation" and summed up its discussions of the origins of evil. *Vestiges,* through its reliance on law, had a profound answer to the problem that had puzzled philosophers for centuries and one that Paley's *Natural Theology* had not solved. "The virtuous man is as liable to such misfortune as the wicked—a fact that would be difficult to reconcile under any other view of the purpose of creation."[16]

Although Hirst had read to the end after the first two weeks in August, he had the book reentered in his name at the Mutual Improvement Society, to abstract what he had identified as its novel argument about organic development. Several pages reported the text without comment and often verbatim, but in closing his discussion of the chapters on transmutation Hirst summed up his own view:

> Such is a general outline of this most extraordinary of theories, a theory evidently the result of a calm, impartial, honest, as well as scientific investigation, divulged with a manly spirit and sustained, in spite of the prejudice and ridicule of the world. Such is ever the fate of one who steps forward so much in advance of his fellow beings. But although believing this theory as likely to be modified in many particulars according to the further discoveries of science, I must say it is the most plausible attempt that ever I heard or saw, to explain the creation and development of the world and its organic inhabitants, by natural law. And at the same time I feel certain that it will have given an impulse to investigation—set it in a proper track, as it were—from which impulse may be anticipated a greatly advanced state of science, together with its attendant blessings.[17]

Hirst had read with devouring attention, more closely in fact than any book up to this point in his life other than the Bible.

14. Brock and MacLeod 1980a, 1 Aug. 1848, f. 257.
15. Ibid., 13 Aug. 1848, ff. 266–67.
16. Ibid., 14 Aug. 1848, f. 267.
17. Ibid., 17 Aug. 1848, f. 272.

Hirst by no means agreed with everything *Vestiges* said. One evening his friend Booth came over to play the flute and piano, and they discussed the book. Booth accepted that the further development of new species probably occurred without divine intervention, according to the *Vestiges* scheme, but doubted whether life could originate from nonliving matter. In contrast, Hirst maintained that if natural law explained species it was likely to do the same for life itself.[18] A few days later Hirst and his friend Roby Pridie, son of the local Congregational minister, read together a "very cleverly written" analysis of the book in an old number of the *British Quarterly Review*.[19] In discussing this review, and another in the same number, Hirst drew the line at the natural origin of human reason and free will. Having just read the chapter on animal intelligence twice, he became convinced that the author "is evidently a materialist or something approaching it—he believes that mind is a function of the brain, governed by as fixed laws as any other part of nature." On this principle, the human soul ought to be explicable as a development from the lower animals, yet the book also stated that man alone might have been "endowed with an immortal spirit."[20] This seemed inconsistent, and Hirst doubted whether man could have developed from the apes.

At the end of August 1848, having finished with *Vestiges,* Hirst borrowed *Explanations,* presumably from the Mutual Improvement Society's library, and completed it four days later. On the evidence of entries in the journal, he had seen only one review of the original work (in the *British Quarterly*), so the sequel gave an opportunity to judge both the criticisms and the author's response. Hirst was impressed: "he certainly sets to rights many of his reviewers' objections." At the same time, though, Hirst thought the new work shifted its ground more than its author was willing to say. What had been widely seen as the principal claim of *Vestiges,* its "particular theory" of development, was now being acknowledged as subsidiary to the larger issue of natural law. This backtracking and "toning down," at least from Hirst's perspective, made the argument "less startling" but also more probable, "capable of directing enquiry into a channel nearer the true one."[21] References to *Vestiges* continue to crop up in the journal over the next year, culminating in an intensive debate among Hirst's friends about Miller's *Footprints.*

Independent Reading

The first thing that strikes a modern reader about Hirst's engagement with *Vestiges*—and the many other books he read—is its extraordinary intensity.

18. Ibid., 9 Aug. 1848, f. 264.
19. Ibid., 19 Aug. 1848, f. 273.
20. Ibid., 23 Aug. 1848, f. 278.
21. Ibid., 3 Sept. 1848, f. 285.

Almost no one reads like this any more. It is the reading practice of a self-improving autodidact, shaped by traditions of Bible-reading among Congregationalists, Presbyterians, and the other denominations of learned, liberal Dissent. Yet what would, a few decades earlier, have been written out as the journey of a soul to Christ has here become a quest for self-identity through an understanding of nature.[22]

Hirst had learned to read at the town school in Heckmondwick. Built in 1809 by subscription, this offered a grounding in basic literacy and arithmetic. When Hirst attended, the school had about 150 scholars, both boys and girls, who paid a small weekly sum for their lessons. A library of several hundred works was housed in the same building, supported by twenty-four subscribers.[23] In the Dissenting traditions that dominated practical education in Yorkshire and Lancashire, the Bible provided a template for how all books should be read: slowly, line-by-line, and with utmost attention to the nuances of the reader's relationship (or lack thereof) with God. Hirst's experience of *Vestiges* embodied ideals at the heart of the exegetical practice of old Dissent: close reading, the consultation of parallel texts, and the need for private judgment. The roots of these techniques are to be found in the Reformation, although they were not distant memories, nor simply the reflections of pious conduct manuals. This is the way Hirst and his friends had learned to read.

These exalted ideals of reading may have their origins in a broadly evangelical tradition, but one book that Hirst hardly ever read was the Bible, even on Sundays. This was most unusual for someone of his background and serious intellectual interests, and he blamed his early schooling, which had made reading the Bible anything but a transcendent experience. The failure of faith was a failure of education in reading:

> To-day I have read my Bible, rather an unusual thing for me—and why? Because my teachers have taught me to profane it. I have been made [to] repeat it when its living passages were meaningless to me; instead of impressing me with its sacredness and making it a sealed book to me until I was by training worthy to open it, I have gabbled it like a parrot, and attached to its living oracles an unworthy, conventional meaning.[24]

This criticism of the dangers of rote learning was itself characteristic of the Dissenting literature on education, which warned of the dangers of introducing

22. For the importance of religious models in the institutions of self-improvement in Halifax and other West Riding towns, see Green 1990.

23. Personal communication, 16 June 1998, from Elizabeth Briggs of the Kirklees District Archives, West Yorkshire Archive Service.

24. Brock and MacLeod 1980a, 6 Oct. 1849, f. 525.

biblical instruction at too early an age. The spiritual development of the individual could proceed only through close engagement with the inspired word.

Like many other young men of his generation, Hirst had turned his quest for spiritual self-realization toward reading in the sciences and in philosophy. The kind of reader he initially aspired to be can be seen in idealized illustrations in useful knowledge compendia. *The Young Man's Best Companion* (1831) was one of a host of texts for boys who wished to educate themselves in "writing, arithmetic, grammar, mensuration, book-keeping, geography, astronomy, history, law, chemistry, algebra, navigation, &c. &c." The frontispiece (fig. 10.2) shows Britannia unveiling the steps to knowledge, beginning with books (including a volume open to geometry); then instruments such as a telescope, globe, and pulley; and finally a sailing ship and the bright sun of commercial

10.2 "Knowledge Is Power." Note the prominence of books on the path to enlightenment, including an open volume on geometry. Frontispiece to *The Young Man's Best Companion* (London: J. Smith, 1831).

success. The message is summed up by the caption "Knowledge Is Power." As the Congregational teacher the Reverend John Pye Smith said, scientific study made good use of a divine gift:

> The cultivation of Natural History and the Sciences will be a dignified means of excluding those modes of abusing time which are the sin and disgrace of many young persons; vapid indolence, frivolous conversation, amusements which bring no good to the mind or the heart, or such reading as only feasts the imagination while it enervates the judgment, and diminishes or annihilates the faculty of command over the thoughts and affections, a faculty whose healthy exercise is essential to real dignity of character.[25]

Hirst's quest for a sense of identity was especially acute, as his formal education had ended at the age of fifteen, three years after his father's death. Hirst was apprenticed to Carter and moved with his mother and sister to Halifax. It was at this time that he had begun to keep a journal of his spiritual progress and everyday activities. He played music, fished, skated, drank, played chess and cards, hunted for fossils, went to church, chapel, and theater, and talked with friends in taverns and local societies.

The railway boom was well under way when Hirst started, which is why his relatives thought engineering a suitable profession, combining as it did trigonometry and fresh-air fieldwork at the leading edge of the Industrial Revolution. His first year was spent surveying the line from Halifax to Keighley. However, by the later 1840s railway building almost completely dried up, and Carter's firm returned to the bread-and-butter work of surveying estates. This was of particular importance after the Tithe Commutation Act of 1838 forced farmers to pay rent on the average value produced by their land, which had to be accurately measured and assessed.[26] This new system of payment created great hardship, which Hirst often witnessed.

With his employer's encouragement, Hirst began to follow an eclectic program of self-education. Most unusually (and probably through an arrangement with the family), Carter allowed his apprentice to read during working hours, although most of the young man's time was occupied with copying documents, surveying, and carrying transits, chains, and theodolites from field to field. Hirst enjoyed the companionship the job provided, but found the work dull, scarcely reflecting his interest in abstract geometry and trigonometry. As early as September 1846 his answer to a mathematical conundrum appeared in the *Family Herald*, a London-based weekly with stories, useful knowledge, and correspondence columns.

25. Smith 1839, 327.
26. Kain and Prince 1985; Moore 1997, 301.

Hirst put reading at the heart of self-improvement: "My studies during the past year may be seen from the books I have read."[27] Every evening he wrote long entries in his journal, which show him as a keen and knowledgeable reader of science. By the time he checked *Vestiges* out of the library, he had already read Lyell's *Principles of Geology,* Humboldt's *Cosmos,* Mantell's *Wonders of Geology* ("the first book of really useful information I ever read through"),[28] Combe's *Constitution of Man,* Jeremiah Joyce's *Natural Philosophy,* Cuvier's *Theory of the Earth,* and numerous other works. He also paid serious attention to theology, though the number of titles was smaller, and he rarely studied his Bible. Books that discussed the theological implications of science were of special interest, including John Pye Smith's *On the Relation between the Holy Scriptures and Some Parts of Geological Science* (1839), Combe's *On the Relation between Religion and Science* (1847), and several of the Bridgewater Treatises.

Hirst read almost no poetry, as he noted when examining Alexander Pope's *Essay on Man* (a work in any case related to his philosophical concerns), and few novels. These included Dickens's *David Copperfield* and *Dombey and Son,* Goldsmith's *Vicar of Wakefield,* as well as Bulwer-Lytton's *The Caxtons,* read in *Blackwood's,* and Kingsley's *Yeast,* which appeared anonymously in *Fraser's.*[29] The story of *Yeast,* which concerned the struggles of a young man to find a creed, was peculiarly appropriate to his own situation, and he approached each new serial in *Fraser's* in keen anticipation that it might be by the same author.

Novel reading was not sinful—nor was dancing or the theater—but its value was debatable. One of Hirst's friends, Haley, gave a paper on it at the Franklin Society, which they later discussed.[30] Hirst rarely commented in any detail on novels in his journal, reserving his close attention for "serious" reading. Different books needed to be read in different ways. At best, as he reflected when in bed with a cold, "stirring fiction" could serve as a medicine by harmonizing with the state of his mind.[31] The right kind of reading was central to self-improvement (fig. 10.3).

Hirst owned only a handful of the titles he read, and his journals do not mention buying secondhand books or clubbing together with others to make purchases. These were often the only options for workingmen who could not

27. Brock and MacLeod 1980a, 31 Dec. 1848, f. 338.

28. Ibid., 12 June 1847, f. 64.

29. See ibid., passim; these titles and many others can be retrieved from the index to this microfiche edition.

30. Ibid., 3 Sept. 1848, f. 285.

31. Ibid., 2 June 1849, f. 414; early examples of this idea (and its converse, that the wrong kind of reading could cause disease) are discussed in Johns 1998a, 580–81; and 1996. For its later history, Mays 1995 and Winter 1998a, 329–30.

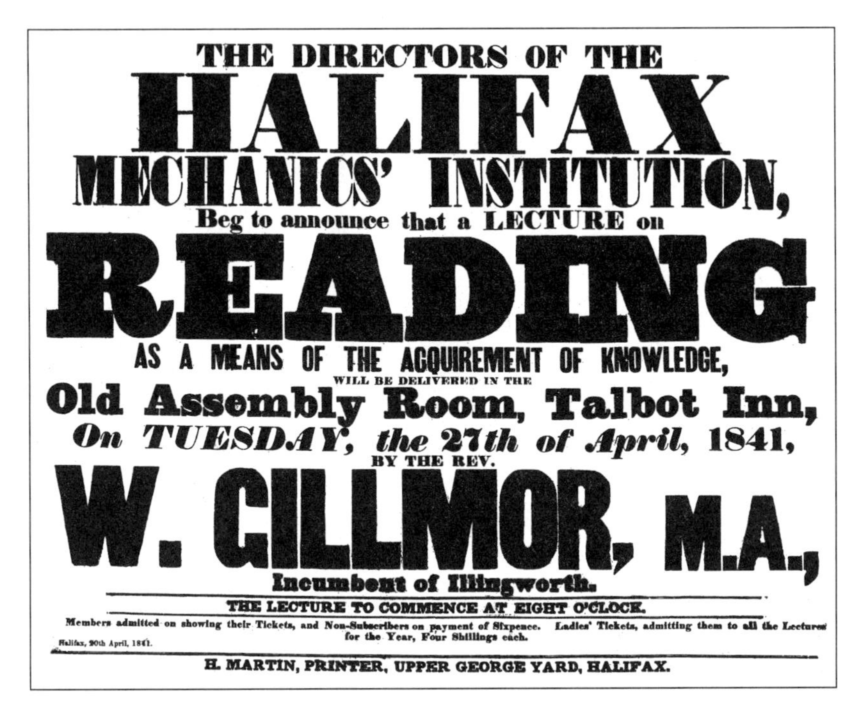

10.3 Public lectures constantly stressed the importance of reading in education. This oversize placard (57 cm × 44 cm) announced a talk in April 1841, delivered by an Anglican clergyman and sponsored by the Halifax Mechanics' Institution. West Yorkshire Archives, Halifax, Calderdale Central Library, HMI: 2.

afford regular library subscriptions. Even on his limited income as an apprentice, Hirst could afford three or four pounds each year to gain access to collections from which he could borrow. The Mutual Improvement Society had been founded in 1846 by a group of workingmen (known as the "Old Muffs") for self-improvement and elementary education.[32] Hirst, with his superior upbringing and education, served as one of their teachers and was a keen supporter and user of the library in their rooms in Waterhouse Street. Besides this collection, he also used the Mechanics' Institution library (with over twenty-seven hundred volumes, founded in 1825) and the circulating library run by John MacCarthur on Jail Lane.[33] Hirst did not seek to join the local Literary and Philosophical

32. "A Meeting of 'Old Muffs,'" Horsfall Turner Collection, Newspaper Cuttings Book, vol. 4, f. 91, Calderdale Central Library, Halifax.

33. Brock and MacLeod 1980a, 12 Oct. 1848, f. 300. The Mechanics' Institution's proceedings and minute books (HMI: 1–7) are preserved in the Calderdale District Archives in the Calderdale Central Library, Halifax. The number of books in its library is taken from a loose circular of 1847

Society, as this would have been both expensive and inappropriate for someone of his age and status. There were many other sources of books in Halifax, including two "public" subscription libraries (one for Anglicans, with ten thousand volumes, the other for Dissenters).[34] However, nothing compared with the riches available to Liverpool and Manchester gentlemen, nor to genteel readers in Oxford, Cambridge, Edinburgh, and London; Sedgwick and Whewell each had more books in their college rooms than did the Halifax Mechanics' Institution or Mutual Improvement Society in their respective libraries.

Newspapers and magazines played an important part in the process of close, comparative reading (fig. 10.4). In local reading rooms and his master's office, Hirst regularly scanned Edward Miall's weekly *Nonconformist,* whose Dissenting outlook and political liberalism he broadly shared; the local conservative weekly, the *Halifax Guardian;* and occasional copies of the liberal weeklies such as *Chambers's Journal,* the *Family Herald, Howitt's Journal,* and the *People's Journal.* He did not read much in the quarterlies, which were expensive and not easily available, although he knew the importance of the *Edinburgh*'s review of *Vestiges* from the discussion in *Explanations.* The sole exception was old numbers of the *British Quarterly Review,* which Hirst's friend Pridie is likely to have obtained from his father, a Congregational minister. Among the monthlies, he read *Fraser's* if it had a good serial. Hirst recognized that all this was what might be expected for someone of his religious and political background—and so, unwilling to limit his horizons to a particular party, he also read the Tory *Blackwood's.* Hirst had nothing but scorn for the radical Chartist and Owenite press, condemning the views taken in the latter as "preposterous."[35]

Hirst's experience shows the dramatic expansion of the range of books available to young men in the middle class and the highest reaches of the artisan class. Nothing like this would have been available to an eighteen-year-old apprentice at the start of the century. Even those lower in the social scale could gain access to a wider range of texts than ever before. *Vestiges* was a common book in the kind of collections young men like Hirst used throughout the country; a survey of catalogues shows it was found in the mechanics' institute libraries of Evesham, Keighley, Liverpool, Manchester, and Warrington, among many others.[36] For working-class autodidacts lower in the social scale, obtaining

in HMI: 2. For the institute's history, see Tylecote 1957, 224–40, and "Opening of the New Hall of the Mechanics Institute," *Halifax Courier,* 17 Jan. 1857, with an illustration.

34. The combined catalogue of the amalgamated Halifax Subscription Library and Halifax Literary and Philosophical Society (Halifax: T. and W. Birtwhistle, 1874) shows a copy of the sixth edition of *Vestiges,* as well as *Footprints* and other related works.

35. Brock and MacLeod 1980a, 13 June 1847, f. 65.

36. Hinton 1979, 256 n. 117.

up-to-date books was a continual struggle, as we have seen in the cases of the ribbon weaver Joseph Gutteridge, the compositor William Chilton, and the schoolteacher Mary Smith.

Access placed important constraints on how books were read. Mary Smith, who also read Carlyle and Fichte, had raced through *Vestiges* in a single night as though it were a novel; but this was at least in part because the book was secretly borrowed from her employer, and had to be back on the drawing room table in the morning. Smith's employer, in turn, had borrowed the volume from a circulating library, which gave him more time for reading but meant that any reactions would have to be recorded in a separate journal. Wealthier readers who owned their own copies could record their reactions directly in marginal annotations. Hirst, in contrast, had to write out large parts of the text in his journal; but this became a form of "mental ownership" that he preferred, even

10.4 All towns and cities had a variety of reading rooms where the latest newspapers and periodicals could be consulted. This figure shows an idealized view of the Working Men's Reading and News Room in York, 1856. The facilities at Halifax appear to have been less extensive, though in 1851 the combined Mechanics' Institution and Mutual Improvement Society took six daily London papers, twenty-two weekly or provincial titles, five quarterly reviews, and eighteen magazines and serials. *British Workman,* Aug. 1856, 78.

after he could afford to buy more books. In copying huge chunks into the journals, he effectively wrote his own *Vestiges,* making his journal into a commonplace book, a private anthology of the kind recommended by John Locke and many authors of advice manuals.[37] The copying of prose did for Hirst what the memorization of poetry did for other readers: it made books more closely part of himself.

Comparative Reading

Such an eclectic program of reading meant that Hirst did not experience books as continuous narratives, but as texts broken up through juxtaposition with other texts. "Reading *Vestiges*" had thus both an internal coherence of its own—partly through the physical experience of turning the pages of a particular volume—but was also open-ended. Close reading is often discussed as if it were an improbably self-contained process, in which the narrative structure of a work shapes the reader's experience. Hirst, like most readers, moved between writings by many different authors in the course of each day—many of them, such as advertisements, placards, and written instructions on medicines—not recorded in his journal. Through the process of juxtaposition, he could structure his own horizon of expectations.[38]

The journal is a record of how Hirst did this. He moved, often as part of a clear plan, between *Vestiges* and other books. These included Paley's *Natural Theology,* Chambers's Information for the People, John Phillips's *Geology of Yorkshire,* Forster's *Life of Goldsmith,* and other works, which he examined in the summer of 1848. He was very explicit about the process, especially in the case of Paley's *Natural Theology,* which became a way of extending the discussion between Paley's grandson and Tyndall. After reading the chapters on astronomy side by side, Hirst concluded that the theory of nebular condensation in *Vestiges* offered a superior explanation, although Paley could be used to show that problems still remained—not least that under the development theory, the sun would at some point cease to emit light. Hirst also noted Paley's caution against evading the need for a Creator by substituting the term "natural law"; but was pleased to see that *Vestiges* "particularly alludes" to this point.[39]

The one comparative reading Hirst resolutely refused to make was between *Vestiges* and the biblical story of creation. This was an explicit decision, a boundary drawing that recognized the troubled relations between scientific and religious writing. After spending a Sunday morning at church, hearing a "very

37. On anthologies and narrative form, see Price 1997.

38. "Intensive" reading of a single book is usually contrasted with "extensive" reading of many. In the present case, the two practices coexist: the intensive reading of a few works occurs against the backdrop of exposure to a much larger number.

39. Brock and MacLeod 1980a, 2 Aug. 1848, f. 258.

good sermon" by the archdeacon on divine omnipotence, and the evening at home in a close reading of *Explanations,* Hirst wrote that true science and true religion would never contradict one another. Correlation between the books of natural and revealed truth would eventually occur, but should not be undertaken prematurely. As with geology, "so with the 'Vestiges'—any contradiction that may occur between its theories and our religious opinions ought only to act as an incentive for testing that theory on its own or scientific merits."[40]

This "testing" became a form of dialogue, so that the arguments of one text were set off against others. Reading one chapter in one work and comparing it with another allowed Hirst to insert fragments of books into a continuous narrative of his own experience, communicated in his letters and conversations and best embodied in his journal. Seen from a wider perspective, Hirst's practice can be understood in the context of the increasing dominance of periodical publication during the 1840s.[41] Hirst's experience of reading complete books like *Vestiges* was structured by his immersion in monthly magazines like *Fraser's* and *Blackwood's,* monthly part works (such as the novels of Dickens), and the secular literary weeklies. Reading was thus a fragmented process, focused more on chapters than on long arguments, and more on comparison and contrast than on single narratives. The whole process, Hirst believed, could be kept under control through a well-regulated critical mind exercising independent judgment.

Periodicals were vital to the process of juxtaposition. Hirst read newspapers and magazines not to agree with them, but to think with them. The literary weeklies offered a very different context for debate than the quarterlies or newspapers, but shared their commitment to intellectual liberality. The *People's Journal* had been one of the strongest supporters of *Vestiges,* complaining of "bullying" tactics and "grossest misrepresentation" by its opponents, whose only interest was in maintaining their own power. "Freedom of speculation in theory," the *People's Journal* noted, "is the natural ally of the advance of useful discoveries in human science, and it becomes us to cherish carefully the one if we regard the other."[42] The *Nonconformist,* on the other hand, condemned *Vestiges* as "erroneous and pernicious."[43] Yet Hirst was soon reading the newspaper's reformist political editorials in terms taken directly from the book. Reflecting on

40. Ibid., 2 Sept. 1848, ff. 285–86.

41. On this issue, see Feltes 1986, Hughes and Lund 1991, Poovey 1988, and several of the essays in Jordan and Patten 1995. Beetham 1990 and Pykett 1990 offer the best overviews. For Hirst, as (one suspects) for most readers, practical problems of time and access were crucial in structuring the text into the discrete units associated with "serialized" reading.

42. W. J. Fox, "On the Progress of Science in its Influence upon the Condition of the People," *People's Journal,* 10 Jan. 1846, 30–35, at 34.

43. *Nonconformist,* 9 July 1845, 490; *Nonconformist,* 13 Aug. 1845, 569.

events during the revolution of 1848, he thought France was attempting to advance too quickly up the progressive scale of civilization. Hirst's comparison was with a growing infant: just as a child needed "the governing care of a father" at certain stages of its development, so did a limited monarchy best suit the French nation. Such an image resonated with many parts of *Vestiges,* especially its quotations from Herschel about the nebular hypothesis. In this way, the journal brought politics and science together in the day-to-day work of writing and reading.[44]

Intensive reading did not just involve assimilating facts or memorizing other people's opinions. Hirst despised those who parroted platitudes or what they had read in reviews. The best periodicals, he thought, offered a range of opinion on different issues, as close adherence to a particular line became repetitive and predictable. Opinions in the different journals to which Hirst had access could scarcely be more different. Within a single periodical, even within a single review, there could be different perspectives to be compared and contrasted. In May 1845 the *British Quarterly* had calmly demolished the development hypothesis as an unsupported generalization, but stressed the need to do this "not in the spirit of bigotry, but in the spirit of men earnest for truth." Yet a note added as an afterthought took a much stronger line, praising John Sheppard's anti-Vestigian tract (which many thought a model of bigotry) for its "learning and acuteness."[45]

Hirst read such articles to sharpen his own opinions and gain unexpected perspectives. Two days after reading the *British Quarterly*'s review, Hirst consulted the essay on James Cowles Pritchard's *Physical History of Man* in the same number. He found the reviewer's reliance on divine intervention unconvincing:

> Now, this appears to me to be a very imperfect solution of the question. He does not believe external circumstances have exerted such changes (in contradiction to known facts), because he considers them accidental, and thus not consistent with the idea of Deity, forgetting or denying that Science is continually tracing such effects to natural laws, and why may not natural laws have produced the differences in the races of men?

Judging by what the reviewer said, Hirst agreed more with Pritchard, that racial types were the products of nature and circumstance, and thought this "in accordance with Divine Writ."[46] The next day, after working at the office, Hirst reread the parallel chapter in *Vestiges* on human origins, finding it "a striking contrast"

44. Brock and MacLeod 1980a, 12 Aug. 1848, f. 266.

45. "Vestiges of the Natural History of Creation," *British Quarterly Review* 1 (May 1845): 490–513, at 513.

46. Brock and MacLeod 1980a, 21 Aug. 1848, ff. 274–76.

to the review, "and after all, a much more plausible theory."[47] He remained equally unshaken after reading an attack on spontaneous generation and evolution in the *Ethnographical Journal,* which Francis Booth loaned him in March 1849 in connection with their continuing conversations. In opposing transmutation, Hirst noted, the author of this article appealed to the "good principle" that the expertise of men like Lamarck and Geoffroy Saint-Hilaire "ought not to bias our minds with regard to its correctness."[48] Authority should not overrule independent judgment.

The range of printed materials newly available to men in Hirst's situation posed what much of the advice literature saw as a potential threat to independent judgment. Critical comparison and juxtaposition of views could easily become superficial, pandering to light conversation—what Tyndall had condemned as "the goose-cackle of society"—rather than serious study. Hirst was particularly impressed by a long letter from Tyndall to a common friend, warning against trying to digest books in too many different subjects, without sufficient critical attention:

> I know one or two most extensive readers who could talk for half a century about various systems of Philosophy, and can tell you the opinions of this and that great man upon such and such subjects; and yet the intellectual power of these readers is truly contemptible. They are merely so many conduit pipes through which information from some other spring finds a passage—throw them into circumstances which demand the exercise of original power, and they get instantly tangled and helpless. This comes from their having contented themselves with driving a retail trade in the opinions of other men, without enquiring into the reasons of these opinions.[49]

Such a system, although calculated to impress, was mechanical and deadening: the reader's mind lost its integrity, becoming part of the machinery of public discussion condemned by Carlyle. Promiscuous readers became "a kind of hamper basket for stowing away the products of braver minds," followers rather than leaders. The remedy, Tyndall suggested, was to gain a general idea of a book on first perusal, and then to focus on novelties and problems, "*noting down your observations in writing*"—something that Hirst already did, but aspired to do more thoroughly. It was also crucial to read on a definite plan. Hirst agreed. He copied Tyndall's letter into his journal, and looking back on the year's reading on the last day of 1848, he noted that "[i]n science I have hovered about the whole field. It is time I should settle on one."[50]

47. Ibid., 22 Aug. 1848, ff. 276–78, at f. 276.

48. Ibid., 29 Mar. 1849, f. 375.

49. J. Tyndall to James Hayran, [1848], Brock and MacLeod 1980a, 4 Sept. 1848, ff. 286–87.

50. Ibid., 31 Dec. 1848, f. 338; for similar recommendations in the Congregational advice literature, see Pye Smith 1839, 327.

Reading and Vocation

The problem was not just which "one" science Hirst would settle on, but what kind of person he was going to be. *Vestiges* and other books helped him to define his beliefs, his relations to others, and above all his sense of vocation. In the Nonconformist tradition, disciplined reading of the Bible had long been the cornerstone of independent judgment, as the individual sinner confronted the divine word. Reading, by exercising the free will granted to men by God, could thereby develop a vocation for life in Christ.

Hirst increasingly conceived his search for spiritual development in terms that some orthodox Congregationalists and evangelical Anglicans found dangerously dissolute. It is appropriate that one of the first occasions when Hirst heard *Vestiges* being discussed was after a lecture by Emerson, for Carlyle and Emerson became Hirst's great heroes, his models for the role of reading in an active life of work. Tyndall interpreted Carlyle's *Past and Present* (1843) as a call to action for new leaders who would guide society in the search for a spiritual understanding of nature and a better life through a knowledge of nature's laws. Careers in business, industry, or engineering need not harden the soul. For Tyndall and for Hirst, as for many young men in northern towns and cities from a Dissenting background, Carlyle and Emerson provided a route to personal spirituality which was free of cant, whether that of evangelical fanatics, Tory bigots, Whig improvers, or Owenite visionaries. Science, within this vision, became a Carlylean vocation appropriate to the industrial age.[51]

Vestiges could be used to ponder the problem of human free will. If astronomy, geology, and other sciences pointed to the conclusion that man was descended from the apes, how could people be morally responsible agents with independent judgment? The problem was at the heart of Hirst's early doubts in August 1848, when first reading the book and some of the reviews. Returning to Halifax on a starlit night after a dancing party, Hirst debated the question of free will with a friend, Dawtry, who believed that all human action was the result of circumstance. Hirst wrote:

> This is the opinion of the "Vestiges," but it is one in which I cannot yet fully agree. Circumstances undoubtedly have a great control over the moral state of an individual, but in order that man may be a responsible being it is necessary that he have some control likewise over those circumstances. Here, in my opinion, is the weakest point in the "Vestiges"—the attempt to prove that mind, like body, has arrived at its present perfection through development. The wide differences between instinct

51. Turner 1993, 131–50; Turner 1981.

and reason appear to me too much to be considered as a progressive step, and the same may be said of man's form.[52]

The individual was not merely shaped by circumstances, Hirst believed, but could control them through moral leadership, educational endeavors, and charity to the less fortunate.

The work of land surveying brought Hirst face-to-face with the issue. Watching a queue of "emaciated faces" in an overseer's office a few days earlier, Hirst witnessed an incident that raised the problem of free will and circumstance in stark terms. A poor woman, who "scraped up her rate" in a pile of copper coins, had been crestfallen to learn that she owed an extra nine pennies. Hirst would have paid for her, but found he was without money. Circumstances had, on this occasion, worked against the promptings of his will, and in consequence he had failed to reap the moral benefit that charitable action would have brought: "how well it would have repaid me, if it had only ensured her a transient pleasure. And yet I could not."[53] The inability to act, Hirst thought, was not the consequence of blind chance, but of his own will; he had not been ready when the time came.

In June 1849, Hirst read *Sartor Resartus* in a copy borrowed from Roby Pridie. At first he found the book "curious" and in "many parts obscure," and even doubted if it could be by Carlyle as was rumored (the title page simply said it was "The Life and Opinions of Herr Teufelsdrockh"). But some passages were of transcendent importance: "'With men of a speculative turn there come seasons, meditative, sweet, yet awful hours, when in wonder and fear you ask yourself that unanswerable question: Who am *I:* the thing that can say "I"?'"[54] Carlyle stripped away the "adventitious wrappages" of clothes-philosophy to the naked biped and spiritual mystery of each man, "our *Me* the only reality." Hirst copied the chapter on "natural supernaturalism" of *Sartor* into his journal in its entirety, finding it the most refined and subtle philosophy he had ever read.[55] A few months later he obtained a copy of the work that had most shaped Carlyle's own thought, Johann Gottlieb Fichte's *Vocation of Man,* which had recently been translated into English.

Hirst became intensely introspective, attempting to follow his philosophical mentors in abstracting "the mysterious 'I'" from the distractions of the world, friends, and his own body. At an evening dance during Halifax summer fair, he reflected on his reading of *Sartor:*

52. Brock and MacLeod 1980a, 27 Aug. 1848, f. 282.

53. Ibid., 23 Aug. 1848, f. 278.

54. Ibid., 24 June 1849, f. 430. For the significance of this kind of reading of *Sartor,* see the discussion of the American case in L. Jackson 1999, 159–62.

55. Brock and MacLeod 1980a, 1 July 1849, f. 459.

> Here again I tried to withdraw myself in thought. Groups of laughing girls and attentive swains were dancing around me. "Look," thought I, "with a cold eye (with pure logic) at that group of dancing automatons, and how inexpressibly foolish their capers appear; yet they are just the persons who pride themselves on commonsense. What miracle, then, can have made them so far forget it? Is it the spirit of music, the electric glance of thy lover's eye, that for once has pierced that hidden soul of thine, buried amongst commonplace wrappages, and to shew thee its power pulls the string of thy human body (like a toy) and makes thee cut those ludicrous capers? Is it that insanity (how pleasant—to be at times insane) that thou laughes[t] at in others when thou art not capable of hearing that all powerful music, of seeing that electric glance that the Universe, that God casts upon us, his children."

Returning home that evening, Hirst read the chapter in *Sartor* called "Pure Reason," which insisted on the need to escape "vulgar logic" and recognize man as "[a] soul, a spirit and divine Apparition."[56]

Only a few months before, Hirst had read the laws of creation in *Vestiges* in the context of a belief in the external reality of God. He was beginning now to think about this in a new way. Nature, as Carlyle and Fichte wrote, was "but the reflex of our own inward Force"; the physical world, as Emerson had said, was the mind precipitated. Free will, which had been the basis of independent judgment for Hirst's faith, now became itself the only faith he had, the only source of truth. Religion, Hirst remarked, had been "reduced to a 'tabula rasa,' and now I am just beginning, I hope, to write upon it words of truth."[57] He became moody and uncertain; his apprenticeship was ending, he no longer wished to follow the profession his family had planned for him, and his old faith was gone. In August he visited Tyndall, who was studying at the University of Marburg, although when his mother died in September he returned to Halifax. The journal became more inward looking, as Hirst began to write his sense of isolation into his records of conversation with friends.

On the evening of Saturday, 24 November 1849, a group of men in their late teens and early twenties met in Hirst's lodgings. Their conversation on a variety of topics was "drawn by a force as powerful as gravitation to Religion, this centre of all things." Pridie, Haley, and Hirst were joined by Booth and John Dyson Hutchinson, two of Hirst's other friends. They began by debating capital punishment, a subject in the newspapers because of the public hanging of Marie and Frederick George Manning for premeditated murder. The discussion moved "by a curious association of ideas"—perhaps involving death, free will, and the problem of evil—to the development hypothesis and Miller's

56. Ibid., 25 June 1849, f. 435.

57. Ibid., 6 Oct. 1849, f. 525; 24 June 1849, f. 432.

Footprints, which had been published earlier in the summer. The journal recorded the conversation in what Hirst admitted was "a faulty record":

> HUTCHINSON. I have just finished reading Miller's work, with which I am well pleased—in my opinion it completely annihilates the theory of the Vestiges.
> TOM [Hirst]. It undoubtedly destroys the evidence of a perceptible progression in Geology, and also says or hints that being a geological theory, it may be considered upset; but is it so in fact, or is geology used as corroborative—are there not many mysterious physiological facts uncombated that may have suggested the theory; when viewed as a whole, does not Nature seem to point this way—in fine, although a partial examination has destroyed a *particular* theory, may not *a* theory of development be still significant?
> BOOTH. Hear, hear!
> HUTCHINSON. But it is not fair to accept in general a truth you cannot partially and particularly establish. As all generalizations are the result of thought on individual phenomena, so in the truth or error of these must a theory stand or fall. Tell me, is not the step between the monkey and the man too much to be conceived? Between the highest type of the former and the lowest of the latter there appears to me a gulf unpassable even by imagination.

For Booth, this comment by Hutchinson recalled the experience of seeing the "Bushmen"—supposedly among the most primitive of all peoples—as part of a traveling show. Their ambiguous position on the scale of nature was heightened by the way they were presented in the illustrated press (fig. 10.5), in which they were portrayed as combining fierce independence with diminutive stature.

> BOOTH. Did you see the Bushmen—they certainly narrowed that gulf to me amazingly.
> HUTCHINSON. No, I did not. But look even at Language—is there not in this a new world opened to man alone?[58]

Faced by the possibility of gradation between the physical characters of humans and other animals, Hutchison retreated to the same ground that Miller occupied—the uniqueness of human reason and the soul.

Throughout the discussion, Hirst remained convinced that however much Miller might hammer its factual errors, *Vestiges* was more successful in dealing with the religious and philosophical issues. Hutchinson went on to say that he believed every word in the Bible, and that language could best be explained as "a divine gift direct from God to Adam in the Garden of Eden." At this, Booth coughed "significantly." Booth, Hirst suspected, was motivated not so much by

58. Ibid., 25 Nov. 1849, ff. 546–47. Speakers' names expanded from initials.

10.5 "The Bushmen Children," as exhibited for Christmas 1845. The group later shown nationally on tour consisted of two men, two women, and an infant. As the *Times* (19 May 1847) commented, "they are little above the monkey tribe, and scarcely better than the mere brutes of the field." "Christmas Sights," *PT,* 27 Dec. 1847, 412–14, at 413.

a search for truth as for logical victory in argument; he had not withstood the effects of strong reading. Hutchinson was equally well-read, but older and classically educated, and "has grappled with opinions manfully."[59]

Although Hirst envied the strength of character that had led Hutchinson to hold his beliefs, he could not adopt them as his own. Drawing implicitly on Carlyle, Emerson, and Fichte, he told the others that he thought truth "is in myself, that there is a common soul animates all men, and that the hundred and one of the many creeds are but the attempts to read this book we carry with us." Hutchinson suspected where this was going, and asked Hirst if he believed in external evidences at all. Did he agree that Napoleon had lived, or that the world had existed before this instant? This led Hirst to one of his introspective passages:

> A pause ensued. I did not like to say I did not. I could not say so, for I am not prepared; yet moments there have been when I have thought that for me it matters little whether they ever did or not; and this indifference carried to a certain excess

59. Ibid.

> would amount almost to practical unbelief. I did not answer, and Booth and Roby carried on the discussion on miracles generally, the difference between the Gospel and other miracles in point of credibility and evidence. But as I scarcely heard them, I shall not transcribe them. Somehow, I could not banish from my mind that they (such subjects) were unimportant, secondary to me; but I do not put aside at once the Divinity of Christ, inspiration, mediation, and a host of other doctrines.[60]

If externals were really so unimportant, if all that mattered was that which could serve as a guide to action, then why not reject traditional faith outright? Why not worship the devil if that became the best path to action? Hirst could not take so great a step. All he could be sure of at this stage was uncertainty and doubt. When Pridie remarked a few minutes later that Booth was "'a sceptical little dog,'" Hirst replied that he should be considered "'a large dog of the same species, I suppose?'" As Hirst put it in a reference to *Sartor Resartus* they would have recognized, much as Hutchinson's comfortable Christian coat might be enviable, he preferred his own, "'even if it be seedy.'"

Only in morals could they all agree, at least after Booth left. As Hutchinson said, "'we must put off corruption, avoid these temptations of a sensual world.'" Hirst affirmed that this was one truth of which he could be certain. Resisting temptation was "'all fine in principle, but devilish hard in practice'" (he had written to Tyndall shortly before of "evil deeds" that may have involved sexual transgressions). Although Pridie was encouraged by the consensus about the Bible's moral principles, Hirst thought their validity derived not from divine inspiration but because the book embodied the best parts of common humanity: "'one nature wrote and the same reads it.'"

The dialogue form used here and elsewhere in the journal is modeled on Hirst's reading of autobiographical novels dealing with intellectual issues, notably *Yeast* and *David Copperfield.* Such works were often cast in the form of diaries. As in these novels, Hirst creates characters who stand for particular positions: Hutchinson, "manfully" grappling with opinions, is akin to one of Kingsley's heroes; Booth is the coolly logical freethinker; Hirst voices doubts expressed in Carlyle and Fichte. The journal draws so closely on the literary formulae of the early Victorian bildungsroman as to read in parts very much like a novel, albeit one written for its author alone.

Moreover, the discussion rehearses debates that the young men have followed in books and reviews; Hutchinson uses arguments from Miller's *Footprints* (and probably the *Edinburgh* critique or perhaps Whewell's *Indications*), while Hirst responds on the basis of the arguments in *Explanations* that he has already recorded in his journal. These literary borrowings are transformed,

60. Ibid., 24 Nov. 1849, ff. 547–48.

however, by their position both within the conversation itself and the prior experience of the various participants. Hirst's fervent agreement about Christian morals grew out of his unspoken consciousness of mortal sin.

The evening, Hirst noted, "was one of the most interesting I have ever spent." He praised the discussion for its "honesty, leniency and sincerity";[61] this was not the witty repartee of fashionable London, an earnest clerical discussion in Ely or Oxford, nor the doctrinal party talk of Edinburgh or Liverpool.

Like other self-improvers of his class and background, Hirst saw philosophical and religious discussion as a masculine activity. All the conversations he thought worth recording in his journal were with men. As one of them said, "when I go to see him we always have a happy time, and discuss subjects of serious import, which, I am sorry to say, are not very popular in mixed company."[62] A few days after the memorable discussion of Hirst and his friends, the Mutual Improvement Society held a joint meeting with its female counterpart. Hirst was disgusted by the "disorderly" behavior and "petty jealousy" of the women:

> I felt many a time a desire to get up and tell them honestly my dissatisfaction; but I thought better of it, and only just brought them back amidst their wanderings and personalities to the point under discussion. It only needs such scenes as these to convince us how much out of place women are in such places. If they want to improve themselves let them do it in God's name without such flummery—if they want to learn to spout at meetings, let it be amongst their own sex.[63]

Hirst thought argument unsuited to woman, who should be "man's companion, his help-mate, invigorating him by her sympathies, her love and her smiles."[64] In the terms defined by Carlyle and Emerson, women were weak-willed and unsuited to leadership through a life of work in the world. *Vestiges,* in contrast, was not effeminate, but what Hirst hailed as the product of a "manly spirit," so that conversation about it reinforced his wish to take his stand as a man.

The End of Reading

Hirst's reaction to *Vestiges* was so strong that he might seem to embody the ideal reader projected in the text. Hirst not only agreed with much of what he had read, but did so as part of a process of self-realization, so that his journal records his own development through the successive chapters. By copying long

61. Ibid., 24 Nov. 1849, f. 550.

62. J. Searle [G. S. Phillips], *Leaves from the Sherwood Forest* (London, 1850), 147–48, quoted in Brock and MacLeod 1980b, 53.

63. Brock and MacLeod 1980a, 3 Dec. 1849, f. 555.

64. Ibid., f. 1130, ff. 484–85; see Brock and MacLeod 1980b, 13.

passages into his private journal, he incorporated the text directly into his sense of who he was and what he was to become. If Jean Ranson is a "Rousseauist" reader of Rousseau in Darnton's classic study, Hirst could be portrayed as precisely the kind of Vestigian reader the *Vestiges* author had had in mind. This, however, involves taking too limited a view of the opportunities afforded by a book.[65]

In the first place, Hirst transformed *Vestiges* by reading it in juxtaposition with other works and in the context of intense discussion with friends. He made books his own, or to put it another way, he pulled them apart to combine them with other works and make them more useful for his own purposes. Like the compiler of a commonplace book or anthology, he looked for "beauties" and passages of particular interest in relation to other works he was reading at the same time. Hirst thereby experienced the book more like the monthly serials and review essays he enjoyed and less as a completed, self-contained narrative to be taken in at a single sitting. The dominance of periodicals was reshaping the fundamental practices of reading.

As a reader, Hirst's needs contrasted with those anticipated by the author. True, Hirst was young and in search of a creed, and came out of religious traditions not all that different from those experienced by Chambers himself. But he resisted the extension of the development theory to man, for that would undermine the notion of free will that was at the heart of his faith. As Hirst came increasingly to adopt perspectives drawn from transcendental philosophy, he conceived of the relations between his individual will, his body, and the external world in a new and very different way from that projected in *Vestiges*, which defined free will as merely a result of the interplay of the phrenological faculties.[66] Hirst's encounter with *Vestiges* and the replies to it thus became part of the process of redefining his own perspective as a reader, the "I" of which *Sartor Resartus* and the *Vocation of Man* spoke so powerfully. Like many young men in Britain and America, particularly those with backgrounds in Congregationalism, Hirst found in these works a way of thinking through who he was.

The issue of free will came to define Hirst's quest for a role in changing the society around him. Halifax in the 1840s presented the liberal middle classes with overwhelming problems in practical political economy. Although built on a steeply sloping hill, the town was badly drained and filthy, with unemployment, destitution, and hunger rife even among the "respectable" poor in the mid-1840s. The population increased dramatically in the middle decades of

65. R. Darnton, "Readers Respond to Rousseau: The Fabrication of Romantic Sensibility," in Darnton 1984, 209–49. Ranson is perhaps too perfect a reader of Rousseau to be an entirely satisfactory example of the disturbing, disrupting effects that reading can have.

66. *V*1, 349.

the century, and over four-fifths of the people were engaged in manual labor. Working on tithe surveys in the surrounding countryside, Hirst became acutely aware of the class divisions and social inequality enforced by the rates, which many found almost impossible to pay in times of destitution.[67] His experience of these conditions encouraged his search for a higher calling.

No single book—and certainly not *Vestiges*—had a transforming effect on Hirst's life. Even his radical self-absorption in Carlyle, Emerson, and Fichte was not a surrender to favorite authors, but a way for Hirst to assert his independence of judgment and to distance himself from the faith in which he had been raised. He could not agree that scientific progress destroyed wonder and mystery; and *Vestiges*, which would have fallen into *Sartor*'s category of mechanized cosmological "Dream Theorems," pointed a way forward for Hirst. Early in 1850 he made the great decision of his life. Following Tyndall's example, and assured of a modest financial independence after his mother's death, he gave up surveying and left England for the Continent to study mathematics, physics, and chemistry at Marburg. Eventually he would obtain a doctoral degree in mathematics, teach at University College, and join the X-Club.

But before this, on the way to Germany (at, of all places, a dinner given by an Anglican clergyman) Hirst met "the great Atheist Holyoake":

> You expect a monster, and overthrower of everything we hold dearest and most sacred; we are surprised to find here a quite human (not demoniacal) being, and that also with some kindness and affection in his nature. He is a man of keen intellect but devoid of all spiritual faculty. In his face so pale and haggard, with dull eyes and compressed, thin lips, and a certain affected working of the facial muscles, especially about the eyes, you can read truly, as on an index page, the contents of the interior—a keen intellectual vision and grasp, but no heart, no higher feelings.[68]

Reading Holyoake's face as an index to his character, Hirst saw no corresponding reflection: "inexpressibly cold and lifeless in his beliefs, which would not at all fit me." Reading *Vestiges* had been a very different experience, more like gazing on the portrait of Tyndall that hung on his wall. "The eyes especially I can look straight into and examine," he had noted in his journal, "I fancy I can read many things there."[69]

Hirst could not agree with everything *Vestiges* said, and thought it tended toward materialism and a denial of free will; but he identified with the author's "manly spirit" and found in its pages much of his own tendency toward independent inquiry. He placed his faith in science; mathematics would be freed from the chain and theodolite.

67. Moore 1997, 300–303.

68. Brock and MacLeod 1980a, 24 June 1850, f. 618.

69. Ibid., 16 Aug. 1850, quoted in Eve and Creasey 1945, xvi.

Like so many others, Hirst had discovered in *Vestiges* new ways of thinking through the problem of vocation. Who, then, did he suspect might have written the work? In February 1848 he and his older brother had a drink with a Scottish traveler, "a very able fellow," in the Upper George Tavern. It was, the man said, "almost beyond doubt" that Robert Chambers was the author.[70] The traveler presumably knew the rumors that had followed in the wake of the article in *MacPhail's.* Given his high opinion of *Chambers's Edinburgh Journal,* Hirst probably took this bit of gossip as a positive recommendation, but basically he cared little about it. The authority of a named author, so important to the aristocracy and the metropolitan literati—and to us—had no discernible effect upon Hirst's approach to the book.

Most contemporaries would probably have agreed. There can be no doubt, though, that the mystery did matter intensely to one particular reader: Robert Chambers. We will now see why he wrote anonymously, how his secret was maintained, and how the narrative voice in *Vestiges* was transformed in response to readers.

70. Brock and MacLeod 1980a, 17 Feb. 1848, f. 193.

CHAPTER ELEVEN

Anonymity

All this is curious—isn't it? Strange to walk about amidst all the jumping jacks, and be pulling grand strings, causing people to wonder, and yet be unseen, unknown all the time—if I really be so now.

ROBERT CHAMBERS to Alexander Ireland, writing secretly from the British Association at Cambridge, 19 June 1845

When Tom Hirst wandered through Halifax, pondering whether streets, milkmaids, and his own body were projections of his inner consciousness, he faced in an acute form a characteristic dilemma of the 1840s. What was the relation between the "I" that expressed who a person was and the rapidly changing settings of urban life?

Being anonymous was both an everyday, taken-for-granted feature of the emerging industrial society and at the same time deeply disturbing. The city is often defined as a place where most people are strangers. It is a dense, highly populated stage, dominated by impersonal interactions of market exchange; individuals tend to have no prior knowledge of one another, no known history, no basis for trust other than that which is either inferred from immediate circumstance or through formal certification. In such a setting, questions of self-identity become questions of self-presentation. No one could ever know the names of the hundreds of people passed on the street in a single day, nor the names of the authors who wrote the words in newspapers and advertisements. Statistics showed that some of those people must be murderers or thieves; some of those authors must be writing with an intent to corrupt. Judging moral and intellectual worth on appearances was fraught with danger. In an increasingly nameless, faceless society, how could one tell?[1]

1. The classic study remains Sennett 1992, which includes a detailed examination of the 1840s. Also important are Auerbach 1990, Barnes 1995; Gallagher 1994; Halttunen 1982; and Morgan 1994, 87–118. Most studies of anonymity deal with individual authors; see, e.g., Welsh 1985, Judd 1995, and Bodenheimer 1994, 119–60 (on George Eliot); Shortland 1994 (on Southey); and Ferris 1991 and Robertson 1994, 117–60 (on Scott). The best general survey is Griffin 1999. Vincent 1999 dis-

The deep anonymity of *Vestiges* reveals just how important this issue was seen to be. In fashionable circles, hunting the "invisible lion" was an attempt to maintain the dominance of the aristocracy over intellectual life. Among the urban missionaries, the search was for the source of a poison that was corrupting the nation's spiritual life. Here was a sensational book that redefined humanity in relation to nature and God, but whose author "bore no bodily shape in the eyes of his fellow-countrymen, and was likely to remain for ever unknown."[2]

The Uses of Anonymity

To the end of her life Camilla Toulmin could never look at a roast leg of lamb without thinking of the mystery surrounding *Vestiges.* As a young contributor to *Chambers's Journal,* she had been invited to dinner at Chambers's house in 1845. A lady at the other end of the table suddenly interrupted a conversation about *Vestiges,* just as their host had begun carving:

> "Do you know, Mr. Chambers, some people say you wrote that book."
>
> Though sitting next my host, I happened to be looking towards Mrs. Chambers, and I saw that she started in her chair and that a frown was on her face. She looked at her husband, but his eyes were bent on the lamb, on which he continued operating in an imperturbable manner, observing—
>
> "I wonder how people can suppose I ever had time to write such a book."
>
> There was silence for a minute, and then I think the subject dropped.

Only when the secret was formally revealed some forty years later did Toulmin understand the reasons for the furtive glance and evasive answer. Despite all the rumors, the secret had been kept "wonderfully well."[3]

Chambers had always been fascinated by anonymity. One of the first essays he ever published, in the *Kaleidoscope*'s opening issue, had satirized the fashion for blank and pseudonymous title pages. The (unsigned) essay notes that secrecy conferred membership in what was effectively a vast family of unknowns. These included "Junius" and the wickedly satirical "Christopher North" of the Tory quarterly *Blackwood's Magazine.* Looming over them all was the Great Unknown, Sir Walter Scott, "the very Prince of Genius and a greater man than even myself."[4] Chambers's first book, *Illustrations of the Author of Waverley,* played with the idea of revealing, or "illustrating," the mysterious author. The veil on

---

cusses secrecy, esp. in relation to the state. The only book-length survey of anonymous and pseudonymous works in English is Courtney 1908.

2. *E*1, 2.

3. Crosland [née Toulmin] 1893, 86–87. Chambers thought Toulmin "a very delightful creature" but cautioned Ireland against marrying her, as she had no income other than that obtained through her writing. See RC to AI, 31 Aug. 1845, NLS Dep. 341/110/191–92.

4. [Chambers], "Concluding Address to the Public," *Kaleidoscope,* 12 Jan. 1822, 12–14, at 14.

11.1 Frontispiece to the second edition of *Illustrations of the Author of Waverley* (1825), written by Robert Chambers.

the framed portrait frontispiece to the second edition (fig. 11.1) was teasingly only half-removed, so the reader was left guessing who was hidden behind.

As a journalist, Chambers had unrivaled practical experience in managing the conventions of invisible authorship.[5] As he wrote in the *Kaleidoscope*, anonymity and pseudonymity had characterized periodical writing "since the invention of that sort of literature," for the voice without gender, class, or status moved among the people at large without being identifiable with any single group. To write under no name or an assumed one, meant "to be absolutely nobody and to live absolutely no where":

5. Vincent (1999, 65–75) briefly analyzes secrecy in relation to anonymous periodical writing. Most of the literature is concerned with the problems of identifying authors; a good survey is Hiller 1978.

> I have been one year the loquacious Tatler and the next the taciturn Spectator; the Rambler and the Idler, with equal propensities to locomotiveness; united the opposite extremes of Leviculus & Gravis, Hermeticus & Flirtillus, Tom Tranquil & Jack Whirler, all in my own form and character; have had no more regard to the decorums of sex than a hacknied actress, in breeches for the hundredth time; have been every thing, yet nothing; every sex and no sex; spoken from heaven in the character of an angel, and howled, with equal complacency, from hell, as Belzebub:—and all to serve you, my dear public.[6]

Anonyms and pseudonyms had thus been part of the persona of the "public" author from the beginning.

Anonymity featured not just on the *Vestiges* title page, but throughout the work, as the authorial voice assumed different guises: the factual reporter of "familiar knowledge"; the master editorial voice of a mass-circulation periodical; the profound philosophical commentator. These roles were all associated with anonymity. They created a sense of the author as a neutral, all-seeing guide, free from human subjectivity, and subtly associated with the "Author of nature" to whom reference is so often made. An invented or assumed name, of the kind associated with political writings and novels, would have been inappropriate. The text did play, however, upon expectations that readers would want to discover who had written it. There was no explicit reference to the lack of an author until the conclusion, when anonymity was tantalizingly reinforced just before the veil was drawn. "For reasons which need not be specified, the author's name is retained in its original obscurity, and, in all probability, will never be generally known."[7]

If that sentence presented readers with a mystery, elaborate precautions were made to keep them from solving it. At first only four people, besides the author, knew the secret (fig. 11.2). Anne Chambers and William Chambers had to be told as a matter of course (though there were later rumors that the secret was "kept from even the wife").[8] Alexander Ireland, the intermediary with the publisher, had recently moved to Manchester, thus allaying any suspicions that the author might be Scottish.[9] Robert Cox, Combe's nephew and editor of the *Phrenological Journal,* was Chambers's closest friend and a trusted confidant.

6. [Chambers], "Concluding Address," *Kaleidoscope,* 12 Jan. 1822, 12–14.

7. *V1*, 387.

8. RC to AI, [12 Feb. 1845], NLS Dep. 341/110/51–52.

9. All correspondence, proofs, and manuscripts relating to the book were supposed to be burned, but Ireland—a keen collector—kept Chambers's letters as a record of events in which he was proud to have participated. The following paragraphs are based on a reading of this correspondence, unsorted and mostly undated, but still largely intact, in NLS Dep. 341/110–13. For Ireland's account of the publication arrangements, see Ireland's introduction to *V12*, xvii.

11.2 Anne Chambers (from Priestley 1908, facing p. 27) and Alexander Ireland in later life (Mills 1899, 139).

Three others were later told: David Page, editorial assistant on *Chambers's Journal* (in February 1845); Neil Arnott, a London physician and author on natural philosophy (in July 1845); and John Pringle Nichol, professor of astronomy at Glasgow (in May 1846).[10] Chambers's other siblings and children, along with good friends such as the Combes, were kept in the dark.

The book was written at a time when a parliamentary inquiry was revealing that government agents had illicitly opened private letters, which put a question mark over the confidentiality of the Penny Post.[11] To avoid being found out (for example, by an evangelical postal clerk), Chambers and Ireland began to employ a system of code words and pseudonyms. *Vestiges* was "opus," Anne Chambers was "Mrs. Balderstone," George Combe was "Jokum," Robert Cox was "Robertus," Ireland was "Alexius," and Chambers himself was "Ignotus," "Mr. Balderstone," "Sir Roger," or "The Unknown." These pseudonyms drew on Chambers's association with Scott (the "Great Unknown") and *Chambers's Journal,* where Mr. and Mrs. Balderstone had appeared as fictionalized alter egos for the Chamberses. Most of the letters were undated and unsigned, often referring to "the author" in the third person.

10. RC to AI, [Feb. 1845], NLS Dep. 341/110/112–13; RC to AI, 31 July 1845, NLS Dep. 341/110/141–42; RC to AI, [May 1846], NLS Dep. 341/110/225–26. With the exception of Arnott, these additional names are not mentioned in Ireland, introduction, *V12*, vii–viii.

11. Robinson 1948, 119–25, 337–52; Vincent 1999, 1–9.

These stratagems meant that a letter that fell into enemy hands could not be used to prove the authorship. There was always a danger that someone would recognize the author's handwriting. To avoid this, Anne Chambers copied the entire manuscript, as well as most changes to the various editions and answers to letters of inquiry sent through the publisher. This was a time-consuming task, which was interrupted when she was away, pregnant, or preoccupied with illness in the family.[12] She does not appear to have agreed with everything in the manuscript she was copying, especially its claim that free will was merely the result of the interaction of the phrenological faculties. As she advised her "dear friend" Alexander Ireland after one of his crises in love, "Do not again talk to me of people having no *free will.* This is nonsense in my opinion, have we not the power of choosing good or evil."[13] Anne Chambers had walked out of Saint Cuthbert's when *Chambers's Journal* was condemned from the pulpit, although unlike her husband she remained throughout her life a regular churchgoer, taking the children to services each Sunday. Her fine Italianate handwriting, familiar only to immediate friends and family, became the graphic mask of the *Vestiges* author, shielding him from public scrutiny. It served something of the function of a female pseudonym—a common device among male writers who wished to speak as unknowns from a realm defined as private and domestic.[14] The characteristic handwriting on documents signed as from "The Author of the Vestiges of Creation" may have encouraged suspicions, which even reached print, that Anne Chambers had written the work herself. However, most people who knew her well enough to recognize her hand acknowledged that this was unlikely, even given her knowledge of the sciences.[15]

For his part, Ireland was told to avoid Churchill's shop and direct contact with the printer or anyone else connected with the production. Documents in Chambers's hand were recopied by Ireland before being forwarded to Churchill, who was to be his sole contact in London. All correspondence and other records were to be burned. The chain of communication was cumbersome but effective: Chambers would tell Ireland to tell Churchill to tell Savill to make a last-minute change in proof; queries in the opposite direction had to filtered through the same intermediaries. Chambers was always looking out for new ways of avoiding detection. All issues relating to *Vestiges* were dealt with behind the imposing facade of No. 1 Doune Terrace in Edinburgh's New Town, the family home

12. RC to AI, [3 Aug. 1844], NLS Dep. 341/110/78–79.

13. A. Chambers to AI, [5 Mar. 1837], NLS Dep. 341/111/5–8.

14. Judd 1995, 259–67.

15. The possibility of Anne Chambers's authorship of *Vestiges* is mentioned in *Proceedings of the Linnean Society of London,* session 1870–71, lxxxiv. For her knowledge of science, see Robert's letters home during his field trips, e.g., RC to A. Chambers, 29 Sept. 1847, Norman mss.

11.3 No. 1 Doune Terrace, in Edinburgh's New Town, where the Chambers family lived after their return from St. Andrews.

after the return from St. Andrews (fig. 11.3). Chambers kept all materials relating to the book in his study, and in a locked desk—just as if he had secret financial dealings or an illicit love affair. It was, as such associations suggest, a highly gendered space: the masculine inner sanctum of the early Victorian private sphere.[16]

Why did Chambers, a key figure in the development of the heroic author and the canon of English literature, wish his involvement with the book to remain such a deep secret? In practical terms, anonymity was a standard, safe course of action in an untried literary enterprise. Many first-time novelists, for example, waited to see if a work sold well enough for a second edition before

16. RC to AI, 5 Nov. 1847, NLS Dep. 341/110/239–40. For evidence about the gendering of domestic reading spaces, see Flint 1993, 102–5; for the role of such spaces in creating male identity, see Tosh 1999.

allowing their name to be on a title page. Chambers had a considerable literary reputation, and if this new venture flopped he had much to lose. Moreover, his name carried no weight in science, however much it might be celebrated in middle-class family journalism. This was not as problematic as it might seem, since what it meant to be "scientific" was much debated. All the same, critics could more easily have dismissed a work under Chambers's name as the product of an inexperienced bungler. Certainly anonymity gave *Vestiges* far more status—particularly aristocratic status—in the critical early months than it otherwise would have had. This was the first book written by Chambers to be noticed by the quarterlies, where his works for the "people" had been cavalierly dismissed. The greatest triumph was when his old nemesis, John Gibson Lockhart, thought *Vestiges* was wonderful.[17]

Anonymity was most effective as a shield against a violent and abusive controversy. There were financial concerns in which William was as much concerned as Robert: an evangelical boycott of their educational publications would have been ruinous. They had agreed never to publish on controversial religious or political subjects. Ultimately, though, Chambers kept the authorship hidden to protect his family and his reputation. Until society progressed, the facade of traditional respectability had to be maintained. He had seen what had happened to Carpenter, Lawrence, and others accused of "atheism" or "materialism." The business might be damaged, the "saints" might attack, but the greatest threat was that his growing family—especially his daughters—might be placed beyond the pale. Years later his son-in-law asked him why his greatest book was shrouded in "impenetrable mystery." Chambers "pointed to his house, in which he had eleven children, and then slowly added, 'I have eleven reasons.'"[18]

The Author as Actor

The imperative for anonymity also allowed Chambers to savor the resulting public confusion as a "treat" to his phrenological organ of Secretiveness. Secretiveness was situated in the middle of the lateral portion of the brain (fig. 11.4; fig. 2.11, faculty number 7), just above Destructiveness; it was often large in actors, who could use it to suppress the natural balance of their own faculties in favor of those of assumed characters.[19] Chambers played his roles with relish and could scarcely wait to find out how his friends would react. He monitored the reception, experimenting with variations in the relation between author and

17. RC to AI, 18 Nov. 1844, NLS Dep. 341/110/43–44.

18. Lehmann 1908, 7, quoting memoirs of Frederick Lehmann, who married one of the Chamberses' daughters in 1852. The main reasons for anonymity are outlined in Ireland, introduction to *V*12, viii, xiii–xvi.

19. Combe 1843, 1:294–311.

reader. He placed himself in situations, often potentially awkward or compromising, that offered opportunities for conversing about the book with those who had no idea that he was the author.

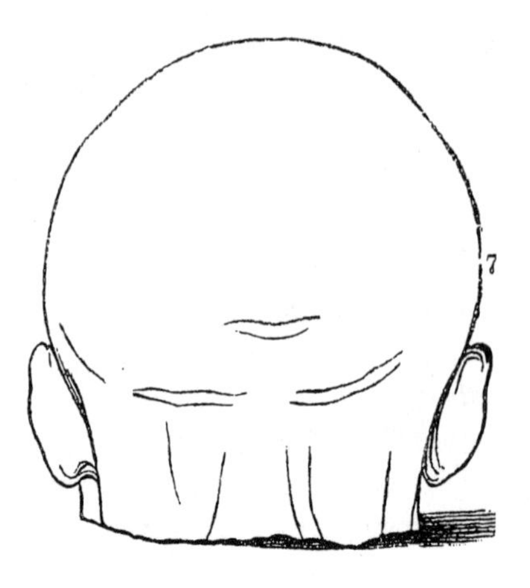

11.4 The phrenological organ of Secretiveness, marked here with a "7." Combe 1843, 1:295.

The one situation to be avoided was lying. Chambers felt he could never do what Vyvyan and other suspects had legitimately done—publicly deny having any part in the authorship. In this, he took a moral stance opposed to his usual model, Walter Scott, who had placed lying about authorship in a special category. As Scott had explained in the general preface to the Waverley novels, "I therefore considered myself entitled, like an accused person put upon trial, to refuse giving my own evidence to my own conviction, and to flatly deny all that could not be proved against me." This legalistic analogy posed its own moral dilemmas, as when King George IV asked point-blank about the Waverley authorship and Scott replied just as blatantly with an outright falsehood.[20] Scott's example was often used to justify lying about anonymous and pseudonymous authorships. As Marian Evans stressed before being revealed as George Eliot, "in such a case one ought to say 'No' to an impertinent querist as one would decline to open one's iron chest to a burglar." For Chambers, the situation was not so simple. He was willing to deny the authorship when strangers inquired about the book in correspondence. As he wrote to the American geologist and novelist Mary Griffith, "I have to acknowledge with best thanks your kind attention in sending me a copy of your Remarks on the celebrated *Vestiges,* even though you have sent it under the false impression that I am the author of that work, a report I can only attribute to my having several times spoken in defence of the author's theory." Yet to equivocate to this extent in public or among friends was unacceptable.[21]

20. In practice, Scott usually softened denial into equivocation, by telling those who asked that authors had the right to protect their secrets even if that meant telling falsehoods. Smiles 1891, 1:474; general preface (1829), in Scott 1986, 349–61, 357. Chambers attributed Scott's attitude to fears that writing for money might be thought "somewhat degrading to the Baronet of Abbotsford;" [Chambers], "Life of Sir Walter Scott," *CEJ,* supplement, 6 Oct. 1832, 289–300, at 296. As shown in Shortland 1994, 59, Robert Southey also denied having written his only novel, *The Doctor.*

21. G. Eliot to J. Blackwood, 1 Dec. 1858, in Haight 1954–78, 2:505; RC to AI, 5 Nov. 1847, NLS Dep. 341/110/239–40; RC to Mary Griffith, 2 Dec. 1846, quoted in Cox 1993, 193.

In such circumstances, to lie about a book that aimed to tell the truth would have negated his self-identity as an author.

Chambers shared the view, common among the early Victorian bourgeoisie, of the need to maintain the appearance of absolute integrity. The ideal derived from traditional codes in which the gentleman embodied truth in his person. Contemporary moralists argued, against Scott, that the secrets of authorship were no exception to the rule that one must not lie. Whewell, in his *Elements of Morality* (1845), took a particularly strong view on this: authors might "baffle curiosity" by evasion or turning the question, but allowing any impression contrary to the truth to develop was bad faith and would leave a "moral stain." The use of language implied certain mutual understandings that should never be violated.[22] Chambers, who by this standard was a moral reprobate, was simply concerned to avoid a direct lie in circumstances where it might be openly contradicted. He could insinuate, equivocate, mislead, tell half-truths; but a public falsehood would have destroyed his character as a man of honor. This, then, was the one role that underpinned all the other parts played by Chambers.

Anonymity in print gave an aura of mysterious power, but in letters and conversations it was a source of fun. Throughout the coming years, Chambers effectively acted the part of an ordinary reader engaged in the game of detecting the author. His secret correspondence is punctuated by miniature dramas, complete with dialogue, that illustrated how he played his role. One remarkable trial in the week of publication had Chambers speaking his anonymous text to close friends to gauge the reaction: he took a copy to Catherine Crowe and read her "the main chapter"—presumably the pivotal discussion of organic development—without admitting that he was the author. Chambers was delighted with her response, but was even more pleased by Combe's praise for this "great new dig into the sides of superstition" and his plan to write to the *Vestiges* author through the publisher. "You cannot imagine how amusing all this was to Ignotus," Chambers confided to Ireland, also asking for any news about the effect of *Vestiges* on "highly endowed minds."[23]

Chambers waited eagerly for word from London. After a few weeks without much news, the scale of the sensation became clear, and it looked as though the veil of anonymity might be withdrawn. A friend from London mentioned

22. Whewell 1845a, 1:253–56. Harriet Martineau felt so strongly about not lying that her servants were even forbidden to use the phrase "not at home" to visitors she did not wish to see. However, this was thought to be somewhat extreme. Whewell could consistently argue that this use of the phrase "not at home" was not technically a lie, as it was conventionally understood to include the possibility that one was at home but did want to let the visitor in. For truth and lies in an earlier period, see Shapin 1994, 65–125.

23. RC to AI, 15 Oct. 1844, NLS Dep. 341/111/ff. 19–20.

that scientific men were full of praise: "'A wonderful work' is the remark of all, while most agree with the author's hypothesis." As Chambers told Ireland in mid-November:

> Your two last communications have given immense pleasure to the author of the opus. Having kept his hopes down at the lowest mark—like a prudent cool Scotchman as he is—he unavoidably feels much elevated by tokens of success so unequivocal, and which have in a manner burst upon him. He capers in thought at the idea of Lockhart's note, considering how that serpent would speak of him at this moment as the author of all he has written besides the opus. This testimony is also valuable as affording a hope that after all the opus may be passable before the world with an author's name at it.[24]

Prudence, according to standard phrenological doctrine, was a characteristic often associated with Secretiveness, which was after all an organ typically large in misers.[25] And it soon became clear that Chambers's initial caution was justified, for although the book was praised, it was clearly proving controversial.

A full four months passed without anyone pointing to Doune Terrace. Chambers rejoiced that the anonymity was "beautifully preserved."[26] As the Newcastle mining engineer Thomas Sopwith confirmed, the mystery was complete: "shadows, clouds, and darkness rest upon it."[27] (As if to illustrate his point, Sopwith had no idea his friend had anything to do with the book.) The delay in associating *Vestiges* with Chambers proved crucial, for when his name did arise from mid-February 1845, it was as only one among a host of possibilities. In all later exchanges over the next forty years, Chambers could "turn" the conversation to other suspects or to the general uncertainty surrounding the authorship.

Rumors that Chambers was the author appear to have emerged in Edinburgh. In conversation with a friend, James McLelland, Chambers had playfully wondered if Nichol might have written *Vestiges*. The game backfired when McLelland and Chambers subsequently went to see Nichol (who was not yet in on the secret) at the observatory in late February:

> McL. said to N. that I had told him of his being the suspected author of the V.; whereat N. *looked a good deal confused,* but quickly said that he knew who was. Mr M'L. said that Dr S. B. [the chemical philosopher Samuel Brown] had found it was a residiary [*sic*] clergyman residing at or near Dundee; whereafter I remarked

24. RC to AI, 18 Nov. 1844, NLS Dep. 341/110/43–44; RC to AI, [Jan. 1845], NLS Dep. 341/110/125–26.

25. Combe 1843, 1:307–8.

26. RC to AI, 8 Dec. 1844, NLS Dep. 341/110/165–66.

27. As reported in ibid. See also RC to A. Chambers, 22 May 1845, Norman mss.

> the strong reason such a man would have for concealment. N. said, "It is no clergyman, but one who, from other literary engagements, has equally good reason to keep on the mark." Meaning me, of course. Yet would you believe it? I looked as unconscious as if I had never heard of the book. I felt afterwards half shocked at my own secretiveness. Was it not delicate, however, of Nichol?[28]

Such incidents were still uncommon enough to be described as "amusing"; Chambers maintained a facade of nonchalance even in his secret letters.

In April, though, Chambers found his name mentioned with increasing frequency when he traveled to London to spy out the situation. He reported that the authorship "had been attributed to a great many persons, *me* among the rest; but this last insinuation, I think, must have been made very faintly. I got through the remarks and conversations with the greatest coolness."[29] Other evidence bears out the impression of a multiplicity of candidates. In March the inventor Charles Wheatstone agreed in conversation with the common view that Vyvyan had probably written most of the book—while thinking "it would have been much wilder" without revisions from someone else.[30] Edwin Lankester, known to be checking proofs, shows that even an insider could judge only on the basis of gossip and internal evidence. His *Athenaeum* review in December 1844 had suggested that the author was "a Scotchman, a large reader, but not an original observer, and one who has mixed little with the men of science of his day."[31] But this was just a guess. Writing in his private diary (and busy revising the work), Lankester remained "entirely ignorant of the author." He continued to think that the book "displays much reading and thought" but no firsthand knowledge of science. However, he believed that "the greatest weight of evidence" pointed to Vyvyan or Lovelace, especially the latter. Rumors of aristocratic authorship had spread beyond fashionable circles.[32]

Back in Edinburgh, Chambers thought the situation might be more serious. Walking home at the end of April after a thrilling astronomical lecture by Nichol, Chambers had asked Crowe if she thought the author of *Architecture of the Heavens* might "from his style" have written *Vestiges.* She replied "'Oh, it is now generally given to you.'" Later that evening, one of Ireland's letters held further bad news: an acquaintance was spreading a report in Manchester that

28. RC to AI, [late Feb. 1845], NLS Dep. 341/110/86–87.

29. RC to AI, [13 Apr. 1845], NLS Dep. 341/113/160–61.

30. As reported in C. Bunbury, diary entry for 5 Mar. 1845, in Bunbury 1890–91, 1:42.

31. [E. Lankester], *Athenaeum,* 4 Jan. 1845, 11–12, at 12; see also [E. Lankester], *Athenaeum,* 13 Dec. 1845, 1190–91.

32. E. Lankester, diary, vol. 9 (1843–45), 3 Mar. 1845, information provided by Peter Bowler, citing the Lankester family papers.

Chambers had been contemplating an anonymous philosophical work, and hence was a likely author. Around this time, too, Hewett Watson, the botanist and phrenologist, addressed a letter directly to Chambers as the author. So did John Forbes, who edited the *British and Foreign Medical Review.* Forbes had initially thought Combe must be the author, but then began to suspect Chambers, whom he confronted directly.[33] On reflection, Chambers decided not to get alarmed. As he wrote to Ireland, "they can but suspect and surmise."[34]

The dilemmas posed by anonymity were most apparent in Chambers's own city and among those closely associated with him. Rumors about the authorship traveled through different channels in Edinburgh from those in England, and tended to be based on personal knowledge. It was far easier for Chambers to manage his self-presentation on the national stage, where he was known primarily through print, than among those with whom he walked and worked every day. Those entrusted with the secret kept quiet, but suspicions about a local author quickly spread. Many people knew Chambers and other phrenologically inclined authors, and made judgments accordingly; some detected traces of Scottish dialect. William Tait believed he had cracked the mystery through conversation, having heard his fellow publisher "speak more than one part of the volume."[35] By the summer of 1845, Chambers knew he was increasingly associated with the book in Scotland. As he told Ireland, "At St Andrews, it is quite concluded upon that Abbey Park was the mint of this strange medal."[36]

But many were not so sure. At a dinner party at the Chamberses' soon after publication, *Scotsman* editor Charles Maclaren named Neil Arnott as the most likely prospect.[37] He picked up the Vyvyan rumor, well known from newspaper reports and letters from London, the following July.[38] But north of the Border neither Vyvyan nor Lovelace was much known or thought of as a strong possibility. Crowe became a favorite local choice. At a dinner at the Chamberses' house in Doune Terrace, the question was discussed with four of the conspirators sitting at the table—Anne Chambers, Robert Chambers, Cox, and Ireland. Many guesses were made and "a brisk fire of conversation ensued." Ireland recalled that one loud-mouthed guest faced Crowe and asked her to confess. "'I have a strong suspicion,' said the questioner, 'that my vis-à-vis . . . is the author of that naughty book. Is it not so? Come now, confess. You cannot deny it.'"[39]

33. RC to AI, [Apr. 1845], NLS Dep. 341/113/164–65; H. C. Watson to RC, 7 Apr. 1847, NLS Dep. 341/98/16; J. Forbes to RC, 17 Jan. 1846, NLS Dep. 341/94, f. 68.

34. RC to AI, [Apr. 1845], NLS Dep. 341/113/164–65.

35. Bertram 1893, 133.

36. RC to AI, 31 July 1845, NLS Dep. 341/110/141–42.

37. RC to AI, [Dec. 1844], NLS Dep. 341/110/22–23.

38. "Edinburgh Review, No. 165," *Scotsman,* 9 July 1845, 2.

39. Ireland, introduction to *V*12, xx–xxi.

Those who were actually in on the secret became skilled at dodging direct confrontations. William Chambers, with his businesslike demeanor and lack of sympathy for the work, found this easy to do, and did not reveal the authorship even in the 1872 biography of his brother.[40] If William had had his way the secret would have gone with them to the grave. Anne Chambers became almost as proficient a dissembler as her husband. On one occasion she visited Tait's shop to ask about a rare book of fairy tales. A "bore" was insisting that the overrated Robert Chambers could not have written *Vestiges*. "'How do you do, Mrs. Chambers,'" Tait said as he came through his private door, and "the bore speedily departed as Mrs. Chambers enjoyed a hearty laugh over his denunciation."[41]

Anne Chambers's husband relied on the integrity of all those in his circle—what he called their "moral tact" or "delicacy."[42] This was the discretion Nichol had displayed in their conversation at the observatory. The ability to keep a secret had long been a defining quality of intimate friendship. In the case of those Chambers told directly, his trust had to be complete, for they were bound to be placed in situations where they would have to equivocate. The book forged a special bond between Chambers and Ireland. The two men came to know one another during their 1837 Irish tour, and Ireland had become close to Anne Chambers, who asked to be considered as a "sister." Her mother had known his mother, and they both already thought of Robert Cox as a "brother."[43] A Unitarian with Chartist sympathies, Ireland managed the advanced liberal *Manchester Examiner* from 1846 and also arranged the lecture tours of Ralph Waldo Emerson in 1847–48, including his appearance in Halifax. He became a keen book collector (with nearly twenty thousand volumes), bibliographer, and supporter of the public library movement. His friendship with Chambers was expressed through the hundreds of letters that passed between "Ignotus" and "Alexius" over two decades. These closed in a variety of ways: "eternally thine," "always yours transcendentally," "always miraculously yours," "yours indescribably," "thine ad infinitum et ultra." Because of his day-to-day work on the book, Ireland was effectively put into situations where he would have to dissimulate too: his Secretiveness, as Chambers put it, was getting lots of exercise.[44]

40. The biography discusses *Vestiges* (Chambers 1872, 274–75), but only to dismiss it as "a work which had caused no little exasperation."

41. Bertram 1893, 133–34. A letter to Anne Chambers of 1846 from Mrs. Charles Kean of New York asked if Robert had "owned himself to be the author"; see Priestley 1908, 48.

42. RC to AI, [4 Dec. 1854], NLS Dep. 341/110/30–31; RC to AI, [late Feb. 1845], NLS Dep. 341/110/86–87.

43. [Anne Chambers] to AI, 3 Jan. 1837, NLS Dep. 341/111/9–11.

44. RC to AI, [Nov. 1844], NLS Dep. 341/113/116–17. Details of Ireland's life are in *DNB;* Mills 1899, 137–44, and *ILN,* 15 Dec. 1894, 738. For his love of books, see Ireland 1884; for his Chartist sympathies, "Edinburgh Complete Suffrage Association," *Fife Herald,* 6 Oct. 1842, 132.

Ireland's silence was crucial, for unlike anyone else outside the immediate family, he held documents that could have instantly unmasked the author. As Chambers remarked when suspicions began to point in his direction, "Proof is out of the question, while you stand faithful."[45]

In his new role as potential suspect, Chambers adopted a repertoire of techniques for presenting himself. The "coolness" he mentioned in writing from London was especially important—not to reveal true feelings, but to maintain a facade of unconcern. The phrenological faculty of Secretiveness made it possible to conceal one's inward emotions as a defense against the assaults of enemies. Chambers became adept at dodging direct confrontations; he was, as Camilla Toulmin noticed, "imperturbable" and kept his eyes on the roast. She also witnessed one of his most common ploys, to claim that the pressure of other business made him an unlikely prospect. Chambers found chance comments in public places the easiest to handle. Asked by a divinity professor from St. Andrews if he had seen "'*that horrible book*'" Chambers was saved from the need to reply by "the bustle of the moment."[46] In crowded surroundings and with relative strangers, conversations could shift ground without raising suspicions.

Secrecy raised problems of intimacy. Keeping "cool" was one thing among strangers; among friends it could create a sense of distance, as when Crowe had to be "turned off with a light observation." This would not have reduced her suspicions, but gently hinted that the topic was not for discussion. In another way, though, existing bonds of intimacy could be strengthened. Friends who were in a position to recognize Chambers as the author knew that only a few could ever be told directly; and although outside this particular confidence, their closeness in other respects put them in a position of trust. At first Combe had had no idea who the author might be, but his growing suspicions were confirmed after criticisms he made verbally to Chambers appeared in a new edition. By the end of June, Combe made this attribution in a letter to the American educationist Horace Mann but asked him not to mention it lest the rumor travel back across the Atlantic. Learning that Sir George MacKenzie planned to announce Chambers as the author in print, Combe successfully pleaded with him not to endanger their common friend.[47] He dissuaded Nichol from exposing Chambers as *Vestiges* author and barefaced plagiarist of *Architecture of the Heavens*. Most strikingly, in print Combe never denied rumors that he had written *Vestiges* himself, although in conversation and correspondence he was more open.[48]

45. RC to AI, [Apr. 1845], NLS Dep. 341/113/164–65.

46. RC to AI, Friday [May 1845], NLS Dep. 341/110/34–35.

47. G. Combe to G. S. MacKenzie, 20 Nov. [1846], NLS mss. 7390, f. 539; G. Combe to H. Mann, 27 June 1845, NLS mss. 7390, ff. 163–66.

48. "I have often found it an advantage to be able to say, *truly,* when asked, that I do not *know* who the author is." G. Combe to J. P. Nichol, 18 Mar. 1848, NLS 7391, ff. 347–50. See also G. Combe,

Thus even those not in on the secret kept their suspicions to themselves to protect a friend. Crowe too was willing to use herself as a blind. At the dinner recalled by Ireland, she refused to answer her questioner. "To our surprise and infinite amusement," Ireland remembered, "the lady did not deny 'the soft impeachment'" but quietly laughed and refused to answer.[49] Although Ireland and Chambers recalled this episode as another manifestation of Crowe's eccentricities, it is more likely that she was actually protecting her host, whom she had long suspected. By allowing suspicion to fall upon herself, she in effect defined herself as a conspirator too.

Although Chambers enjoyed playing two roles at once, to be present at great public occasions simultaneously as a celebrated journalist and as the invisible "Mr. *Vestiges*" could be a bizarre experience. In June 1845, he traveled with Anne to Cambridge to report the British Association meeting and sample reactions to his book. They bumped into William Carpenter as the audience was leaving the Senate House:

> As we came out, Carpenter came up along side of us, and said, "so the poor Author of the ——— has got it in all directions tonight!" I said, "Yes, it would be curious if he was present to hear it all." To which he replied, with a nudge in the elbow and a laugh, "Some people say he is one who is not far from me at this moment," evidently meaning me. I laughed too, and only remarked, "What you say might have two meanings," to which he added, "But I have denied it out and out," and then we were divided by the crowd.[50]

Carpenter was only guessing, for he had seen Anne Chambers's handwriting at Churchill's office.[51] This rather convenient, theatrical denouement is typical of the way in which circumstance could be grasped to avoid the need for a direct answer. The exchanges move toward a moment of confession, avoided only when the two suspects are swept back into the anonymous crowd—which, by implication, consists of potential authors.

The Property of an Author

The deep anonymity of *Vestiges* brings to light the conventions that we use to separate readers and authors. For any act of reading to become part of the historical record, it needs also to be an act of authorship and leave written traces. We have already seen how readers wrote their own *Vestiges* as they copied

"To the Author of the 'Vestiges of the Natural History of Creation,'" 1 Mar. 1845, NLS mss. 7390, ff. 66–71.

49. Ireland, introduction, *V12*, xx–xxi.

50. RC to AI, 19 June 1845, NLS Dep. 341/110/137–38.

51. Priestley 1908, 48.

passages into diaries or summarized their views in published essays or letters to friends.[52] Some went further. James Bertram recalled how a few aspiring Vestigians "smiled and smirked their friends and even themselves into the belief that 'they really had a finger in the pie.'" One reader took a copy of Thomas Monck Mason's *Creation by the Immediate Agency of God* and annotated its critique of *Vestiges* in Chambers's name (fig. 11.5). Another went the rounds of Edinburgh's publishers with the manuscript of a pamphlet reply to his critics, "by the Author of the 'Vestiges of the Natural History of Creation.'"[53] For readers to achieve this degree of identification with a work was thought to require either imposture or insanity. As Bertram said, the only charitable view to take of the would-be pamphleteer was that he was "a lunatic."

Chambers could be defined as the author—rather than as madman, plagiarist, or ordinary reader—through the network of the ties linking his actions with those involved in making the book. His special status was the result of a complex array of tacit agreements about production and circulation. These agreements had a legal foundation in copyright law, which defined the author as owner of a literary property for a specified period of time. The Copyright Act of 1842 gave protection for forty-two years or until seven years after the author's death. This was an important moment in the long process of defining the author, built on the foundations of the 1710 act, which had for the first time connected the author's life span and the right to his or her works.[54]

This definition of the author was by no means watertight. The first application of International Copyright did not take place until 1846[55] and in any event was never applied to the Dutch and German translations. Until the 1880s, most American imprints of British titles were pirated, without payment or permission. All of the *Vestiges* editions published in the United States appeared in this way. Wiley & Putnam went so far as to advertise the book under Vyvyan's name (fig. 11.6), though they never interfered with the title page. Moreover, almost all the American editions included hostile critiques unsanctioned by the author, usually the scathing review in the *North British*. As Chambers commented when he heard about this, "the publisher, not willing to stop sales, clapped a

52. The *Expository Outline* in the *Atlas* took this process close to the limit, with passages from one anonymous author (Laing) being distinguished only by typography from those of another (Chambers).

53. Bertram 1893, 134. Such cases of imposture could become serious, as when Joseph Liggins claimed to be the still-unidentified George Eliot: Bodenheimer 1994, 129–34, 137–42; Welsh 1985, 128–31.

54. Feather 1994; see also Rose 1993 and Woodmansee 1994 on the role of copyright in defining the author.

55. Feather 1994, 172–73; for the debates about the act, in which Chambers was involved, see Barnes 1974.

70 GEOLOGICAL CONCLUSIONS.

working of a Divine will. And such exceptions are everywhere to be met with. The regularity observable in the conditions of the celestial creation is not that regularity which we have a right to expect from the operation of inflexible causes, but may very well consist with the working of that Power which does nothing in vain, and among the distinguishing characteristics of whose productions the extremes of uniformity and variety are equally prominently displayed.

GEOLOGICAL CONCLUSIONS. PRE-ADAMITE CONDITION OF THE EARTH.

I now pass on to the examination of the remaining sections of the volume, in which the same principle of natural causation is sought to be extended to the rest of the visible works of creation, and more especially to the production and conditions of subsistence of the various organic beings inhabiting the earth.

The first class of facts by reference to which this doctrine of an organic creation by natural law is attempted to be sustained, presents us with a statement of particulars respecting the primitive condition of the terrestrial globe and its pre-adamite inhabitants, professedly compiled from the researches of modern geologists, and intended to represent the actual state of that science according to the views of those by whom it has been reduced into its present form. +It is not my intention to follow the author

+ Why does he avoid the ~~larger~~ half of my book, by far the most important & conclusive half too, — Cowardly, sneaking, paltry, stony Mason. R. Chambers

Not a bad pun for you, Mr Chambers

11.5 Pretending to be Robert Chambers, a student at the University of Edinburgh has annotated Thomas Monck Mason's anti-Vestigian polemic *Creation by the Immediate Agency of God* (1845). "Why does he avoid the larger half of *my* book, by far the most important & conclusive half too,— Cowardly, sneaking, paltry, *stony Mason. R. Chambers.*" Another reader has responded: "Not a bad pun for you, Mr Chambers." EUL, SC 6838.

VESTIGES OF THE CREATION.

Vestiges of the Natural History of Creation. By Sir Richard Vyvyan, Bart., M. P., F. R. S., &c. One vol. 12mo. well printed. Price 75 cents.

CONTENTS.—1. The bodies of space, their arrangements and formation—2. Constituent materials of the earth and other bodies of space—3. The earth formed; era of the primary rocks—4. Commencement of organic life; sea plants, corals, &c.—5. Era of the old red sand-stone; terrestrial zoology commences with reptiles; first traces of birds—5. Era of the oolite; commencement of mammalia—6. Era of the cretaceous formations—7. Era of the tertiary formation; mammalia abundant—8. Era of the superficial formations; commencement of the present species—9. General considerations respecting the origin of the animated tribes—10. Particular considerations respecting the origin of the animated tribes—11. Hypothesis of the development of the vegetable and animal kingdom—12. Maclay system of animated nature; this system considered in connexion with the progress of organic creation, and as indicating the natural status of man—13. Early history of mankind—14. Mental constitution of animals—15. Purpose and general condition of the animated creation—16. Note conclusory.

"This is a remarkable volume—small in compass—but embracing a wide range of inquiry, from worlds beyond the visible starry firmament, to the minutest structures of man and animals. The work is written with peculiar and classical terseness, reminding us very much of the style of Celsus. . . . . . We have dedicated a large space to this remarkable work, that may induce many of our readers to peruse the original. The author is, decidedly, a man of great information and reflection."—*Medico-Chirurgical Review.*

"This is a very beautiful and a very interesting book. Its theme is one of the grandest that can occupy human thought—no less than the creation of the universe. It is full of interest and grandeur, and must claim our readers' special notice, as possessing, in an eminent degree, matter for their contemplation, which cannot fail at once to elevate, to gratify, and enrich their minds." —*Forbes' Review.*

"A neat little volume of much interest. Judging from a brief glance at the contents of the volume, the author has produced a work of great interest, and one which, while it affords the reader useful instruction, cannot fail to turn his mind to a very profitable channel of reflection."—*Commer. Adv.*

"A small but remarkable work. It is a bold attempt to connect the natural sciences into a history of creation. It contains much to interest and instruct, and the book is ingenious, logical, and learned."—*Newark Adv.*

"This work discovers great ingenuity and great research into the mysteries of nature. It is a noble work, and one which no intelligent person can read without finding a fresh impulse, communicated to his thoughts, and gaining some higher impressions of the Creator's power, wisdom, and goodness."—*Albany Argus.*

"A novel and remarkable work, which will speedily attract the attention of all inquisitive readers. There is much that is new and ingenious in the book. The author, whoever he is, is a man of varied philosophical and literary attainments, and master of a style in conveying his thoughts, so pure, simple, and modest, that his treatise will be everywhere widely read."—*N. Y. Morning News.*

11.6 International copyright and the definition of the author. The failure of copyright agreements to control piracy of British titles in the United States meant that an anonymous book could be advertised under the name of an author without his or her consent. In this case, Wiley & Putnam of New York have advertised Sir Richard Vyvyan as author of *Vestiges.* Advertisement bound in the 1845 second American edition of *Vestiges.*

clerical drag to the machine and thus satisfied his conscience."[56] Who was the "author" of a pirated edition with a self-destructing introduction?

Copyright legislation was only one sign of the special position accorded to authors in the emerging industrial economy of print. The true meaning of the text, as well, was increasingly identified by readers as an author's literary property. The act of reading involved a tacit agreement to respect the author's intentions. In normal circumstances, these could be judged through a whole range of clues, both in the text and outside it. Such information was of course

56. RC to AI, 10 July 1847, NLS Dep. 341/112/11–12.

missing in the case of *Vestiges,* which thus carried an unusual interpretative burden. It is not surprising, then, that Chambers soon complained of being misread and misunderstood: his literary property was being violated. *Vestiges* had been an experiment in the ability of an anonymous work to carry Chambers's intentions, but almost no one read it as expected.

Struggles over the meaning were most acute in relation to questions of expertise. Chambers had anticipated clerical opposition, so that the attacks of the "stupid and dishonest" Reverend Hume and the antics of the Scottish Association for Opposing Prevalent Errors came as no surprise.[57] (Indeed, priests who had anything positive to say always surprised him, so that he was astonished to have a supportive letter from the Reverend George Crabbe of Suffolk, son of the poet and author of an orthodox Anglican work on natural theology.)[58] Chambers was appalled, however, to find that "able men, of marvellous industry, and unimpeached zeal for science" would trip over themselves in condemning a work they had barely read.[59] The issue first arose during the Liverpool debate. Chambers, infuriated, responded as publicly as circumstances allowed in the long letter to the *Liverpool Journal* signed "G. S. of Manchester." The pseudonym blocked suspicions that might have arisen had the letter appeared, like *Vestiges,* without any signature at all. What appeared as an impartial intervention from "a lover of fair discussion and its proper progeny,—truth," was actually a reply from the secret author.[60]

Arriving at the Cambridge British Association meeting, the Chamberses expected the worst. As if to symbolize their isolation from the centers of power in science, they stayed in a bed and breakfast lodging on Jesus Lane, while most of the association's leaders had college rooms. Still, they enjoyed Mozart in an afternoon service at King's, attended evensong at Trinity, walked in the college gardens and countryside, and met many friends.[61] Sitting in the front

57. RC to AI, [12 Feb. 1845], NLS Dep. 341/110/51–52. Combe told Chambers about the association and its attack on *Vestiges* and the *Constitution:* "There is honour for that author & me!" G. Combe to RC, 11 Mar. 1847, NLS Dep. 341/98/7.

58. RC to AI, [late 1844–early 1845], NLS Dep. 341/110/163–64.

59. *E1*, 34–36. RC to AI, [12 Feb. 1845], NLS Dep. 341/110/51–52. He had originally encouraged Churchill to publish in time for the September 1844 British Association meeting, which he thought would provide favorable publicity for the work; see RC to AI, 12 Sept. 1844, NLS Dep. 341/110/153–54.

60. G. S. [Chambers], "Discussion on the 'Vestiges of the Natural History of Creation,'" 22 Feb. 1845, *Liverpool Journal,* 5. Two weeks before, Chambers had replied in much the same way to an astronomical critic in the *Manchester Guardian;* see J. G. S. [Chambers], "Motion of our Solar System," *Manchester Guardian,* 8 Feb. 1845, 5, replying to H. H. Jones, "Motion of the Solar System," *Manchester Guardian,* 29 Jan. 1845, 8.

61. T. Sopwith, diary entry for 22 June 1845, Robinson Library, University of Newcastle, vol. 40, f. 88.

row at Senate House, they had heard the great astronomer John Herschel—supposedly a man of wide and philosophic views—deliver a devastating attack on *Vestiges.*[62] Elsewhere at the meeting, Sedgwick previewed his forthcoming demolition job in the *Edinburgh Review:* the one cleric Chambers might have expected as an ally, on the basis of his opposition to Cockburn, had become his most damning critic. Soon afterward the Chamberses left for London, disgusted by the snobbery, clerical privilege, and clubby familiarity that the British Association seemed to ratify. He could only dismiss the response as willful misinterpretation. Scientific readers seemed to be cloaked in the same black ignorance as the clergy.

Chambers henceforth began to construct his role as an author distinct from, and in opposition to, what he began to call "the scientific class." The public failed to appreciate science, he argued in reporting on the Cambridge meeting for *Chambers's Journal,* not because speculation was feared, but because it was desired. The report cited a taxonomic paper on entomology and James Joule's announcement of the mechanical equivalent of heat as just two examples the crabbed timidity that avoided all larger conclusions:

> Turning to a graver matter, we cannot but sympathise a little in the disappointment so often felt by the public with regard to matters brought forward at the Association. There are many valuable isolated facts, but there never is such a thing as a comprehensive view of nature in any of her departments. . . . Common minds see no sense in this wasting of life in the establishment of new species of moths and ascertainment of new laws presiding over the heating of bodies. They call for something either offering great practical advantages, or opening up new and better views of the relations in which we stand to the great agencies external to ourselves. It is all very well to decry rash generalisation; but while science professes to show merely a collection of bricks towards the building of a house, it will never get any credit as an architect, and the popular apprehension of the value of its researches will be dubious and obscure.[63]

Chambers had hoped for support from scientific men, having written *Vestiges* in such a way that they could rally around it, just as they had stood up for Carpenter's *Principles.* But he had been wrong: these were readers his book had failed to address. They were held down, as he later noted, by the same clerical

62. RC to AI, 31 May 1845, NLS Dep. 341/110/189–90; RC to AI, 19 June 1845, NLS Dep. 341/110/137–38.

63. [Chambers], "The Scientific Meeting at Cambridge," *CEJ,* 2 Aug. 1845, 65–69, at 69; see the similar passage in *E*1, 177–78. Combe, in contrast, thought Herschel's address was "like an Oasis in the desert. He is the only man among all the existing philosophers who appears to me to rise up to a just conception of the relation of Science to man's nature." G. Combe to H. Mann, 27 June 1845, NLS mss. 7390, ff. 163–66.

"Patent Safety Drag" that destroyed elementary education by providing infant readers with only Bibles and religious books.[64]

As Chambers walked through the courts of Trinity and Kings, he had reflected on the power of aesthetics in protecting what he saw as outmoded institutions and Anglican privilege. Cambridge's architectural splendors seemed part of a preordained plan, an impression of organic continuity underlined by the neogothic revivals under construction. Whewell, the master of Trinity College, believed the design of a true city gave no evidence for "progressive development."[65] Chambers experienced a resurgence of his romantic antiquarianism:

> We bethought us of the power of even external beauty in protecting the institutions here established. The most daring innovator might come hither full of eagerness to meddle with systems unquestionably representative rather of past centuries than the present, and we can conceive him so captivated through the agency of the mere aesthetics of the place, that, like Alaric awed by the venerableness of the Roman senate, into which he had intruded, our innovator would shrink from his task, and leave the business of reform to ruder hands.[66]

Like the barbarians who conquered Rome, Chambers was tempted, but only momentarily, to "leave the business of reform to ruder hands." *Vestiges* pointed to a new order of knowledge. The false temple of science would be destroyed, and a new one—open to the people under the supervision of publishers, printers, and literary men—built in its place. As Chambers wrote to Ireland during a dispute about the appointment of a bishop, "How silly all these . . . controversies must appear in a few years, when whole masses of the people are tearing at the front pillars of the temple itself, and howling the dogs of bigotry into insignificance!"[67]

Reading for Revision

Chambers had already begun refashioning his authorial role. The extensive program of revisions through eleven editions was an attempt, only partly successful, to recast the narrative voice of *Vestiges* in a more effective register, to limit "misreadings" by hostile critics. Of all reading practices, this kind of editorial intervention could be among the most intensive: as reviser, Chambers was his own closest reader. And unlike those of most other readers, his annotations and

64. [Chambers], "The Scientific Festival at Oxford," *CEJ*, 31 July 1847, 65–67.

65. Whewell 1846, 16.

66. [Chambers], "The Scientific Meeting at Cambridge," *CEJ*, 2 Aug. 1845, 65–69, at 65.

67. RC to AI, 4 Jan. 1848, NLS, Dep. 341/112/5–6; the furor over the appointment of R. D. Hampden as bishop of Hereford, Chambers felt, exemplified the kind of minor issue that would be dwarfed by the coming changes.

additions could become part of the printed text. All manuscript traces of this kind—for example, interleaved copies of each edition—were destroyed to preserve the secret, but we do know that Chambers worked in seclusion in his study, using his large library of scientific books and private collection of replies and responses. He compiled factual corrections from reviews, comments from casual conversation, and reports from expert readers.

Incorporating these changes required Chambers to assimilate the closely focused, "studied" reading practices common in the learned world, very different from those required for editing a family weekly. He had tried to anticipate these standards, but recognized that he had failed. "I thought all these things had been made exact," Chambers explained after an early review complained of errors, "—for I took the greatest care—but some minds are more microscopic than others—and your microscope makes a merciless critic."[68] Chambers proved a quick learner of these "microscopic" reading practices; as the University of Edinburgh natural philosopher James David Forbes noted, through five editions and a sequel the author "has shewn himself a very apt scholar, & has improved his knowledge & his arguments so much since his First Edition that his deformities no longer appear so disgusting."[69]

More familiar than these academic practices, certainly to someone used to editing reference works, was keeping the book up to date. Outmoded facts or theories could be as damaging as mistakes. Almost immediately, for example, Chambers realized that making so much of William Sharp Macleay's circular system of classification had been ill judged. The first edition had used this scheme to show that nature had an order that could be explained only by law. All organisms were in nestled groups of fives (fig. 11.7). Opinion on the validity of this "quinarian" scheme had turned dramatically among men of science a year or two before, and so the system was largely dropped from the third edition.[70] Similarly, the announcement of the resolution of the great nebula in Orion by Lord Rosse's Leviathan telescope required major changes. ("There should be thanksgivings in all the churches," Chambers told Ireland privately.)[71] However, the application of natural law to the heavens did not necessarily fail; it simply meant that no evidence for a developmental sequence could be observed. In subsequent editions, the nebular hypothesis was intro-

68. RC to AI, [late Nov.–early Dec. 1844], NLS Dep. 341/113/70–71.

69. J. D. Forbes to W. Whewell, 8 Jan. 1846 (copy), St. Andrews University Library, J. D. Forbes papers, letterbook 4, pp. 60–63. Chambers reflects on the difficulties of mastering these practices in RC to AI, [Apr. 1845], NLS Dep. 341/110/64–65.

70. Compare *V1*, 236–74, with *V3*, 244–61; the changes are summarized in Ogilvie 1973, 370–439. The wider shift in opinion is outlined in Ospovat 1981, 107–8, 112–13.

71. *E2*, 199; RC to AI, [May 1846], NLS Dep. 341/110/221–22. The largest changes appear in *E2*, 199–205, and *V6*. See Ogilvie 1975 and S. Schaffer 1998.

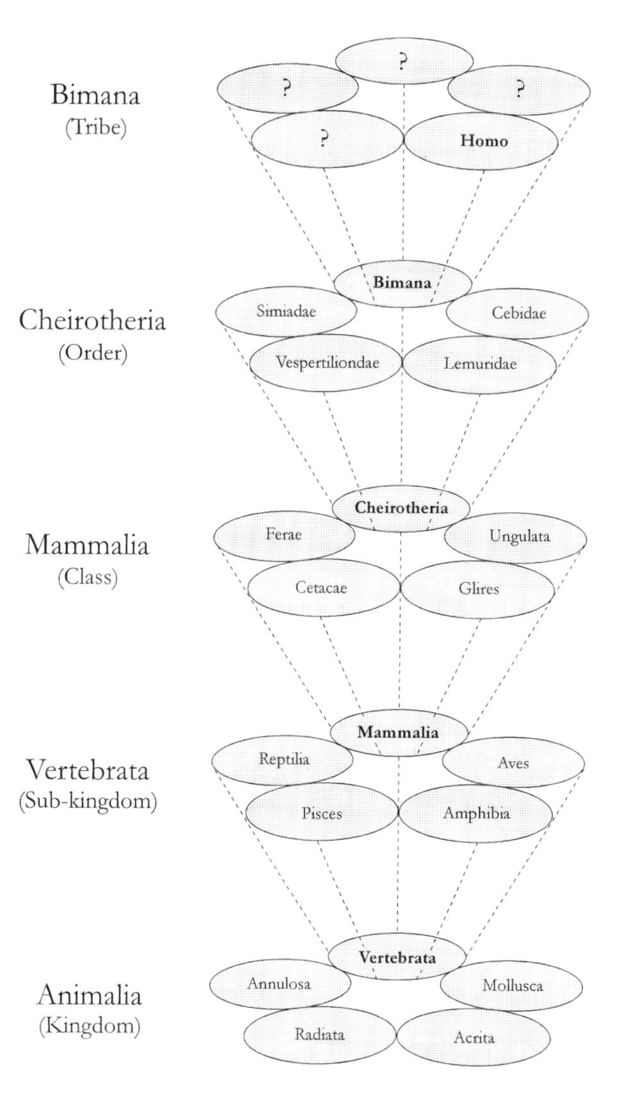

11.7 Diagram illustrating the position of humans in the circular classification of early editions of *Vestiges*. Starting at the bottom of the chart, the whole animal kingdom is divided into five groups, each of which in turn consists of five interconnected groups, and so forth. A system of elaborate analogies and parallels gave the system a satisfying regularity. Note, however, that our species is alone within in its particular circle, thus leaving room for the "nobler type of humanity" still to come. Drawn from information in *V*1, 262–76.

duced much later in the first chapter, giving priority to the order and regularity of the heavens.

Chambers's more assertive, oppositional role as an author also began to be reflected in the text. The first edition had depended on the narrator being seen as an authoritative guide through the facts of science. As the volume of criticism increased, the narrative voice was gradually transformed. The conclusion had always mentioned that the evidence for development was provisional, but the second edition of December 1844 was more apologetic, admitting that the work was "in some measure crude and unsatisfactory, even overlooking errors of detail justly attributable to my own defective knowledge." This forestalled potential critics, but made it harder to imagine the unknown author as a master spirit of science. A footnote was added to this passage in the third edition, explained that "a few alterations and omissions" had been made to avoid minor points being used as the basis for "sweeping objections."[72] Still, the overall tone remained that of the calm, Olympian voice.

It was only after the British Association and the big quarterlies weighed in against the book that frustration became apparent. *Explanations* and subsequent editions of *Vestiges* tackled the problem of misreading directly, noting that the argument "has been generally misunderstood" and "in a great measure misapprehended in its general scope."[73] The "scientific class" was censured for narrow views, timid prejudice, and lack of candor. The danger was that "ordinary readers" would be misled into thinking that the whole work was riddled with error. The first edition had implicitly asked men of science to become fellow laborers in a great cause. Now the appeal was redefined: "No, it must be before another tribunal, that this new philosophy is to be truly and righteously judged."[74]

So while remaining contrite about "isolated facts," the narrative voice of *Vestiges* became bolder and more speculative. On some extremely controversial issues the book stood its ground. Thinking machines, dogs playing dominoes, plants growing like frost crystals, mammals hatching from birds' eggs, apes turning into men: all the editions of *Vestiges* were filled with propositions that, if true, would transform the basis on which science was conducted. A notable example involved the *Acarus crossii,* the tiny mites that had crawled from the electrical experiments of Andrew Crosse and William Henry Weekes. Because these creatures straddled the crucial boundary between matter and life, they remained in the text, although it was also emphasized that the truth or falsity of the argument did not depend on them.[75] A note added to the third edition also pointed out that the *British and Foreign Medical Review* had opposed any easy dismissal of spontaneous generation.[76] Over and over, the later editions stressed that *Vestiges* should be read not as a new theory of the origin and development of life, but as an argument for a fixed order of natural laws.

The most dramatic assertion of the author's right to speak on matters of theory involved replacing the circular with a genealogical classification. This appeared in the fourth edition and was developed in *Explanations* and all editions

72. *V*3, 381; *V*2, 392.

73. *E*1, 3, 1.

74. Ibid., 179.

75. *V*3, 193, and esp. *V*10, 138; Ogilvie 1973, 300–307, sums up the changes; for their significance in the argument, see Secord 1989b.

76. *V*3, 193; [W. Carpenter], "Natural History of Creation," *British and Foreign Medical Review* 19 (Jan 1845): 155–81, at 170–72.

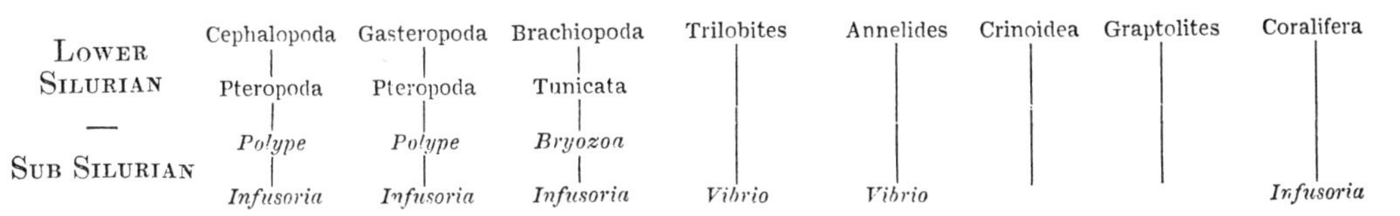

11.8 In the genealogical classification proposed in the later editions of *Vestiges,* the groups ordinarily used by naturalists were reinterpreted as "grades" characteristic of a particular developmental phase. This chart from the seventh edition traces the early history of the main invertebrate groups of the Lower Silurian. Note that the Cephalopoda and Gasteropoda passed through the same developmental sequence. All the groups originate independently through spontaneous generation; different outcomes are the result of environmental circumstance, so there is no single chain of being. Italics are used for those groups that had not yet been found in the fossil record. *V7*, 162.

of *Vestiges* from the fifth. The basic idea was simple. Previous taxonomists had taken parallelisms of structure, or organic affinities, as the basis of natural order. For example, each of the invertebrate classes was seen as separate and distinct: all polyps were grouped together. But *Vestiges* now argued that this view was incorrect. The grouping together of similar species on the basis of affinities needed to be replaced by a system of genetic lines. The polyps did not belong together, but were a "grade" through which the true lines of development had passed many times and in different parts of the world (fig. 11.8). The lines of connection the new classification drew in the natural order were vertical, extending through time, rather than horizontal, between similar groups existing at any one time. As the narrator put it, "The classification which this system implies may be said to transverse to all ordinary classifications."[77]

This scheme was revolutionary: old classifications would have to be thrown out, museums reorganized, textbooks rewritten from scratch. In his letters to Ireland, Chambers was supremely confident. Not least, he had the sanction of David Page, "an acute naturalist" in whose judgment he had every confidence. On the other hand, Chambers recognized that Carpenter would not approve his new views, but remained certain that "*mine are right.*" As he noted just before the fourth edition appeared, it was "the greatest burst of light on the Animal Kingdom that has happened since Cuvier's time."[78] Above all, Chambers hoped to create a classification that readers would understand. As the third edition of *Vestiges* noted, the "present confused scene" in natural history, with its technical terms and a lack of organizing principles, baffled "all but life-long

77. *E1*, 71.

78. RC to AI, n.d., NLS Dep. 341/110/130; memo in RC to AI, 29 Mar. 1845, NLS Dep. 341/113/168–69.

attention." A genealogical classification would be "simple and beautiful," "instructive and an aid to memory." Rather than being accessible only to specialists, nature would be open to all. A true classification would be a people's classification, accessible even to children.[79]

As the debate went on readers could feel they were no longer dealing with a complete unknown. Still shrouded in secrecy, "The Author of *Vestiges of the Natural History of Creation*" became a voice with a history. Continuity of title in the revised editions created expectations of continuity in authorial identity, and when *Explanations* was published as a sequel to *Vestiges,* it was explicitly identified as "By the Author of that Work." In this way, the link with *Vestiges* began to function as a pseudonym, just as "By the Author of *Waverley*" or "By the Author of *Frankenstein*" had served to label work by earlier authors who wished to conceal their real names.[80] In some contexts (such as the Liverpool newspapers), Chambers could employ pseudonyms ("G. S.") without any connection with *Vestiges,* thereby allowing him to enter the lists under a guise of neutrality. Complete anonymity also continued to have its uses too, although in the case of *Chambers's Journal* his known association with the periodical meant that his comments, despite being unsigned, always had to be indirect.

The sense of authorial presence was most apparent in the tenth edition (1853), which included a preface explaining how *Vestiges* came to be written. This creates a voice at once so authoritative and so personal that historians have regularly used it as a transparent record of Chambers's intentions.[81] Its purpose, like all such framing devices, is tactical. Even through the cloak of anonymity, the tone is confidential, inviting the reader to share in secrets. The rhetoric is akin to some of the clandestine letters to Ireland, but it is written in the third person, as though by a common friend of author and reader. *Vestiges* has now come of age, and its history deserves to be recorded for posterity. Its author is presented as a man of faith and a profound discoverer, whose every step on the path to truth derives from a conviction in "the Divine Governor of the world." The two issues are intertwined at every point:

> After long cogitation, the idea at length came unpromptedly into his mind—and therefore so far was an original idea—that the ordinary phenomenon of reproduction was the key to the genesis of species. In that process—simple because familiar to us, but in reality mysterious, because we only can look darkly and

79. *V*4, 278.

80. Griffin 1999, 880–83.

81. This important preface was first noted in Hodge 1972a, 134–35, who specifically used it as a "precious, strangely neglected" commentary on authorial intentions. As Fiona Robertson (1994, 142–60) points out, critics have often followed a similar procedure with Scott's notes to his collected Magnum Opus.

> adoringly through results to the inscrutable Agent—we see a gradual evolution of high from low. . . . all in unvarying order, and therefore all natural, although all of divine ordination.

The narrative of discovery serves to validate the author's originality, breadth of vision, and benign intentions, which are contrasted with the critics' blinkered prejudice and illiberality. Having taken readers into his confidence, the author of the preface invites them, in effect, to become the work's first supporters: "It has never had a single declared adherent—and nine editions have been sold." The extent of the opposition is here exaggerated to encourage an act of heroic independence, as readers are told to think for themselves as individuals, and to choose facts and logic over unreasoning obloquy.[82]

As review after review had pointed to the *Vestiges* author's lack of experience, Chambers had become convinced that the best way forward would be to transform the practice of science. Establishing a people's science thus involved creating new forms of field and laboratory practice. This was a strategy used in *Explanations,* with its masculine analytical structure and marshaling of facts, but he planned to do more. As he told Ireland, "Perhaps I may yet surprise the scientific world with a rather formidable addition to its stock of knowledge."[83] Beginning on the St. Andrews golf links (fig. 11.9), Chambers measured the heights of ancient sea terraces throughout Britain and Europe. He spent several days at Glen Roy, examining the celebrated Parallel Roads (the subject of Darwin's first fieldwork after the *Beagle* voyage) and claimed to have "solved" the problem. His conclusion, that traces of a gradual lowering of the sea could be found throughout Britain, was summarized in *Ancient Sea-Margins,* published in 1848 by W. & R. Chambers. This, as he exulted to Ireland, would "give reputation" and "astound the big wigs!"[84]

Ancient Sea-Margins appeared with Chambers's name on the title page. Since Chambers knew by this time that many people suspected him of authoring *Vestiges* anyway, publishing scientific work under his own name could become a way of demonstrating that he was not just a closet speculator. He published papers in the *New Edinburgh Philosophical Journal,* the British Association *Reports,* and the Royal Society of London's *Proceedings.*[85] In some respects this strategy worked; Chambers began to "pass" at the British Association and the Geological Society, with his work on terraces and glaciation being referred to (albeit grudgingly) by the leaders of metropolitan science.

82. *V*10, v–x.

83. RC to AI, 23 Sept. 1846, NLS Dep. 341/110/197–98.

84. RC to AI, 19 Oct. 1847, NLS Dep. 341/110/247–48; RC to AI, 10 July 1847, NLS Dep. 341/112/11–12.

85. Royal Society of London 1867–1925, 1:868, 7:366, gives a list of his papers.

Scientific authorship was only one of the ways that Chambers attempted to recast his identity in the mid- and late 1840s. Moving into the big house at Doune Terrace was another, as was adopting a less direct connection with the day-to-day work of the business. To signify Robert's emergence into the ranks of standard authors, in 1847 W. & R. Chambers issued a seven-volume set of his collected works. Four of the volumes reprinted his best essays from *Chambers's Journal,* thereby asserting their permanent literary value and his own dignity as an author. But Chambers was not simply assuming the ways of a gentleman: an implicit rejoinder to Lockhart explained why he had abandoned

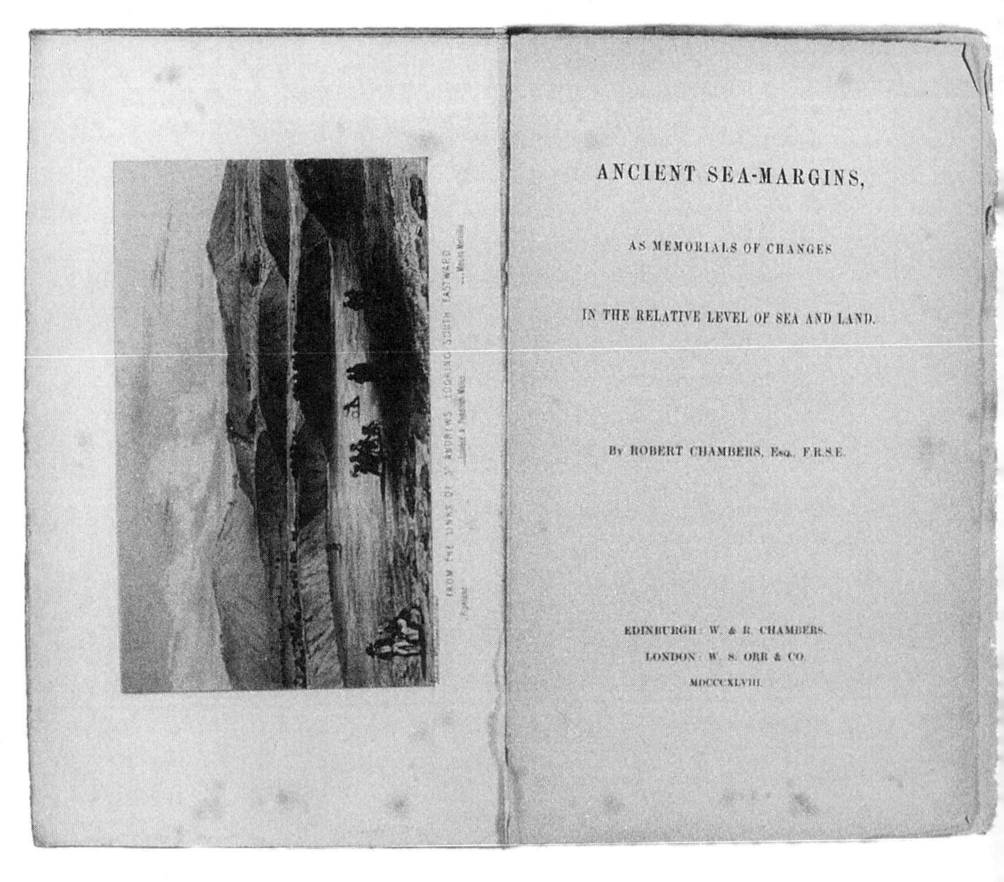

ANCIENT SEA-MARGINS,

AS MEMORIALS OF CHANGES

IN THE RELATIVE LEVEL OF SEA AND LAND.

By ROBERT CHAMBERS, Esq., F.R.S.E.

EDINBURGH: W. & R. CHAMBERS.
LONDON: W. S. ORR & CO.
MDCCCXLVIII.

11.9 Title page and frontispiece of *Ancient Sea-Margins,* published by the family firm in 1848. Chambers, an ardent golfer, began his fieldwork on the famous "links" at St. Andrews in the early 1840s, and the lithographed view illustrates the terraces to the southeast.

antiquarian studies to become "the essayist of the middle class"—the class into which he had been born and was proud to stay.[86]

This was the audience Chambers hoped to reach with the people's editions of *Vestiges*. "Now does the time approach for illuminating the masses on this grand subject . . .," he wrote to Ireland. "What we may hope to see coming from all this for the good of mankind!"[87] The apotheosis of the author as hero of middle-class culture came early in 1848 before a cheering crowd in Glasgow. "It was very remarkable to me," Chambers exulted, "the other night at the Glasgow Athenæum soiree, that Jokum and myself were received with the greatest applause after Dickens. . . . Presbyteries might have pined at the sight."[88] Such triumphs, like the repeal of the Corn Laws and the sale of *Vestiges*, suggested that reform would succeed, despite the bleak prospects in Edinburgh.

Vestiges of the Author

Chambers's emergence as a literary celebrity had the unintended consequence of making it easier to peg him as author of *Vestiges*. Ada Lovelace was rarely suspected after the spring of 1845, and Vyvyan's categorical denial in October of that year had effectively ended his candidacy. As the work became less fashionable, those who knew the world of journalism began to hunt for the likely combination of knowledge and literary skill. The private secrets of a public author had a way of getting into the newspapers, as the Edinburgh election debacle showed only too well. The collected essays, now attributed to Chambers for the first time, were a gold mine of corresponding passages, as was *Ancient Sea-Margins*. Many of the parallels cited by his enemies were close. After all, the first words of the new book were "Taking observed facts for our data," emphasizing the empirical basis of the speculations to come. As an American reviewer put it, the author "strikes off at once, like a bold swimmer, into the heart of his subject"—just as in *Vestiges*.[89]

As his name topped the list of suspects, Chambers relied more than ever on the etiquette surrounding anonymity. Speculating about secret authorships was perfectly acceptable over dinner, in letters, or even in the newspapers, but not when extended to demanding public declarations. Unless the circumstances were exceptional, an author had the right to remain anonymous, and remarks were usually managed so as not to demand a reply. Breaches of these rules were rare and long remembered. The woman who confronted Chambers at dinner,

86. Chambers 1847, 1:iv.

87. RC to AI, [June or July 1846], NLS Dep. 341/110/201.

88. RC to AI, 4 Jan. 1848, NLS Dep. 341/112/5–6; also Gibbon 1878, 2:241–42.

89. [F. Bowen], "Recent Theories in Geology," *North American Review* 69 (July 1849): 256–59, at 257; Chambers 1848, 1.

Toulmin noted, spoke with "singular impropriety"; the man who challenged Crowe, Ireland recalled, was "a noisy, obtrusive *gobe-mouches,* with a strident voice, and zealous in the pursuit of truth in season and out of season."[90]

The accusations in *Macphail's* and the *Manchester Courier* in November 1847 were far more worrying, for Chambers could now reasonably be expected to issue a public denial, "if deny I can, and this I must not do." Still, no one had any conclusive proof. He pointed out to Ireland that the textual parallels were what might be expected from any works dealing with the same subject; and *Ancient Sea-Margins* actually argued for a different position on the fall of sea levels than that taken in *Vestiges* and most of the existing literature. The *Courier*'s story about multiple copies of *Vestiges* being found in the library at Doune Terrace was pure fiction, as no more than one was ever there at a time. "The one consolation," Chambers told Ireland, "is that the utmost rage of bigotry cannot now visit me with absolute poverty."[91] The real force of the "rage of bigotry" became apparent after the "horrid botheration" of the election for lord provost the following year. With accusations about *Vestiges* hanging around his neck, he had little choice but to abandon hopes of public office. "The Free Church was a formidable opponent in this case," he told Ireland, "and some scoundrel of that kidney published an anonymous libel against me, insinuating that I was the author of the V. of C. Wasn't it detestable?"[92] The barest hint of a connection with *Vestiges* could not be combined with political office, not when Edinburgh was dominated by what Chambers termed "the present fatal Christianity."[93]

In England, the most potent rumors about Chambers came from the world of metropolitan journalism. Here people tended to know Chambers on a commercial basis and were less aware of just how much was at stake in Edinburgh. Carpenter was the main source of gossip, especially at soirées and British Association meetings.[94] Such "inside" knowledge could be cited only in conversation and could indicate either the truth of what Carpenter said or a wish to cover his own authorial tracks. There were other sources too, of varying degrees of credibility. William Howitt told fellow radical journalist Samuel Smiles that he had seen a copy of the book corrected in Chambers's hand in the Doune Terrace drawing room.[95] Another of Chambers's friends, the barrister and author Samuel Carter Hall, suggested that Chambers "wrote a part and superintended the publication," but believed that most of the text was by Leitch Ritchie, a later

90. Crosland 1893, 86–87; Ireland, introduction to *V12*, xx–xxi.
91. RC to AI, 5 Nov. 1847, NLS Dep. 341/110/239–40.
92. RC to AI, 21 Nov. 1848, NLS Dep. 341/112/2–3.
93. RC to AI, 4 Apr. 1849, NLS Dep. 341/112/34–37.
94. Priestley 1908, 48.
95. Ibid., 48–49.

editor of *Chambers's Journal.*[96] And it was through the atheist compositor William Chilton's connections with one of the commercial men of science (perhaps Carpenter?) that freethinkers began to link Chambers with *Vestiges* too.[97]

Yet those who spoke most confidently about the authorship, even from the evidence of their own eyes, could never be certain. The page proofs seen by Howitt might seem to convict Chambers; it was rumored, however, that the real author had posted duplicate proofs to the half-dozen people most likely to be suspected, in hopes that an incriminating packet would be left out and seen by a visitor.[98] A prodigious range of suggestions were made in the pubs and clubs of Fleet Street. The Chartist Thomas Cooper—who had been scouting out Paternoster Row for a publisher for his prison poem—later recalled that Arnott, Chambers, and Lovelace had been accused, but the authorship remained an enigma to him.[99] Those most likely to have concrete evidence were too closely implicated themselves to voice suspicions in print.

The most serious attempt to break the anonymity was almost exactly a decade after publication of the first edition, when one of the secret holders exposed Chambers as the author. David Page, a geologist and literary man who had subedited *Chambers's Journal* and written a successful series of geology primers for W. & R. Chambers, had quit the firm when they refused to make him a partner. Page told what he knew after giving a lecture at Dundee, just north of St. Andrews in Scotland. He explained that after being asked by William Chambers to review *Vestiges* for *Chambers's Journal* in late 1844, he became convinced that Robert had written it. When some weeks later Page happened to mention errors that could have been avoided, Robert let him in on the secret and sent proofs for correction. Page felt no guilt in having helped out in this way: he "had done no more than what many men were in the habit of doing for others, and what he himself more recently had done" for the Reverend John Anderson's anti-Vestigian *Course of Creation.* However, the tenth edition led Page to speak out, for he was disgusted by its "tone and spirit," especially the egotism of its autobiographical preface. He was also tired of being accused of the authorship himself.[100]

96. Hall 1883, 2:292. Hall, who regretted the "Free-thinking" character of the book, claimed not to have read it.

97. W. Chilton to G. J. Holyoake, 1 Feb. 1846, Holyoake Correspondence no. 155, Co-operative Union, Manchester; also printed in Royle 1976, 141–42.

98. Hall 1883, 2:292 n. There is no evidence that Chambers did this. Another rumor that Chambers later did encourage, that he had been involved in the book's production, provided an explanation of the handwriting Carpenter had seen at Churchill's shop.

99. Cooper 1878, 10; his literary contacts and (temporary) abandonment of Christianity are mentioned in Cooper 1872, 259–78.

100. "Who Wrote the 'Vestiges of Creation?'" *Dundee, Perth and Cupar Advertiser,* 24 Nov. 1854, clipping in CUL Add. mss. 7652.II.L.23. Page was first told in February 1845; "I have taken up a

Page had bragged vaguely all along about knowing the author, but a formal announcement was potentially disastrous. A report of his statement was published in the *Dundee, Perth and Cupar Advertiser,* which Page forwarded to Sedgwick and the widely circulated *Athenaeum,* which ran the item in "Our Weekly Gossip" for 2 December 1854. At least one reader who had been presented with a copy of the book pasted this notice inside.[101] Chambers, really alarmed this time, wrote to Ireland:

> You will have seen the Athenæum of Saturday, containing a notice from D. Page regarding the authorship of a certain book. It is a shocking piece of malice from a man to whom we showed much kindness, but who finally quitted us in wrath because he found he could not get into a partnership quite so fast as he wished. I always hoped that a wish to maintain his own honour would keep him silent; but being deficient in moral tact, this has not entirely availed with him.

Robert and William decided to let the matter rest with "such uncertainty as may yet be felt to hang upon it." They asked Churchill to issue no further editions. Chambers also began to admit in private conversation that he might have had something to do with the proofs, which kept the authorship under wraps, while not accusing Page of being a liar.[102]

There were reasons to be optimistic. In publishing circles, Page was known to be an angry ex-employee who had struggled with hack journalism and lecturing after quitting his job.[103] When working for W. & R. Chambers, he had seen nothing irreligious in the origin of humanity through natural causes, but had since returned to his roots in the Established Church of Scotland. In Scottish journalism Page was already notorious as the "Judas editor" for having switched overnight from a Tory to a Whig paper. This was not a man who could be trusted.[104] Above all, it was bad manners to break a confidence when an author wished to remain unknown. This left some hope that most newspapers would not carry the story. In the event, they did not, and Page's betrayal had nothing like the impact the brothers feared. Had he spoken out in 1847, at the time of the *Macphail's* essay, a potentially fatal weapon would have been placed in the

---

scientific confidant in Mr Page, who is now labouring in the cause beside me. His delicacy hitherto shows he is worthy of my confidence." RC to AI, [Feb. 1845], NLS Dep. 341/110/112–13.

101. "Our Weekly Gossip," *Athenaeum,* 2 Dec. 1854, 1463–64; a clipping of this is in James David Forbes's copy of the first edition (St. Andrews University Library, For QH.363.C4.)

102. RC to AI, [4 Dec. 1854], NLS Dep. 341/110/30–31.

103. RC to AI, 27 May 1851, NLS Dep. 341/112/70–71.

104. Campbell, n.d., chap. 6; for Page's beliefs on the creation of man, see *Fifeshire Journal,* 21 Sept. 1848, 4, an address to the St. Andrews Horticultural Society. This is attacked (with Page being mentioned) in Miller 1849, 17.

hands of the saints. Not only was the tone of intellectual controversy changing by the mid-1850s, but Page's claim did not come as much of a surprise; he fixed the authorship, as the *Athenaeum* noted, upon "a gentleman who has been generally credited with the work." The damage that a direct accusation could do had largely been done.[105]

Allegations might be false, parallel passages might be coincidence. Some of the book's best-informed critics thought Chambers an unlikely suspect. In 1849 the *North British* (the quarterly associated with the Free Church) commissioned David Brewster to review Chambers's *Ancient Sea-Margins* and Miller's *Footprints of the Creator.* Presumably the expectation was that this would fix the *Vestiges* authorship once and for all. But Brewster refused to rise to the bait. He praised *Ancient Sea-Margins* as a fine contribution to science, and used *Footprints* to damn *Vestiges.* He wrote the review at an inn where Chambers happened also to be staying, and so the two men discussed the *Vestiges* authorship over dinner. "I battled against his idea that the work is atheistical," Chambers reported to Ireland, "but with little success. It was, however, gratifying to find that he acquits me of all suspicion of having written that atrocious book. It was queer having such a critique in process next room to me." *Vestiges* was too "horrible" to have been written by a decent family man.[106]

The Character of Truth

A book is bound up with its author's identity in ways that are uneasy and intimate. Chambers was a casual reader discussing *Vestiges* with friends; the conspiratorial Ignotus, whose letters were to be burned; a secret participant in newspaper debate; the legal owner of copyright; a journalist initiated into practices of close reading, revision, and scientific authorship; a closet speculator who had stumbled into the field. Perhaps Chambers was too busy, too reverential, or too stupid to have written *Vestiges;* then again, perhaps he had helped with the proofs. Vilified as Mr. Vestiges from pulpits and in the press, he could assume the character of a middle-class hero and martyr to the cause of progress.

The early Victorian bourgeoisie stressed the need for playing different roles appropriate to circumstance, and elaborated a subtle range of distinctions between public and private. That there was an inner, private essence to the self was central to contemporary notions of character. In his self-proclaimed role as "essayist of the middle class," Chambers discussed the issue as a matter of "insides" and "outsides." As one of his essays noted, external appearances of dress, speech, and demeanor were essential to preventing the confusion of identities

105. "Our Weekly Gossip," *Athenaeum,* 2 Dec. 1854, 1463–64.

106. RC to AI, 14 Jan. 1850, NLS Dep. 341/112/66–67; the review in question appeared as D. Brewster, "Hugh Miller's *Footprints of the Creator,*" *North British Review* 12 (1850): 443–81.

in large cities, where "most are strangers." One could visit a family's house on many occasions and still never have any idea of the true character of the inhabitants. Only prolonged exposure and intimate knowledge would uncover what the essay termed "the *real* insides." Men could simulate virtue with conformity of opinion and appropriate expressions.[107] Dissimulation worked because the world was willing to rely on externals. What the elaborate safeguards surrounding *Vestiges* show, though, is the work it took to create the separation between public and private. These domains overlapped at virtually every point—as indeed they did throughout Victorian life.

If appearances counted for so much, how was a stable identity to be achieved? *Vestiges* stated its creed as a concluding benediction: "I believe my doctrines to be in the main true; I believe all truth to be valuable, and its dissemination a blessing."[108] The character of the book, and by implication the character of its unknown author, rested on its claim to truth. As many critics pointed out, publishing anonymously sat rather uncomfortably in this context. If truth mattered so much, why should its author remain shrouded in deceit and dissimulation? The book, as Page said, "stands bastardised by the moral cowardice that shrinks from avowing its paternity."[109] Yet the *North British Review* declared that for what really mattered, anonymity was no disguise. Whatever his or her name, the author was a speculator, a defiler, a materialist, "imperturbable on his throne of fire-mist." Real men of science did not rampage uncontrolled through virgin nature, but had to "strive to unveil" her mysteries, always guarding the "vestal fire" of truth. An "intellectual priesthood" had to defend the purity of science just as Jesus had cleansed the Temple in Jerusalem by driving away the money changers and speculators.[110]

This fusion of religion and credibility was, for Chambers and his circle of reformers, a failure of character in its own right, which is why they attacked the science of the saints with such passion. The truth had to remain anonymous precisely because organized religion had kept the people in intellectual infancy. Auguste Comte had said that all societies passed through three stages: the theological, the metaphysical, and the scientific. Only in the final, positive stage—as *Vestiges* predicted—would complete frankness and openness prevail.[111] Truth could then appear under an author's name, and not as an anonymous

107. "Outside and Inside," reprinted in Chambers 1847a, 2:207–12.

108. *V*1, 388. See remarks on his deathbed, reported in Ireland, introduction to *V*12, x–xi.

109. Page 1861, 209.

110. [D. Brewster], "Vestiges of the Natural History of Creation," *North British Review* 3 (Aug. 1845): 470–515, at 473; [D. Brewster], "Explanations," *North British Review* 4 (Feb. 1846): 487–504, at 494.

111. *V*1, 355.

voice from the crowd. Identity should not be defined through an individual's spiritual state in relation to formal religion, as the urban missionaries would have it: instead, each man and woman had to be evaluated on the basis of character.

Not only *Vestiges,* but everything Chambers wrote was engaged in forming the character of readers, instilling virtues of temperance, thrift, hard work, and honesty. This is why Chambers refused to tell a falsehood about the authorship that might have been discovered. It would have threatened what one critic has called "the theatricality of sincerity itself."[112] If outright denial had been an option, then distancing his role as a famous author from his secret role as creator of *Vestiges* would have been easier. But in his position, a public lie would undermine the defining quality of the moral self.

From Chambers's perspective, every person had to be judged as an individual, according the balance of his or her mental faculties. In contrast, the character of society as a whole was subject to the same numerical regularities as the weather. "Man is now seen to be an enigma only as an individual," *Vestiges* said; "in the mass he is a mathematical problem." Following contemporary statisticians, Chambers lost no opportunity of pointing out the law-bound consistency of such quirky social facts as the annual number of crimes committed in France and drunks arrested in London. He was a pioneer in encouraging insurance schemes among the middle and working classes, and wrote a tract on the subject that sold thousands of copies.[113] The national character could advance only if the people understood that society fell under the dominion of natural law. Statistics was not about counting souls for Christ, but the people's moral progress.

At no time during the nineteenth century had definitions of "the people" been more contested than in the 1840s. Europe was embroiled in revolution, trade had slumped, and a revitalized Chartist movement was sweeping the country. *Ancient Sea-Margins* was published in the spring of 1848, during the most turbulent months of the Victorian era. Edinburgh witnessed two nights of rioting and over 130 were arrested in disturbances in Glasgow.[114] As Chambers told Ireland a few days after the Chartists marched on Kennington Common in London: "Reform in the finances, the redress of some of the blunders of the reform bill, secular education, sanitary enactments, these are the things

112. Auerbach 1990, 114.

113. *V*1, 331. On insurance, see "Messrs Chambers's Soirée," *CEJ,* 8 July 1843, 197–99, at 198; [R. Chambers], "Social Economics of the Industrious Orders," in Chambers and Chambers 1842, 2:305–20; see also Scholnick 1999. On statistics, see "Regularity of Occasional Things," *CEJ,* 18 Sept. 1841, 273–74.

114. Saville 1987, 89–90.

now required, if we would avoid things we have less mind to."[115] If the advocates of physical force Chartism could be kept down, the electorate could be widened gradually to include the working classes, but only if education (the key to the moral force campaign) was freed of the "Patent Safety Drag" of theological dogma. There was no need to oppose religion outright—Chambers condemned that as the mistake of socialists and communists, who "fix upon themselves the disrepute of want of religion."[116] But rational reform, underpinned by education in the laws of nature, was essential if Britain was not to follow Continental Europe into chaos.

Chambers's vision of an anticlerical apocalypse, with an enlightened people storming the temple, was just one among the imagined futures in which science occupied a central place. The roles of scientific practitioners were changing, more than Chambers suspected, potentially more than at any time since the seventeenth century. Their uses of *Vestiges* reveal the landscape of potential futures with vivid clarity. What place would makers of knowledge have in the emerging industrial society? It is to these readers that we now turn.

115. RC to AI, 28 Apr. 1848, NLS Dep. 341/112/7–10; unlike Chambers, Ireland was a supporter of universal suffrage.

116. On the clerical drag, see [Chambers], "The Scientific Festival at Oxford," *CEJ*, 31 July 1847, 65–67; RC to AI, 10 July 1847, NLS Dep. 341/112/11–12; RC to A. Chambers, [24 June 1847], Norman mss.; RC, memorandum book, 21 Apr. [1848], NLS Dep. 341/42.

PART FOUR

Futures of Science

CHAPTER TWELVE

The Paradoxes of Gentility

Hard cash, paid down, over and over again, is an excellent test of inherited superiority.

CHARLES DARWIN, *The Variation of Animals and Plants Under Domestication* (1868)

WHO WAS A GENTLEMAN? One way to find out was to smell his library. A collection gathered by a parvenu smelt like a tannery. It could knock you down with the odor of leather polish and binder's glue, showily bound volumes that would never be read.[1] A genteel library was more likely to be acquired over generations or for straightforward practical use. Other hints were provided by clothes. Take the gentleman of science in figure 12.1, who is wearing typical formal evening dress: black waistcoat, double-breasted black tailcoat, white cravat, linen shirt with high collar and frilled opening. Over this he has an academic gown, as he is a master of arts at a Cambridge college. The uniformity of male fashion, compared with the extravagance of previous decades, indicates moral probity, respectability, and lack of self-interest.[2] The sitter's expression, too, bears this out: an "open" face with nothing to hide. The globe, to his left, signals his identification with the entire range of knowledge.

That early Victorian science was dominated by gentlemen has become a commonplace of writing about the period. The old misconception that dilettantish "amateurs" impeded the making of knowledge has been recognized as a retrospective creation. Instead, many of the leading figures are now seen as gentlemen of science, devoted to the serious pursuit of knowledge as a vocation, but not for pay.[3] In contrast, the typical figure of the later nineteenth century is

1. [Disraeli] 1807, 2:11.

2. For a survey of male fashion, Buck 1984, 185–201; Harvey 1995, 31–38, discusses the meanings of black. On gentility, see Vincent 1999, 37–45, and Collini 1991.

3. The classic work is Morrell and Thackray 1981; see also Morrell 1971, Porter 1978a, and Rudwick 1985.

12.1 The ideal of the Christian philosopher: Sir John Frederick William Herschel, Bart. F.R.S. Steel engraving, made in honor of his British Association presidency in 1845, after the oil portrait by H. W. Pickersgill. [Timbs] 1846, frontispiece.

characterized as a paid professional—a "scientist"—working in a government, industrial, or academic laboratory.

These changes, however, are often described with an inevitability that vanishes on close examination. It is easy to forget that gentlemanly science was an ideal as much as a reality; relatively few of those engaged in science were in fact genteel by birth. Gentlemanly science was never so dominant as is usually assumed, nor did its disappearance appear predestined. No one mentioned in this book ever called him- or herself a "scientist" (a term tentatively coined by William Whewell in 1833) and most despised it as an ugly neologism. As Sedgwick wrote in the margins of a volume in which Whewell stressed the need for such a word, "better die of this want than bestialise our tongue by such barbarisms."[4]

The status of the practitioner of knowledge was at a point of maximum flux in the 1840s. Examined more closely, the form of the portrait in figure 12.1 shows how the brutal realities of a commercial society were creating new opportunities for science. Originally an oil painting visible only within the walls of St. John's College in Cambridge, it was reproduced in 1835 as an expensive mezzotint, and a decade later in the crude, steel-engraved version shown here—the frontispiece to a compilation of facts by a science journalist. The man of science it portrays is Sir John Frederick William Herschel, the son of William Herschel, a German émigré bandsman whose astronomical discoveries had brought wealth and status. For all groups, from aristocratic grandees to Grub Street hacks, there was everything to play for. *Vestiges* became a touchstone for predicting what the role of the maker of knowledge would be.

The Portrait of the Gentleman

Ideals of gentlemanly character can appear deceptively akin to those of modern scientific researchers, and the parallel between them has often been made to give contemporary relevance to historical cases.[5] But of all the scientific ideals of the past, that based on gentility ought to be the most alien. Ideals of gentlemanly science were embedded in a vision of a hierarchical society, masculine authority, and government by an aristocratic elite. To be a gentleman of science was, ideally, a matter of blood, of being male and born to an established family. Clothes and facial expression, like the smell of a library, were only outward (and often unreliable) signs of inner qualities. And of course, money mattered too, in that the ranks of the genteel were continuously replenished by those with newly acquired fortunes. A gentlemanly style of life implied a certain leisure and independence of income.

Most important, the ideal gentleman was a man of honor, whose word could be trusted. Commercial men and working people were considered less reliable because they were (in different ways and to different degrees) in positions of dependence. Women, with their supposedly weak constitutions and susceptibility, were often portrayed as too trusting to be trustworthy. The highest accolade

4. Morrell 1990 offers a sensible review of the literature on professionalization. On "man of science" and "scientist," see White forthcoming and Ross 1962; Sedgwick's comment is reported in J. Wyatt 1995, 6. Whewell proposed the term at the 1833 Cambridge meeting of the British Association for the Advancement of Science. It was first used in print in [W. Whewell], "Mrs Somerville on the Connexion of the Sciences," *Quarterly Review* 51 (Mar. 1834): 54–68, at 59–60.

5. See esp. Rudwick 1985, 14–15, and Shapin 1994, 409–17. Although its limits are acknowledged in principle, in practice the parallel has tended to assimilate earlier forms of scientific inquiry to modern ones.

was to be identified as a Christian gentleman, in whom moral integrity rested on an unobtrusive but strongly held faith. Indeed, one of the most important changes taking place in the 1840s was that the term "gentleman" increasingly referred as much to character as to birth. Association with such qualities made gentility an important resource among all classes of society. This was especially true for practitioners of the sciences, where virtues of disinterestedness, piety, and independence were paramount. Gentility, as Steven Shapin has said of the early modern period, "was a massively powerful instrument in the recognition, constitution, and protection of truth."[6]

The British Association for the Advancement of Science afforded an ideal opportunity for reasserting the right of Christian philosophers to demolish any pretensions *Vestiges* might have to being part of true science. For over a decade, the association had symbolized a scientific alliance between the metropolitan gentlemen and the ancient universities. With presidential addresses, discussions of papers, and after-dinner toasts, the British Association displayed the ideals of gentlemanly scientific practice in action (fig. 12.2).[7] In June 1845 the association met in Cambridge, and the president was John Herschel.

From Herschel's point of view, *Vestiges* had to be destroyed. It had the outward show of the science supported by the British Association, and was read that way in the metropolitan salons; but it came from an unknown and hence potentially illegitimate source. It cited authoritative works of leading geologists and astronomers, including those by Herschel himself; but it did so along with phrenology, quinarian circles, and spontaneous generation. Lockhart summed up the dilemma his friends among the scientific "clique" faced:

> It (*The Vestiges*) is a work of extraordinary knowledge and ability, dashed with apparently mad extravagance—the effect, however, bringing into form all the elements of infidelity scattered thro' the modern tomes of Geology, Botany, Zoology, etc., and written so clearly and powerfully that the *savans* anticipate an outburst of orthodox wrath upon their whole clique. . . . You should see this book. And Sir John Herschell [*sic*] ought to answer it, and if you advise I shall ask him.[8]

Unless the sensation could be stopped, scientific authority would be cut loose from its foundations in sound religion and polite culture.

Herschel did not choose to review *Vestiges* for Lockhart's *Quarterly*, but instead used an even more prominent platform, his British Association presidential address. Speaking in the Senate House, he invoked values shared between

6. Shapin 1994, 42. On gentility and character, see Collini 1991.

7. Morrell and Thackray 1981. The best survey of *Vestiges* from the perspective of the gentlemen of science is Yeo 1984.

8. J. G. Lockhart to J. W. Croker, n.d., in Paston 1932, 49.

the ancient universities and the metropolitan savans. He drew upon his own "stirring and deep-seated remembrances" of Cambridge—personal recollections, but ones in which "*self* has no place." Herschel commended the "sobering discipline of mind" and "moral influence" that mathematical studies provided, and surveyed the university's greatest triumphs. He then turned to concerns closer to his own interests and those of other metropolitan gentlemen: the establishment of a global network of physical observatories; the earl of Rosse's Leviathan telescope; and the proper method for pursuing science. On the latter subject, he pointed to the differences between his own philosophy and that of Whewell, a contrast that in many ways embodied the differences between London and Cambridge science. Yet conflicts were minimized to show the pursuit of science as a "brotherhood." "Let interests divide the worldly," Herschel said, "and jealousies torment the envious! We breathe, or long to breath, a purer empyrean." All disciplines, all men of whatever background, could unite in understanding of "the wonderful works of God."[9]

Vestiges offered precisely that "propensity to crude and over-hasty generalisation" that Herschel expected the "sound and thoughtful and sobering discipline" of Cambridge mathematics to combat. He avoided mentioning the book, not wanting to add to the sensation, but his target was clear.[10] He claimed that the mathematics in the opening pages was flawed, and that the nebular hypothesis was but "a matter of pure speculation." Without an observable cause, progressive development was just as unexplained and "*miraculous*" a process as the immediate creation of all species and all individuals. A "mere speculative law of development" was no law at all, but a vague generalization without explanatory power. The address was appallingly delivered (Herschel had a cold and was a feeble orator), but this did not matter, for his words were reported in newspapers throughout the country.[11] The nation's most prestigious man of science had declared against the book.

Herschel's celebrity rested on his *Preliminary Discourse,* a widely read conduct manual for proper scientific practice. The book showed how those engaged in the study of nature could partake of the virtues of the Christian gentleman—calm, devout, dispassionate—whatever their station or degree. In his famous address on the reading habits of the working classes, Herschel had spoken of the "gentle, but perfectly irresistible coercion in a habit of reading well directed

9. J. Herschel, "Address," in *1845 British Association Report* (London: John Murray, 1846), xxvii, xliv. On Herschel's cosmology, see Cannon 1978 and Hoskin 1987.

10. Herschel, "Address," xxvii–xxviii. A footnote reference was added to the published report (xxxix).

11. For Herschel's delivery, J. Romilly, journal entry for 19 June 1845, CUL Add. mss. 6823, p. 39; for the quotation, Herschel, "Address," xxxix, xlii.

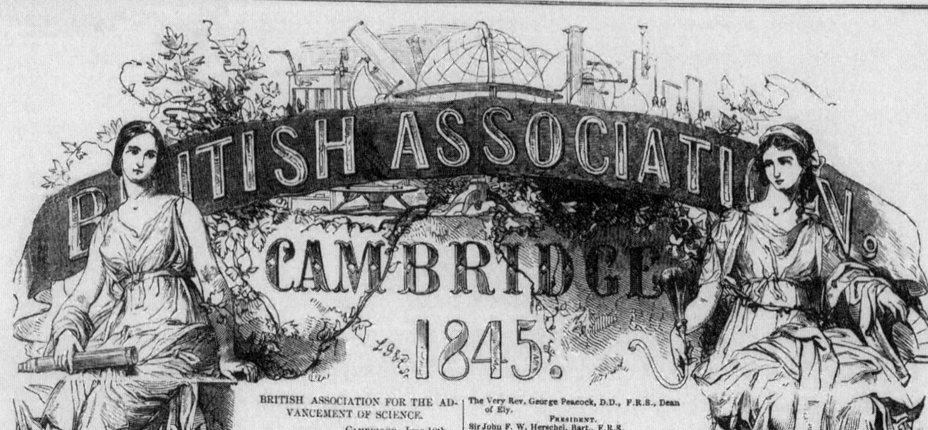

BRITISH ASSOCIATION FOR THE ADVANCEMENT OF SCIENCE.

CAMBRIDGE, June 18th.

(From our own Correspondent.)

The first day of this, the Fifteenth Meeting of this celebrated body, did not bid fair for a numerous attendance during the succeeding ones. It opened inauspiciously as regarded the weather, the rain commencing at a very early hour, and continuing to descend in torrents till an advanced period of the afternoon. The Town Hall presented almost as melancholy an appearance as the streets and the College grounds, few being adventurous enough to venture out in such weather. Not many of the distinguished persons who were expected arrived during the day, this being looked upon more as one of preliminaries, than general business. The Council met on the previous night, on the subject of arrangements, and amongst these it was discussed whether or not Medical Science should in future be made subject of inquiry, or form a portion or section of the Institution. A pretty general feeling seemed to exist amongst the members that it would be better to leave it out altogether, for the reason—that such members of the medical profession as could give interesting papers were too much occupied with, and chained, as it were, to, their practice, to do so. It was left, however, an open question for the meeting of the General Committee to adjudicate upon, and no decision was pronounced by the Council.

The following is the authentic list of the new Officers of the Association:—

TRUSTEES, PERMANENT.

John Taylor, Esq., F.R.S, Treas. G. S.
Roderick Impey Murchison, Esq., F.R.S.
The Very Rev. George Peacock, D.D., F.R.S., Dean of Ely.

PRESIDENT.

Sir John F. W. Herschel, Bart., F.R.S.

VICE-PRESIDENTS.

The Earl of Hardwicke.
The Bishop of Norwich, President of the Linnean Society.
The Rev. J. Graham, D.D., Master of Christ's College.
The Rev. G. Ainslie, D.D., Master of Pembroke College.
G. B. Airy, Esq., F.R.S., Astronomer Royal.
Rev. A. Sedgwick, F.R.S., Woodwardian Professor of Geology.

GENERAL SECRETARIES.

Roderick Impey Murchison, Esq., F.R.S.
Lieut.-Col. Sabine, F.R.S.

ASSISTANT GENERAL SECRETARY.

Professor John Phillips, F.R.S., York.

GENERAL TREASURER.

John Taylor, Esq., F.R.S., 2, Duke-street, Adelphi, London.

LOCAL SECRETARIES.

Wm. Hopkins, Esq., F.R.S.
Professor Ansted. F.R.S.

LOCAL TREASURER.

C. C. Babington, Esq., F.L.S.

The Council Room in the Town Hall was appointed for the reception of members, and tables were placed on either side for the secretaries, officers, clerks, &c., with a temporary little post-office at the upper end, for the letters and communications addressed to the members. At about half-past two the General Committee met for the election of sectional officers, and other business, the Very Reverend the Dean of Ely, Professor Peacock, late President of the Association, in the chair. The Secretary read the report of the proceedings of the various committee meetings during the past year, which were confirmed. He then went into a statement concerning

THE RECEPTION ROOM, AT THE TOWN HALL.

12.2 *Opposite:* Opening page of the *Illustrated London News*'s special report on the Cambridge meeting of British Association for the Advancement of Science: two female figures, representing the sciences and holding scientific equipment, float above an all-male reception in the newly opened Town Hall. (Women, however, were admitted to most of the association's gatherings.) The high ceiling in this and many other contemporary illustrations emphasizes the character of the building as a public space. *ILN*, 21 June 1845, 393.

over the whole tenor of a man's character and conduct." Manners took "a tinge of good breeding and civilization, from having constantly before one's eyes the way in which the best bred and the best informed men have talked and conducted themselves."[12] The *Preliminary Discourse* had been written with precisely this aim, which made its author's declaration against *Vestiges* particularly compelling.

The onslaught continued during the rest of the week. The section meetings aimed to show specialist science in action, which left evening assemblies for addressing wider cosmological issues. Rather than the unintended fiasco of a newspaper war, this was the way in which true gentlemen conducted their business. Speaking to a mixed audience, Roderick Murchison lectured on his recent geological travels in Russia. He tried to end the confusion (so apparent after the Liverpool fiasco) about the difference between geological progression and progressive development. He pointed to numerous instances of degradation, where the "most elaborately formed and highly organized" forms were "the first created of their classes." This claim was buttressed by the entire geological column: it was hard to argue with a speaker who could thrust two huge tomes on Russia before his audience and say that "every portion of geological evidence sustained the belief that each species was perfect in its kind when first called into being by the Creator." Murchison's geology was a showpiece of imperial science. Its antievolutionary lessons were then reinforced by his old friend Sedgwick, who summed up his forthcoming *Edinburgh* article. Whatever their differences—and Sedgwick profoundly opposed what he saw as the premature expansion of the Silurian empire in the older fossil-bearing rocks—the two men could agree on the need to combat the "desolating pantheism" of *Vestiges*. As Bishop Samuel Wilberforce had said to the cheering audience, only a rash man would stand up in Cambridge to prove any connection between irreligion and science.[13]

As the British Association's leaders knew all too well, a book claiming sanction from their results was being read simultaneously in backstreet freethinkers' halls and at Buckingham Palace. The reading habits of the nation seemed out

12. "An Address to the Windsor and Eaton Public Library and Reading Room," in Herschel 1857, 13.

13. "Scientific Meeting at Cambridge," *Cambridge Chronicle*, 28 June 1845, 2; Secord 1986a, 110–201, on Murchison and Sedgwick's geology.

of control. Metropolitan savants and clerical professors responded by drawing together to rout the common enemy. Their union combined global imperial interests with those in defending the faith. The result was not a "Cambridge network," as historians have called it,[14] but a tactical alliance between different but overlapping interests. As a metropolitan savant, imperial astronomer, Cambridge graduate, and philosophical author, Herschel embodied in his own person the Christian gentleman of science. *Vestiges,* temporarily, made that ideal stronger than it already was.

Science in the Salons

The admission of scientific practitioners into elite Society had been one of the ways that the aristocracy and urban gentry adapted to the reform era. Men of science exploited their status as lions. Many were younger sons and professional men, whose principal entrée into high Society was their interest as intellectual celebrities: Herschel was the inheritor of an illustrious scientific name, not an ancient estate. They could lobby for support of special projects, obtain peers as speakers at annual meetings, and engage aristocrats as practitioners in their own right. Some notable fixtures on the social calendar were organized specifically to bring men and women of rank and talent together. These grand soirées, and dozens of smaller dinner parties, were not part of an "alternative" social world, an "intellectual aristocracy" defined in distinction to Society proper—this developed only in later decades. Rather, the authority of this form of science was located within the wider political and religious authority of the aristocracy and gentry.[15]

An indication of how expertise functioned in these circles is available from conversations recorded by Sir Charles James Fox Bunbury, whose aristocratic connections included some of the most famous Whig families in the country. Bunbury typified the connection between social rank and the new sciences. He had become interested in botany while at Cambridge, and afterward studied the flora of the Cape of Good Hope during a year's residence. In 1844 he married Frances Horner, a daughter of the Whig factory inspector Leonard Horner. At the suggestion of his new brother-in-law, Charles Lyell, Bunbury commenced the study of fossil plants, a subject in which Britain had few specialists. A modest, retiring man, Bunbury was still a beginner, and he was reading standard works and encyclopedia articles while working at the coal plants in the museum of the Geological Society of London.[16] In common with many of his status, he detailed his reading and conversations in a diary.

14. Cannon 1978, 29–71.

15. Davidoff 1973.

16. Diary entry for 17 Feb. 1845, in F. Bunbury 1890–91, *Middle Life,* 1:35.

Arriving in London from their country mansion in Suffolk, the Bunburys stayed at the Horners' Bedford Place house in Bloomsbury. At an evening party in February 1845, Bunbury sat near the botanist Robert Brown of the British Museum. *Vestiges* was the common topic of conversation. Asked about the development of oats into rye—a key Vestigian fact—Brown "smiled sarcastically, and remarked that such a transmutation might be *very convenient*" for the author of *Vestiges,* although he "would not express decidedly any opinion on the subject." Bunbury was also left with the impression that the elderly botanist (who carried genteel scientific reserve to an extreme) "evidently disbelieves" in the Crosse experiments. "The account I have heard of the book is," Bunbury wrote in his diary after further informal discussion in the Geological Society's museum, "that every really scientific man who has examined it, finds it shallow and unsound in his own particular department."[17]

Focusing on errors was a conversational tactic also used by Murchison, who damped down enthusiasm for the book in the soirées and clubs. A number of leading parliamentarians were warned that it was full of mistakes, and was probably by Vyvyan or Ada Lovelace.[18] The integrity of science was vitally important to Murchison's own social ambitions, for his father was an army surgeon with a fortune gained under dubious circumstances on the Indian subcontinent. His book on Russian geology was a princely production subsidized by the czar and costing eight guineas, more than twenty times as much as *Vestiges.* The larger meanings of creation were only just touched upon in its pages, but behind the scenes Murchison was extremely active "in scaring away that night mare," as he told Hugh Miller.[19]

George Bellas Greenough also occupied a prominent place in polite scientific circles. A radical and former M.P. in his sixties with a large fortune, he lived in a large mansion in Regent's Park, purpose-built to hold his natural history collections and library. Greenough took extensive notes on *Vestiges,* listing facts, authors cited, and felicitous phrases in seven pages of tiny handwriting. This practice, as befitted a gentlemanly collector, disaggregated books into

17. Diary entries for 17 and 19 Feb. 1845, in ibid., 1: 35, 37.

18. In Australia, the gentlemanly naturalist William Sharp Macleay (whose circular system of classification featured largely in the first two editions of *Vestiges*) warned Robert Lowe (later Viscount Sherbrooke) that the book was "incorrect as to facts and therefore valueless, however attractive it may be in style." The episode is recalled in W. S. Macleay to R. Lowe, May 1860, in Martin 1893, 2:204–7, at 205. See also Finney 1993, 97–99.

19. Murchison, de Verneuil, and Keyserling 1845, 9*. See also Thackray 1979; Secord 1986a, 45, 120–25; Rudwick 1985, 395–97. On Murchison's social origins, see Stafford 1989. For his opposition to *Vestiges,* see R. Murchison to R. Owen, 2 Apr. 1845, in Owen 1894, 1:254; J. C. Hobhouse, diary entry for 17 Mar. 1845, BL Add. mss. 43747, f. 82; and R. Murchison to H. Miller, 26 Sept. 1849, NLS mss. 7527, no. 33.

"specimens." A sharp critic, he was skeptical about the book's merits, and questioned whether the facts of premature births supported a hypothesis of evolution through embryological development—did human babies delivered at seven months really show the characteristics of a wolf? Greenough discussed his reading with the Unitarian literary man Henry Crabb Robinson, who agreed that the book was "an edifice very skilfully constructed so that it will stand, though the materials are mere rubbish."[20] Again, factual criticism provided a way of demolishing a book without being impolite.

Greenough, Murchison, and other wealthy gentlemen orchestrated opportunities for bringing science to the center of fashionable society. Lord Northampton held large conversaziones in his capacity as Royal Society president, as did Lord Rosse once he took office.[21] Here curiosities of science and technology—figure 12.3 shows a printing machine in the salon—could be displayed along with new books such as *Vestiges.* Some sense of what these occasions were like is given by descriptions of the great soirées held by the Murchisons in spring 1845 as part of their campaign to become Sir Roderick and Lady Murchison. Prepublication copies of the Russian volumes would have been on display in their Belgravia mansion, together with a monumental vase of Siberian aventurine presented by the czar.[22] Men of science brought fossils, electrical apparatus, astronomical illustrations, and rare zoological specimens that could then be examined by the assembled company. Objects of research became conversation pieces and brought discoveries to the attention of the fashionable world.

Vestiges would certainly have been discussed at these soirées. The surgeon Gideon Mantell showed specimens of live infusoria to "many intelligent persons" through an achromatic microscope (fig. 12.4).[23] Continental work on these tiny creatures was just becoming well-known in Britain, and had been cited in *Vestiges.* In the following year Mantell published his own introductory guide to the "invisible world," drawing on his experiences in presenting these attractive forms (fig. 12.5). The notes dwelt upon Crosse's experiments to create life through electricity, the nature of the primordial cell, and the relation

20. H. C. Robinson, diary entry for 24 Mar. 1845, in Morley 1938, 2:652; Greenough, "Remarks on the Vestiges of Creation," History of Science Collections, Cornell University Libraries. For Greenough's place in science, see Rudwick 1985 and Wyatt 1995.

21. Hall 1984, 77–78.

22. Secord 1986a, 43–45; Geikie 1875; "Varieties," *Literary Gazette,* 8 Mar. 1845, 158; and for Murchison's political connections, Stafford 1989. *PT,* 15 Mar. 1845, 173, for the Russian vase. Greenough's diary (University College London, Greenough Papers, 7/48) gives some idea of the dense schedule of dinners and soirées for those at the heart of London science.

23. G. Mantell, journal entry for 7 May 1845, in Curwen 1940, 194. On Mantell's social position, see Dean 1999, esp. 214.

12.3 The pictorial newspapers that began to appear in the 1840s often illustrated fashionable soirées for their middle-class family readerships. This woodcut shows Prince Albert in conversation with Lord Northampton at one of his regular Royal Society receptions. These gatherings, like the Royal Society itself, were for men only. William Carpenter, Robert Edmond Grant, Henry Hallam, William Jerdan, Edwin Landseer, Charles Lyell, Richard Owen, and John Timbs (subeditor of the journal in which this illustration appeared) mixed with the aristocracy at these affairs. Besides Prince Albert, the star attraction was the double-action printing machine on the right-hand side of the picture; this was kept in operation during the evening, producing ten to twelve thousand impressions each hour. *ILN,* 20 Feb. 1847, 117.

between life and organization. They show how *Vestiges* would have been discussed over the microscope. As befitted his modest social status, Mantell avoided being dogmatic, cautioning only that its theories were not conclusive and the evidence as yet "insufficient." In contrast, when writing anonymously in the *American Journal of Science,* he complained that "all its errors are swallowed by the upper classes, to whom every thing boldly asserted and in captivating style is gospel."[24]

24. Mantell 1846, 24, 25, 86, 95–98; [G. A. Mantell], "Vestiges of the Natural History of Creation," *American Journal of Science* 49 (Apr. 1845}: 191. For a detailed account of Mantell's reaction, see Dean 1999, 205–13.

12.4 Like the Murchisons' soirées, this "scientific conversazione" was open to both women and men. Note the microscopes and the diagrams of microscopic life, and the changes in fashion by the mid-1850s, with men now sporting facial hair and women wearing crinolines. Detail from "Scientific Conversazione at the Apothecaries' Hall," *ILN,* 28 Apr. 1855, 405.

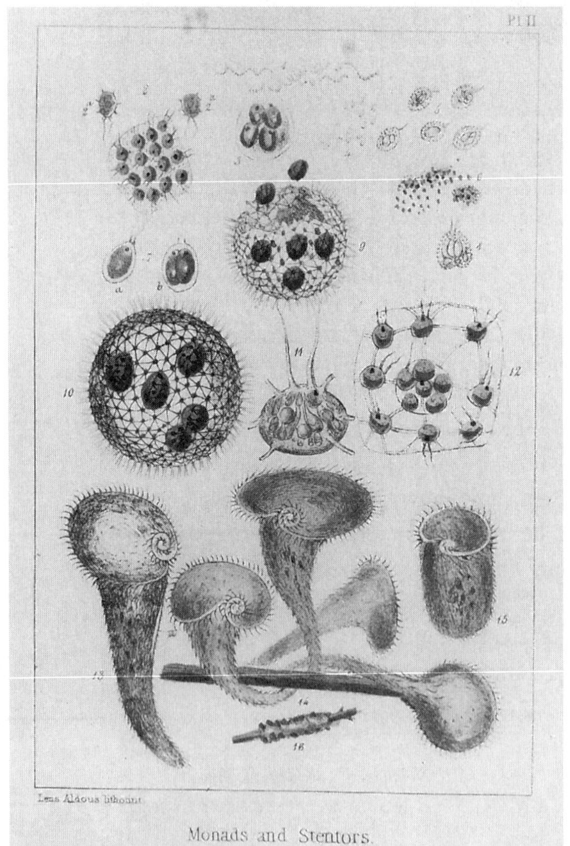

12.5 "Monads and Stentors." The simplest forms of life, shown at the top of this plate by Gideon Mantell, had been mentioned in *Vestiges* and could serve as an opening for discussing the book at a conversazione. Mantell 1846, plate 2.

The role of the gentlemen of science in all the conversations discussed up to this point is strictly limited: they point to "blunders," "heresies," "rubbish," and other factual problems. The book, as Bunbury noted, was "described as abounding with errors as to the facts of science" and "ingeniously reasoned, though on unsound data."[25] Herschel lost patience while reading his presentation copy and lamented living in an age marked by "the appetite for the trashy."[26] The book, in more than one sense, was in bad taste. Pointing to mistakes was one of the least obtrusive ways to intervene in conversation. An opinion could be expressed, and claims to expertise established—but without seeming to dictate truth. In this respect, novels, histories, and biographies all posed fewer problems to drawing-room culture than did works of science. Like any propositions claiming truth, no matter how well handled, *Vestiges* was potentially too dogmatic. Most awkwardly, it relied upon material that might be thought to require specialist knowledge. Expertise had to be dealt with carefully lest it upset the equilibrium upon which the culture of conversation depended. Men of science, whose knowledge of specific facts might be unquestioned, often lacked the social status to participate fully in discussions on sensitive religious topics. Why should experience with the hammer, retort, or dissecting knife overrule the authority of the high aristocracy, particularly when the origins and destiny of the human race were being discussed? It was easy to sound overbearing or pedantic. Anonymity exacerbated the problem, as forcible criticism could prove offensive: one even might be talking to the author. The ultimate nightmare was to be a bore.[27] Although the reviewing process softened these difficulties—breaking up the text, pointing out elegant extracts—they still had to be faced.

Scientific clerics were given greater latitude for touching on religion and morals in polite conversation, although they too needed to handle the situation carefully. In the judgment of most observers, Sedgwick failed in the *Edinburgh* but was effective in the salons, where he had a reputation as one of the liveliest of talkers. *Vestiges,* he said, was "Monmouth Street Philosophy."[28] Monmouth Street was the center of London's rag trade, a source (as *Sartor Resartus* pointed out) not only for old clothes but the paper used to make books. In *Sketches by Boz,* Dickens had referred to Monmouth Street's market stalls as the "burial place of fashion," as a byword for flimsy finery and gaudy magnificence. On this analogy, *Vestiges* had a flashy exterior to make it sell, but its substance was

25. Diary entry for 5 Mar. 1845, in Bunbury 1890–91, *Middle Life,* 1:43.

26. J. Herschel to A. Sedgwick, 15 Apr. 1845, RSL Herschel Letters, vol. 15, ff. 426–27.

27. Such fears related to older images of the scholar as pedant: see Shapin 1991; 1994, 116–25.

28. R. Murchison to R. Owen, 2 Apr. 1845, Owen 1894, 2:254.

stale, shabby, and secondhand. There was no better strategy for rendering a book unfashionable than by associating it with Monmouth Street.

Implicit codes of propriety also meant that genteel women, whose social positions were independent of their intellectual accomplishments, could discuss the implications of *Vestiges* more readily than could some of the British Association lions. While Mantell, Bunbury, and Brown were condemning shoddy facts, Florence Nightingale and Lady Ashburton were speculating about the future of humanity. Women, it was often said, were experts in the moral realm. Yet even in the highest circles there were limits to what could be said. Sir John Cam Hobhouse reported a dinner conversation with Ada Lovelace (whom he suspected was a witch) that involved issues relevant to the final chapter of *Vestiges.* Lovelace questioned the validity of "the common argument in favour of a future state" derived from our intimations of immortality. However, "she would not say so much to any one, except very privately," lest she deprive them of "the consolations of belief." Her own thoughts were "only doubts"—to be dogmatic would be "presumptuous." Even so, Hobhouse was taken aback, for these were "subjects which few men, and scarcely any women, venture to touch upon."[29] Lovelace's behavior was simultaneously identified as gendered, aristocratic, and eccentric. Certainly any gentleman of science would have recognized that reference to such matters would have been out of place. As Thomas De Quincey said, "real good breeding . . . shows itself far less in what it prescribes, than in what it forbids."[30]

Science in Clubland

Where, then, were the suitable places for gentlemen of science to read and discuss controversial books? It might be expected that the learned societies in Somerset House on the Strand would provide an important focus for deliberations about *Vestiges.* After all, copies were sent to the libraries of the Linnean, Geological, Astronomical, and Royal societies, and the communal ideals of gentlemanly science achieved their fullest expression in their meeting rooms. Here controversy could be conducted without boring those who did not care for the laws of heat flow or the classification of trilobites. Presenting and discussing papers in these circumstances was a ritualized extension of "serious" male conversation in clubs, just as Parliament was itself the greatest club of all. Almanacs and wall charts with the dates for the year's scientific meetings were published annually, which show that science moved with the rest of the London literary and political season.[31]

29. Diary entry for 3 June 1846, in Hobhouse 1911, 6:175.

30. [T. de Quincey], "Conversation," *Tait's Edinburgh Magazine,* Oct. 1847, 678–81, at 679.

31. See Brock and Meadows 1998, 45–47; Rupke 1994, 55–59, offers one of the few available discussions of science in clubs. Masson 1908, 211–56, offers a survey of the clubs by a contemporary,

12.6 Imaginary conversation piece with geologists and palaeontologists discussing fossils. The main row of figures from left to right includes Roderick Murchison, Richard Owen, Henry De la Beche, Adam Sedgwick, Charles Lyell, William Fitton, William Buckland, [unidentified], William Smith. John Phillips is in the foreground leaning over a fossil; the figure with the ear trumpet at the top of the painting is probably George Bellas Greenough. Oil painting by Thomas Henry Gregg, c. 1838.

Debates about new findings were often called "conversations," especially at the Geological Society, which was famous for them. Given the small numbers who participated, the atmosphere of male intimacy, and the prohibitions on press reports, the term was appropriate. The oil painting in figure 12.6 suggests how the geologists could be portrayed as a small group of independent gentlemen engaged in scientific talk. As the *British Critic* noted, the cultivators of geology "resembled more one of the schools of philosophers of the ancient world, where conversation and action were the means of communicating knowledge, than a sect of these modern times, in which the press is supposed to be the

---

and Timbs 1866 reviews their history. A study of nineteenth-century clubs along the lines of Clark 2000 is much needed.

source of all information, and popular circulation the criterion of all progress."[32] Or as Darwin told an old Cambridge friend, "Geology is at present very oral."[33]

The natural history of creation, however, was not part of the formal proceedings of scientific meetings.[34] These were reserved for what gentlemen of science considered as the proper subject of their practice: reports of original research in the field, observatory, laboratory, and dissecting room. Such meetings, for example, were where the Storeton footprint fossils and the resolution of the great nebula in Orion were discussed. They were where Darwin presented his *Beagle* discoveries, including his theories of coral reef formation and crustal uplift. Even at the Geological Society, which was celebrated for lively discussions, questions about the origins of man and the earth were excluded. The annual presidential address did provide an opportunity to speak to wider issues; but even there, *Vestiges* was never mentioned.

This practice did not rule out dealing with the book implicitly, either in technical papers (such as those on fossil fish) or in the presidential addresses. Lyell, who was Geological Society president in 1849 and 1850, used this platform to renew his attack by lumping evolutionary development with geological progress. As he had pointed out to the Reverend Hume and in his *Principles of Geology,* if no progress from simple to complex could be read from the strata, then transmutation could not be true. Lyell piled up dozens of examples to show how animals and plants of "high" status could be found low in the geological record. In all this, he never mentioned *Vestiges,* nor did he refer to it in his books. What really alarmed him was the sheer number of copies being read— "a larger public than ever, in an equally short time, bought a scientific book." As a "popular" work, *Vestiges* was outside the confines of scientific discussion, in a way that even Lamarck's "romance" had not been.[35] In his address, Lyell could present the Geological Society as the guardian of public taste.

The dangers of straying from these limits are illustrated by what happened when *Vestiges* became the subject of a lecture at a rather different kind of London club. The British and Foreign Institute had been founded in 1843 with

32. [W. Whewell], "Lyell—*Principles of Geology,*" *British Critic* 9 (Jan. 1831): 180–206, at 180.

33. C. Darwin to J. M. Herbert, [3 Sept.? 1846], in Burkhardt et al. 1985–99, 3:338. The importance of oral communication networks in geology is stressed in Rudwick 1985, 18–30, and in Secord 1986a, 14–24.

34. For the limits to gentlemanly specialist debate, see Desmond 1989, 294–334; Rudwick 1985; Secord 1986a; and esp. Morrell and Thackray 1981. As Burkhardt (1972, 33) notes, a similar situation prevailed in relation to Darwin's *Origin.*

35. Lyell's fear of the "popularity" of *Vestiges* is expressed in his species journals from the 1850s; see Wilson 1970, 84, 180, 190. On the address, see Wilson 1998, 333–38; Bartholomew 1973, 290–92.

Prince Albert's patronage, and the press pictured it as ideally suited for ladies and gentlemen visiting the metropolis. Early in 1845 the institute's director, James Silk Buckingham, decided that *Vestiges* would make an ideal start to a series of verbal book reviews. It was, Buckingham said, "one of the most remarkable Books of the year, at once scientific, metaphysical, original, daring, and sublime."[36] He quoted long extracts on the nebular hypothesis and morals, but stressed that he did not agree with other aspects, which might have been harder to reconcile with the "elegant" and "respectable" image he wished to foster.

The institute, despite the facade of genteel success, soon proved a failure—as did Buckingham's scheme for lecture-reviews. Creation was a suitable topic for a mixed audience in a way that would not have been the case at the specialist societies; but the setting was all wrong. Only "geese" and "spoons," contemporary slang for simpletons, would attend, and no one wanted the tastelessly over-the-top *Transactions* distributed to all members. Such actions mimicked the learned societies without their originality, substance, and opportunities for manly combat. *Punch* ridiculed Buckingham's one-man show as "The British and Foreign Destitute": pretentious, underfunded, and appealing only to bluestockings and traveling salesmen.[37] Two contrasting images of reading at the institute are shown in figures 12.7 and 12.8. The first, from a eulogistic article in the graphic press, depicts earnest gentlemen consulting the institute's "well-stored" library, while small groups engage in serious discussion.[38] The second, from Thackeray's *Vanity Fair* (1847), shows an institute member—the much-ridiculed antiquarian George Jones—reading in the same building. "Yes," the narrator says: "I can see Jones at this minute (rather flushed with his joint of mutton and half-pint of wine), taking out his pencil and scoring under the words 'foolish, twaddling,' &c., and adding to them his own remark of *quite true.*'" Thackeray parodied any "lofty man of genius" who applied the wrong technique of reading in the wrong place to the wrong kind of book.[39]

Silent Voices

The failure to locate gentlemanly science within drawing room culture has had unfortunate consequences. The views of many geologists, astronomers, and anatomists are considered somehow dishonest or contradictory; and historians often implicitly criticize their subjects for an inordinate fondness for rank.

36. J. S. Buckingham, "Soirée. First presentation and review of books, with verbal descriptions of their contents," *Transactions of the British and Foreign Institute* 1 (8 Jan. 1845): 86–97.

37. [J. Leech], "Bubbles of the Year: the British and Foreign Destitute," *Punch Almanack for Sept. 1845*, Jan. 1845; Altick 1997, 638–42.

38. "The British and Foreign Institute," *PT*, 23 Mar. 1844, 177, 181.

39. Thackeray 1847–48, 5. Altick 1997, 639–41, discusses the antiquarian George Jones, whose identification of the native Americans in books of 1843 and 1844 were widely discussed and mocked.

12.7 The ideal of reading in the library of a learned club: "The Library of the British and Foreign Institute," *PT,* 23 Mar. 1844, 181.

12.8 Jones at his club: Thackeray's satire on reading at the British and Foreign Institute. Thackeray 1847–48, 5.

But gentlemanly science was practiced in a community whose very existence depended on assumptions about breeding. Those who aspired to a conception of scientific practice based on research, supported by ideals of vocation and an aristocracy of birth, appreciated the virtues of tact.[40]

Science in these circles was embedded in codes of gentility, which meant that claims to legislate over nature were unlikely to succeed. As long as the

40. The approach adopted in Morrell and Thackray 1981 has clear affinities with the revisionist account of aristocratic politics in Mandler 1990.

gentlemen of science expressed their views in the appropriate manner, which often meant sticking to their expertise except when speaking in confidence, they could believe what they wished on religious and political issues. By regulating their talk and expressing certainty only for specific "facts" in their special "departments," men of science could be both polite and authoritative at the same time—something that was not always easy to do. Beyond that, they were no more than ordinary participants in the conversation that defined polite society. The modest social origins of many men of science meant that silence was the most effective way of exercising authority. They spoke to larger issues where consensus already existed; hence the emphasis on providence, progress, and adaptation in natural theology and in philosophical conduct manuals such as the *Preliminary Discourse.* Neutrality was necessary if science's claim to absolute truth were not to conflict with the demands of civility. To say more would have been inappropriate and boring. In these circumstances, silence could be a form of power.

The case of the anatomist Richard Owen illustrates how these issues were negotiated. The most effective way to understand the much-debated problem of his attitude to *Vestiges* is to examine his strategies for self-presentation. The son of a West India merchant, Owen had pushed his way to the top of contemporary science and wished to be admitted to the best circles as an insider and a gentleman. This was no easy feat, requiring an exceptional understanding of the possibilities for pursuing science within genteel culture. Owen's perspective was defined not by adherence to abstract doctrines such as idealism or functionalism but around issues of who ought to be allowed to speculate and in what circumstances.[41]

Owen knew that his acceptance into the best circles rested first on his unrivaled mastery of the collections at the Royal College of Surgeons and the Zoological Society; second on his patronage by the prime minister Sir Robert Peel and other social superiors; and only then on the independence of his character. Having no fortune, and rising as a surgeon through a series of apprenticeships, Owen could never take his position in Society for granted. As Buckland had warned Peel, there was always the chance that Europe's greatest anatomist would be "obliged to descend to the Condition of a Bookseller's Hack"—a very real possibility before Parliament had voted a pension.[42]

These difficulties in career making left Owen with an exceptionally fragile

41. Rupke 1994, the standard biography of Owen, stresses these theoretical allegiances in the context of an account mainly devoted to his museum work; see also Rupke 1993. Desmond 1989 puts Owen's philosophical anatomy in a political setting.

42. W. Buckland to R. Peel, 12 Jan. 1842, in Rupke 1994, 52. On Owen as hack and pensioner, see Desmond 1989, 354–58; Rupke 1994, 52–53.

sense of status, which is evident in his letter of acknowledgment sent to the *Vestiges* author through the publisher:

> I beg to offer you my best thanks for the copy of your work . . . which I have perused with the pleasure and profit that could not fail to be imparted by a summary of the evidences from all the Natural Sciences bearing upon the origin of all Nature, by one who is evidently familiar with the principles of so extensive a range of human knowledge. It is to be presumed that no true searcher after truth can have a prejudiced dislike to conclusions based upon adequate evidence, and the discovery of the general secondary causes concerned in the production of organized beings upon this planet would not only be received with pleasure, but is probably the chief end which the best anatomists and physiologists have in view.[43]

This letter has created considerable controversy, for it conflicts in both tone and substance with letters written to Whewell only a few months later.[44]

The dilemma is resolved once we know that Owen thought that Sir Richard Vyvyan was probably the *Vestiges* author.[45] By this time, Owen would have been introduced to Vyvyan, who had invited him to discuss the *Harmony of the Comprehensible World* three years before. He deftly qualified his praise in writing to the *Vestiges* author, but he could not offend a powerful patron by appearing condescending in any way. Not least, his plans for a national museum of natural history needed to be maneuvered through Parliament. Moreover, he believed that Vyvyan had every right to contribute to philosophic conversation. Owen was not trimming, truckling, or revealing his "real" views in acknowledging *Vestiges*. He was being polite. The two men later became friends, Owen being invited to Vyvyan's Cornish estate and in turn providing Vyvyan with tickets to see the exhibition of the so-called "Aztecque" children, whose remarkable crania were relevant to transcendental anatomy.[46]

Owen had to be tactful. Most of his letter sticks to factual issues ("a few mistakes where you treat of my own department of science, easily rectified in your second edition"). For all his enthusiasm for "the discovery of the general

43. R. Owen to the "Author of 'Vestiges,' &c.," [1844], in Owen 1894, 1:249–52.

44. The discrepancy was pointed out in Brooke 1977, and the extensive literature is best surveyed in Richards 1987.

45. Diary entry for 19 Feb. 1845, in Bunbury 1890–91, *Middle Life,* 1:37. This reference was first noted by M. J. S. Hodge (see Brooke 1977, 144 n. 20), although its significance has not been appreciated. Corsi 1988, 267, suggests that Owen might have suspected Prince Albert of being the author, but there is no evidence that he held this uncommon view.

46. R. Vyvyan to R. Owen, 6 Apr. 1849, 22 Jan. 1851, 31 July 1851, 12 Oct. 1852, 13 Oct. 1852, 13 July 1853, BL Add. mss. 39954, ff. 140, 197, 219, 267, 270, 296–97. For Owen's interest in the crania, see Richards 1994, 400–404.

secondary causes concerned in the production of organized beings upon this planet," Owen never told his anonymous correspondent what his own views were.[47] He did not pass judgment, for to do so would have overstepped the bounds of propriety in dealing with a member of the aristocracy. Rather, he put his expertise at the service of a social superior and acknowledged *Vestiges* as a legitimate voice in an ongoing discussion.

Taking visitors around the collections of the Royal College of Surgeons, Owen was equally circumspect. In January 1848, he told one such guest that there were at least five more plausible mechanisms for introducing a new species than the theory of progressive development "servilely" copied by *Vestiges* from the French. "'What are they?'" the visitor eagerly replied, but Owen declined to say, preferring instead to allow his museum to speak. During the remainder of their tour, he pointed out specimens—perhaps relating to the alternation of generations—that suggested at least one of the possibilities. Writing later about this encounter, Owen described his visitor as "the (reputed) author of 'Vestiges.'" Two months after the *Macphail's* and *Manchester Courier* furor, this would almost certainly be understood as a clear reference to Chambers. Owen, at any rate, took obvious pleasure in this particular visitor's failure to understand facts thrust under his nose. "He saw nothing of their bearing," Owen concluded, "and I shall refrain from publishing my ideas on this matter till I get more evidence."[48]

Owen chafed at the limits of his independence. Independence was the mark of the true gentlemen, so that for someone in his position to have been able to express opinions freely would have signaled the complete integration of gentlemanly culture and the culture of science. Only when that happened, Owen believed, would English science cease to be narrow and insular. He had expressed such views often, most strongly to Carl Gustav Carus at a dinner for the king of Saxony early in 1844.[49] He did not expect scientific men to pontificate on nature's laws, but rather that they might state their views without fear of theological attack. At a smaller dinner party, Owen told Bunbury that the secrecy surrounding *Vestiges* came as no surprise, because such views would "bring odium"

47. R. Owen to the "Author of 'Vestiges,' &c.," [1844], in Owen 1894, 1:249–52.

48. R. Owen to J. Chapman, [Jan. 1848], in Owen 1894, 1:309–10. Chapman, at any rate, thought he must mean Chambers: "I presume the 'reputed Author of the "Vestiges"' who paid you a visit was Robert Chambers? I think there is now pretty strong evidence to fix the paternity upon him." See J. Chapman to R. Owen, 13 Jan. 1848, NHM, Owen Correspondence, vol. 7, f. 26. For Owen's interest in alternation of generations in relation to this episode, see Desmond 1982, 34–37, Rupke 1994, 225–30. Owen later suspected Combe (see chap. 14), but the evidence is scanty for his views from 1845 through the middle of the next decade.

49. Carus 1846, 60–62, 93–94.

not only from the "vulgar," but from "men of science."[50] This was, of course, an indirect way of agreeing with the book's general aims, however shoddy its facts might be. Implicitly, he was criticizing university-based clerics like Sedgwick for speaking dogmatically, although he knew he could not afford to alienate them, for their approval depended on doctrinal soundness.[51] For Owen to advocate any specific scheme for the continuous "coming-into-being" of new creatures would threaten his acceptance by at least some clerics, who could in turn scupper his chances in good Society (and would lead to De Morgan–style jokes).

Owen was in an impossible position. He had expressed his secret sympathy with the search for a developmental law, but recognized that getting his clerical patrons to contemplate any form of organic descent would now be far more difficult.[52] He could only hope that *Vestiges* would die a quick death. Owen's Cambridge friends could not so readily leave the book alone, and Sedgwick traveled to London to gain "Owen's sanction as to the *facts*" of the embryology in his *Edinburgh* essay "paragraph by paragraph." Owen gave what was wanted, but refused to allow his name to be used in print in any way.[53] He developed his most profound generalization, the vertebrate archetype, from suspect sources in transcendental anatomy, particularly Carus and Lorenz Oken (fig. 7.9), and to make these safe the archetype had to be presented within a strongly expressed argument from design. No wonder he welcomed Church's temperate review of the controversy in the *Guardian* and the new freedom for scientific enquiry being encouraged at Oxford.[54]

By the end of the decade, the strategy based on these new possibilities began to fall apart. When Owen hinted at the Royal Institution in 1848 (fig. 12.9) that archetypal development might be a real, causal process—albeit directed by God—he was suspected of heretical sympathies.[55] Ralph Waldo Emerson,

50. Diary entry for 5 Mar. 1845, in Bunbury 1890–91, *Middle Life*, 1:42–43. S. P. Woodward, who worked closely with Owen at the British Museum for many years, called him a "'Vestigian'" who believed in transmutation through the force of conditions. S. P. Woodward to C. Lyell, 6 Mar. 1863, EUL Lyell 1/6481.

51. W. Buckland to R. Peel, 12 Jan. 1842, BL Add. mss. 40499, f. 256; Desmond 1989, 351–58; Rupke 1994, 47–55.

52. Desmond 1982, 29–37; Richards 1987; Brooke 1977; and Rupke 1994, 220–35, discuss the difficulties in detail. The literature on this subject has a tendency to lump together Cambridge and Oxford as "Oxbridge," when the problems facing the two universities in relation to the sciences were very different.

53. A. Sedgwick to M. Napier, 13 May [1845], BL Add. mss. 34625, ff. 207–9; A. Sedgwick to M. Napier, [28 May 1845], BL Add. mss. 34625, ff. 233–35; R. Owen to W. Whewell, 14 Feb. 1844 [1845], in Brooke 1977, 142.

54. J. B. Mozley to A. Mozley, 4 Apr. 1846, in Mozley 1885, 176–77.

55. The best analysis of this incident is Richards 1987, 161–67, building on work in Brooke 1977 and Desmond 1982, 42–48.

12.9 Richard Owen lecturing on extinct vertebrates to a packed fashionable audience at the Royal Institution. Note the large drawings of fossils, and the books and papers on the desk. Richard Doyle, "A Scientific Institvtion," *Punch,* 28 Sept. 1850, 136.

who sat in the audience, commented that Owen's face was "a powerful weapon," with its "surgical smile" and "air of virility, that penetrates his audience." For an Englishman Owen had gone far in "scientific liberalism," but then "indemnified himself in the good opinion of his countrymen, by fixing a certain fierce limitation to his progress, & abusing without mercy all such as ventured a little farther; these poor transmutationists, for example."[56]

Men of science of secure gentlemanly status—notably that "poor transmutationist" Charles Darwin—negotiated these difficulties with much greater ease. Darwin did not have to give public lectures nor take the views of aristocratic

56. Emerson, notebook for 1848, in Sealts 1973, 10: 525, 527. Writing to a friend soon after *Vestiges* appeared, Emerson welcomed it as "a good approximation to that book we have wanted so long" and tentatively attributed it to Vyvyan (R. W. Emerson to S. G. Ward, 30 Apr. 1845, in Rusk 1939, 3:383). Three years later Emerson met Chambers, and by then was sure that the book was his (R. W. Emerson to L. Emerson, 22 Feb. 1848, in Rusk 1939, 4:19).

metaphysicians seriously. Sedgwick and Lyell, Owen's potential patrons, could be treated as mentors and colleagues. It is not too surprising that Darwin was never sent a presentation copy of the first edition of *Vestiges,* for he was known as an outstanding practical geologist and travel writer, not a speculator on creation. Suggestions that Darwin was the author were shots in the dark. When most people thought of the Darwin name in connection with evolution, they did so because of his celebrated grandfather. In 1844 only a handful of close friends and family knew that the author of the *Naturalist's Voyage* no longer believed in species fixity.

How Darwin handled such delicate topics in conversation is revealed in Bunbury's diary. In the autumn of 1845, Bunbury lunched with Darwin at the Horners'. They had established "a great intimacy" some years before, so that Darwin could now admit that he was a believer in transmutation "to some extent." He also stressed that this was not "exactly according to the doctrine either of Lamarck or of the 'Vestiges.'"[57] As when talking with Lyell and Owen, Darwin did not reveal the specifics of his evolutionary mechanism. Only the naturalist Reverend Leonard Jenyns and the young botanist Joseph Hooker appear to have been given the opportunity to read his theory in draft (and Jenyns did not take it up). For all his caution, Darwin was able to go further than a man like Owen who depended on patronage and a salary.[58] He could afford to keep his opinions to himself.

The Naturalist in the Study

Soon after his return from the *Beagle* voyage, Darwin had gained a reputation as the beau ideal of the scientific gentleman. The older generation had welcomed him partly because there were so few new recruits with the financial independence to pursue the study of nature as a vocation. Science was short of genteel blood. Most of those who entered science in the 1830s and 1840s had to find ways of doing it for money. As one of them, the young geologist Andrew Ramsay, remarked, "Darwin is an enviable man. A pleasant place, a nice wife, a nice family, station neither too high nor too low, a good moderate fortune, & the command of his own time." Scientifically inclined aristocrats like Bunbury envied men like Darwin too, for not being burdened with a great estate or the

57. Diary entry for 23 Nov. 1845, in Bunbury 1890–91, *Middle Life,* 1:77; Burkhardt et al. 1985–99, 3: xiv, 237 n. 5. For general accounts of Darwin's reaction, see Desmond and Moore 1991, 320–23, and Browne 1995, 457–70.

58. Darwin left his species essay to be published in the event of his death, with suggestions as to potential editors. "Professor Owen w^{d} be very good," he told his wife Emma, "but I presume he w^{d} not undertake such a work." C. Darwin to E. Darwin, 5 July 1844, in Burkhardt et al. 1985–99, 3:45; Browne 1995 gives an insightful account of Darwin's connections with genteel society. On his management of the boundary between public and private, see Rudwick 1982.

12.10 Darwin's old study at Down House. Note the files by the desk; the scientific portraits (including Lyell); and the book-lined shelves of his working library reflected in the mirror. CUL Darwin archive.

expectation of continuing a political dynasty. "Nature certainly intended me for quiet times and a private station," Bunbury wrote. "I have sometimes wished that I was not an eldest son."[59]

Darwin manifested the ideals of gentlemanly science most completely in his study at Down (fig. 12.10). His procedures as a reader give a good idea of the practices characteristic of the literary men of his class and station.[60] For such readers—Herschel, Lyell, Sedgwick, and many others—books were not for ostentatious display, but tools for use. Darwin split them in half, ripped pages out of pamphlets, and never had anything rebound. His library was, in many ways, an extension of his self-presentation: simple, manly, and devoid of pretense. The vast mass of printed paper in his study attested to his single-minded identity as a gentleman of science. It was, as much as the great library at Hol-

59. C. Bunbury, diary entry for 21 Jan. 1837, in Bunbury 1890–91, *Early Life,* 224; A. C. Ramsay, diary entry for 13 Feb. 1848, IC, Ramsay Papers, KGA Ramsay 1/10, f. 26r.

60. The next few paragraphs are based upon an examination of books in Darwin's library in Cambridge, together with the letters and reading notebooks printed in Burkhardt et al. 1985–99, and Francis Darwin's unusually full recollections of his father's reading habits in Darwin 1887, 1:150–53. I am also indebted to the introduction in Di Gregorio and Gill 1990, esp. 1:xii–xvii, and the brief analysis of his readings of natural history journals in Sheets-Pyenson 1981a.

land House (fig. 5.5), a testimony to the intellectual seriousness of the reforming Whigs.

Darwin's genteel independence was clearest in his ability to pursue a major project for years without journalistic deadlines or lecturing commitments. He had access to all the great London research libraries, with those of the Athenaeum, British Museum, and Geological Society of particular relevance. These were, quite simply, not available to most readers at this time, nor did public libraries with specialist collections come into being until later in the century. Darwin also accumulated a personal library of several thousand books, but was very selective and bought only those likely to prove valuable to his work in specific ways. Money was not a major obstacle. Everything was aimed toward maximum efficiency in constructing and elaborating his theories.

Reading became a process of extraction. A small notebook recorded most of the titles Darwin had read and (starting from the opposite end) those he was planning to read.[61] Books to be read at home accumulated on a special shelf. Darwin judged books quickly, sometimes dismissively ("rubbish"), to decide whether they were worth reading thoroughly or needed only to be skimmed. The process of intense, "studied" reading was, as already seen in Sedgwick's case, developed from traditions of scholarly annotation and commentary. Working with a pencil, Darwin marked relevant or striking passages, added marginal comments, and kept a running index of page numbers at the back. The book would then go to another shelf; before being put away, he would make a rough abstract of important points on separate sheets, with numbers coded to the book. These sheets could then be filed in portfolios relating to particular subjects, to be reunited with the book itself after use. In this way, hundreds of thousands of printed pages could be processed with a minimum of effort.

Books not used for work were read in a very different way. With the exception of the daily newspaper, which Darwin read lying on a couch, all novels, histories, and other "non-scientific" works were read aloud to him, usually by his wife Emma. A self-improving reader like Hirst, say, juxtaposed and compared genres to make a meaningful whole. Darwin made sharper demarcations—some books were read for extraction, others for relaxation or amusement.

In the earlier stages of the species work this had not been the case, at least to the same extent. Initially Darwin had kept reading notes in the same secret notebooks that also included theoretical reflections on metaphysics, natural history observations, and information from friends. This was an outgrowth of his meticulous record keeping on the *Beagle*, and was broadly similar to the notebooks kept by Lyell and other naturalists. The notebooks, as a form of highly

61. These notebooks, with most titles identified, are published in Burkhardt et al. 1985–99, 4:434–573.

abbreviated diary, recorded the sequential development of his theory and his changing sense of personal identity. After moving to Down in 1842, however, Darwin's methods changed to the portfolio-based system.[62] In practical terms he was compiling an anthology of passages and references whose order could be endlessly revised. This was more flexible for locating individual facts, but less useful for thinking through major changes to his argument. However, he had written out a sketch of his theory, the structure of his planned book was becoming clear, and his health was not good. In the event of his death, another naturalist had to be able to retrace his reading if the great work was to be prepared for publication. Darwin sometimes spoke of his library as an extension of his identity, which with all its annotations would live on after he was gone.[63]

Darwin probably learned about *Vestiges* from an advertisement in the *Times* or *Athenaeum*. He made a note to read it, and copied down the publisher and price. When next in town for a Geological Society council meeting on 20 November 1844, he read the new work in the bustling, flea-infested British Museum library (fig. 12.11).[64] His techniques of close study were as typical among gentlemen of science as his questions were unusual. Darwin read scientific books almost exclusively in relation to the species issue and by the early 1840s with little interest in metaphysics and theology. His reading of *Vestiges*—the dominant reading today—was at this time an idiosyncratic one. Large parts of what moved other readers, such as the stirring account of the nebular hypothesis or the future of humanity, were quickly skimmed. Darwin approached the text not as a sweeping cosmological narrative but as a botched version of his own manuscript.

In many ways, Darwin had been scooped: here was a book advocating a natural origin for species in a framework of material causation and universal law. Only that summer he had outlined his own theory in a manuscript essay, which when copied out occupied 231 pages. The first part outlined a mechanism for species origins through individual competition and natural selection, very different from the developmental model of *Vestiges.* The second part argued for species mutability on the grounds of embryology, distribution, and classification. This was, at least in general terms, much closer to the new work. Another author had brought transmutation onto the public stage—and, what was worse, had drawn out all the religious and moral implications.[65]

62. The form of Darwin's note taking in relation to other contemporary diaries and commonplace books has as yet been little examined. For the notebooks and the shift, Barrett et al. 1987, esp. the introduction by Herbert and Kohn.

63. C. Darwin to E. Darwin, 5 July 1844, in Burkhardt et al. 1985–99, 3:43–44.

64. "Books to be read," CUL, DAR 119: 19v, in Burkhardt et al. 1985–99, 4:449; for the meeting, see Burkhardt et al. 1985–99, 3:396; for the library, Harris 1998, 154–58; for the fleas, Miller 1973, 157.

65. Desmond and Moore 1991, 316–20; Browne 1995, 445–47. On the essay and part 2 of Darwin's species work, see Ospovat 1981, esp. 87–114. The essay was published in Darwin 1909, 55–255.

12.11 The British Museum Reading Room, where Darwin read *Vestiges* and many other books. The room was famous for overcrowding, heat, a confusing catalogue, and the notorious "museum fleas." British Museum, Department of Prints and Drawings, 1938-11-12-1.

In letters to close friends, Darwin immediately distanced himself from the book, both to safeguard his priority and to show he was doing something entirely different. As he responded to Hooker, whose initial impression was more positive, "the writing & arrangement are certainly admirable, but his geology strikes me as bad, & his zoology far worse."[66] Darwin's reading notes picked up on what most naturalists were already condemning as one of its weakest features, its dependence on Macleay's circular system of classification. He would need to go into this question after all, and he filed away a note to this effect in the portfolio on classification. A defense, Darwin decided, could be built around the practical difficulties "*sound* anatomical naturalists" faced in putting organisms into groups; these difficulties made a circular arrangement unlikely.[67]

Darwin never paid down hard cash for *Vestiges*. He may have planned to buy a copy, as he marked the price in his list of books to be read, but it was temporarily unavailable when he went up to London, and he obviously decided it was not worth the money. Nor does he seem to have given the work a careful, "studied" reading for several years. Presumably he felt that the early *Vestiges*

66. C. Darwin to J. Hooker, [7 Jan. 1845], in Burkhardt et al. 1985–99, 3:108; see also 104.

67. Browne 1995, 461–62; Ospovat 1981, 113; Burkhardt et al. 1985–99, 3: 109 n. 5; CUL, DAR 205.5: 108.

editions were too full of mistakes to be a reliable storehouse of facts, and the broader philosophical issues were already familiar. Only after receiving a presentation copy from the publisher in 1847 did Darwin mark it up in his usual way. This was the sixth, heavily revised gentlemen's edition, and some of his annotations concern minor facts that might be useful for his own work.[68] However, Darwin continued to be less interested in any new information it might contain than in learning about the expectations readers would bring to his own work. Reading reactions to *Vestiges* became a way of imagining the reception of his own, unpublished essay. He put himself in the shoes of the *Vestiges* author, working out strategies for his own writing. As another of his notes put it, "The publication of the Vestiges brought out all that cd be said against the theory excellently if not too vehemently."[69]

For this purpose, following the *Vestiges* controversy was more important than reading the book itself. Darwin ignored the theological antidotes, especially after Hooker warned that Bosanquet's pamphlet was "not half so *nice*" as the work it combated. Unlike many readers, Darwin dismissed such attacks as worse than useless. However, he did read Miller's *Footprints,* Powell's *Essays, Explanations,* and several of the main critical essays with care (fig. 12.12).[70] The review that disappointed him most was Sedgwick's. Among the women at Down House, or so Darwin later recalled, it caused considerable distress, for the *odium theologicum* "much pains all one's female relations & injures the cause."[71] From his own perspective Darwin was unimpressed. He searched for things to praise ("a grand piece of argument against the mutability of species") but found the review forced, dogmatic, and predictable. As he wrote later, reassuring Lyell about the soundness of his own views, "I read it with fear & trembling, but was well pleased to find, that I had not overlooked any of the arguments, though I had put them to myself as feebly as milk & water—."[72] Darwin now knew the worst, and it was not so bad as expected.

Reading the reviews allowed Darwin to sharpen the distinction between his own theory and its competitors. *Vestiges* was reduced to a simplified version in which new organic forms sprang from old ones with no intermediate stages.

68. C. Darwin to J. D. Hooker, [18 Apr. 1847], in Burkhardt et al. 1985–99, 4:36; this copy is in the Darwin Library, with annotations transcribed in Di Gregorio and Gill 1990, 1:163–65, and partially in Di Gregorio 1984, 51.

69. From notes on Sedgwick 1850, copy in Darwin Library; transcribed in Di Gregorio and Gill 1990, 1:750.

70. Burkhardt et al. 1985–99, 4: 478, 493, 470; J. D. Hooker to C. Darwin, [28 Apr. 1845], in Burkhardt et al. 1985–99, 3:184.

71. C. Darwin to W. Carpenter, 3 Dec. [1859], in Burkhardt et al. 1985–99, 7:412–13.

72. C. Darwin to C. Lyell, 8 Oct. [1845], in Burkhardt et al. 1985–99, 3:258. On Darwin's reading of this review, see Egerton 1970.

215

Explanations of the Author of the Vestiges of Creation.

p 47 — Ansted's Geology the p 60. — ammoniacal matter in rocks below the Silurian

p 57. A Cestracion below the small fish of the Aymestry limestone. —

p 81. said (no authority) in embryo state of frog & crocodile, have the bicuneate form of the vertebrae of old Saurians

p 90 Mayer (Report of Roy Soc I) says in Emeu a ~~pouch~~ pouch form in certain organs approaching to Marsupial structure.

p 118 insists that no reason that parent form shd perish before descendants (I must allude to this)

Ly {But the history of Fossils shows that such usually is the case, as might be expected of being beaten by better adapted offspring.

12.12 Darwin, like other gentlemen of science, kept careful reading notes on books he did not buy. In this sheet of references to *Explanations,* he remarks on curious facts (including those in other works) and on strategies for composing his own theory ("I must allude to this"). CUL: DAR 205.9(2): 215.

One author on pigeons, for example, argued that development could occur only if completely new species could be found hatching in modern dovecotes; Darwin answered in the margin with an emphatic "no."[73] He was contemptuous of the attempt in the later editions of *Vestiges* to construct a genealogical system of classification—"The idea of a Fish passing into a Reptile (his idea) monstrous." The main lesson (which Darwin never really learned) was to play down progress in nature. Making notes on the sixth edition, he reminded himself not to mention terms like "higher" and "lower," but rather to speak of organisms being "more complicated." In fish, for example, the supposedly "lower" cartilaginous forms were actually "higher" in the sense that they were more affiliated to the reptiles.[74] Similarly, he commented that Sedgwick had "written against law of development higher & higher with which I have nothing to do."[75] To Darwin, the idea that the Galapagos Islands had no mammals because the gestating process had not yet progressed to perfection was absurd: "really I need not allude to such Rubbish."[76]

As Darwin had told his cousin, he was "much flattered and unflattered" (mostly the latter) whenever this "strange unphilosophical, but capitally-written book" was attributed to him.[77] He had already decided to wait before publishing, but *Vestiges* reinforced his conviction that the factual grounding of his theory would take years. This meant measuring up to the highest standards of gentlemanly science, and convincing men like Lyell and Herschel. It demanded becoming a skilled observer of variation in nature, differential survival, and problems in classification. Taking the luxury of time available only to an independently wealthy gentleman, Darwin put off publishing and began a study of barnacles that was to take another eight years. He did not want to become another "Mr. Vestiges."

The Hindrance of Pride

The covert cut and thrust found in Lyell's and Herschel's addresses, in which the real target was never named, was typical of the specialist societies. Certainly any attempt to identify the author in print, as was done in evangelical circles, would have been viewed as a distasteful violation of privacy. In correspondence

73. Annotation to Dixon 1851, 76, Darwin Library, transcribed in Di Gregorio and Gill 1990, 1: 200; see also Secord 1981, 185.

74. For Darwin's annotations, see Di Gregorio and Gill 1990, 1:163–64, and Egerton 1970, 180–81.

75. From notes on Sedgwick 1850, copy in Darwin Library; Di Gregorio and Gill 1990, 1:750.

76. Annotation to *V*6, 367, copy in Darwin Library; Di Gregorio and Gill 1990, 1:164. On the Galapagos issue, see Hodge 1972b, 148–49. An extensive literature on Darwin's difficulties with progress includes Bowler 1976, 117–29; Ospovat 1981, 211–27; Richards 1992; Ruse 1996.

77. C. D[arwin] to W. D. Fox, [24 Apr. 1845], in Burkhardt et al. 1985–99, 3:181.

and conversation, however, the gentlemen of science were eager to track down the invisible lion who was trespassing on their territory. As Charles Babbage asked Herschel:

> Have you read the "Vestiges"
> of which every body talks of
> which many are suspected
> but which nobody owns.[78]

After many months of detective work, most had hunted down the author to their satisfaction. It was too late to make a difference in the salons, where the sensation was over; but the preserves of scientific practice could still be protected. By the summer of 1845, *Vestiges* began to be labeled a "popular" work, to distinguish it from "real science" based on laboratory and field experience. This became especially urgent after the leading suspect began to stake a claim as a practical man of science though his studies of ancient sea terraces. Now, a presumed author could be criticized, in his own person, for sloppy *Vestiges*-like reasoning and bad fieldwork, but without violating the genteel codes of honor surrounding anonymity.

The stakes in developing an alternative research practice became extremely high. At the Geological Section at the Oxford British Association meeting in 1847 (fig. 12.13), the gentlemen geologists heard Chambers emphasize the parallelism of nine main marine terraces, ranging up to 386 feet above sea level, as evidence for land elevation (or oceanic subsidence) over large areas of the planet. The paper concluded by drawing on Darwin's authority for uniform elevation at many different points throughout the globe. When Chambers sat down, the elite of science took revenge. Buckland opened up the attack as session chairman, denouncing the paper as "audacious theorizing." Many of the supposed "beaches," John Phillips suggested, were produced by the erosion of softer beds, not by marine deposits. Sedgwick thought it "extremely improbable" that elevation had occurred so uniformly all over Britain, let alone Norway or America. The government geologist Henry De la Beche and the natural philosopher James David Forbes both emphasized the difficulty of determining what were really raised beaches. Lyell was willing to be more speculative, believing that the American evidence suggested that ancient beaches could sometimes be traced across entire continents. Yet he too stressed the need for caution.[79]

78. C. Babbage to J. Herschel, 8 Apr. 1845, Harry Ransom Research Center, John Herschel Papers.

79. *Athenaeum*, 3 July 1847, 714.

12.13 The Geological Section at the British Association at Oxford in 1847. *ILN*, 3 July 1847, 9–10, at 9.

Darwin, who was in the audience with his species essay in his pocket, was in an awkward position. His *Beagle* fieldwork had been called upon to buttress a practice he saw as fatally flawed, even worthless. Yet he agreed with most of Chambers's conclusions, save the possibility of a global fall in sea level. Darwin thus became the only speaker reported as having anything favorable to say, noting that the North American plains and the South American pampas were indeed almost perfectly horizontal, just as Chambers had claimed. In general, though, Darwin thought the work leading up to *Ancient Sea-Margins* displayed the same failings as *Vestiges*. As he wrote Lyell, "If he be, as I believe, the Author of the Vestiges this book for poverty of intellect is a literary curiosity."[80]

The Oxford discussion had been, by design, a not-so-indirect assault on cosmic evolutionary speculation. As one listener noted in his diary, Chambers "certainly pushed his conclusions to a most unwarrantable length, & got roughly handled on account of it." Lyell stressed in conversation afterward that he had attacked Chambers "purposely that he might see that reasonings in the

80. C. Darwin to C. Lyell, [16 June 1848], in Burkhardt et al. 1985–99, 4:152; Desmond and Moore 1991, 347–48. Darwin's presence at the session is revealed in *Athenaeum*, 3 July 1847, 714.

style of the Author of the Vestiges would not be tolerated among scientific men."[81] In contrast, the High Church *Guardian* condemned the "great sparring match" as disgraceful, Chambers being overwhelmed by sheer numbers. "The reputed authorship of the 'Vestiges of Creation,'" its reporter wrote, "which has been commonly ascribed to him, did not probably tend to diminish the violence of the storm by which he was assaulted."[82] An entire way of doing science was being rejected. On the following Sunday, Bishop Wilberforce preached in the Church of Saint Mary the Virgin on "Pride a Hindrance to True Knowledge." Nothing better illustrates the way in which Christian morality underpinned gentlemanly science than the rapturous reception this sermon received from the assembled savants. Pride, Wilberforce said, led to premature theorizing and the inability to approach female nature like a gentleman. The speculator "grows to deal boldly with nature, instead of reverently following her guidance. He seals his heart against her secret influences. He has a theory to maintain, a solution which must not be disproved . . . and once possessed of this false cypher, he reads amiss all the golden letters around him."[83]

Not everyone agreed. An anonymous report in *Chambers's Journal* lamented that Wilberforce's sermon seemed aimed "against the spread of knowledge" and charged it with being "of a discouraging nature"—a product of the clerical "Patent Safety Drag" that destroyed elementary education. The article acknowledged that the bishop spoke well, but "a certain element of masculinity was wanting in the visage." "In literature and in delivery the discourse was very masterly," the report went on to note; "from beginning to end, not one word, or look, or gesture amiss. But the impression left on the mind was, upon the whole, of a discouraging nature. Once more the drag."[84]

The British Association's leaders knew that their vision of the future was not universally shared—otherwise no one would have read *Vestiges* after the summer of 1845. Their authority, as we will see, was even more fragile in the wider sphere of debate fed by steam printing machines and science shows.

81. A. C. Ramsay, diary entry for 24 June 1847, IC, Ramsay Papers, KGA Ramsay/1/8, f. 58v.

82. *Literary Gazette,* 10 June 1848, 392.

83. Wilberforce 1847, 19.

84. [Chambers], "The Scientific Festival at Oxford," *CEJ*, 31 July 1847, 65–67.

CHAPTER THIRTEEN

Grub Street Science

It ran about town for a long time, knocking at every scientific man's door; but the answer invariably was, "No thank you, my good book; we don't want anything in your way." Sometimes it would call upon the "Fast Man" of some light Review, or the editor of some heavy Quarterly, but with no better success. It was always denied, and threatened with the police "if it did not carry its rubbish elsewhere." The *Vicissitudes of the Vestiges of Creation* would make quite a pathetic little book.

[HORACE MAYHEW], "The Book That Goes A-begging," *Punch* (11 December 1847)

THE BRITISH, IT WAS SOMETIMES SAID, were "the 'Staring nation.'"[1] Science was part of a commercial culture of exhibition, reflected in the glittering prose of journalism, in lecture demonstrations, panoramas, museums—and in the evolutionary narrative of *Vestiges.* The revolution in communication was transforming opportunities for making money from the display of knowledge. Those being paid to do science defined their status by choosing among a vast array of venues. Money could be made through authorship, editing, reviewing, specimen dealing, industrial consulting, instrument making, museum curating, lecturing, and showmanship.

Together these opportunities created what will here be called commercial science. This is a more useful term than "professional science," which has positive connotations that were largely absent from the paid pursuit of science in the first half of the nineteenth century; for a young person of good family, the career path it involved could be dangerously close to a "job" in both senses of the word. A more attractive possibility is to use the term "Low Science," based on an analogy with Robert Darnton's work on the French Revolution; but this

1. Bulwer-Lytton, *The Siamese Twins* (1831), quoted in Altick 1978, 1.

creates a misleading dichotomy with "High Science" and is too easily elided with the dismissive catch-all "popular science."[2]

Since the eighteenth century, lecturing had been a mainstay for those who pursued science for pay. By the 1840s, a few venues dominated the metropolitan scene, such as the Royal Institution. A teaching position at King's or University College, although dismally paid, could also anchor a career. The hospital schools provided opportunities and a few independent medical schools struggled on, although many had failed in the previous decade.[3] And, as the next chapter will show, the gentlemanly ideals dominant since the seventeenth century were beginning to be challenged by state-funded institutions and innovations in education. Tom Hirst's later career is a case in point; on returning from Marburg in 1853, he took over a post at Queenwood College in Hampshire, teaching boys geometry. It was a fitting symbol of the growing importance of science that this novel social experiment—a technical college for farming, engineering, and surveying—should occupy what had been the failed Harmony Hall of the Owenites.[4]

Writing remained by far the most significant way of making money from science, and the most important changes involved new roles for authorship. Nigel Cross has discussed the emergence of the "common writer," who produced magazine stories and novels for the circulating libraries. The rewards were appalling—rarely enough to live on—but such authors were beginning to occupy a more secure social status. One effect of increased security was to marginalize women within certain arenas of literary production. With the burgeoning culture of printing and publishing in London, science writing could be combined with lecturing, museum work, or (for the fortunate few) a small independent income. The economics of authorship in genres other than the novel has been little examined, but the number of hungry writers on science is striking.[5] As Richard Owen said, changing circumstances of scientific production produced real hardship. "All these evolutions are attended with concomitant casualties . . .," he told the charitable Royal Literary Fund "and require corresponding expansions of means of relief."[6]

2. Sheets-Pyenson 1985. My use of the term "commercial science" derives from the work of Brock (1980) and Sheets-Pyenson (1981b) on commercial science periodicals. For paid science in the early nineteenth century, see Allen 1985 and Porter 1978a.

3. Desmond 1989, 382–87; Hays 1983.

4. "Queenwood College Revisited," essay 17 in Brock 1996.

5. Cross 1985 is outstanding on the career of authorship, but focuses largely on fiction and poetry. On the role of women as authors in the second half of the century, see Tuchman 1989 on novelists and several of the essays in Gates and Shteir 1997, esp. those of Lightman and Gates.

6. R. Owen, Royal Literary Fund Address, 1859, in Cross 1985, 58.

A nostalgic obsession with gentility has blinded historians to the fact that for most practitioners science was a demanding, even desperate way of making money. Owen was not the only scientific man to be granted a Civil List pension on the grounds that this would prevent him from becoming a slave to the publishers.[7] Real hacks—and there were far more of them than salaried men like Owen—might make only a few pounds from a title selling thousands of copies, if the copyright been sold outright. Celebrated authors and lecturers, too old to work, ended their days in abject poverty and starvation. Hirst, like many who occupied teaching posts, suffered a nervous collapse, while the government geologist Andrew Ramsay lost an eye and Thomas Henry Huxley had a breakdown. For some, it was too much: the trilobite expert John William Salter, out of work and insane, jumped off a steamer into the Thames and drowned; the geological lecturer George Richardson went bankrupt and ended his own life; the country's leading writer on veterinary science, William Youatt, lost his job at the Zoological Gardens and committed suicide by eating a bun sprinkled with prussic acid.[8]

The commercial middlemen of science were among the most influential readers of *Vestiges*. The book offered occasions for paid reviews and public lectures; for stressing the value of technical expertise; and for raising public interest through controversy. It became a touchstone for defining the role of the journalist, lecturer, museum curator, and government expert. Such definitions concerned the polity of knowledge, the relations between expertise and commerce, and the acceptable range of public involvement in scientific decision making. For anyone pursuing science for cash, the issue came down to the price of knowledge in the marketplace. To understand why *Vestiges* raised these issues so acutely, we need first to return to the text.

The Exhibitionary Text

Vestiges incorporated the tensions surrounding the pursuit of knowledge for pay. The evolutionary narrative—drawing on the novel, the encyclopedia, and mass-circulation journalism—was created from kinds of writing most closely associated with new kinds of authorial role. A rhetoric of spectacular display was built into the text, which addressed its readership through a common experience of science as visible, clear, and accessible.

One way of understanding this is to see how the book, even without illustrations, drew upon a powerful and widely distributed set of visual images of

7. W. Buckland to R. Peel, 12 Jan. 1842, in Rupke 1994, 52.

8. On Hirst, Brock and MacLeod 1980b, 24; on Ramsay, Geikie 1895, 345; on Huxley, Desmond 1997, 27–28; on Salter, Secord 1985; on Richardson, Curwen 1940, 224; and on Youatt, Crosland 1898, 321.

natural wonders. We are used to seeing the Victorian world through the printed word, but the visual field of contemporaries was dominated by shows, exhibitions, and pictorial representations.[9] *Vestiges* was read not only as a book, but *seen* as a museum of creation. The opening paragraph could rely upon an understanding of the solar system as "familiar" because Nichol and other lecturers had toured to paying audiences for decades. The first sentence—with its reference to the sun circled by planets circled by satellite moons—is a verbal orrery, calling up the central image of astronomical display. The reader's relation to the culture of display is invoked most directly in connection with the calculating engine, the principles of which Babbage had demonstrated in his house during his unsuccessful quest for further state funding. By the time that *Vestiges* appeared, the engine had become the centerpiece of the scientific instrument gallery at King's College on the Strand (fig. 13.1).[10]

The rest of *Vestiges* could be read as a guided tour of science spectaculars in London, Liverpool, Manchester, and other cities.[11] Visitors to these shows could see monstrous embryos, electric telegraphs, phrenological casts, scenic panoramas, fossil bones, tattooed men, the remains of lost civilizations, and the Eureka, a machine that could compose Latin verse. In the "Caverns" of the Regent's Park Colosseum, visitors could wonder at a sequence of four murals illustrating "The Geological Revolutions of the Earth," including the Creation, the age of the iguanodon and megalosaurus, the Garden of Eden, and the Deluge.[12] One could examine fossilized bones of extinct creatures at the British Museum and the Royal College of Surgeons (fig. 13.2). In the wake of the creatures that had crawled out of Andrew Crosse's electrical experiments in 1837, there was a particular fascination with gigantic images of cheese mites, and machines capable of generating living beings. In Pall Mall, visitors could pay a shilling to see the "Eccaleobion" (fig. 13.3), which exhibited "[t]he wonderful process of the Development of Organization" through machinery, "one of the most magnificent spectacles of infinite power and wisdom which the human mind can contemplate."[13] The facts of sex, embryological gestation, and

9. The importance of exhibitionary culture to the sciences is stressed in Altick 1978, Morus 1998, Morus, Schaffer, and Secord 1992, Richards 1991, and Winter 1998a. The essays in Christ and Jordan 1995 offer a sampling of recent work on Victorian visual culture.

10. On the display of Babbage's engine, see Schaffer 1996. For the move to King's, see *ILN*, 1 July 1843, 5; and Hyman 1981, 192–93.

11. Altick 1978 offers a tour around the main sites in London.

12. "The Gallery of Natural Magic, an Entirely New Exhibition of the Colosseum, Regent's Park," Bodleian Library, John Johnson Collection, Scientific Instruments, box 1. Gideon Mantell had had at least one large scene, "Reptiles Restored," on show in his museum in Brighton in the 1830s; Dean 1999, 123.

13. Altick 1978, 371–72, discusses this and other mechanical hatcheries.

13.1 From July 1843 Babbage's calculating engine was on display in the gallery of scientific instruments at King's College on the Strand. In this woodcut, the engine—just to the left of center—is a principal object of attention. "Prince Albert Opening George the Third's Museum, King's College," *ILN,* 1 July 1843, 5.

generation—so central to *Vestiges'* argument—were also on show in the West End, although these had to be discreet so as not to offend public morals. Dr. Kahn's Anatomical Museum at Piccadilly had eight hundred "Superbly Executed" models, open at separate times to women and men.[14] Some of the most popular exhibitions brought peoples from Africa, the Arctic, and North America under the gaze of the curious. To those who had seen them, *Vestiges* called up such imperial experiences with dramatic immediacy: Hirst's skeptical friend Francis Booth was among the thousands whose first sight of the Bushmen "narrowed that gulf" between monkey and man "amazingly" (fig. 10.5).[15]

The most important place for contemplating that "gulf" was the Royal Zoological Gardens in Regent's Park. In 1843 Queen Victoria had watched in horrified fascination as the orangutan, Jenny, made and drank a cup of tea: "He

14. Mason 1994a, 190–91.

15. T. A. Hirst, journal entry for 24 Nov. 1849, Brock and MacLeod 1980a, f. 547; on the Bushmen, Altick 1978, 279–81.

THE ILLUSTRATED LONDON NEWS

No. 179.—Vol. VII. FOR THE WEEK ENDING SATURDAY, OCTOBER 4, 1845. [Sixpence.

ROSAS AND BUENOS AYRES.

From the mingling of the Spanish and Indian race there has arisen on the wide Plains and Pampas of the Rio de Plata, a nomad and half savage people, possessing "much cattle," and being, as horseman and herdsman, without their match in the world—the Guachos. They are proud of their wild freedom and rude equality; bold to rashness; cunning, as most savages are, used to hardship and endurance; and, never sparing themselves, they are as destitute of the "quality of mercy" to others. To this race belongs Don Juan Manuel de Rosas, the almost absolute Master of the Argentine Republic. His reign has continued now for some years.

Among the races of Spanish blood, Constitutional Government seems an impossibility; all is either the violence of one strong hand and will, or the confusion of anarchy. Buenos Ayres under Rosas is an example of the former; to the latter, the Republic of Mexico seems to be rapidly declining. At the present moment more than usual interest attaches to Rosas and his policy since his blockade of Monte Video has brought him nto collision with England and France, who, uniting, have seized on the blockading force by which Rosas attempted to stop the navigation of the Rio Plata, and impede the commerce of the vast regions on its banks. The trade of Monte Video is a growing one, and civilised nations have an interest in not permitting it to be crushed in its infancy by the savage despot. Rosas is now at open variance with England and France As there must be something in the man who thus ventures to "pluck lions by the beard," some notice of his career may not be uninteresting at the present juncture. It will show the sort of ruler with whom we have to deal.

Rosas exhibits in his conduct, all the peculiarities of the race from which he has sprung; they have made him popular among the Guachos, and to his untiring energy and courage, he owes his position as ruler. Among a nation of horsemen, the boldest rider is sure to be honoured, and in this accomplishment Rosas was without a peer. By surpassing the rest in strength and activity, he first gained an ascendancy over his people; he kept it by the force of a strong mind that knew how to flatter their feeling of equality, and yet enforce regulations checking their disorders, particularly the crime of assassination; he prohibited the carrying the knife in the girdle on the Sabbath—the day of festivity, and consequent broils—under the penalty of severe corporal punishment. It is said that once, in forgetfulness, he violated his own order, and insisted on undergoing the punishment for the violation, to the intense delight of the Guachos. But his exertions in repressing assassination did not proceed from any aversion to bloodshed; when engaged in war, or inflicting political vengeance, he is a perfect demon—cold, ruel, and atrociocus, massacreing in mass with the utmost indifference. He is enormously wealthy, possessing an estate covering about seventy-four square miles, and feeding three thousand head of cattle. This territory he ruled like a Prince, commanding a kind of army of his own, raised from among his admiring Guachos, and trained by himself, to serve as a force against the Indian tribes of the Pampas, who made frequent incursions on the grazing grounds of the settled districts, and were as skilful reivers and cattle stealers as ever were our Scotch borderers; but the Rob Roys of the Pampas found in Rosas an enemy as cunning as themselves, stronger by discipline, bold in the conflict, and, in the victory, cruel and unsparing; all prisoners were massacred. On one occasion, after a skirmish, eighty were shot at once, and their bodies flung into a pit together. This was the school in which were developed those military talents he has since exhibited in a wider sphere. In conversation, he is said to be grave and earnest; in his dealings, cunning, showing himself as great a master of dissimulation and as treacherous as Ferdinand the Seventh; as of that most legitimate Monarch, his smile is reported to be of evil omen, and he on whom it falls most blandly, frequently finds himself in a dungeon, or has to be sought in a grave.

In a country where the Governments are as often convulsed by revolutions as the soil by earthquakes, there were opportunities enough for such a man to rise, and he was not slow in taking advantage of them. At the close of the war in the Banda Oriental between Buenos Ayres and Brazil, General Lavalle, the commander of the Buenos Ayrean troops, made an attempt to seize the principal authority in these republics—always the prize of the most successful soldier. Dorrego, the President, applied to Rosas, then master of a considerable force, and celebrated by his wild warfare with the Indians, for assistance. It was given, but as yet Lavalle was too strong; Rosas was defeated in an engagement, and Dorrego, having been taken prisoner, was shot, according to the invariable Spanish practice in such cases. But one defeat was not destruction to a man with such talents for a guerilla warfare as Rosas; he

THE HUNTERIAN MUSEUM, AT THE ROYAL COLLEGE OF SURGEONS.—SEE NEXT PAGE.

[COUNTRY EDITION.]

13.2 *Opposite:* Front page of an issue of the *Illustrated London News,* with a woodcut of the Hunterian Museum of the Royal College of Surgeons, showing spectacular specimens of South American fossil mammals; the accompanying article discusses General Rosas, the "savage despot" of the Argentine Republic. *ILN,* 4 Oct. 1845, 209.

Extraordinary Novelty!!!

EXHIBITION

OF THE

ECCALEOBION,

WHEREBY

LIFE

IN

COUNTLESS THOUSANDS

OF ANIMAL BEINGS,

From a WREN to an EAGLE,

IS PRODUCED BY MACHINERY.

THIS EXTRAORDINARY MACHINE

Which in the ungenial climate of England, realizes the

Greatest Wonder of Egypt!

Both in the ancient and modern world has already given existence to

Many Thousand Chickens & other Birds

It stands upon a table, and animal life in its most perfect and healthful state is made to burst into existence in the presence of the visitors. It is capable of containing Two Thousand Eggs, and of hatching about One Hundred daily, or Forty Thousand per annum.

The wonderful process of the

Developement of Organization

As discoverable by the assistance of powerful instruments, and the no less striking and interesting phenomena that accompanies it, renders this Exhibition one of the most magnificent spectacles of infinite power and wisdom which the human mind can comtemplate, attractive alike to every class in society.

At 121, PALL MALL

(Opposite the Opera Colonnade).

A Pamphlet, explanatory and historical, and treating of the mystery of Life and its co-relatives Mind and Instinct, as developed by

THE ECCALEOBION

ONE SHILLING.

Open from Ten until Six o'Clock daily.

ADMISSION ONE SHILLING.

W. S. Johnson, "Nassau Steam Press," Nassau Street, Soho.

13.3 "Extraordinary Novelty!!! Exhibition of the Eccaleobion." Printed handbill, late 1830s.

is frightful & painfully and disagreeably human."[16] The eponymous hero of Henry Cockton's comic novel *Valentine Vox* (1840) was a ventriloquist who secretly projected his voice to disrupt ordinary social situations; at the zoo, Vox made the orangutan speak in the character of a philosopher (fig. 13.4). Jenny died soon after the novel appeared, but the arrival of a chimpanzee from Sierra Leone made the issue timely again, so that crowds thronged the zoo just as the *Vestiges* sensation reached its peak in the spring of 1845. The *Pictorial Times* featured a wood engraving (fig. 13.5) and warned against seeing any approach to

16. Blunt 1976, 38; on the zoo's great apes, see Desmond 1989, 290–94; Desmond and Moore 1991, 243–44.

the human in her physical or mental character. Unlike chimps previously on display, docile because raised in captivity, the mature animal illustrated "the true savage character," "unreasoning and headlong." She fought with her keeper, forced the bars on her cage, and even injured herself.[17] The similarities with humans were deeply disturbing. The *Edinburgh*'s review of *Vestiges* ironically hailed this "glorious specimen," calling it a "satire on humanity" that suggested that the gap between humans and monkeys might be narrowing, but not in the direction usually assumed: "if monkeys be not passing into men, it is plain there are men in plenty who are passing into monkeys."[18]

Some of these would-be monkeys used *Vestiges* as a guidebook to what they saw at the metropolitan shows. In the summer of 1846 Florence Nightingale and a friend visited the museum at the Royal College of Surgeons, where they traced anatomical connections between different species of the flightless birds of New Zealand. Owen had described these strange birds without drawing any evolutionary implications, but Nightingale took a bolder view. The newly discovered fossil skeleton of the giant moa, larger than an ostrich, had the same "two little bits of wings" as its diminutive modern representative, the apteris, which was about the size of a sparrow. As she told her cousin at school, "the thing which was most curious of all, was to see how the species ran into one another, as *Vestiges* would have it."[19]

As the illustrations suggest, the 1840s witnessed the first great flowering of the graphic newspaper press.[20] The new journals carried on the traditions of the *Mirror of Literature* (f. 1823), the *Penny Magazine* (f. 1832), and the *Saturday Magazine* (f. 1832) in new directions. Titles that started publication ranged from middle-class humor magazines such as *Punch* (f. 1841) to penny working-class weeklies such as *Reynolds's Miscellany* (f. 1846), linking an increasingly characteristic "popular" language for science to the conventions of romantic melodrama. In science coverage, the most important newcomer was Herbert Ingram's *Illustrated London News*, founded in 1842. Sold at sixpence, it soon had a circulation of forty-five thousand, mainly among upper-middle-class families with children.[21] From 1843 it was challenged by the *Pictorial Times*, which at the same price offered a more critical (and protectionist) political line and editorial

17. "The Chimpanzee, Regent's Park Gardens," *PT*, 20 Aug. 1845, 140.

18. [Sedgwick], "Natural History of Creation," *Edinburgh Review*, July 1845, 82, 1–85, at 65; see also "Lyell's Travels in America," *English Review* 4 (Oct. 1845): 72–99, at 74.

19. F. Nightingale to W. Shore Smith, 19 July [1846], BL Add. mss. 46176, f. 15; Goldie 1983, fiche 2, A9, 283.

20. The best general survey is Fox 1988; Anderson 1991 provides an important analysis of some of the leading titles, and Sinnema 1998 discusses the most famous of the papers (the *Illustrated London News*) in connection with issues of national identity.

21. "The Newspaper and Periodical Press of London," *London Journal*, 19 July 1845, 328.

13.4 “Aesop Eclipsed.” The speaking orangutan at the Zoological Gardens. Illustration by Thomas Onwhyn, from Cockton 1844a, facing p. 380.

13.5 To stress her inhuman ferocity, this chimpanzee (in contrast to previous zoo favorites) was not named in the accompanying article. “The Chimpanzee at the Royal Zoological Gardens,” *PT*, 3 Aug. 1845, 140.

comments and illustrations to match. With a circulation of ten thousand, the *Pictorial Times* lost money and ceased publication in 1848. When a ton of back issues was sold off—the largest remainder ever to hit the market—it reached an even wider audience on the streets.[22]

The production of the new magazines was itself a commercial spectacle. Two hundred sandwich men marched down London's streets to advertise the first number of the *Illustrated London News.*[23] Graphic portrayal of news events was made possible by the railways, telegraph, techniques for printing woodcuts of unprecedented size and quality, and the demand for pictures engendered by panoramas and shows. Much of the vast output of periodical production related to science and engineering: comets, zoological novelties, curious cacti, British Association celebrities, the latest technological developments. George Cruikshank's *Comic Almanack* for 1846 nervously joked that authorship might soon go on display as a mechanized form of commodity production, with a Eureka-like "New Magazine Machine" producing essays, stories, and reviews to order (fig. 13.6).

Places of display were major features of the mass-circulation illustrated newspapers, but are referred to only tangentially in *Vestiges.* A philosophic overview of creation could not be associated with Leicester Square or Piccadilly, the geographical heart of the London shows. Babbage's engine, one of the few public demonstrations explicitly mentioned in the text, was the prestige instrument of a celebrated natural philosopher. Street shows, panoramas, menageries, tableaux vivants, and similar displays were potentially out of control and too uncertain in status. Similarly, when specialist naturalists like Owen did draw on material from such places in their works—which was often—they rarely mentioned the source in print. Scientific examinations of the rare, exotic, and monstrous had to be distinguished from the public's prying gaze.

Exhibitions of animal sagacity, which did feature in the discussion of mental phenomena in *Vestiges,* were especially controversial. The book refers to the trained dogs of Adrien Léonard, which "play at dominoes, and with so much skill as to triumph over biped opponents, whining if the adversary play a wrong piece, or if they themselves be deficient in a right one."[24] Léonard's pointers, Braque and Philax, were seen by thousands in London during the summer of 1841 and were carefully described in a pamphlet. Gideon Mantell found they played dominoes "most admirably." They impressed a skeptical *Athenaeum* reporter, whose account led Elizabeth Barrett to teach her lapdog Flush arith-

22. "The Newspaper and Periodical Press of London," *London Journal,* 30 Aug. 1845, 431; Mayhew 1861–62, 1:289, mentions the remaindering.

23. *ILN,* 14 May 1842, 16.

24. *V*1, 338; see Léonard 1842.

13.6 The death of the author in the application of mechanism to periodical journalism: "the handle is turned, and the fountain-pens immediately begin to write articles upon everything." George Cruikshank, etching for *The Comic Almanack,* 1846, in Cruikshank 1870, 2:120.

metic and spelling.[25] Léonard's findings were of a kind used by Darwin and other researchers into animal behavior, but they were also easily mocked as cheap showmanship.

It was partly to avoid the taint of commercial spectacle that the first nine editions of *Vestiges* had no illustrations, for depicting the full range of phenomena would have brought home the book's character as a cornucopia of natural wonders. (The woodcuts for the tenth edition maintained some sense of discipline by being confined to anatomy, physiology, and palaeontology.) Even without illustrations, however, the vivid analogies and skillfully managed rhetoric could be used to raise the possibility that this was not a high-minded treatise, but a cheap trick. From this perspective the book was *too* well written: this

25. Léonard 1842; G. Mantell, diary entry for 31 July 1841, in Curwen 1940, 146; "Our Weekly Gossip," *Athenaeum,* 24 July 1841, 560; "Educability of Animals," *CEJ,* 16 Apr. 1842, 97–98, at 97; E. Barrett to M. R. Mitford, [27–28] Nov. 1843, in Kelley et al. 1984–92, 8:65.

was style without substance, shiny brilliance without the solidity of true metal. Because it drew on the showy, spectacular, and extraordinary, *Vestiges* could be accused of appealing to the same audiences that flocked to the gaudy displays of the West End. Any hint (as *Blackwood's* put it) that P. T. Barnum might have "the *Vestiges of Creation* in his eye" had to be avoided.[26]

Preachers of Science and Medicine

The consumption of popularized science, as the ubiquitous analogy went, was like eating. For its critics, bad science posed a threat to the nation's spiritual body as potent as that which flour laced with chalk (or worse) posed to its physical health—a much debated issue in the mid-1840s. Hostile evangelicals compared *Vestiges* with quack nostrums and adulterated food. "It is full of insidious poison—," the celebrated science lecturer John Murray wrote, "a pitfall for the unwary youth and the superficial reader—because tinselled with the brocade of *ad captandum* language."[27] "Infidelity," the Reverend Stowell had warned at Exeter Hall, "is no longer administered in the suffocating doses of Socialism, but in homeopathic globules, which poison, but do not alarm."[28]

This theme was widely canvassed in the reviews. The article in the Free Church–sponsored *North British* praised Whewell's *Indications* for its "plain and even dry" style, in contrast to the "flowery and seductive" *Vestiges*. The Edinburgh *Torch* compared the unknown author—"some reputed Cornish knight" (i.e., Vyvyan)—to an "insignificant gnat" sucking up the discharge from festering wounds, "and then pouring it out again a sweetened, but pestilent honey." That the public had "bolted and swallowed" so many editions of so disgusting a thing was a sign of "the childlike tastes of the foolish multitude."[29] The cheap editions only made things worse. As *Fraser's Magazine* complained, they were designed for those "who swallow without discrimination the poison of any nostrum presented to them by the quack in the name of science; and believe, with implicit credulity, any deception practiced upon them by the juggler."[30]

The accessible language of these attacks, however, points to paradoxes in the enterprise of providing science for the people. As the naturalist Edward Forbes advised a would-be lecturer, presentations needed to be "as *striking* as possible,

26. [W. E. Aytoun], "Revelations of a Showman," *Blackwood's Edinburgh Magazine* 77 (Feb. 1855): 187–201, at 189.

27. John Murray, "The *Vestiges*," *Mining Journal*, 5 Feb. 1848, 67. Radway 1986 discusses the continuing power of the analogy with eating.

28. Stowell 1849, 153.

29. "The Delusions of the Day," *Torch* 1 (10 Jan. 1846): 21–23, at 23; [D. Brewster], "Whewell's *Indications of the Creator*," *North British Review* 4 (Feb. 1846): 364–79, at 364.

30. "Geology versus Developement," *Fraser's Magazine* 42 (Oct. 1850): 355–72, at 359.

that is, as *remarkable to that part of the audience* (more than half) *which consists of those who know nothing about the science of the matter, as to those who know and understand.*"[31] Reviews targeted, just as *Vestiges* did, audiences that had no familiarity with practical scientific work. Supporters of the book complained that its assailants used verbal sophistry to appeal to popular prejudice. The Unitarian *Prospective* noted "that some are desirous of securing immunity to their own speculations, by a cheap display of eloquent zeal against all who dare to go beyond their measure."[32] For the reviewer in *Fraser's*, on the other hand, those who supplied science to the multitude had a responsibility "to supply it pure; unadulterated by the palatable and sparkling, but poisonous mixtures, which have been of late but too successfully mingled with it."[33] Defining "popular science" meant defining what was meant by "the people," what was meant by "science," and what their relations should be.

Vestiges posed this problem in an acute way for the flamboyant, one-eyed Robert Knox, who had introduced a generation of students at Edinburgh to philosophical anatomy. After his disgrace in the notorious Burke and Hare resurrectionist scandal, he reemerged in London as an itinerant lecturer on anatomy, physiology, and racial science.[34] Knox dismissed *Vestiges* as the work of a dull Saxon, "a popular writer, an adept at plagiarism and at arrangement":

> Progress towards what? The idea has been thrown out by a utilitarian mind, an unconscious disciple of Paley; a nibbler at philosophy, who scarcely understood the thing. He wished to give *a reason* for everything: a Saxon, no doubt, and so he thrusts himself unwittingly into the councils of the Great First Cause. And so it ever is with the half-educated; the utterly ignorant, the *canaille,* flee at once in all arguments to a first cause. . . . With them all is mystery, a *lusus naturæ,* a visitation of Providence, a direct interference; with them the Deity is ever present; he has no power to bestow secondary laws on matter, with them attraction has no real meaning; every animal required a distinct creation. A material Jove still thunders.[35]

The success of *Vestiges* brought home the appalling nature of his own fall from grace. As a "savage radical" and "moral anatomist," Knox believed in the iron-clad operation of natural law and a theory of descent. At the same time, he denied organic transmutation as usually understood. New species arose through

31. E. Forbes to J. Percy, [Jan. 1845], in Wilson and Geikie 1861, 384.

32. [F. Newman], "Vestiges of the Natural History of Creation," *Prospective Review* 2 (Jan. 1846): 33–44, at 37.

33. "Geology versus Developement," 355.

34. Rehbock 1983, 31–55; Desmond 1989, esp. 388–89; Richards 1989b.

35. Knox 1850, 170, 423.

sudden changes in the "generic animal" or embryo, and not by gradual, progressive changes in the mature organism.[36] From the early 1840s, Knox was in the galling position of lecturing to audiences that confused his sophisticated theories with those of *Vestiges.* Embittered and struggling, he recognized that the forces that had hunted him down ("furious in support of orthodoxy and Cuvier") were now his allies in exposing the half-baked puerilities of "a pseudo-philosophical work."[37]

Science lecturers often had passionate views on religion; many were evangelicals and followed a lecture circuit much as Nonconformist preachers might have done. Unlike freethinkers, who tended to advocate science rather than doing it, evangelicals were often keen practitioners, earning badly needed cash while expressing their beliefs. For example, at a pitch in Whitechapel, a working-class Methodist naturalist offered penny views through his microscope to impart moral lessons about divine creation and the distinctions between man and beast. Creation was "Adam's library," which God had bidden man to read.[38]

The lecturing calendar rotated around Easter week, when the theaters were open only for science shows. The distance between pulpit and platform was short, and the audiences were often identical. One of the perennially popular performers was John Wallis. His astronomical lectures were undertaken on a contract basis, with a fee—often as much as forty guineas—charged for a course.[39] It was a debilitating schedule, which he later abandoned despite the need to feed a large family. "I know that God is holy and just, and good, in all his dispensations," he lamented, but "my worn out constitution cannot rally."[40] In his prime Wallis created stage spectaculars that displayed God's creation as "sacred evidence of His universal presence." Wallis confessed his awkwardness away from the podium, but felt angry enough to write a pamphlet against *Vestiges* and the nebular hypothesis as "sophistical speculations." This "bold and specious conjecture," Wallis argued, had been favorably received in scientific quarters, and so he focused on the mathematical and physical problems of the Firemist. Truth was his greatest weapon against the infidel, whose arguments led only to negations, difficulties, and doubts that made understanding causation in nature impossible. "Nothing that is true," Wallis said, "can possibly be of any service to atheism."[41] This pamphlet—which seems to have had a limited

36. Richards 1989b reassesses Knox's turn toward racial science in the context of his professional decline. Lonsdale 1870 surveys his disastrous career.

37. Knox 1850, 170–71.

38. Richmond 1843, 29. See Mayhew 1861–62, 3:83–88, on street microscopy.

39. The London lecturing scene is best described in Hays 1983, 91–119 (for Wallis, see 99).

40. J. Wallis to E. W. Brayley, 19 Apr. 1852, quoted in E. W. Brayley to Royal Literary Fund, 31 May 1852, archives of the Royal Literary Fund, case file no. 1297.

41. Wallis 1845, 6–7, 28.

circulation—was his only publication, and his views spread primarily though his lectures.

No inhibitions about appearing in print troubled the chemical lecturer John Murray, who wrote over a hundred books and pamphlets on subjects ranging from miners' lamps to plant morphology. Murray, a Scotsman, toured the country from his base in Hull and had spoken for decades in London at the Surrey Institution on Blackfriars Road. Even before *Vestiges* appeared, Murray had refuted all the doctrines of "that silly and frivolous romance" both in print and on the platform. He had railed against the nebular hypothesis, progressive geology, Lamarckian transformism, and the spontaneous generation of life through electricity. In February 1845, he wrote (probably following the Reverend Hume's example) to leading men of science, including Ansted, Lyell, and Nichol.[42] He then drew upon their replies in letters to the London-based weekly *Mining Journal. Vestiges,* he warned, was "full of insidious poison—a pitfall for the unwary youth and the superficial reader." He objected most of all to the way that *Vestiges* (like other geological works) meddled with cosmology and creation, whose true basis was scriptural. "For all this romance," he concluded, parodying the work's final sentence, "there is not a tittle of proof."[43]

The most sustained attack from the showmen of science was from Thomas Monck Mason. An impoverished Irish gentleman in his forties, with a degree from Trinity College Dublin, Mason had a checkered career. He had lost a huge fortune as lessee of the Opera House in Covent Garden—the case would be in Chancery for twenty-three years—had composed operas, and was an outstanding flautist.[44] By the mid-1840s he was best known as a balloonist, whose latest aeronautical invention was displayed at the Adelaide Gallery (fig. 13.7).[45] His response to *Vestiges* was a 182-page book, *Creation by the Immediate Agency of God, as Opposed to Creation by Natural Law,* which appeared in November 1845. Philosophy and the sciences, Mason lamented, were being used to disparage the "infallible authority" of the Bible, and so his book tore apart the factual evidence of *Vestiges.* Like many Irish Protestants, Mason harbored reservations about geological claims concerning the age of the earth and the history of creation.[46] He did not, however, base his case on such controversial grounds;

42. D. Ansted to J. Murray, 21 Feb. 1845, EUL Gen. 1971/2/1; C. Lyell to J. Murray, 24 Feb. 1845, EUL Gen. 1971/2/30; J. P. Nichol to J. Murray, 10 Mar. 1845, Gen. 1971/2/33. On Murray's opposition to the Crosse experiments in spontaneous generation, see Murray 1837 and Secord 1989b, 370; for his views on plant morphology, see Murray 1845.

43. J. Murray, "Geology," *Mining Journal,* 4 Dec. 1847, 578; J. Murray, "The *Vestiges,*" *Mining Journal,* 5 Feb. 1848, 67.

44. Thomas Monck Mason, case file 1715, archives of the Royal Literary Fund.

45. On the Adelaide Gallery, see Morus 1998, 75–98; Altick 1978, 377–82.

46. Mason 1845, 14–15.

13.7 The best place to see the work of Thomas Monck Mason, balloonist, failed entrepreneur, and evangelical theologian, was in displays of his inventions in the Adelaide Gallery at the Lowther Arcade off the Strand. The gallery combined orchestral entertainment and tableaux vivants (just visible on the far stage) with exhibits of science and technology. The gallery's union of "amusement" with "instruction" became increasingly characteristic of commercial science during the mid-1840s. Mason's balloon is perhaps the large domed object on the ground floor. "The Adelaide Gallery, Easter Week," *PT,* 13 Jan. 1844, 29.

for the sake of argument, his book accepted the premises of geology and proceeded to show that they failed to support a developmental law. Mason also set to one side the most compelling argument of all from his own perspective, the total incapacity of natural laws to explain the immortality of the soul.[47] Nonconformists and Low Church Anglicans alike praised Mason's book, which became the standard source of ammunition for evangelicals engaged in pastoral work, missionary visits, and sermons.

In the household of Emily and Philip Gosse of the Plymouth Brethren in suburban Islington, science writing was just one of the aspects of everyday life bound to evangelical religion. Philip Gosse had decided "to look to literature for a livelihood" in 1843, although he sometimes lectured at the main London venues.[48] For the most part Gosse adopted indirect tactics against skeptics,

47. Mason 1845, 7.

48. Freeman and Wertheimer 1980, 3. In the context of a thorough bibliography, this book gives a wealth of information on the economics of natural history authorship. See also Gosse 1890.

writing books such as *The Ocean* (1845) for the Religious Tract Society and *The Aquarium* (1854) for John Van Voorst to convey enthusiasm for God's creation, to replace novels as appropriate reading for the middle classes, and to support his family. For Gosse, science was one of the few allowable genres of writing that would also sell. He confronted challenges to Scripture in *Omphalos: An Attempt to Untie the Geological Knot,* published by Van Voorst in 1857. This argued what many evangelicals had long believed, that geologists had made a terrible mistake in assuming that fossils implied a long history for the world. The *Vestiges* author was singled out as beyond the pale. "Coolly bowing aside His authority," Gosse remarked, "this writer has hatched a scheme, by which the immediate ancestor of Adam was a Chimpanzee, and his remote ancestor a Maggot!"[49] Gosse had expected to do well from *Omphalos,* for he had four thousand copies printed, priced at 10s. 6d. each. Some three-quarters had to be pulped.[50]

Other evangelicals engaged in commercial natural history also spoke out. The veterinarian William Youatt defended a scriptural view of creation in his celebrated manuals on domestic animals; the Reverend Edmund Saul Dixon, an ill-paid Anglican rector in Norfolk, argued against transmutation in his practical handbooks on poultry and pigeon breeding.[51] The Quaker Edward Newman, printer of Wallis's pamphlet and publisher of the *Zoologist* and *Phytologist,* condemned *Vestiges* in both. These were monthly magazines costing seven pence a copy and capitalizing on the middle-class fashion for natural history (fig. 13.8). Although the *Phytologist,* unlike the *Zoologist,* was never financially successful, they became the first long-lasting commercial periodicals devoted to specific branches of natural history.[52]

Newman knew his market well, for he had first worked as a wool dealer (like Hirst's father) and then as the owner of a rope-making establishment. A deeply religious man, he had long denied claims that natural history led to atheism. Using the familiar metaphors of consumption, he lamented that *Vestiges* slipped poison into a food that was otherwise scripturally sound, "wholesome and delicious." As he wrote in the *Zoologist,* its references to divine action were a sham: the anonymous author was "only catching at a straw to save himself from drowning in the sea of avowed atheism." The book was "absurd, unnatural, and illogical," "flatly contradicting the sacred truths."[53] A later discussion in the

49. Gosse 1857, 27.

50. Freeman and Wertheimer 1980, 59–61.

51. Youatt 1855, 1; Dixon 1848, x–xiii; see also the attack on "the Vestigiarian theory of Creation" in [E. S. Dixon], "Poultry Literature," *Quarterly Review* 88 (Mar. 1851): 317–51, at 332.

52. Allen 1996, 115–18, and Sheets-Pyenson 1981b discuss Newman's publishing activities and the history of the two magazines.

53. E. Newman, "Notice of the Natural History of Creation," *Zoologist,* 1845, 3: 954–63, at 954–55, 963.

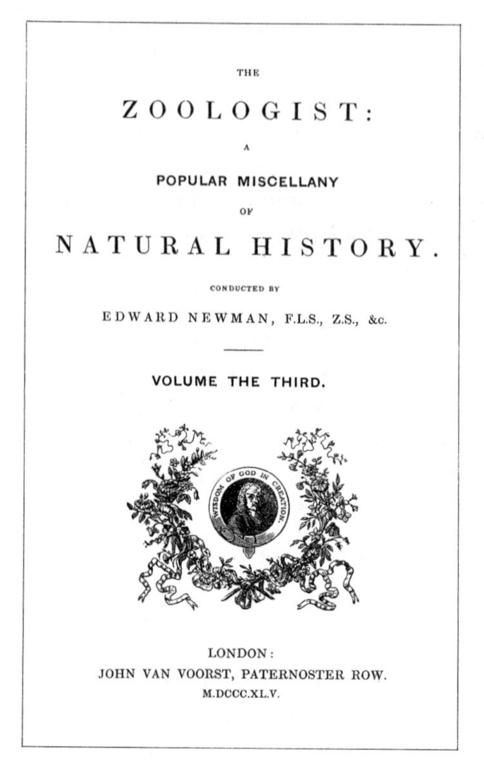
THE

ZOOLOGIST:

A

POPULAR MISCELLANY

OF

NATURAL HISTORY.

CONDUCTED BY

EDWARD NEWMAN, F.L.S., Z.S., &c.

VOLUME THE THIRD.

WISDOM OF GOD IN CREATION.

LONDON:
JOHN VAN VOORST, PATERNOSTER ROW.
M.DCCC.XL.V.

13.8 Title page of Edward Newman's *Zoologist.* Note the portrait of the seventeenth-century naturalist and divine John Ray. The motto, "Wisdom of God in Creation," is adapted from the title of his classic work of natural theology.

Phytologist dismissed it as the work of an impostor, akin to outpourings of Andrew Jackson Davis, the illiterate Poughkeepsie Seer from New York whose "Divine Revelations" had just been issued in Britain.[54]

Newman used rhetoric as unsparing as anything in the *Nonconformist* or the *Record.* But no matter how strongly he felt, as a Quaker he also believed in reasoned debate. Newman thus printed his searing condemnation over his own name, stripped of the mask of editorial anonymity. This was a dramatic step: out of the scores of contemporary attacks on *Vestiges,* his were the only signed periodical reviews. All the rest were anonymous.

The *Phytologist* put a similarly even-handed policy into practice, also printing letters that strengthened the botanical evidence for transmutation.[55] These were by Hewett Cottrell Watson, known for his interests in plant distribution, phrenology, and transmutation. Watson, who had moved from Edinburgh to Thames Ditton, was an uncompromising democrat and no Christian. But

54. E. Newman, "Notice of "The Principles of Nature, her Divine Revelations, and a Voice to Mankind," *Phytologist* 3 (1848): 149–57, at 149.

55. H. Watson, "On the Theory of 'Progressive Development,' Applied in Explanation of the Origin and Transmutation of Species," *Phytologist* 2 (1845): 108–13, 140–47, 161–68, 225–28.

the *Phytologist* was open to his views; in fact, the *Phytologist*'s editor George Luxford disagreed with Newman's condemnation of progressive development. Luxford was a botanical writer, compositor, and publisher who had learned his science while apprenticed to a printer. Luxford not only worked with Newman (the two men were at one point partners), but also subedited the quarterly *Westminster,* where he reviewed *Vestiges* and *Explanations* in 1847.[56]

Between men like Luxford and Newman, there was no agreement about doctrines, only that the issues deserved to be discussed. At the same time, though, their own status—and that of the periodicals they edited—was uncertain. Newman acknowledged that defending the Bible might appear "altogether foreign" to the purpose of a scientific journal focussed on classification and observation.[57] Luxford and Newman's own books on botany and zoology usually steered clear of controversy; this was not what collectors of seaweeds, ferns, or insects wanted to read. The great dynasty of commercial natural history, the Sowerby family, also avoided commenting on debatable questions. Writing from a background in evangelical Anglicanism, James de Carle Sowerby took extensive notes on *Vestiges,* concerned to reconcile progressive geology with the scriptural record; but he published nothing on this subject, seeing his job as running a business.[58]

Mr. Vestiges and Mr. Punch

Strict evangelicals condemned as unsafe many aspects of the "light and transient literature of the day." Few doubted that middle-class heroes such as Dickens and Thackeray had literary talent. But their bias was often radical and secular. Some journalists, such as G. M. W. Reynolds, were self-proclaimed freethinkers. Others had suspected sympathies with radical publishers, particularly through common interests in copyright legislation. From an evangelical perspective, the greatest talents of the age were being wasted on satires of the faithful, or on fiction, rather than being consecrated to Christian salvation.[59] The evangelicals, in their turn, were targeted as canting hypocrites in novels, the theater, and comic journalism.

The leading humor magazine was of course *Punch,* founded in 1841 and with a weekly circulation of some forty thousand in 1845; its famous dinners, at which the staff fought over politics, tone, and content, were at the center of literary bohemia. A half-dozen young writers wrote (anonymously) most of the

56. [G. Luxford], "Natural History of Creation," *Westminster and Foreign Quarterly Review* 48 (Oct. 1847): 130–60.

57. Similar views appear in the main Quaker weekly the *Friend,* as quotations from works by Hugh Miller and Hitchcock.

58. James de Carle Sowerby, ms. notes on *Vestiges,* NHM, General Library, Sowerby Collection, item C112.

59. For the situation up to 1830, see Rosman 1984.

jokes. Initially it was, as Thackeray said, "a very low paper . . . and a great opportunity for unrestrained laughing sneering kicking and gambadoing."[60] One of the leading contributors, Douglas Jerrold, was a known infidel (and ate his peas with a knife), and it was partly through his influence that *Punch* was intermittently radical during its first decade. Unlike, say, Disraeli's *Tancred, Punch* made only the mildest attempts to ridicule the foibles of progressive development, despite the sensation *Vestiges* was causing among its readership. No mention was made of the book until November 1845, when a brief paragraph noted that the next edition would be dedicated to "the immortal Widdicomb" for facts concerning the moon on display at the Colosseum being "a slice of the one he recollects when he was Master of the Ceremonies at the Ampitheatre at Rome."[61] John Widdicomb was ringmaster at Astley's Amphitheatre and Vauxhall Gardens; to connect him with *Vestiges* was to link the book with pantomime and popular farce.[62] This was a world that the *Punch* staff—especially the subeditor, Horace Mayhew, who wrote these lines—knew extremely well. Horace (nicknamed "Ponny") was a writer of theatrical burlesque, and the younger brother of the bankrupt Henry Mayhew, who had founded the magazine and still wrote for it on occasion.[63]

There were further innocuous references early in 1846, when *Vestiges* was used to parody two of *Punch*'s favorite targets. Another staff regular, Tom Taylor, suggested that parliamentary records would provide "Vestiges of the Natural History of Peerage Creation," showing ministers' tracks to be "analogous to the footmarks of animals of the rat species."[64] And as part of *Punch*'s campaign for free trade, a paragraph by Percival Leigh noted that

> Protection originated in a mist, which, however, was not a fine mist, but an intellectual fog of singular density. Its vestiges are apparent in the imperfect state of Agriculture, which no doubt would have been improved by competition; also in a crippled and shackled condition of Commerce. Pauperism and the Union Workhouse are vestiges of Protection, which are also observable in the ducal skull; with whose thickness it has an evident connection.[65]

60. W. M. Thackeray to Mrs. Carmichael-Smyth, 11 June 1842, in Ray 1945–46, 2:54. On *Punch,* see Altick 1997 and Spielmann 1895.

61. [Horace Mayhew], "The First Man of the Day," *Punch,* 22 Nov. 1845, 226; see also "'Vestiges of the Natural History of Creation.'—Messrs. Widdicomb and John Cooper," *Punch,* 20 Mar. 1847, 122.

62. For Widdicomb, see Altick 1997, 278, 535.

63. The authors of individual contributions have been identified from the account book for the years 1843–48 in the *Punch* Library in London. On Horace Mayhew, see Spielmann 1895, 327–30.

64. [Tom Taylor], "Vestiges of the Natural History of Peerage Creation," *Punch,* 25 Apr. 1846, 186.

65. [Percival Leigh], "Vestiges of the Natural History of Protection," *Punch,* 14 Feb. 1846, 83.

Protection is juxtaposed with the Fire-mist, a fossil cranium is compared with the duke of Wellington's: whatever humor such jokes had depended on the assumption that politics and science are different things. *Vestiges* itself is let off easily. *Punch's* benign attitude contrasts with its heavy-handed humor about Buckingham's "British and Foreign Destitute," and the British Association for the Advancement of Science, which it parodied as the "Meeting for the Advancement of British Cookery," with sections devoted to "Beefology," "Tartology," and "Meat-pieology."[66]

Punch identified itself with the broad range of liberal, middle-class opinion against pretense, snobbery, and self-righteousness. In the cartoon shown in figure 13.9, the "Vestiges of Creation" are fashionably dressed female bodies with the heads of ducks, strolling through a London park.[67] As a letter to the *Times* had recently said, the Serpentine and Belgravia had once been a lagoon of the Thames. If so, surely the ancient denizens must have promenaded then just as they do now? It is the late summer of 1859, so crinolines are an established fashion (they had not been invented in the 1840s). A simpering woman addresses the ducks with deference, failing to notice anything amiss.

> The slimy reptile here, no doubt,
> Wriggled and crawled in greed or malice:
> Now see the Courtier creep about—
> Near as he dares to yonder Palace.

The humor is not directed primarily against *Vestiges* or science; rather, evolutionary development is used to parody high Society, with its "toads," "serpents," and "cackling ducks." A "duck," in contemporary slang, was someone who could not meet his or her financial obligations—clearly the finery on display here is a cover for fraud.

The bohemian sympathies underlying these attitudes is clear in the full-column article on "The Book That Goes A-begging" (fig. 13.10), in which *Vestiges* is batted about from author to author and sits unwanted on a foundling hospital doorstep.[68] This again was by Horace Mayhew. Abusing sham charities, brainless newspapers, and starving Irish, this might seem no more than a typical mix of *Punch* themes, and simply one more manifestation of the *Vestiges* sensation.

66. [G. À Beckett], "Meeting for the Advancement of British Cookery," *Punch,* 12 Oct. 1844, 168. Paradis 1997, 154–55, points out that the farcical humor *Punch* applied "is less the engagement of scientific materials than loss of capacity to engage them."

67. "Vestiges of Creation," *Punch,* 3 Sept. 1859, 100; for helpful comments, see Paradis 1997, 143–45. An analogous parody of fashionable life is the cartoon in John Leech's "Rising Generation" reproduced here as figure 5.4.

68. [Horace Mayhew], "The Book That Goes A-begging," *Punch,* 11 Dec. 1847, 230.

VESTIGES OF CREATION.

"The Serpentine, and the whole of Belgravia, were formerly a Lagoon of the Thames."—*Sir S. M. Peto in the Times.*

WHAT, all Belgravia grand and fine,
Was once a mess of marsh and lakes!
PROFESSOR OWEN, be it thine
To prove it in a brace of shakes.

Tell doubters that they need not sneer,
Nor set their puddle-minds in storm;
For all the ancient life is here,
And only changed in outward form.

The slimy reptile here, no doubt,
Wriggled and crawled in greed or malice:
Now see the Courtier creep about—
Near as he dares to yonder Palace.

If tadpoles in the marsh were black,
There is one CONINGSBY can tell
Belgravia's Tadpoles swim in track
Where Tapers guide them to Pall Mall.

If the old lake was rich in toads,
Look out, and you 'll be sure to meet 'em;
If not, it is because such loads
Of people here delight to eat 'em.

With cackling ducks the old lagoon
At times, perchance, alive was seen:
Our Ducks come out each afternoon,
And chatter in their Crinoline.

Lay serpents in the wet nooks twined?
We still can point them out at need:
Search any street, and you shall find
Some home empoisoned by their breed.

Doubtful if Thames were ever den
Where the old monsters made their feasts,
But if we'd Mega-Theria then,
We still can show a few great Beasts.

Adjutants, or Gigantic Cranes,
Croaked o'er the marsh with voices hard.
The first at yonder barracks trains,
The Cranes are loud in CUBITT's yard.

Just as "in earth there is no beast
But 's rendered in some fish of sea,"
One would not say we 'd lost the least
Of that old marsh's family.

13.9 "Vestiges of Creation," *Punch,* 3 Sept. 1859, 100.

THE BOOK THAT GOES A-BEGGING.

The *Vestiges of Creation* has been offered this week to another celebrated author, and again refused. This poor book is doomed apparently to be "The Disowned" of literature. No one will have anything to do with it. It has been left at every author's, like the packets of sealing-wax which respectable beggars carry from house to house, and leave with the printed directions, "If not wanted, to be returned." But not a single author will keep it. LORD BROUGHAM kept it longer than anybody else. He thought, probably, the book could not do him much harm, and so it was allowed to remain for several weeks with his name and address upon it. The rumour immediately was circulated through every paper:—"We can state with the greatest confidence, and upon most unquestionable evidence, that LORD BROUGHAM is the author of the *Vestiges of Creation*." His Lordship has been the author of so many strange things in his day, that the rumour was readily believed, and the book sold another edition of five-and-twenty in consequence. But LORD BROUGHAM thought he had absurdities enough of his own to answer for, so he rose one morning in one of his magnificent passions, and flung the book out of window. It ran about town for a long time, knocking at every scientific man's door; but the answer invariably was, "No, thank you, my good book; we don't want anything in your way." Sometimes it would call upon the "Fast Man" of some light Review, or the editor of some heavy Quarterly, but with no better success. It was always denied, and threatened with the police "if it did not carry its rubbish elsewhere." The *Vicissitudes of the Vestiges of Creation* would make quite a pathetic little book. Its travels would vie in romance with those of OMOO. It has journeyed all over England, it has even penetrated into Scotland, coming back poorer than it went. It has not yet visited Ireland. Probably it thinks that that fossil country might take its title as an insult. It is still a wanderer on the face of literature. Every man's reputation is turned against it. We fear there is no rest for this pilgrim of books but the butter-shop. We don't mind taking it in, a vestige at a time, with a pound of the best Dorset or Stilton, always providing that the stereotyped paragraph is not constantly repeated by some malevolent critic, and copied by every spiteful newspaper, that "We can state with the greatest confidence and upon the most unquestionable evidence, that the celebrated *Mr. Punch* is the author of the *Vestiges of Creation*."

Seriously, however, the destitution of this friendless little literary orphan is a most deserving case for the benevolent. We propose that a certain sum be subscribed in this wealthy metropolis, to pass it on to its own parish. But then again, there is this difficulty: which is its parish? for it does not know its father, and seemingly it never had a home. Heigho! we can only say that "It's a clever book that knows its own author!" Poor *Vestiges of Creation!* Hast thou no strawberry-leaf on thy frontispiece? no stain or blot about thee, by which thy parentage can be recognised? Unhappy foundling! Tied to every man's knocker, and taken in by nobody; thou shouldst go to Ireland! There thou wilt find plenty of kind fathers to own thee and adopt thee! Now we think of it, we are really lost in wonder that no Irishman has yet declared himself the author of the *Vestiges of Creation!* It does not say much for the book, or else the thing would have been claimed long ago, directly it had been known that the authorship of it was a profound mystery.

13.10 Authors bat about the unwanted *Vestiges; Vestiges,* like a packet of unwanted sealing-wax, left at Lord Brougham's door-knocker; and *Vestiges* weeping unwanted outside the door of a foundling hospital. [Horace Mayhew], "The Book That Goes A-begging," *Punch,* 11 Dec. 1847, 230.

Yet readers who bought the magazine off the newsstand in mid-December 1847 would have understood the specific meaning of the joke. *Punch* had triumphed over all other forms of caricature—and competing comic papers—by tying inoffensive humor into the latest news. In this case, the targets were *Macphail's Edinburgh Ecclesiastical Journal* and the *Manchester Courier,* which had just claimed that Chambers had written *Vestiges.* These reports had indeed been picked up both by "malevolent critics" and "spiteful newspapers." "The Book That Goes A-begging" was *Punch*'s counterattack, part of a longstanding campaign against what its writers saw as religious bigotry. The article noted that *Vestiges* "has been offered . . . to another celebrated author, and again refused." Of course, Chambers had *not* refused, nor could he without lying in public. His friends were trying to throw readers off the scent by making the authorship once again "a profound mystery."[69] *Punch* was ideally suited to such a purpose, for it made the entire question seem ridiculous. Middle-class journalists, sheltering what *Punch* called "a wanderer on the face of literature," suspected they were defending one of their own.

Creation in the Marketplace

Common writers, freelance lecturers, and piecework professors participated in a social world poised between the sordid degradation of Grub Street and the attractive (but almost always illusory) hope that authorship might be a route to the status best embodied in the flash gentility of Dickens. As the most widely discussed work on science ever published, *Vestiges* did create opportunities for making money; but these too brought out the moral dilemmas involved in the pursuit of science for cash. Churchill paid Edward Forbes to review the book in the *Lancet;* Wallis's pamphlet was, in part, an advertisement for his lectures and part of his claim for charity; *Omphalos* was a passionate expression of Gosse's faith and also an abortive attempt to earn money.

Reviewers benefited from the *Vestiges* sensation in much the same way as from dozens of other books. The literary weeklies that covered science all relied on a small cadre of contributors who in turn depended on reviewing for a regular part of their income. The book's unique status also created opportunities for exceptionally well-paid work in periodicals that featured science on

69. There is no evidence about what Mark Lemon, Douglas Jerrold, and other *Punch* regulars thought about the authorship. Chambers's friendship with them is clear in C. E. S. Chambers, "Robert Chambers's Commonplace Book," *Chambers's Journal,* 19 Oct. 1901, 737–40, at 738–39, and also in the effusive report of the Burns festival in [Douglas Jerrold], "The Burns Festival.—Repentant Scotland," *Punch,* 24 Aug. 1844, 81. J. Murray, "The Vestiges," *Mining Journal,* 5 Feb. 1848, 67, reviewed the question in the wake of the *Punch* article and mentioned Vyvyan, Nichol, William Chambers, Ada Lovelace, Brougham, only to conclude "Stat nomnis umbra" ("the name remains in shade").

a less regular basis. The best pay (sixteen to twenty guineas per sheet of sixteen pages) was offered by the established quarterlies.[70] The struggling metropolitan geologist David Ansted would have sent in his review for possible inclusion in the *Edinburgh* in full knowledge of these attractive rates. Although Napier did not take the article, some other paying editor almost certainly did. Sedgwick, his old teacher, could afford to be squeamish about taking money offered for a review, but Ansted needed every penny. He was a King's College professor, Geological Society secretary, and editor of its new *Quarterly Journal,* besides authoring several books. Sedgwick feared his former assistant would kill himself: "I had rather be ground under a millstone," he told a friend, "than attempt his work."[71]

No women, so far as I am aware, reviewed *Vestiges,* and despite their involvement in other aspects of the debate (after all, Ada Lovelace, Harriet Martineau, and Catherine Crowe were among the chief suspects for the authorship) no female author seems to have published anything about the controversy until the 1850s. Women made up about 20 percent of all writers, but most wrote novels, poems, and stories. Until much later in the century, women who did publish on science tended to produce manuals and children's books.[72] Mary Shelley, as we saw earlier, had proposed an elementary work on geology to John Murray, and many evangelicals wrote for the Religious Tract Society. A few, like Anna Jameson, Eliza Lynn, and Marian Evans, were leaders in otherwise male fields of history, philosophy, and criticism—as indeed was Shelley herself—but they had reputations as "strong-minded women" who had to define themselves in relation to fields dominated by men; almost none who wrote for money had the opportunity or the education to pronounce critically on new scientific books. Mary Somerville, who might seem an exception, only wrote one review in her whole life, and that was for the genteel *Quarterly.*[73]

Over the longer term, *Vestiges* exercised its most important influence by providing a template for the evolutionary epic—book-length works that covered all the sciences in a progressive synthesis. This eventually became a field cultivated by women authors like Lyell's former secretary, Arabella Buckley, but until the 1870s the vast majority of ambitious science surveys continued to be written by men.[74] Such books offered the same cosmic range as the offending work, but without the developmental law that rendered *Vestiges* suspect. Ansted, for example, set to writing *The Ancient World; Or, Picturesque Sketches of Creation* for

70. Cross 1985, 120.

71. A. Sedgwick to J. Phillips, 20 Jan. 1845, CUL Add. mss. 7763 (copy).

72. Shteir 1996, 1997 and Fyfe 2000; see also Cross 1985, 164–68, the essays in Gates and Shteir 1997, and esp. Tuchman 1989.

73. [M. Somerville], "Astronomy—The Comet," *Quarterly Review* 105 (Dec. 1835): 195–233.

74. On Buckley, see Gates 1997.

John Van Voorst, stressing the lack of "regular gradation" and explicitly denying transmutation.[75]

The controversy also provided an occasion for updating and revision. One of the most significant projects of this kind was the sixth edition of Mantell's *Wonders of Geology* in 1848, commissioned (for £105) by the rising star of cheap publishing, Henry Bohn. *Wonders* had been among the leading surveys of geology, with material on the nebular hypothesis and human origins. Mantell had meanwhile been scathing about the spread of *Vestiges* in the salons, had opposed it in writing to Hume, and had welcomed Herschel's British Association address. But after the unknown author's revisions and *Explanations,* Mantell became slightly more charitable. A new footnote cited the Congregationalist teacher John Pye Smith's view that the "primordial elements" might well have been endowed with properties that could give rise to life. This position had become increasingly important for those with backgrounds in liberal Nonconformity, although for Mantell the problem was the lack of proof. Otherwise the development theory would "explain many obscure physiological phenomena" and bring the laws of life and matter into accordance. He pointed, for example, to the anatomical parallels between humans and fossil monkeys, but refused to extend these to the arena of mind and spirit.[76]

Other publishers exploited the revived market for general books on creation, many titles showing the continuing salability of evangelical, Bible-based science. These sold extraordinarily well: Stephen Watson Fullom's *The Marvels of Science, and their Testimony to Holy Writ* (1852) was into an eighth edition after two years, with a cheap edition from Routledge for the railway market in 1860; the Reverend Henry Christmas's *Echoes of the Universe: From the World of Matter and the World of Spirit* (1850), combining astronomy, geology, demonology, and angelology, reached a seventh edition in 1863. One could not make a middle-class living if the copyrights to such books had to be sold (publishers paid from nothing to £250 for an elementary nonfiction text) but they were useful supplements to other sources of income. The tenth edition of *Vestiges* condemned such titles as shameless hypocrisy: "Professing adversaries write books in imitation of his, and, with the benefit of a few concessions to prejudice, contrive to obtain the favour denied to him."[77]

Most of these competing works avoided narrative structures that could be read as linking astronomy, geology, and the human sciences in a causal se-

75. Ansted 1847, 55, 102. Other books incorporated elements of the *Vestiges* title to attract interest, see, for example, Kemp 1852 and 1854 and [Urquhart] 1849.

76. Mantell 1848, 1:47–48. The revisions to this edition are outlined in Dean 1999, 224–29.

77. *V10*, ix.

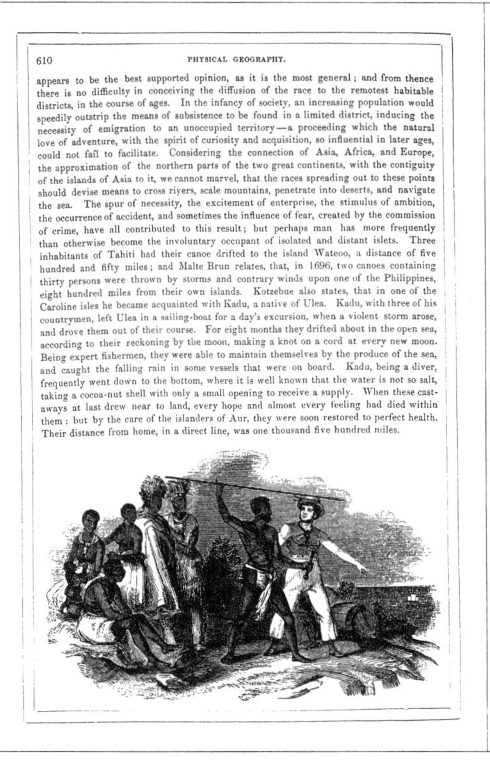

610 PHYSICAL GEOGRAPHY.

appears to be the best supported opinion, as it is the most general; and from thence there is no difficulty in conceiving the diffusion of the race to the remotest habitable districts, in the course of ages. In the infancy of society, an increasing population would speedily outstrip the means of subsistence to be found in a limited district, inducing the necessity of emigration to an unoccupied territory—a proceeding which the natural love of adventure, with the spirit of curiosity and acquisition, so influential in later ages, could not fail to facilitate. Considering the connection of Asia, Africa, and Europe, the approximation of the northern parts of the two great continents, with the contiguity of the islands of Asia to it, we cannot marvel, that the races spreading out to these points should devise means to cross rivers, scale mountains, penetrate into deserts, and navigate the sea. The spur of necessity, the excitement of enterprise, the stimulus of ambition, the occurrence of accident, and sometimes the influence of fear, created by the commission of crime, have all contributed to this result; but perhaps man has more frequently than otherwise become the involuntary occupant of isolated and distant islets. Three inhabitants of Tahiti had their canoe drifted to the island Wateoo, a distance of five hundred and fifty miles; and Malte Brun relates, that, in 1696, two canoes containing thirty persons were thrown by storms and contrary winds upon one of the Philippines, eight hundred miles from their own islands. Kotzebue also states, that in one of the Caroline isles he became acquainted with Kadu, a native of Ulea. Kadu, with three of his countrymen, left Ulea in a sailing-boat for a day's excursion, when a violent storm arose, and drove them out of their course. For eight months they drifted about in the open sea, according to their reckoning by the moon, making a knot on a cord at every new moon. Being expert fishermen, they were able to maintain themselves by the produce of the sea, and caught the falling rain in some vessels that were on board. Kadu, being a diver, frequently went down to the bottom, where it is well known that the water is not so salt, taking a cocoa-nut shell with only a small opening to receive a supply. When these cast-aways at last drew near to land, every hope and almost every feeling had died within them; but by the care of the islanders of Aur, they were soon restored to perfect health. Their distance from home, in a direct line, was one thousand five hundred miles.

GEOLOGY

Geology—a discourse concerning the earth—the signification of the two Greek words from which the term is derived—ranks next to astronomy in the interest and grandeur of its revelations. It treats of the elementary substances of which the crust of the globe is composed; the structure of the masses of rock which form its rind; the order in which they occur; the agencies that have operated in their production; and of the fossilised organic remains with which they abound. This is an investigation which requires the aid of several branches of natural knowledge—of mineralogy, of chemistry, of mathematical dynamics, of zoology and comparative anatomy, and of botany: but a number of illustrious men have appeared, in modern times, in whom these various accomplishments have centred, who have succeeded in gaining an intelligent apprehension of the arrangements and combinations of the accessible parts of our planet—have unfolded forms of organic life, now withdrawn from the sphere of living existence, which flourished in succession upon the surface of the globe, in ages long anterior to the era of man's creation upon its soil—and have demonstrated the story of its past career to be a longer chapter than commonly imagined, marked

13.11 Nature as tour rather than story. An account of human migration—the final page of "Physical Geography"—faces the opening of the section on "Geology." Copious wood- and steel-engraved illustrations were a major selling point for works such as Thomas Milner's *Gallery of Nature,* which was frequently reissued later in the century. Milner 1846, 610–11.

quence. This is evident in the Reverend Thomas Milner's *Gallery of Nature* (1846), probably the decade's best-selling scientific part work (fig. 13.11). The *Gallery* was structured as a tour of creation, not a story of development. On natural law, Milner (a former Congregationalist minister) took broadly the same position that Mantell's *Wonders* had done, quoting John Pye Smith to the effect that powers divinely implanted in matter could produce "all the forms and changes of organic and inorganic natures."[78] But such issues were introduced as digressions. As indicated by his subtitle (*A Pictorial and Descriptive Tour through Creation, Illustrative of the Wonders of Astronomy, Physical Geography, and Geology*), Milner wanted readers to view nature as travelers, not as seekers after narrative suspense. He never exploited the storytelling possibilities of serial

78. Milner 1846, 192.

publication that Dickens and other novelists were developing in part issues of fiction.

Milner, although only thirty-eight, would soon give up his position as a Congregationalist minister in Northampton due to ill health; he later paralyzed his right arm with writing.[79] He earned little from his books, although these were extremely popular for birthday gifts, family reference, and school prizes. The *Gallery*, issued on behalf of a consortium through the London firm William S. Orr & Co., could be purchased in weekly parts at threepence or monthly ones at a shilling. Readers of one part were tempted into purchasing the next through the striking illustrations.[80] As a complete eight-hundred-page volume, the *Gallery* was packed with 250 woodcuts, 8 landscape engravings, and 6 double-page astronomical diagrams.

The flowering of the market for such books showed how religious belief could be integrated into the concerns of commercial science. The connection was concretely embodied in Van Voorst's warehouse on Paternoster Row, which served as a depository for Bibles and natural history books. Countering the infidel uses of science, authors could serve God and Mammon at the same time.

The Contradictions of Commerce

The moral dilemmas of commercial science became especially acute when the connection with *Vestiges* was direct. With so much in the book emanating from recent anatomy and physiology, many traced the author to the hospital schools near Churchill's Soho shop, where transcendentalist anatomies had sprouted for decades. A damning notice in *Wade's London Review*, a short-lived monthly, identified the author as a novice hospital lecturer:

> His mental character, acquirements, and mode of thinking bespeak the *medical profession* as his worldly status, and the dogmatical and flowery style of his composition would bespeak him a hospital-lecturer of barely mature age for his office. His learning is that of the lecturer—shewy rather than profound—and effective rather than systematic: it is learning "got up," not science investigated; it is knowledge taken at second-hand and upon trust, not elaborated in his own mind nor fully comprehended as to its exact demonstration.[81]

As *Parker's London Magazine* said, progressive development "has its origin in the dissecting-room, and is not the offspring either of astronomy or geology."

79. Thomas Milner, case file 1385, archives of the Royal Literary Fund; Milner 1846, 192.

80. *Publishers' Circular*, 16 Dec. 1844, 373.

81. "Vestiges of the Natural History of Creation," *Wade's London Review* 1 (May 1845): 382–404, at 384.

Young medical men were notorious for dissolute habits both in body and mind, as suggested in *Punch* cartoons and comic novels.[82]

Those in scientific commerce with a higher opinion of *Vestiges* attributed it to medically trained authors known to be associated with Churchill.[83] Edward Forbes was thought likely, given his wide-ranging interests and his anonymous praise of the book in the *Lancet.* Perhaps embarrassed by his initially favorable response, he refused to help with the revisions and instead devoted his energies to attacking the book and tracing its author. Forbes's status as a gentleman whose financial expectations had been dashed gave the issue a particular urgency. Before tagging Chambers as author in December 1845, he had suspected first Vyvyan and then the young Scottish moral philosopher Alexander Bain.[84] Edwin Lankester, as a devout Congregationalist struggling to make ends meet, was also accused of the authorship, but was less fastidious in taking on the job of revising. Lankester condemned progressive development at Exeter Hall and as an anonymous reviewer in the *Athenaeum:* but behind the scenes he earned thirteen guineas for correcting the text.[85]

William Carpenter too acted a double life in relationship to *Vestiges.* "The *ideas* are quite familiar to me," he told Ada Lovelace, whose children he was tutoring, "and I long ago expressed them in a general form. Some of the reasoning is loose and unconclusive; but I very much admire the tone and spirit of the book." Soon after reading the first edition, Carpenter suggested that they collaborate with "Faraday, or some other luminary" and erect "an Acarus-producing apparatus in our workshop" at Lovelace's Ockham Park estate.[86] This, however, was never done. Not only did the earl of Lovelace believe that his wife's intimacy with the tutor had exceeded propriety, but Carpenter consistently ranked himself as a gentleman of science rather than a servant in an aristocratic household.[87] He resigned during the late spring of 1845, and plans for experiments were abandoned. In anonymous journalism, Carpenter identified with the wider aims of *Vestiges,* as is clear from the *British and Foreign Medical*

82. "The Natural History of Creation," *Parker's London Magazine* 1 (Feb. 1845): 95–104, at 97. On medical students, see Desmond 1989, 12, 167; also Cockton 1844b, facing p. 115.

83. Carpenter 1888, 34.

84. E. Forbes to A. C. Ramsay, 22 Nov. 1844, IC, Ramsay Papers; E. Forbes to [J. S. Blackie?], Jan. [1845], NLS mss. 2642 f. 227. For Forbes's career, see Wilson and Geikie 1861, Rehbock 1983, and Mills 1984. Forbes's assistance in revision is proposed in RC to AI, 1 Mar. 1845, NLS Dep. 341/110/155–56.

85. Lankester 1848; [E. Lankester], review of *Vestiges, Athenaeum,* 4 Jan. 1845, 11–12; authors' ledger of J. Churchill, Reading University Library, mss. 1393/385, p. 290 (2d ed.), p. 304 (3d ed.), p. 338 (4th ed.).

86. W. B. Carpenter to A. Lovelace, 19 Nov. 1844, Bodleian Library, Dep. Lovelace Byron 169, ff. 218–23.

87. Moore 1977, 193–210.

Review and an eighteen-part series in the weekly Unitarian *Inquirer.* However, all his scientific patrons loathed the book. So Carpenter would disown *Vestiges* at soirées and British Association meetings, only to praise it anonymously or slip into Churchill's office to improve the text—not least, it brought in money, over thirty-five pounds in all.[88]

Many thought that the most celebrated science lecturer of the period, John Pringle Nichol, had written the book, which at the very least was indebted to his writings in language, intended readership, and general aims. Nichol held the astronomy chair at Glasgow, but needed paying audiences for his books, newspaper articles, and lectures. He disliked *Vestiges* even before reading it, as he admitted to fellow lecturer John Murray in March 1845. In a letter to a former student, the young natural philosopher William Thomson, Nichol dismissed it as "a foolish book" that had attracted "still foolisher reviews," which raised the specter of "Atheism &c." The best that could be done now was to find "something sufficiently probable to be accepted as a cosmogony."[89] The situation could scarcely be more serious, for Nichol had gone bankrupt four years earlier.[90] *Vestiges* destroyed the delicate compromise he had engineered in his *Architecture of the Heavens,* so that nebular hypotheses had become linked with controversies about the status of the soul.

Worst of all, *Vestiges* forestalled Nichol's plans for a multivolume cosmological work on the heavens, earth, and man. Nichol had told Chambers about his scheme during their Irish tour in 1837, and believed that his erstwhile friend had stolen the idea. Long quotations from letters he had written for *Chambers's Journal* had been silently incorporated into the text. As he complained in 1848 to Combe, "I think I can promise you as curious a history of the unblushing & wholesale appropriations of another mans plans & thoughts, as probably ever occurred." The issue needed "to be decided by some code of Morals," and Nichol invited Combe to serve as judge.[91] His suspicions about the authorship had been confirmed when, two years earlier, he had heard the secret from

88. [W. Carpenter], "Natural History of Creation," *British and Foreign Medical Review* 19 (Jan. 1845): 155–81; [W. Carpenter], "On the Harmony of Science and Religion," *Inquirer,* 22, 29 Mar.; 5, 12, 26 Apr.; 3, 10, 17, 24, 31 May; 7 June; 5, 12, 19, 26 July; 2, 9, 16 Aug. 1845: 182–83, 198, 214, 230–31, 262–63, 277–78, 293–94, 308–9, 325, 341, 358, 422–23, 437–38, 454–55, 470–71, 485, 502–3, 520. Authors' ledger of J. Churchill, Reading University Library, mss. 1393/385, p. 304 (3d ed.); mss. 1393/386, p. 347 (10th ed.).

89. J. P. Nichol to W. Thomson, 1 Feb. 1846, CUL Add. mss. 7342.N.28; J. P. Nichol to J. Murray, 10 Mar. 1845, EUL Gen. 1971/2/33.

90. Smith and Wise 1989, 74–75; *Times,* 30 Mar. 1842, 3.

91. J. P. Nichol to G. Combe, 25 Feb. 1848, NLS mss. 7296, ff. 76–82; G. Combe to J. P. Nichol, 18 Mar. 1848, NLS mss. 7391, ff. 347–50. Nichol also spoke privately to the newspaper editor Charles Mackay about the grievance (Mackay 1887, 1:177–80).

Chambers himself, who asked for advice on astronomical revisions. Although Nichol checked proofs of the second edition of *Explanations* and never exposed the secret, his own literary plans were in ruins.[92] All subsequent editions of the *Architecture* were sharply distinguished from *Vestiges,* which it never mentioned by name.

As a commercial author Nichol could not afford, and had no wish, to draw down the saints' wrath. Lord Rosse's long-anticipated resolution of the Orion nebula into stars in March 1846 provided a way out. Nichol began to focus on the broader and more conjectural question of the structure, size, and fate of the entire universe (fig. 13.12). He developed a new view of nebular condensation in which gravitation brought individual stars together. Nebulae were dense stellar clusters, not traces of the primordial Fire-mist.[93] Nichol's lectures and journalism became even more grandly sublime, but in a way that allied him with the observational triumphs of aristocratic astronomy, rather than the noisy debate surrounding *Vestiges.*

Mental Pregnancies

The most remarkable attempt to resolve the pressures of scientific commerce was Robert Hunt's allegorical novel *Panthea, the Spirit of Nature,* published in 1849 by Reeve, Benham and Reeve of the Strand. The noble hero, Julian Altamont, is torn between "the conflicting views which beset our philosophy."[94] Repelled by the instructors chosen by his parents, Julian comes under the influence of the aged, insidious Laon Ælphage and his beautiful daughter Æltgiva, votaries of Panthea, the Spirit of Nature. Guided by Laon, Julian surveys the wonders of life on the globe and the distribution of organic beings. Then, taught by Panthea to see with his spiritual eye, the young nobleman apprehends the mysteries of creation itself, from primordial chaos to the progress of the human race. Although attracted to pure contemplation, Julian chooses the love of a practical woman and a life of useful work—but she is beautiful and it is a life informed by the visions he has witnessed.

Drawing on the transcendental philosophies of Coleridge, Fichte, and Oken, Hunt also built his reading of *Vestiges* into Julian's vision of creation. An experimental literary form enabled him to recognize the limits of science, while

92. RC to AI, [May 1846], NLS Dep. 341/110/225–26; RC to AI, [late May or early June 1846], NLS Dep. 341/110/229–30.

93. This view is developed in Nichol 1848; for his relations with the Edinburgh evangelicals, see "The Limitations Essential to Every Speculation in Cosmology," *Free Church Magazine* 3 (Jan. 1846): 7–14, and esp. 3 (Apr. 1846): 101–5, which contains Nichol's first considered reaction to the Orion resolution. My account of Nichol's revisions is indebted to Lane 1995.

94. Hunt 1849, v; see also Millhauser 1959, 153–54.

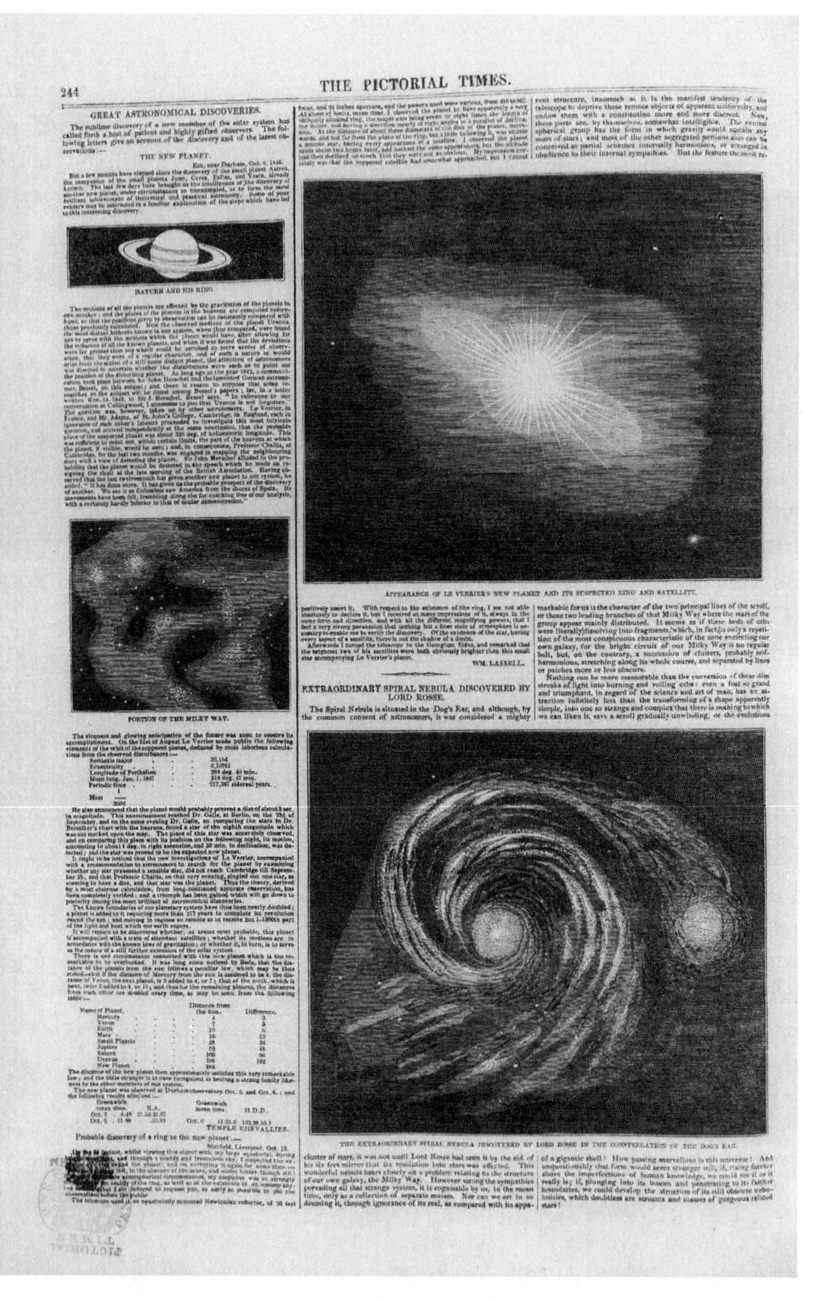

244 THE PICTORIAL TIMES.

GREAT ASTRONOMICAL DISCOVERIES.

The sublime discovery of a new member of the solar system has called forth a host of patient and highly gifted observers. The following letters give an account of the discovery and of the latest observations:—

THE NEW PLANET.

Esh, near Durham, Oct. 8, 1846.

But a few months have elapsed since the discovery of the small planet Astrea, the companion of the small planets Juno, Ceres, Pallas, and Vesta, already known. The last few days have brought us the intelligence of the discovery of another new planet, under circumstances so unexampled, as to form the most brilliant achievement of theoretical and practical astronomy. Some of your readers may be interested in a familiar explanation of the steps which have led to this interesting discovery.

SATURN AND HIS RING.

The motions of all the planets are affected by the gravitation of the planets to one another; and the places of the planets in the heavens are computed beforehand, so that the positions given by observation can be constantly compared with those previously calculated. Now the observed motions of the planet Uranus, the most distant hitherto known in our system, when thus compared, were found not to agree with the motions which the planet would have, after allowing for the influence of all the known planets, and when it was found that the deviations were far greater than any which could be ascribed to mere errors of observation, that they were of a regular character, and of such a nature as would arise from the action of a still more distant planet, the attention of astronomers was directed to ascertain whether the disturbances were such as to point out the position of the disturbing planet. As long ago as the year 1842, a communication took place between Sir John Herschel and the lamented German astronomer, Bessel, on this subject; and there is reason to suppose that some researches on the subject will be found among Bessel's papers; for, in a letter written Nov. 14, 1842, to Sir J. Herschel, Bessel says, "In reference to our conversation at Collingwood, I announce to you that Uranus is not forgotten." The question was, however, taken up by other astronomers. Le Verrier, in France, and Mr. Adams, of St. John's College, Cambridge, in England, each in ignorance of each other's labours proceeded to investigate this most intricate question, and arrived independently at the same conclusion, that the probable place of the suspected planet was about 325 deg. of heliocentric longitude. This was sufficient to point out, within certain limits, the part of the heavens at which the planet, if visible, would be seen; and, in consequence, Professor Challis, at Cambridge, for the last two months, was engaged in mapping the neighbouring stars with a view of detecting the planet. Sir John Herschel alluded to the probability that the planet would be detected in the speech which he made on resigning the chair at the late meeting of the British Association. Having observed that the last twelvemonth has given another new planet to our system, he added, "It has done more. It has given us the probable prospect of the discovery of another. We see it as Columbus saw America from the shores of Spain. Its movements have been felt, trembling along the far-reaching line of our analysis, with a certainty hardly inferior to that of ocular demonstration."

PORTION OF THE MILKY WAY.

The eloquent and glowing anticipation of the future was soon to receive its accomplishment. On the 31st of August Le Verrier made public the following elements of the orbit of the supposed planet, deduced by most laborious calculations from the observed disturbances:—

Semiaxis major	36,154
Eccentricity	0,10761
Longitude of Perihelion	284 deg. 45 min.
Mean long. Jan. 1, 1847	318 deg. 47 min.
Periodic time	217,387 sidereal years.

Mass $\frac{1}{9300}$

He also announced that the planet would probably present a disc of about 3 sec. in magnitude. This announcement reached Dr. Galle, at Berlin, on the 23d of September, and on the same evening Dr. Galle, on comparing the stars in Dr. Bremiker's chart with the heavens, found a star of the eighth magnitude which was not marked upon the map. The place of this star was accurately observed, and on comparing this place with its position on the following night, its motion, amounting to about 4 deg. in right ascension, and 30 min. in declination, was detected; and the star was proved to be the expected new planet.

It ought to be noticed that the new investigations of Le Verrier, accompanied with a recommendation to astronomers to search for the planet by examining whether any star presented a sensible disc, did not reach Cambridge till September 29, and that Professor Challis, on that very evening, singled out one star, as seeming to have a disc, and that star was the planet. Thus the theory, derived by a most abstruse calculation, from long-continued accurate observation, has been completely verified; and a triumph has been gained which will go down to posterity among the most brilliant of astronomical discoveries.

The known boundaries of our planetary system have thus been nearly doubled; a planet is added to it requiring more than 217 years to complete its revolution round the sun; and moving in regions so remote as to receive but 1-1360th part [illegible]

[illegible]

Name of Planet.	Distance from the Sun.	Difference.
Mercury	4	3
Venus	7	3
Earth	10	6
Mars	16	12
Small Planets	28	24
Jupiter	52	48
Saturn	100	96
Uranus	196	192
New Planet	388	

The distance of the new planet then approximately satisfies this very remarkable law; and the little stranger is at once recognised as bearing a strong family likeness to the other members of our system.

The new planet was observed at Durham observatory Oct. 3. and Oct. 6.; and the following results obtained:—

[illegible]

TEMPLE CHEVALLIER.

Probable discovery of a ring to the new planet:—

Starfield, Liverpool, Oct. 12.

[illegible]

The telescope used is an equatorially mounted Newtonian reflector, of 20 feet focus, and 24 inches aperture, and the powers used were various, from 316 to 567. At about 8½ hours, mean time, I observed the planet to have apparently a very obliquely situated ring, the major axis being seven or eight times the length of the minor, and having a direction nearly at right angles to a parallel of declination. At the distance of about three diameters of the disk of the planet, northwards, and not far from the plane of the ring, but a little following it, was situate a minute star, having every appearance of a satellite. I observed the planet again about two hours later, and noticed the same appearances, but the altitude had then declined so much that they were not so obvious. My impression certainly was that the supposed satellite had somewhat approached, but I cannot positively assert it. With respect to the existence of the ring, I am not able absolutely to declare it, but I received so many impressions of it, always in the same form and direction, and with all the different magnifying powers, that I feel a very strong persuasion that nothing but a finer state of atmosphere is necessary to enable me to verify the discovery. Of the existence of the star, having every aspect of a satellite, there is not the shadow of a doubt.

Afterwards I turned the telescope to the Georgium Sidus, and remarked that the brightest two of his satellites were both obviously brighter than this small star accompanying Le Verrier's planet.

WM. LASSELL.

APPEARANCE OF LE VERRIER'S NEW PLANET AND ITS SUSPECTED RING AND SATELLITE.

EXTRAORDINARY SPIRAL NEBULA DISCOVERED BY LORD ROSSE.

The Spiral Nebula is situated in the Dog's Ear, and although, by the common consent of astronomers, it was considered a mighty cluster of stars, it was not until Lord Rosse had seen it by the aid of his six feet mirror that its resolution into stars was effected. This wonderful nebula bears closely on a problem relating to the structure of our own galaxy, the Milky Way. However strong the sympathies pervading all that strange system, it is cognisable by us, in the mean time, only as a collection of separate masses. Nor can we err in so deeming it, through ignorance of its real, as compared with its apparent structure, inasmuch as it is the manifest tendency of the telescope to deprive these remote objects of apparent uniformity, and endow them with a constitution more and more discreet. Now, these parts are, by themselves, somewhat intelligible. The central spherical group has the form in which gravity would sustain any mass of stars; and most of the other segregated portions also can be conceived as partial schemes internally harmonious, or arranged in obedience to their internal sympathies. But the feature the most remarkable for us is the character of the two principal lines of the scroll, or those two leading branches of that Milky Way where the stars of the group appear mainly distributed. It seems as if these beds of orbs were literally dissolving into fragments, which, in fact, is only a repetition of the most conspicuous characteristic of the zone encircling our own galaxy, for the bright circuit of our Milky Way is no regular belt, but, on the contrary, a succession of clusters, probably self-harmonious, stretching along its whole course, and separated by lines or patches more or less obscure.

Nothing can be more memorable than the conversion of these dim streaks of light into burning and rolling orbs: even a feat so grand and triumphant, in regard of the science and art of man, has an attraction infinitely less than the transforming of a shape apparently simple, into one so strange and complex that there is nothing to which we can liken it, save a scroll gradually unwinding, or the evolutions of a gigantic shell! How passing marvellous is this universe! And unquestionably that form would seem stranger still, if, rising farther above the imperfections of human knowledge, we could see it as it really is; if, plunging into its bosom and penetrating to its farther boundaries, we could develop the structure of its still obscure nebulosities, which doubtless are streams and masses of gorgeous related stars!

THE EXTRAORDINARY SPIRAL NEBULA DISCOVERED BY LORD ROSSE IN THE CONSTELLATION OF THE DOG'S EAR.

13.12 "Great Astronomical Discoveries." In this page from the *Pictorial Times,* an unattributed letter from John Pringle Nichol about Lord Rosse's discovery of the spiral nebula is combined with William Lassell's account of a satellite around Neptune ("Le Verrier's new planet"). As Nichol said, "How passing marvellous is this universe! And unquestionably that form would seem stranger still, if . . . we could develop the structure of its still obscure nebulosities, which doubtless are streams and masses of gorgeous related stars!" *PT,* 17 Oct. 1846, 244.

using his "fancy" to penetrate to the deeper wonders of nature: the chapter which most closely echoed *Vestiges* was called "The Vision of the Mystery." Speculative ideas about the cosmos could provide the basis for an epic prose-poem: the dreamlike *Panthea* could voice "truths which would not be received in plain spoken, common sense English," as Hunt told his colleague Andrew Ramsay.[95]

Panthea aimed to resolve the dilemmas of commercial science. Much of Hunt's day-to-day work, like that of many young scientific men, was dull and repetitive. Born in 1807 in Devon as the posthumous son of a naval officer, Hunt had been trained for a career as a surgeon, which he abandoned for pioneering photographic experiments and folklore studies after inheriting a small property. However, he needed other work in order to make a living, so from 1845 he combined an appointment as keeper of mining records with miscellaneous journalism. None of this paid well, but it brought in money. Sales of the *Poetry of Science,* he hoped, would enable him to pay off a £150 debt.[96]

Panthea took a higher path: as Hunt told Ramsay, it was "a Pilgrim's Progress—a new Christian in pursuit of scientific truth." The story embodied "the struggle between Intelligence and sensuality in the onward progress of an individual mind towards the acquirement of the highest truths."[97] He had contemplated such a work for years:

> like a boiling pot my brain is hot and troubled—and all the elements of my dream in confusion—such is always the condition of my mental pregnancies—I conceive—The thought quickens—and then it must have its five months roll in the darkness of my intellectual womb by and by the hour of birth comes—and what is the child—Alas! I am rarely satisfied—[98]

The novel was appreciatively reviewed, but little read: the publisher cut the price from 10s. 6d. to 6s. to move the stock.[99]

Panthea became one of many attempts in the next few years to create literary forms capable of expressing the wider meanings of science.[100] Charles Mackay,

95. R. Hunt to A. C. Ramsay, 25 May 1847, IC, Ramsay Papers, KGA/Ramsay/8/506/36.

96. R. Hunt to A. C. Ramsay, 11 Nov. 1848, IC, Ramsay Papers, KGA/Ramsay/8/506/42.

97. R. Hunt to A. C. Ramsay, 30 Oct. 1848, IC, Ramsay Papers, KGA/Ramsay/8/506/41.

98. R. Hunt to A. C. Ramsay, 11 Nov. 1848, IC, Ramsay Papers, KGA/Ramsay/8/506/42.

99. Low 1864, 387; for a favorable review, see *Westminster and Foreign Quarterly Review* 52 (Jan. 1850): 607–8. See also R. Hunt to A. C. Ramsay, 29 Nov. 1849, IC, Ramsay Papers, KGA/Ramsay/8/506/54, which cites a woman reader who hoped that Hunt would write a novel about humans next time.

100. The most famous is the "Dream Land" chapter in the Reverend Charles Kingsley's *Alton Locke* (1850), in which the protagonist recapitulates the history of the human race, from the "lowest point of created life" (336) to the origins of civilization. This vision was not evolutionary, but its progressive gestations would have been read in relation to *Vestiges,* a work for which Kingsley had "a special dislike." C. Kingsley to P. H. Gosse, 4 May 1858, in Gosse 1890, 280–83, at 282.

who worked for the *Illustrated London News* from 1848, wrote an epic poem *Egeria* (1850) with cosmological themes so close to those of *Panthea* that Hunt privately accused him of plagiarism.[101] A similar allegorical drama was pursued in *Thorndale, or the Conflict of Opinions* (1857), written by the journalist William Henry Smith and published by Blackwoods. One character in the book's philosophical dialogues is a utilitarian who advocates a "law of indefinite progress." Smith had criticized *Vestiges* anonymously in *Blackwood's Edinburgh Magazine*, but was well aware of its sources and took them seriously in his novel.[102]

Cosmological novels and epic poems of science rarely sold well. Some, like the anonymous two-volume *Nimshi: The Adventures of a Man to Obtain a Solution of Scriptural Geology, to Gauge the Vast Ages of Planetary Concretion, and to Open Bab Allah—The Gate of God* (1845) simply fell dead from the press. To write such a work at all showed that an author had escaped from the clutches of the market (and perhaps, as reviewers wondered, from an asylum).[103] Smith composed *Thorndale* in almost complete isolation, having moved from London to the Lake District and depending on journalism to supplement a small independence. Hunt and Mackay lived in the metropolis, but had relatively secure employment and reputations. Writing a philosophical opus indicated aspirations beyond cataloging fossils and writing newspaper reviews. Many of these works have strong echoes of *Sartor Resartus*, and express hopes of fulfilling a Carlylean sense of vocation through science. "May Armageddon fly away with me," Hunt told Ramsay after forgetting about some charts, "but I am sometimes rapt—and being exalted into the Empyrean how can I think of maps of this gross Earth."[104]

101. R. Hunt to A. C. Ramsay, 22 Aug. 1850, IC, Ramsay Papers, KGA/Ramsay/8/506/64.

102. [W. H. Smith], "Vestiges of the Natural History of Creation," *Blackwood's Edinburgh Magazine* 57 (Apr. 1845): 448–60; Smith 1857. On *Thorndale*, see Millhauser 1959, 154–55.

103. Reviewers, as in the *Atlas*, 24 Sept. 1845, 622, thought the work resulted "from too exclusive self-communion, and little collision with other and more stable intellects." The author is listed in the British Library Catalogue as Samuel Hanna Carlisle.

104. R. Hunt to A. C. Ramsay, 30 Oct. 1848, IC, Ramsay Papers, KGA/Ramsay/8/506/41.

CHAPTER FOURTEEN

Mammon and the New Reformation

In literature a man may write for magazines and reviews, and so support himself; but not so in science. . . . A man who chooses a life of science chooses not a life of poverty, but, so far as I can see, a life of *nothing*, and the art of living upon nothing at all has yet to be discovered.

THOMAS HENRY HUXLEY (1851)

But of the Future Man what do we know, and how wildly may we not predicate!

ELIZA LYNN LINTON (1859)

WHAT WAS THE ROLE OF THE SCIENTIFIC PRACTITIONER TO BE? We have seen how readings of *Vestiges* were embedded in gentlemanly science, commercial science, and the culture of the exhibition. Reading also offered opportunities for contemplating the creation of new sciences and new roles. Two novel and controversial projects for transforming the place of the scientific practitioner gained prominence in the 1840s and 1850s. Small increases in the number of posts—especially in the government Geological Survey—created expectations that science might be pursued as a regular career, akin to medicine, the military, or the priesthood. At the same time, periodicals of "advanced thought" began to use the sciences to bring about what Combe and others were calling "the New Reformation," with religious faith regrounded in an understanding of nature's laws. These were simultaneously utopian fantasies and practical projects for reform, which became immediately bound up with readings of *Vestiges*.

These visions of the future can best be understood by examining their stresses and fault lines. Was expert criticism of facts a sign of narrow specialism, or the only real license for public comment on the moral issues posed by science? Was the best-qualified reviewer a man of letters or a man of science? Or was it possible to be both? What was the place of women in projections of the scientific writer? Definitions of "science" and "journalism" were sharply contested, as part of wider struggles about democracy, education, and the press between the Chartist era and the passage of the Second Reform Bill in 1867.[1] If knowledge was power, it was because of the changing relations of science to the state, to the educational system, and to the growing audience for books, periodicals, and lectures. All writing on the phenomena of nature—from taxonomic description to abstract philosophy—was bound up with these issues. Whatever else they did, those engaged in scientific practices needed to position themselves as authors in a competitive economy.

Utopias never materialize, yet the futures projected by the protagonists in this chapter have often been taken as destined for success. The realist novel, the rejection of dogmatic religion, and the making of science as a paid profession have a painless inevitability about them that is hard to resist. George Eliot, Thomas Henry Huxley, and others who risked everything in the early 1850s found themselves just a decade later on their way to dominant positions in the mid-Victorian scene. Because their works are read today, their readings of *Vestiges* shine in the canonical glow surrounding *Middlemarch* and *Man's Place in Nature.* Only in recent biographies does Huxley burst on the London streets as a pushy surgeon just back from the South Seas, and George Eliot appear as Marian Evans, a young freethinker from Coventry planning to board at 142 Strand, home of the radical *Westminster Review.*[2] What did the future of science look like to them?

Scientific Young England

When Huxley returned in 1850, power within science continued to be wielded by practitioners who had come of age in the 1810s and 1820s. This was an unstable situation, and their leadership was no longer secure. The supply of genteel blood in the sciences had been failing for over two decades, so that typical recruits were the middle-class men like Tyndall, Lankester, Hirst, Carpenter, and Ansted, sons of manufacturing or professional families, from Scotland or the English provinces, of some education but without independent incomes.[3] A few, like Edward Forbes, had had gentlemanly expectations dashed through financial failure. These men read *Vestiges* aware that their positions in society depended on the balance between status and expertise.

1. This chapter is informed by general accounts of the rise of professional middle-class careers in Britain, notably Altick 1962, Burn 1964, Gross 1969, Heyck 1982, and Perkin 1969.

2. I am especially indebted to Desmond 1994, Desmond and Moore 1991, Ashton 1991 and 1996, Bodenheiner 1994, and Hughes 1998.

3. The best starting points in a large literature are Turner 1993, esp. 171–200; Heyck 1982; Desmond 1982; Porter 1978a.

The most obvious new source of salaried posts was at the Geological Survey, which had been founded through a strategic alignment of metropolitan utilitarian interests with provincial agricultural and mining ones. In 1845 a parliamentary act established the Survey's independence, and the number of posts began to increase. Yet in its cramped Museum of Economic Geology in Whitehall, De la Beche's outfit was confessedly temporary and not the haven it seems in retrospect. Worried parents viewed a scientific career as tantamount to joining the circus. Jobs with the Survey and related institutions were salaried and of higher status than freelance lecturing, hand-to-mouth journalism, or commissioned natural history work; but holders of the new posts fully shared in the uncertainties of commercial science. Most posts paid a pittance, so that several had to be combined to scrape together a career that even approached the lower bounds of gentility.[4]

The young men feeding off the scraps of science formed a subset of the bohemian world of pubs, clubs, and masculine camaraderie sketched by Thackeray in *Pendennis* (1848–50). They met every month in a convivial dining club, the Metropolitan Red Lions, at the Cheshire Cheese in Fleet Street. The original Red Lions had been set up in 1839 as a protest against "dons and donnishness," because British Association meetings were stuffy, genteel, and old-fashioned. Red Lion gatherings in London provided what dining clubs like the Shakespeare Club, the Museum Club, and the Fielding Club offered for literary men; indeed, there was a considerable overlap of membership. There were songs and poems, with much growling, drinking, and flourishing of coattails.[5] This was an all-male world, dominated by men who had remained bachelors into their thirties due to their uncertain prospects.

The world of scientific bohemia is best recorded in the diaries of the Scotsman Andrew Ramsay, who had joined the Survey in 1841, much to his mother's dismay—for she was shocked by the pittance he received as a salary. Ramsay and his friends would emerge from their tiny digs wearing the carefully preserved formal dress needed for a dinner at the Murchisons' Belgravia mansion, a Royal Society soirée, or a Friday evening discourse at the Royal Institution. Or they might stroll to the theaters, an oyster bar, the dance halls, or the West End's tableaux vivants ("mere nakedness," Ramsay complained, "without anything like art").[6]

4. Secord 1986b, Porter 1978a, McCartney 1977.

5. Wilson and Geikie 1861, 247–50, 388; Gay and Gay 1997, 431–36; Gardiner 1993; Masson 1908, 211–56.

6. Diary entry for 16 Jan. 1847, IC, Ramsay Papers, KGA Ramsay/1/8, f. 13r. Geikie (1895) publishes some passages from the diary.

Ramsay and his colleagues saw themselves as an all-male "brotherhood of the hammer." Women tended to be depicted with the chivalric conventions of sentimental melodrama. Those close to their own status whom they might meet at formal occasions were portrayed as dreamy unobtainable maidens, or as closely observed "specimens" ("very pretty" but "by no means uncommon" as noted in the original caption to figure 14.1). More often the Survey men met unattached lower-middle-class women at the West End dance halls. The Adelaide Gallery had been one of London's premier sites for popular science (Thomas Monck Mason had exhibited there), but in 1846 was transformed into

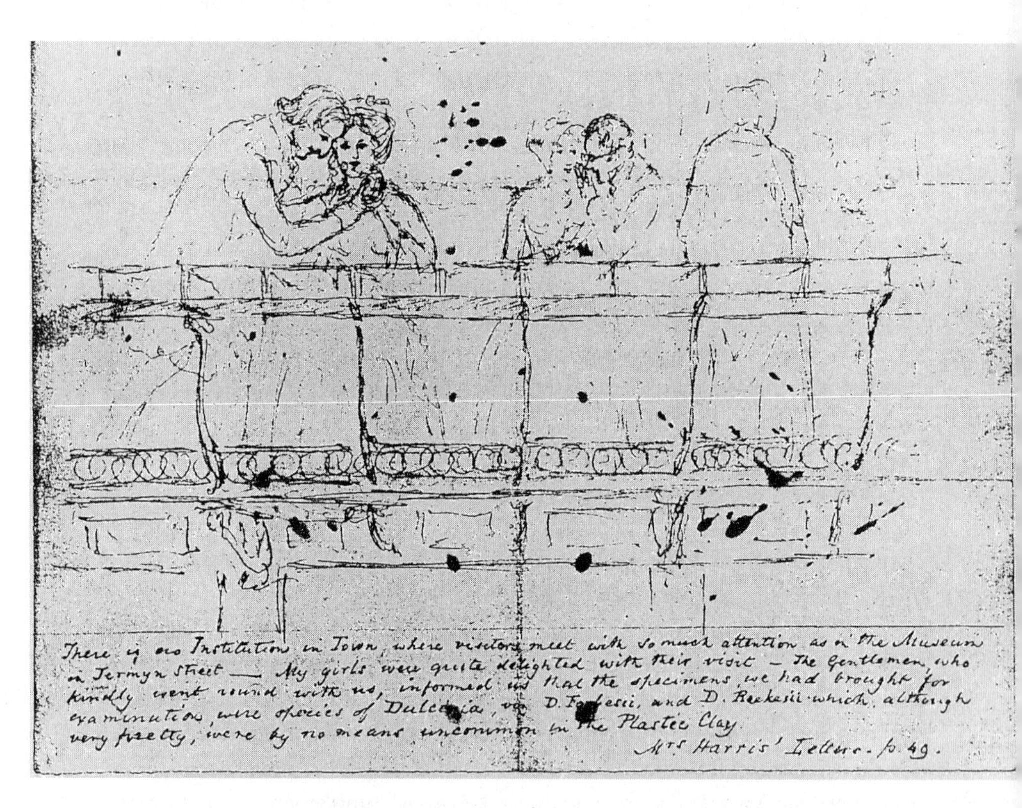

14.1 In this informally circulated drawing, two Survey men, probably Edward Forbes and Trenham Reeks, show female "specimens" around the Museum of Practical Geology, while their chaperone examines an exhibition case. "There is no Institution in Town, where visitors meet with so much attention as in the Museum on Jermyn Street—My girls were quite delighted with their visit—The Gentlemen, who kindly went round with us, informed us that the specimens we had brought for examination, were species of Dulcinia, viz D. Forbesii, and D. Reekesii which although very pretty, were by no means uncommon in the Plastic Clay. Mrs Harris' Letters p. 49." The "Plastic Clay" was a reference to the strata underlying London.

a casino. "I had a long chat with a little French demoiselle," Ramsay wrote after one of many visits, "& found Jukes busy with a big Edinburgh lass, who for a fille de jou seemed marvellously gauche & innocent."[7] The talk was probably not about science, but Ramsay and a friend did meet one young woman, "the most lovely creature in the world," who on the way home "began to launch out on Anatomical & physiological subjects in a most modest & intelligent manner to our infinite surprise & great admiration." They also admired the "nice little library" in her rooms at Euston Square, but declined her invitation to stay for tea. Like many young men, Ramsay went to the dance halls to talk, flirt, and drink, and not (so far as the diaries show) for sexual liaisons.[8] The following Saturday might find him conversing—though not about physiology—with Lady Shelley or Lady Murchison at a formal reception. Ramsay and his friends in "Scientific Young England" occupied an unstable place on the social map: neither securely within genteel society, nor in the hand-to-mouth world of Grub Street.

This can be seen in the case of Ramsay's friend Edward Forbes, the founder of the Red Lions, who scrabbled together a scientific career after his father lost his fortune. Trained in natural history and medicine, he became professor of botany at King's College, curator to the Geological Society, and (from 1844) palaeontologist to the Geological Survey. None of these positions paid well enough to maintain the style of life into which he had been born, so Forbes supplemented his income by anonymous writing for the periodicals and newspapers. He signed contracts for books with John Van Voorst, whose series on British zoology gave work to many naturalists. He wrote genial, uncontroversial reviews for the *Westminster,* viewing this as another way to keep science in the public eye and money in his purse.[9]

Forbes's anonymous notice in the *Lancet* reveals the ambiguities *Vestiges* posed for a reader in his position. He admired the book's spirit, which he welcomed for opening up connections between the sciences "to the general reader" rather than limiting them to "the timid coteries of the so-called scientific world." For those laboring away among "the *facts* of science," *Vestiges* was "a breath of fresh air to the workman in a crowded factory."[10] But (as Churchill had been alarmed to see) Forbes went on to criticize numerous details. Alarmed by the book's runaway success, he distanced himself from *Vestiges* in letters, at soirées, in the concluding lecture of a Royal Institution series, and in the

7. Diary entry for 16 Jan. 1847, IC, Ramsay Papers, KGA Ramsay/1/8, f. 13r.

8. Diary entry for 23 Mar. 1847, IC, Ramsay Papers, KGA Ramsay/1/8, f. 32v. For the reputation of these entertainments, see Mason 1994a, 98–103.

9. Wilson and Geikie 1861, Mills 1984. For the range of Forbes's publications, see Rehbock 1979.

10. [E. Forbes], *Lancet,* 23 Nov. 1844, 265–66.

periodicals.[11] Reviewing the controversy a few years later in one of his regular articles in the *Literary Gazette,* Forbes affected to believe that "the imaginative and elegant" *Vestiges* was on its deathbed, but he remained worried by continuing sales.

> The clear, pleasant, racy, self-sufficient style of the Vestigian captivated, when the dry, heavy, technical disquisitions and manuals of professors in science disgusted. The naturalists were taught the good lesson . . . to give the public the results of their researches . . . in plain, readable, and comprehensible language, and not to keep the philosophy of their science to themselves; for if they do so, others, unqualified for the task, will impose a sham philosophy on the people, who like to have a reason for their belief, and to be assured of the causes of things.

Reading had an indelible effect on the "unscientific brain": once absorbed, ideas became "an article of faith" almost impossible to dislodge. The diffusion of falsehood could be explained by the failings of scientific men as authors and the public as readers. Forbes still had hopes of the *Vestiges* author, whom he expected to see "calmly confessing his sins, and publicly recanting his faith in the transmutation of species."[12]

Forbes's serene tone exemplifies his belief that the manners of gentlemanly science should carry over into the new world of commerce; but younger men fighting to establish a new status for experts were less charitable. Joseph Hooker, recently returned from Sir John Ross's Antarctic voyage, briefly joined the Survey in 1845 as part of his campaign to obtain employment as a botanist. As he told Darwin in a letter, he found the book fun to read but full of errors and impossible to take seriously:

> I have been delighted with *Vestiges,* from the multiplicity of facts he brings together, though I do [not] agree with his conclusions at all, he must be a funny fellow: somehow the book looks more like a 9 days wonder than a lasting work: it certainly is "filling at the price".—I mean the price its reading costs, for it is dear enough otherwise; he has lots of errors. . . . After all what is the great difference between *Vestiges* & Lamarck, whom he laughs at.

For all Hooker's disclaimers, the book did remind him to tell Darwin of evidence from his travels that a mother's mental state might be imprinted to her

11. Examples of his comments (besides those already cited) include Wilson and Geikie 1861, 383–84; [E. Forbes], review of C. Lyell, *A Manual of Elementary Geology, Literary Gazette,* 24 Jan. 1852, 79–80; and E. Forbes to R. Jameson, 17 Feb. 1852, EUL mss. 1429/4, f. 64.

12. [E. Forbes], review of A. Sedgwick, *A Discourse on the Studies of the University of Cambridge, Literary Gazette,* 4 Jan. 1851, 5–7, at 5.

offspring; there might, he indicated, be more in such "nursery stories" than met the eye. But Hooker remained unconvinced about the development of species; by the summer of 1845 he thought (prematurely, as it turned out) that *Vestiges* was "already defunct."[13]

The young men of the new government-sponsored institutions, through their connections with journalists and literary editors, had been responsible for some of the first negative reviews. Hooker's dismissive attitude was shared by other Survey men, most of whom had been in their thirties when *Vestiges* appeared. Lyon Playfair, Ramsay's roommate and the Survey's chemist, found it full of faults. The Survey geologist Joseph Beete Jukes, the son of a Midlands manufacturer, thought it "crotchety."[14] Robert Hunt—while willing to treat progressive development as an imaginative fantasy in *Panthea*—warned readers in his fact-packed survey of *The Poetry of Science* (1848) that *Vestiges* could not be trusted.[15] Ramsay condemned it too, and edited out references that might seem favorable from official publications. In the proofs of a local geological memoir, one of the Survey men had described ammonites "as *susceptible of physical changes.*" Ramsay recognized that his colleague had meant that the range of these fossils in time had been influenced by changes in physical conditions, but worried that this would be misread. In a government-sponsored publication, organisms could not be seen as subject to evolutionary development.[16]

Of course, the dangers of speculation were a constant complaint from those experienced in the field and laboratory. But now the reasons were different. Unlike Brewster, Sedgwick, and Wilson, many of the young guard were indifferent to religion, so that separating theological concerns from scientific ones was not an emotional wrench. Ramsay rarely went to church, Playfair had been repelled by "arid Scotch orthodoxy" at an early age, and Jukes was privately a freethinking skeptic. De la Beche was conventionally an Anglican, but his private diaries reveal an anticlerical deist. Hooker was a Tory Anglican, but made little reference to such matters in his everyday affairs. Among the Survey men the only staunch Christians were Forbes and Sowerby's son-in-law, the palaeontologist John W. Salter—an evangelical Anglican who distributed tracts on street corners and carried his Bible on field trips. Salter's self-righteousness grated so badly with his colleagues that on one memorable occasion Jukes

13. J. D. Hooker to C. Darwin, 30 Dec. 1844, and Hooker to Darwin, 5 July 1845, in Burkhardt et al. 1985–99, 3: 103, 211. For Hooker's early career, see Huxley 1918.

14. On Playfair, RC to AI, [13 Apr. 1845], NLS Dep. 341/113/160–61; on Jukes, J. B. Jukes to C. M. Ingleby, 12 July 1848, in Jukes Browne 1871, 385.

15. Hunt 1848, 342, 366–67.

16. A. C. Ramsay to T. Reeks, [30 Dec. 1856?], British Geological Survey, GSM 1/420.

kicked him. Even for Jukes though, the point was not that science disproved religion—but that the two were different things.[17]

Speculation threatened the integrity of the new roles that the Survey and related institutions were attempting to foster. The new generation of careerists, emulating De la Beche and Airy, wished to establish science as the province of a paid elite independent of the mass audience. The shabby genteel world of commercial science, with its Grub Street connections and lack of security, was no longer satisfactory for ambitious men—now in their mid- to late thirties—who wanted to marry, settle down, and pursue research without scrambling for lecture shows and journalism. They wanted a new kind of intellectual aristocracy, in which social standing and evaluation of character would depend entirely on expertise. Publishers and editors would be in subservient positions, with scientific practice clearly demarcated from the communication of its results. The place that knowledge should have in the new order was embodied in the magnificent four-story Museum of Practical Geology, under construction from 1848 and opening its doors to the public in 1851, the year of the Great Exhibition (fig. 14.2). Tellingly, the entrance fronted on elegant Jermyn Street: the obvious alternative, Piccadilly, was too closely associated with the mermaids, dwarfs, and monsters of the West End science shows.[18]

The new building epitomized a new kind of science. Throughout the 1840s, the young government geologists had been immersed in tracing the details of the limestones, sandstones, and grits on the one-inch maps of the Ordnance Survey. This demanded hard-won specialist skills in surveying, including measuring six-inch-to-the-mile sections with theodolite and chain—a level of detail unknown in previous geological work. Thousands of fossils had to be collected for identification, cataloguing, statistical analysis, and display.[19] Technical monographs and memoirs had to be prepared to make the results available to other experts. Similar regimes were also being introduced in other government institutions, such as the precision observational astronomy of the Royal Observatory at Greenwich and the analysis of imperial coal supplies at the Royal College of Chemistry on Oxford Street. The knowledge maker in these places had little in common with Harriet Martineau's image of the inquirer into nature as "a free rover on the broad, bright breezy common of the universe."[20]

17. Reid 1899, 19 (Playfair); J. B. Jukes to A. H. Browne, 10 Apr. 1843, in Jukes Browne 1871, 404–5 (Jukes); Secord 1985 (Salter); McCartney 1977, 15–19 (De la Beche).

18. Forgan 1999; Yanni 1999, 51–61; Altick 1978, 498–99; Geikie 1895, 182–85.

19. Secord 1986a, 202–14.

20. Martineau 1877, 1:116; on precision astronomy, see Schaffer 1988; on chemical analysis, see Bud and Roberts 1984.

14.2 The spectacle of state expertise: engraved view of the interior of the Museum of Practical Geology on Jermyn Street. *Builder,* 28 Oct. 1848, 522.

Science in Literary Bohemia

For the men of the Survey, science achieved its power by being a neutralized form of expertise, taking the great problems of belief off the agenda. Though less shocking than outright infidelity, such studied indifference was a more radical break with the past. From the perspective of Jermyn Street, science needed to be purged not just of a particular form of doctrinal religion, but of speculative cosmology of any kind.[21] These standards of expertise were unusual and controversial, in some ways increasingly so in the more tolerant intellectual debates of the 1850s. Reviews of the tenth edition of *Vestiges,* other than those in the

21. As Joseph Hooker said, "These parsons are so in the habit of dealing with the abstractions of doctrines as if there was no difficulty about them whatever, so confident, from having the talk all to themselves for an hour at least every week . . . that they gallop over the course when their field is Botany or Geology as if we were in the pews and they in the pulpit. Witness the self-confident style of Whewell and Baden Powell, Sedgwick and Buckland." From J. D. Hooker to A. Gray, 1857, in Huxley 1918, 1:477–78; for discussions of this issue, see Turner 1993, 171–200, and Desmond 1982.

religious press, were generally positive. Thus the *Critic* gave two notices, the first a half-column eulogium and the second a substantial survey of the contents. The *Lancet* too praised the tenth as "by far the most interesting edition" of "this remarkable work" with "very excellent wood-cuts." With the book having "taken its stand amongst the philosophical works of the age," the reviewer could only wonder why it remained anonymous.[22]

There were commercial authors who took on any subject as long as it paid, although most had a few (antiquities, travels, biographies, natural history) around which most of their work clustered. Such writers came with a general grasp of their topic, an intimate knowledge of their audience, and a ticket for the British Museum library. Thackeray satirized this kind of reading in *Pendennis:* after a quick scan and a thoughtful moment over a cigar, the young critic might signify his "august approval" of Macaulay or Herschel, "as if the critic had been their born superior and indulgent master and patron."[23] The praise of *Vestiges* in the newspapers derived in part from impatience with the niceties of expertise. The new generation of journalists had been deeply impressed by the book and sympathized with the view—expressed so eloquently in *Explanations*—that the larger questions of science needed to be addressed. Marmaduke Blake Sampson, who worked at the Bank of England, also wrote for liberal magazines and newspapers like the *Spectator,* and became city editor for the *Times.* Active in the London Phrenological Society, he enthused about *Vestiges* in a letter to the unknown author, which he also forwarded to Combe. Progressive development offered "an escape from the fearful doctrines of the destinies of the soul at present promulgated by religious teachers," although mesmerism was for him more satisfying and scientific.[24]

The metropolis became a magnet for dissident middle-class radicals who increasingly labeled intellectual life in other settings as parochial. At the age of thirty-one, Marian Evans, the Coventry-born translator of Strauss, moved to London after a wrenching break with her father and the evangelical faith in which she had been raised. Evans read voraciously on mesmerism, physiology, and development (including *Vestiges*) and had her head cast by a London phrenologist. Combe found her brain extraordinary; she was "the ablest woman whom I have seen."[25] A fellow lodger remembered her as one of the "Insurgents": by the fire in a dark room, "with her hair over her shoulders, the easy chair half sideways to the fire, her feet over the arms, and a proof in her hands."

22. *Lancet,* 25 June 1853, 582; *Critic,* 1 July 1853, 336, 1 Oct. 1853, 510–11.

23. Thackeray [1848–50] 1994, 444.

24. M. B. Sampson to G. Combe, 11 Feb. 1845, NLS mss. 7277 ff. 35–38; on Sampson, see Gibbon 1878, 2:134, and de Giustino 1975, 101, 146.

25. Ashton 1996, 3, 89–90.

Eliza Lynn, who also boarded for a time at the Strand, found the newcomer provincial and gauche.[26]

Another exile from the Midlands was Herbert Spencer, a railway engineer who became subeditor of the *Economist* and published comprehensive schemes of cosmic development. Tyndall came up from Queenswood in 1853, becoming Faraday's successor at the Royal Institution, and Hirst followed seven years later, to a mathematics lectureship at University College.[27] By the late 1840s, Baden Powell had a reputation for heterodoxy that made him an irrelevance in the reform of Oxford science, and he associated instead with the metropolitan radicals. Like many others, he wrote to the *Vestiges* author, mainly to claim priority. His *Essays on the Spirit of Inductive Philosophy, the Unity of Worlds, and the Philosophy of Creation* (1855) gave *Vestiges* the first wholehearted support it had ever had in a book, stopping only at the origin of man.[28] Although Powell's *Essays* went into a second edition, they were too abstruse to sell really well.[29]

In the expansive world of metropolitan journalism, *Vestiges* became a useful tool in an ongoing campaign. Several weeklies tied the flag of progressive development to their mastheads. One was the liberal *Atlas* (f. 1826), with its lengthy and sometimes critical supplement of *Vestiges;* another was the *Critic* (f. 1843), which combined the romantic Toryism of Disraeli with an enthusiasm for mesmerism and Continental literature. Here is what it had to say about *Vestiges* at the end of 1845:

> For our own part, without subscribing to the theories of the author, or even admitting all his facts, we rejoice in the appearance of his book, and the attention it has attracted. Scientific men have a great tendency to dogmatism. . . . The use of such daring and dashing intruders as "The Vestiges," lies in this,—that it prevents the sages from going to sleep under their laurels, and flattering themselves that no more remains to be discovered. . . . The author of "The Vestiges" may be as wrong in his conjectures and facts as was asserted in the clever article in the *Edinburgh Review,* but that does not disentitle him to the thanks of those who think that to stimulate inquiry is to ensure progress, and that he has done more to this end than any other writer of our day will be acknowledged by his most bitter opponents.[30]

26. Ibid., 5; Linton 1899, 94–99. W. H. White, *Athenaeum,* 28 Nov. 1885, 702.

27. Eve and Creasey 1945, 43–58; Brock and MacLeod 1980b, 14.

28. Corsi 1988, 272–85; Powell 1855; Powell to the author of *Vestiges of the Natural History of Creation,* in *V*12, xxx–xxxi.

29. As Owen Chadwick (1971, 555) said, "For his impact upon the world at large he might have been writing in Latin"; the *Essays* did, however, occasion a number of reviews and comments in presidential addresses at scientific societies.

30. Review of *E*1, *Critic,* 20 Dec. 1845, 676–78, at 676; see also *Critic,* 10 Apr. 1847, 277; *Critic,* 1 July 1853, 336; *Critic,* 1 Oct. 1853, 510–11.

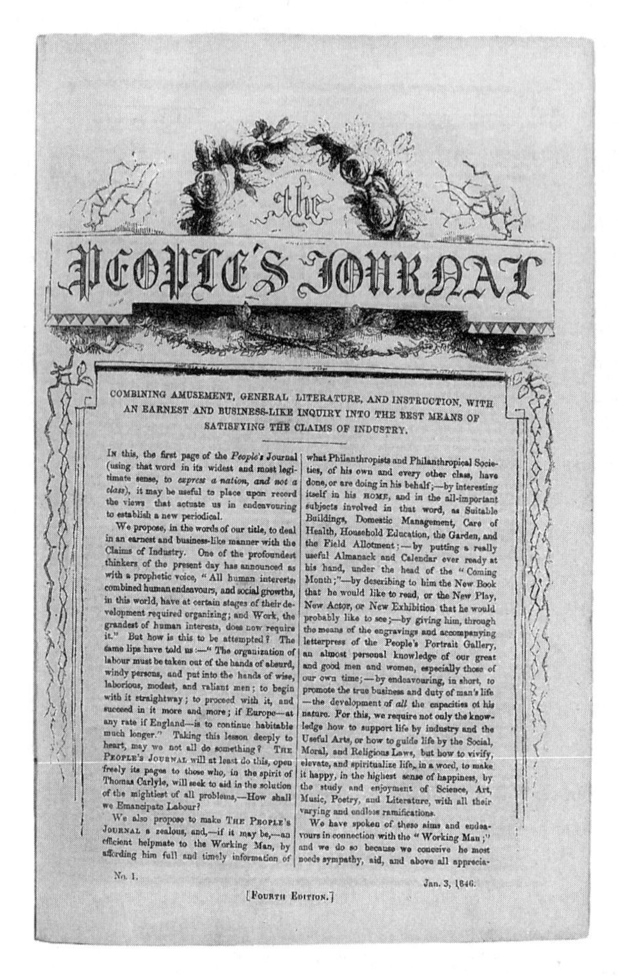

The People's Journal

COMBINING AMUSEMENT, GENERAL LITERATURE, AND INSTRUCTION, WITH AN EARNEST AND BUSINESS-LIKE INQUIRY INTO THE BEST MEANS OF SATISFYING THE CLAIMS OF INDUSTRY.

In this, the first page of the *People's* Journal (using that word in its widest and most legitimate sense, to *express a nation, and not a class*), it may be useful to place upon record the views that actuate us in endeavouring to establish a new periodical.

We propose, in the words of our title, to deal in an earnest and business-like manner with the Claims of Industry. One of the profoundest thinkers of the present day has announced as with a prophetic voice, "All human interests, combined human endeavours, and social growths, in this world, have at certain stages of their development required organizing; and Work, the grandest of human interests, does now require it." But how is this to be attempted? The same lips have told us:—"The organization of labour must be taken out of the hands of absurd, windy persons, and put into the hands of wise, laborious, modest, and valiant men; to begin with it straightway; to proceed with it, and succeed in it more and more; if Europe—at any rate if England—is to continue habitable much longer." Taking this lesson deeply to heart, may we not all do something? The People's Journal will at least do this, open freely its pages to those who, in the spirit of Thomas Carlyle, will seek to aid in the solution of the mightiest of all problems,—How shall we Emancipate Labour?

We also propose to make The People's Journal a zealous, and,—if it may be,—an efficient helpmate to the Working Man, by affording him full and timely information of what Philanthropists and Philanthropical Societies, of his own and every other class, have done, or are doing in his behalf;—by interesting itself in his home, and in the all-important subjects involved in that word, as Suitable Buildings, Domestic Management, Care of Health, Household Education, the Garden, and the Field Allotment;—by putting a really useful Almanack and Calendar ever ready at his hand, under the head of the "Coming Month;"—by describing to him the New Book that he would like to read, or the New Play, New Actor, or New Exhibition that he would probably like to see;—by giving him, through the means of the engravings and accompanying letterpress of the People's Portrait Gallery, an almost personal knowledge of our great and good men and women, especially those of our own time;—by endeavouring, in short, to promote the true business and duty of man's life—the development of *all* the capacities of his nature. For this, we require not only the knowledge how to support life by industry and the Useful Arts, or how to guide life by the Social, Moral, and Religious Laws, but how to vivify, elevate, and spiritualize life, in a word, to make it happy, in the highest sense of happiness, by the study and enjoyment of Science, Art, Music, Poetry, and Literature, with all their varying and endless ramifications.

We have spoken of these aims and endeavours in connection with the "Working Man;" and we do so because we conceive he most needs sympathy, aid, and above all apprecia-

No. 1. Jan. 3, 1846.

[Fourth Edition.]

14.3 First page of the weekly *People's Journal,* a progressive journal targeting the workingman and grounded in the traditions of philosophical radicalism. This was among the periodicals avidly read by Tom Hirst and other self-improvers. *People's Journal,* 10 Jan. 1846, 1.

The *Critic* claimed to support *Vestiges* as part of its aim to speak neither for publishers nor authors, but for readers. It advertised itself as being conducted by a group of independent literary gentlemen, "not having themselves any connection with the *business of literature.*" In fact most of its authors were not very different from those who wrote for the other literary weeklies—young men about town who earned a portion of their income by commenting on current affairs.[31]

The importance of *Vestiges* for the weekly periodical press is indicated by the *People's Journal,* which advocated progressive, humanitarian causes (fig. 14.3). The first issue printed an address to the working classes by the celebrated Unitarian preacher, the Reverend W. J. Fox, who identified opposition to *Vestiges* with attempts to uphold class interests by restricting the spread of knowledge:

31. *Newspaper Press Directory* 1846, 55. Many of the contributors were connected with the law; see Watkins 1982, 89–90.

> But those who find advantages—those who retain power—by keeping the human mind in leading-strings, by acting on its timidity and exciting its apprehensions—those whose highest objects of ambition is dominion over the consciences of their fellow-creatures—they, like black magicians, raise up their hideous phantoms to guard their enchanted ground, and call forth foul shapes to fill the mind of man with doubts and fears, in order that it may not venture to pierce through the circle of ignorance in which they should confine it, and ascend upwards to its own appropriate sphere of knowledge. Now, science seldom takes a decided step in advance, but there is a clamour made as though there were danger in it to the souls of men.[32]

Fox claimed that in misrepresenting *Vestiges* as infidel materialism, the reviewers had been like "a bullying counsel," bigoted and unfair. In addressing working people (though most readers were probably middle-class) these journals attempted to sow prudence, self-help, temperance, and Christian charity. "No living beings," the writer Mary Howitt reassured her sister, "can estimate Divine Revelation higher than we do." The aim, in part, was to counter the crudities of the sensationalist press with a sharper, livelier mix of articles than that provided by *Chambers's Journal*.[33]

The most daring and successful of the new radical weeklies was the *Leader* (f. 1850), which advocated secular education, Chartism, and international republicanism. Finances were provided by the Reverend Edmund Larken, the clergyman at whose table Hirst had met Holyoake. Holyoake, who continued to edit the *Reasoner*, also worked as the *Leader*'s office manager, an indication of how former working-class "infidels" were making common cause with middle-class journalists. George Henry Lewes, a novelist, actor, and author, handled the review columns (fig. 14.4). The Carlyles fondly called him "the ape."[34] As Eliza Lynn later remembered, he was the most audacious talker she knew, with "neither shame nor reticence in his choice of subjects, but would discourse on the most delicate matters of physiology with no more perception that he was transgressing the bounds of propriety than if he had been a learned savage."[35] However, Lewes keenly felt the pressure of convention: contemporary censorship was "more domesticated, more searching and constraining." In a lithographed circular, he invited Combe, Chambers, and a few others to join in a confidential society that would allow them to express "their real opinions

32. W. J. Fox, "On the Progress of Science in its Influence upon the Condition of the People," *People's Journal,* 10 Jan. 1846, 30–35. See also G. I. Cowan, "Development," *People's Journal,* 29 Apr. 1848, 255–57.

33. Howitt 1889, 2:39–40. On magazines of "popular progress," see Maidment 1984.

34. Ashton 1991, 73. Besides Ashton's excellent biography, see Desmond 1982, 28–31, which associates Richard Owen with the Lewes and Chapman circle. The antagonisms, however, seem at least as important: see Rupke 1994, 207–9, 215.

35. Linton 1899, 18.

14.4 Unrestrained talk among the "advanced thinkers" in London's literary circles tended to be a masculine affair. This etching of 1842 shows the aging bohemian Leigh Hunt, his son-in-law George Henry Lewes, his son Vincent Hunt, and the artist William Bell Scott engaged in conversation. W. Minto 1892, 1: facing p. 130.

with all the frankness which security can give."[36] This was the kind of place where the veil of anonymity could be lifted from *Vestiges.*

The *Leader* attempted to embody something of the same conversational daring in print (its provisional title had been "The Free Speaker").[37] Its reviews and editorials argued that *Vestiges* was too tame and flawed by dependence on a preordained metaphysical law: but "delightful" and still the best available defense of the application of law to nature and society. No mere charms of style, nor the counterfeit glitter of deadly poison, could have sold so many editions. Exaggerated claims for the literary qualities of the work, Lewes wrote, were attempts to draw attention from the real source of its popularity: "an uneasy unrest in the minds of men, a painful suspicion of the validity of what acknowledged Teachers choose to avow. Choose your horn!"[38]

The "acknowledged Teachers" targeted by the *Leader* included Sedgwick, whose fifth edition of the *Discourse on the Studies of the University of Cambridge*

36. G. H. Lewes to G. Combe, 8 Feb. [1850], NLS mss. 7309, ff. 78–79; G. H. Lewes to RC, 8 Feb. [1850], NLS Dep. 341/98/76; Ashton 1991, 90–91.

37. Ashton 1991, 86–111, is especially helpful on the founding of the *Leader;* see esp. 89–91.

38. [G. H. Lewes], "Development Theory and Mr. H. Miller's Book," *Leader,* 4 May 1850, 138; see also "Lyell and Owen on Development," *Leader,* 18 Oct. 1851, 996; "The Development Hypoth-

was reviewed in 1850. Asked by his publisher, John W. Parker, to provide new introductory material for his sermon of 1832, Sedgwick allowed the book to grow uncontrollably in three years into a "brain-monster" about popery, transmutation, transubstantiation, and academic reform. *Vestiges* was the main target.[39] As with the *Edinburgh,* Sedgwick never had his entire text before him, but this time there was no restraining editorial hand. His philosophical testament bloated into a bibliographical curiosity, with the original 94-page sermon lost in 442 pages of preface and 228 pages of appendices. Even supporters were disappointed, Brougham backhandedly thanking him for "the somewhat amorphous—at least oddly-proportioned—book."[40]

Lewes launched into this flabby target. As the *Leader* sneered, "This slovenly and inexcusable construction" contained a host of curious facts, but was "intemperate and illogical," a worthless tissue of "flippant profundity":

> Has it intrinsic value? Frankly, it has little. Professor Sedgwick has a respected name in geology, and friends speak of him as a man richly endowed by nature. . . . He is in truth a great master of the Vituperative Syllogism; he reasons with epithets. He is fond of discrediting opponents by coupling some offensive epithet with their opinion. Yet this same Professor is found complaining of "brawling and ignorant declamation," and declaring that the author of the *Vestiges* (usually considered a singularly mild and courteous writer) "braves out a bad cause by insulting language, and a confident tone of superior intelligence." Certainly, if ever man deserved that to be said of him, Professor Sedgwick is the man.[41]

There was a clear sense, and not just among free-living literati, that the Anglican natural theologians had lost—and the author of *Explanations* had won—the war of scientific manners. Sedgwick's anger and impatience had reduced the credibility of his position. Moreover, doctrines that had been in the vanguard of intellectual life were increasingly identified as the peculiar property of the evangelical party. Not surprisingly, the *Discourse* was more popular in Pres-

esis of the 'Vestiges,'" *Leader,* 13 Aug. 1853, 20 Aug. 1853, 27 Aug. 1853, 10 Sept. 1853, 784–85, 812–14, 832–34, 883–84, quotation on p. 784. See also G. H. L[ewes], "A Precursor of the *Vestiges,*" *Fraser's Magazine* 56 (Nov. 1857): 526–31.

39. A. Sedgwick to C. B. Phipps, 9 Sept. 1850, in Clark and Hughes 1890, 2:186; for the genesis of the book, see Clark and Hughes 1890, 2:187–93.

40. Quoted in Clark and Hughes 1890, 2:193. Some old friends took a more positive view, and even the aging geologist William Daniel Conybeare hinted that not much would be changed even if the development theory proved true: "as to Vestiges of Creation I dont care for them. Were the theory true, it would only appear to me an involved series of consequences contemplated & adjusted ab initio by the same Creative & Provident intellect to which reason quite as much as faith . . . impels us to ascribe everything." W. Conybeare to A. Sedgwick, 15 Dec. 1851, CUL Add. mss. 7652.II.Q.40a.

41. [G. H. Lewes], "Philosophy at Cambridge," *Leader,* 7 Sept. 1850, 566–67, at 566.

byterian Edinburgh than in London Society. In the late 1840s, Napier was asking men like Lewes to write for the *Edinburgh,* not Sedgwick or Brewster.

When Chapman bought the *Westminster* in 1852, developmental progress became the coping stone of its editorial philosophy. The prospectus, largely written by Marian Evans, proclaimed that "[t]he fundamental principle of the work is the recognition of the Law of Progress" (fig. 14.5). Chapman's house and publishing premises on the Strand became a focal point for uniting the scattered forces of liberalism, from James Martineau's new vision of Unitarianism to Combe's programs for education and prison reform. A committed vanguard could undermine the compromising marriage between Church, aristocracy, and intellect that had held together the country since the first Reform Act.[42]

The contributors in the first year—Combe, W. R. Greg, Harriet Martineau, John Stuart Mill, Francis Newman, Herbert Spencer—had longstanding interests in extending the boundaries of science. Chapman was no exception, as in 1848 when he had initiated a correspondence with Richard Owen about the formation of new species. This was not out of general curiosity, but because Chapman wanted Owen to introduce the British edition of Andrew Jackson Davis's mesmeric visions about life on Jupiter and Saturn. Owen refused, but Chapman remained keenly interested in the problem, which typified the belief in advanced circles that the boundaries of science would be radically transformed.[43] Even before Chapman's takeover, the *Westminster* had consistently supported progressive development; in 1847, the then-editor William Hickson took the unusual step of arranging for the sixth edition of *Vestiges* to be reviewed to make amends for its mildly critical notice of the first. George Luxford, as subeditor, even asked Churchill for advice from the author. *Vestiges* might or might not give correct or competent answers, but as the review said, there was no work better adapted "to give a right direction to the philosophical investigation of the highest subjects of human interest."[44]

Many in Chapman's circle were more equivocal in their praise, preferring to trace the development of their own ideas from less controverted (and usually Continental) sources. Spencer had first heard of *Vestiges* at the age of twenty-

42. The full text of the original prospectus is in Moore 1988, 432–35, and Eliot 1990, 3–7. On the *Westminster,* see Rosenberg 1982; Postlethwaite 1984; Desmond 1982, 30–31; Cashdollar 1989, 69–72, and the large literature on the early writings of George Eliot.

43. J. Chapman to R. Owen, 13 Jan. 1848; 11 Mar. 1848, NHM Owen Correspondence, vol. 7, ff. 26–28. The letter was first printed in Owen 1894, 1:309–10, and is discussed in Desmond 1982, 33–37, and Rupke 1994, 225–30.

44. [G. Luxford], "Natural History of Creation," *Westminster and Foreign Quarterly Review* 48 (Oct. 1847): 130–60, at 130; see also [G. Luxford], "Natural History of the Human Species," *Westminster and Foreign Quarterly Review* 49 (Apr. 1848): 284–86. For the correspondence behind the scenes, see RC to AI, [Apr. 1847], NLS Dep. 341/110/237–38, 53–54, which includes materials sent to "The Editor of the Westminster Review, care of Mr. George Luxford."

THE WESTMINSTER REVIEW

Is designed as an instrument for the development and guidance of earnest thought on **Religion and Theology, Social Philosophy, Politics, and General Literature;** and to this end the Editor seeks to render it the organ of the most able and independent minds of the day.

The fundamental principle of the work is the recognition of the Law of Progress. In conformity with this principle, and with the consequent conviction that attempts at reform—though modified by the experience of the past and the conditions of the present—should be directed and animated by an advancing ideal, a comparison of the actual with the possible is steadily maintained, as the most powerful stimulus to improvement. Nevertheless, in the deliberate advocacy of organic changes, it is not forgotten, that the institutions of man, no less than the products of nature, are strong and durable in proportion as they are the results of a gradual development, and that the most salutary and permanent reforms are those which, while embodying the wisdom of the time, yet sustain such a relation to the moral and intellectual condition of the people, as to ensure their support.

In regard to **Religion and Theology** the Review unites a spirit of reverential sympathy for the cherished associations of pure and elevated minds with an uncompromising pursuit of truth. The elements of ecclesiastical authority and of dogma are fearlessly examined, and the results of the most advanced Biblical criticism are discussed without reservation, under the conviction that religion has its foundation in man's nature, and will only discard an old form to assume and vitalize one more expressive of its essence.

The Review gives especial attention to that wide range of topics which may be included under the term **Social Philosophy.** It endeavours to form a dispassionate estimate of the diverse theories on these subjects, to give a definite and intelligible form to the chaotic mass of thought now prevalent concerning them, and to ascertain both in what degree the popular efforts after a more perfect social state are countenanced by the teachings of politico-economical science, and how far they may be sustained and promoted by the actual character and culture of the people.

In the department of **Politics** careful consideration is given to all the most vital questions, without regard to the distinctions of party; the only standard of consistency which is adhered to being the real, and not the accidental relations of measures—their bearing, not on a ministry or a class, but on the public good.

In the department of **General Literature** the criticism is animated by the desire to elevate the standard of the public taste, in relation both to artistic perfection and moral purity, and to guide the reader in selecting for himself only those books which are of intrinsic value.

An endeavour is made, by a careful analysis and grouping of each quarter's productions, at once to exhibit the characteristics of the individual works reviewed, and to supply a connected and comparative History of Contemporary Literature.

JOHN CHAPMAN, 8, KING WILLIAM STREET, STRAND.

14.5 Prospectus for the *Westminster Review,* Oct. 1857; a shortened version of the statement largely written by Marian Evans and first published by John Chapman in 1852. (Bound in *Cambridge Essays, Contributed by Members of the University* [London: John W. Parker, 1857].)

four, while he was working as a railway engineer in Birmingham in 1845. After talking with a Liverpool gentleman who spoke highly of it, he recommended the book to a friend and for possible purchase by the Birmingham Mechanics' Institute. Five years later, after moving to London and writing his own essays on the "development hypothesis" for the *Leader* and *Westminster*, he was much more dismissive.[45] The common view among those well read in German-language physiology was that the book offered a serviceable overview calculated to make complex ideas accessible, even as it suffered from inaccuracy and anonymity. The most likely author, it was generally felt in these circles, was Chambers, whose literary talents were admired even as his capacity for rigorous philosophy was questioned.[46] "There are faults in that delightful work," Lewes wrote in the *Leader*, "errors both in fact and philosophy; but compared with the answers it provoked, we cannot help regarding it as a masterpiece."[47]

Evans, whose Strauss translation had been vilified at Exeter Hall, recognized the value that reading *Vestiges* could have in helping others to escape the clutches of cant. Her *Westminster* essay on "evangelical teaching" made quick work of the Reverend John Cumming's dismissal of what the Presbyterian leader had denounced as "'that very unphilosophical book.'" Cumming "tells us that 'the idea of the author of the "Vestiges" is, that man is the development of a monkey, that the monkey is the embryo man, so that *if you keep a baboon long enough, it will develop itself into a man.*'" Cumming's qualifications as a judge could be measured, Evans wrote, by his implied claim that the nebular hypothesis had been put forward in *Vestiges* for the very first time.[48]

Because of Marian Evans's later fame, her entertaining dismissal has been taken as the last word on Cumming's millenarian brand of evangelical teaching. But in the 1850s this was not the case. Cumming was an influential preacher whose works sold tens of thousands of copies; Evans was a social outlaw living with a married man. After she moved in with Lewes, old friends like the Chamberses and Combes would have nothing to do with her. Her own position as an author and editor was marked by the difficulties of being a woman.[49] These were less acute than those involved in entering the all-male world of the

45. Compare with Spencer's dismissal of the book in reporting his first conversation with Lewes in 1850 (Spencer 1904, 1:348), with H. Spencer to E. Lott, 18 Mar. 1845 (Spencer 1904, 1:266–69, at 269).

46. E.g., R. Chapman to R. Owen, 13 Jan. 1848, NHM Owen Correspondence, vol. 7, f. 26.

47. [G. H. Lewes], "Lyell and Owen on Development," *Leader*, 18 Oct. 1851, 996.

48. [G. Eliot], "Evangelical Teaching: Dr. Cumming," *Westminster Review* 64 (Oct. 1855): 436–62, at 450, in Eliot 1990, 54.

49. David 1987, 159–224, and for the context in changes in the publishing industry, Tuchman 1989. Ashton 1996 and Hughes 1998 provide moving accounts of Evans's radical position and her ostracism from society.

Red Lions and "the free-and-easy jolly boys" at the Geological Survey: forging paid careers in science meant, quite simply, pushing women out.[50]

For Evans, such tensions were released through writing fiction. Her early stories and novels, written under the carefully guarded male pseudonym George Eliot, applied the methods of scientific natural history to ordinary, everyday human life. The characters in *Scenes of Clerical Life* (1857) and *Adam Bede* (1859) were closely observed, set in clear relation to one another, and shown in accordance with inner qualities of character and the outer circumstances of time and place. This stress on accurate observation combined her love of Scott's historical fiction with a remarkable knowledge of physiology, phrenology, and evolutionary philosophies of progress. She had been immersed in such works since the early 1840s. That at least some of these writings were also indebted to Scott's novels made the connection a natural one. The result was a new form of realist fiction, very different from the comic exaggerations of Dickens or the glittering fantasies of Lynn Linton or Bulwer-Lytton. Developmental evolution thus shaped the realist novel not through Darwin's plots, but through those in Combe's *Constitution*, Spencer's *Westminster* essays, and *Vestiges*.[51]

Ignotus in the Metropolis

Sexual propriety was not the only issue that threatened to tear radical London apart. Tactics in religious controversy were another, as demonstrated when Chapman published Harriet Martineau and Henry George Atkinson's *Letters on the Laws of Man's Nature and Development* (1851), which offered an uncompromising blend of atheism, mesmerism, and materialism. Although Evans admired their courage, she found the book "studiously offensive."[52] Respectability remained very much an issue. The Howitts and other authors whose writings had already found a place in middle-class homes had to be especially careful. The *Leader*'s prospectus would appear so inflammatory in Edinburgh, Combe told Lewes, that their common friend Chambers would be able to subscribe only if copies were hand-delivered "secretly by a confidential agent."[53]

50. Richards 1989b; for the "jolly boys," see A. C. Ramsay to H. T. De la Beche, 12 May 1847, NMW De la Beche Papers.

51. Shuttleworth 1984, Paxton 1991, and A. S. Byatt, introduction to Eliot 1990, are among the best analyses of these issues; see also Postlethwaite 1984, which (contrary to its aim) illustrates the difficulty of extracting a unified worldview from the *Westminster* circle. The attention devoted to Darwin in Beer 1983 and Levine 1988—while illuminating the later novels—has tended to reinforce the tendency to make evolutionary connections primarily through the *Origin*.

52. Morley 1938, 2:707, quoted in Ashton 1996, 82. Webb (1960, 299–302) discusses the *Letters* and contemporary reactions, as does Postlethwaite (1984, 141–55), although her argument that it was a "prototypical expression of their Victorian world-view" (142) is surely mistaken.

53. G. Combe to G. H. Lewes, 30 Dec. 1849, NLS mss. 7392, ff. 19–21.

In London things were less fraught, although the *Vestiges* secret had to be kept there too. When Chambers traveled to the metropolis in April 1845 to hear the gossip, his contacts were in the world of radical journalism. He met fellow editor G. M. W. Reynolds, dined with the horticultural author Jane Loudon and her daughter, and heard speeches at a Covent Garden rally that stressed that the Corn Laws violated the laws of nature. A quiet Saturday evening was spent with the Howitts, the Sopwiths, and Playfair, whom Chambers knew through family connections in St. Andrews. There was much talk about *Vestiges,* and Playfair echoed the common refrain that no book since *Waverley* had made a greater sensation.[54] On a visit later that summer, Chambers was introduced to Leigh Hunt, who spoke moderately about *Vestiges* before the company, but with great warmth once they were alone. "What an idea have we here," Chambers commented, "of the vast amount of liberality of mind there is in this country, smothered under the terror of reigning dogmas!"[55]

With the help of his old friend Thomas Sopwith, Chambers also began "breaking ground amongst the scientific people." Connections between literary and scientific bohemia made the move relatively easy to make; with Sopwith as his sponsor, he had joined the Geological Society in December 1844. On 17 April the Red Lions invited Chambers to the Cheshire Cheese for one of the wildest dinners the club ever held.[56]

At these gatherings, Chambers's name came up only occasionally as a possible *Vestiges* author. Carpenter would have conveyed his suspicions; but the Survey men preferred to follow a trail of facts. This was a form of detective work peculiarly suited to their "microscopic" reading practices and more convincing to them than any rumors. After making many other guesses, at an informal gathering at Down House in late 1845 Forbes pointed out a similarity in an error between *Vestiges* and writings in which Chambers was known to have a hand. Further confirmation came when *Chambers's Journal* extracted an account of the Kerguelen's Land cabbage (fig. 14.6) from a monograph by Hooker, and used it to support the marine origin of all plants. Hooker learned

54. RC to AI, [13 Apr. 1845], NLS Dep. 341/113/160–61; see also Thomas Sopwith, diary entry for 12–15 Apr., Robinson Library, University of Newcastle.

55. RC to AI, [July 1845], NLS Dep. 341/110/62–63.

56. Wilson and Geikie 1861, 388, mentions Chambers's presence at the dinner; for the other details, see RC to AI, [Apr. 1845], NLS Dep. 341/113/56–57; RC to AI, [13 Apr. 1845], NLS Dep. 341/113/160–61. His membership certificate (no. 1439) at the Geological Society was put out on 6 Nov. 1844, with support from Buckland, Horner, and S. Pratt—and Sopwith "from Personal Knowledge." The election took place 18 Dec. 1844. Further documents dealing with Chambers's visits to London are published in C. E. S. Chambers, "Robert Chambers's Commonplace Book," *Chambers's Journal,* 19 Oct. 1901, 737–40.

14.6 Facts as clues: the botanist Joseph Hooker used an article in *Chambers's Journal* about the Kerguelen's Land cabbage—a plant described in his *Flora Antarctica*—to trace the *Vestiges* authorship to Robert Chambers. The article claimed that the plant had developed from a seaweed under the influence of new conditions. Hooker 1844–47, plates 90–91.

soon after that Darwin had received a presentation copy of *Vestiges* shortly after meeting Chambers for the first time.[57]

Such clues were important because reputation among the young professionals depended so heavily on scientific credentials. Labeled as the work of a self-identified "essayist of the middle class," *Vestiges* lost status. Embarrassed by Chambers's attempts to instruct them, Survey men such as Ramsay dismissed his fieldwork as "puerilities about terraces." At the 1847 British Association meeting, Ramsay could only agree with Wilberforce and the gentlemanly geologists. When Chambers remarked to him afterward that the bishop's sermon was "levelled against the spread of knowledge," Ramsay had no sympathy:

57. C. Darwin to J. D. Hooker, [10 Feb. 1846], in Burkhardt et al. 1985–99, 3:287–89, at 289; J. D. Hooker to C. Darwin [1 Feb. 1846], in Burkhardt et al. 1985–99, 3:281–84, at 283–84. [Chambers], "The Kerguelen's Land Cabbage," *CEJ*, 31 Jan. 1846, 76–77, extracted from Hooker 1844–47, 238–41.

"most people seemed to think it showed the proper way to pursue knowledge. Chambers will go down to Scotland with the feeling of a martyr."[58]

As a veteran of the Scottish wars between evangelical Protestantism and the sciences of progress, Chambers had been amazed to discover that the Survey men had no religious objections to the book. "Strange to say," he had reported to Ireland after first encountering the Red Lions, "there is hardly a word against it on theological grounds, but the scientific men find many faults."[59] He realized that men like Ramsay, Playfair, Hunt, and Forbes would be crucial to ridding science of dogma. They formed what he told an American correspondent was "a little school" of men "who appear likely to elicit new truths of considerable importance."[60] He would do anything possible to push their careers forward, and regretted that the fiasco of the lord provost election had destroyed any chance of influencing university appointments in Edinburgh for these "worthy fellows."[61]

In the years that followed, Chambers had worked hard to meet the demands of the young guard, both in his work on terraces and in the revisions to *Vestiges*. In 1851 he began to prepare the illustrated edition proposed by Churchill. Drawing again upon Carpenter's assistance, he was confident of revising to the highest standards, with "facts more conclusive and more guarded against challenge, and arguments less assailable."[62] Impatient, he wrote to Ireland:

> What makes me now eager to come out is, that I have prepared new controversial chapters on Hugh Miller's Footprints of the Creator and Sedgwick's preface to the Discourse on the Studies at Cambridge, not merely defending every position of my own, but punishing the insolent foes in what I think a very effective manner. I anticipate that these chapters will give quite a fresh interest to the book. It is surely a great pity to allow the enemy to go on year after year unanswered. . . .

58. A. C. Ramsay, diary entry for 27 June 1847, IC, Ramsay Papers, KGA Ramsay/1/8, f. 59v. Five years later, Ramsay took a much more positive view of Chambers's studies of glaciation, perhaps because his own work was being attacked for being too speculative: see A. C. Ramsay to RC, 19 Nov. 1852, NLS Dep. 341/87/49.

59. RC to AI, [13 Apr. 1845], NLS Dep. 341/113/160–61; see also T. Sopwith, diary entries for Mar.–Apr. 1945, Robinson Library, University of Newcastle, vol. 39.

60. RC to M. Griffith, 2 Dec. 1846, quoted in Cox 1993, 193.

61. Chambers blamed his failure to obtain the post on a late start, not on accusations of the *Vestiges* authorship that had actually precipitated his withdrawal: he knew well enough that continued association with the book was unlikely to enhance his status with his London friends. (Hunt, for his part, was pleased at a promise that his *Poetry of Science* would be featured in *Chambers's Journal*, and thought Chambers's sorrow at being unable to help further "a most noble regret.") R. Hunt to A. C. Ramsay, 18 Nov. 1848, IC, Ramsay Papers, KGA/Ramsay/8/506/43; RC to [R. Hunt], 16 Nov. 1848, APS, Letters of Scientists, 509/L56.

62. RC to Churchill, [1851], NLS Dep. 341/111/28.

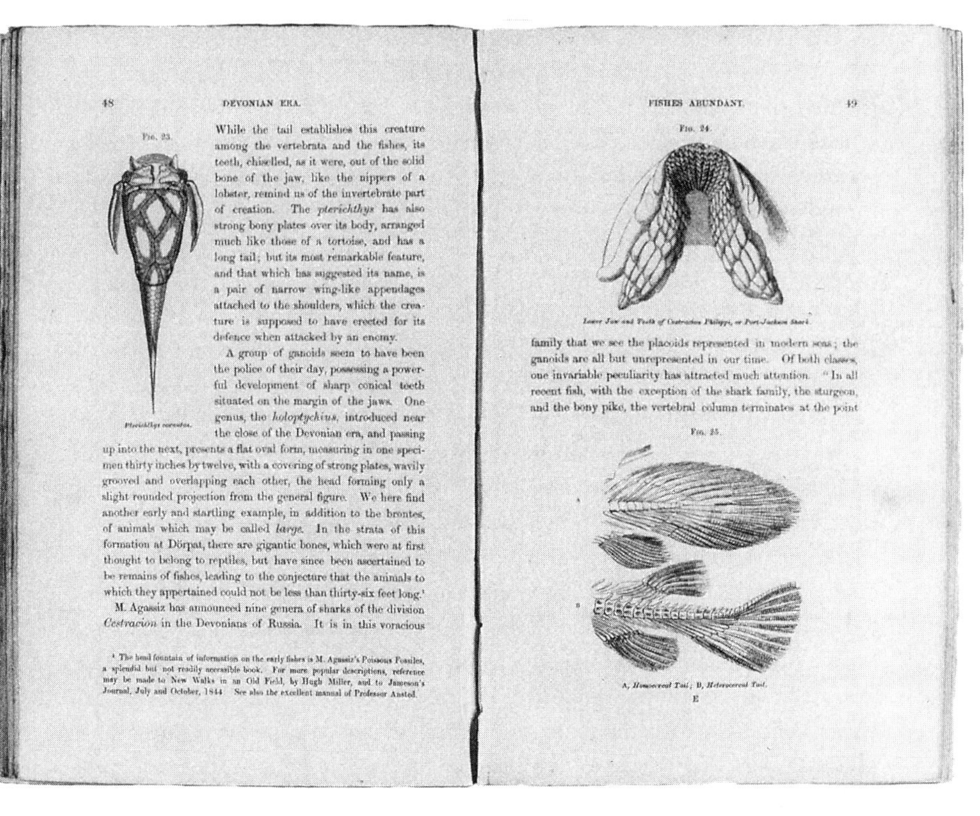
48 DEVONIAN ERA.

Fig. 23.

Pterichthys cornutus.

While the tail establishes this creature among the vertebrata and the fishes, its teeth, chiselled, as it were, out of the solid bone of the jaw, like the nippers of a lobster, remind us of the invertebrate part of creation. The *pterichthys* has also strong bony plates over its body, arranged much like those of a tortoise, and has a long tail; but its most remarkable feature, and that which has suggested its name, is a pair of narrow wing-like appendages attached to the shoulders, which the creature is supposed to have erected for its defence when attacked by an enemy.

A group of ganoids seem to have been the police of their day, possessing a powerful development of sharp conical teeth situated on the margin of the jaws. One genus, the *holoptychius*, introduced near the close of the Devonian era, and passing up into the next, presents a flat oval form, measuring in one specimen thirty inches by twelve, with a covering of strong plates, wavily grooved and overlapping each other, the head forming only a slight rounded projection from the general figure. We here find another early and startling example, in addition to the brontes, of animals which may be called *large*. In the strata of this formation at Dörpat, there are gigantic bones, which were at first thought to belong to reptiles, but have since been ascertained to be remains of fishes, leading to the conjecture that the animals to which they appertained could not be less than thirty-six feet long.[1]

M. Agassiz has announced nine genera of sharks of the division *Cestracion* in the Devonians of Russia. It is in this voracious

[1] The head fountain of information on the early fishes is M. Agassiz's Poissons Fossiles, a splendid but not readily accessible book. For more popular descriptions, reference may be made to New Walks in an Old Field, by Hugh Miller, and to Jameson's Journal, July and October, 1844. See also the excellent manual of Professor Ansted.

FISHES ABUNDANT. 49

Fig. 24.

Lower Jaw and Teeth of Cestracion Philippi, or Port-Jackson Shark.

family that we see the placoids represented in modern seas; the ganoids are all but unrepresented in our time. Of both classes, one invariable peculiarity has attracted much attention. "In all recent fish, with the exception of the shark family, the sturgeon, and the bony pike, the vertebral column terminates at the point

Fig. 25.

A, Homocercal Tail; B, Heterocercal Tail.

E

14.7 The production values of the tenth edition of *Vestiges* stressed its status as a standard scientific work. Note the woodcut of *Pterichthys* (fig. 23), a typical armored fish of the Old Red Sandstone, and of the difference between homocercal and heterocercal fish tails (fig. 25). According to Agassiz and the author of *Vestiges,* fish with homocercal tails were of the higher and more perfect type. *V*10, 48–49.

> Hugh Miller's book is actually in its fourth edition unanswered, though from end to end mere rubbish as far as the points in controversy are concerned.[63]

Chambers was especially proud of the sixty-seven-page appendix with "proofs in the language of the original authorities"—Agassiz, Ansted, Bronn, Carpenter, Heer, Mill, Murchison, Owen, Phillips, Stewart, Watson, and many others. Combined with his own field experience, this anthology of favorable testimony could "blow both Sedgwick and Miller out of the water."[64] When the tenth edition appeared in 1853, he thanked the publisher for "a most successful effort in the preparation of a handsome volume with illustrations" (fig. 14.7).[65]

63. RC to AI, "Monday morning" [1852], NLS Dep. 341/113/14–15; the reference is to *V*10, i–lxvii, esp. pp. xxxiii–lix.

64. RC to Churchill, [1851], NLS Dep. 341/111/28.

65. RC to Churchill, [June 1853], NLS Dep. 341/113/120.

The Chamberses relished the ferment of London life, and from the late 1840s they spent two to three months a year in town. Robert was regularly invited to the Red Lions (fig. 14.8) and also attended Geological Society meetings. Anne Chambers was no sofa-inclined dilettante. As the mother of eleven children, an accomplished hostess, and an exceptional musician, she struck Thackeray as a formidable example of the "strong-minded woman" when they met at a dinner given by one of the publishers of *Punch.* "I was afraid of Mrs. Chambers I confess," he half-jokingly wrote to Carpenter, "but of no (other) man in company—."[66] There seemed a world of like-minded liberal friends in London. Robert's younger sister Janet had married a subeditor who worked for Dickens's *Household Words;* she, like Anne, became a noted hostess.[67] The contrast with Scotland was becoming painful. From the Chamberses' perspective, the "Athens of the North" had degenerated into a provincial backwater ruled by Calvinist fanatics. A turning point came with the "horrid botheration" of the lord provost election, when Robert abandoned hopes of public office.[68]

Chambers continued to pursue new frontiers of knowledge through experiments on table turning and other spirit manifestations. In the early 1850s he began compiling a major work on the history of superstition, which attempted to give such manifestations a physical basis. In 1853 he witnessed a table move along a floor without being pushed; the cause, he suspected, was an electrical power connected with the medium's mental action: "no spirits in the case."[69] However, less than a year later Chambers became convinced that the spirits were both real and dangerous. Under their influence, the family's old neighbor, Catherine Crowe, imagined that the world's sufferings would be relieved if she walked naked through the Edinburgh streets holding her card case in one hand and her handkerchief in the other. She had to be rescued by her friends from this "terrible condition of mad exposure."[70] Chambers became a cautious believer in the reality of spirits, and later burned his book manuscript. He attended séances and began to abandon his private skepticism about the reality of an afterlife. Testimony, in this case that of his own eyes and that of many friends, needed to be accepted even when it flew in the face of physical principles. Facts that supported *"unpopular theories"* were too often ignored or ridiculed

66. W. M. Thackeray to W. B. Carpenter, 24 May [1846], in Harden 1994, 1:171.

67. Corsi 1988, 274 n. 4.

68. RC to AI, 21 Nov. 1848, NLS Dep. 341/112/2–3.

69. RC to AI, [Apr.–May 1853], NLS Dep. 341/113/145–46. Chambers's interest in these phenomena was first documented in Millhauser 1959, 175–76; see also Oppenheim 1985, 272–78. Winter 1998a, 262–67, discusses table turning and related phenomena.

70. RC to AI, 4 Mar. 1854, NLS Dep. 341/112/115–16, and Kemble 1882, 2:83.

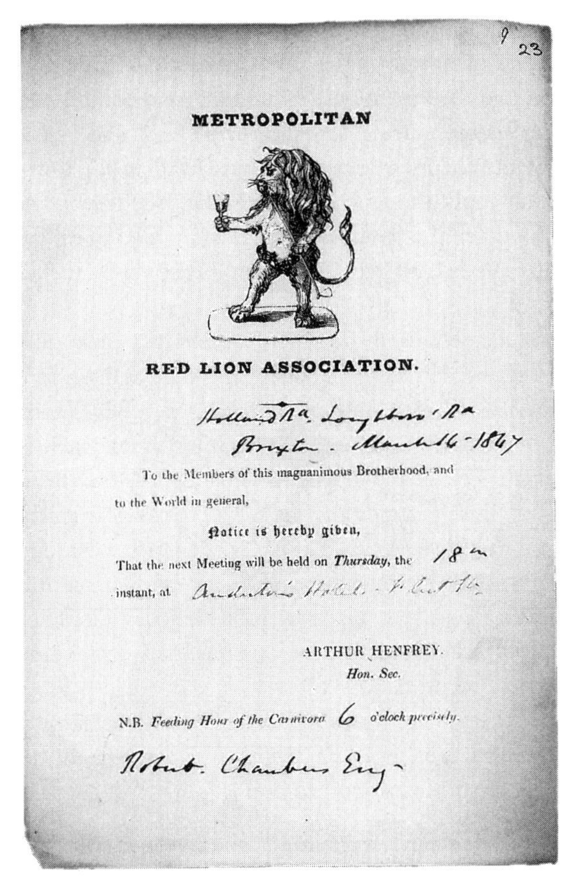

METROPOLITAN

RED LION ASSOCIATION.

Holland Rd. Loughboro' Rd
Brixton. March 16-1847

To the Members of this magnanimous Brotherhood, and to the World in general,

Notice is hereby given,

That the next Meeting will be held on *Thursday*, the 18th instant, at Anderton's Hotel

ARTHUR HENFREY.
Hon. Sec.

N.B. *Feeding Hour of the Carnivora* 6 o'clock *precisely.*

Robert Chambers Esq

14.8 From the mid-1840s, Chambers began to socialize with the young career men of science. Circular for the Metropolitan Red Lion Association, enclosed in a letter of 18 March 1847 from Arthur Henfrey, addressed to Robert Chambers through Andrew Ramsay at the Craig's Court Museum. NLS Dep. 341/109/23.

even when reported by thousands of reliable witnesses. Spiritual phenomena suffered from the same skepticism that condemned the developmental law of progress. Chambers expressed these views in *Testimony: Its Posture in the Scientific World,* a pamphlet issued early in 1859.[71]

Even at this late date, one phenomenon that remained resistant to proof through testimony was the *Vestiges* authorship. Chambers was still by no means the only candidate, nor did any reliable source seem willing to state categorically who was. This is shown by a furious debate about the issue that blew up in the newspapers in the spring of 1859.

Working in the British Museum library, Owen had noticed that *Vestiges* had been catalogued under "Chambers." Supposing that this might reflect a formal

71. [Chambers] 1859; RC, entry for 2 June 1853, memorandum book, NLS Dep. 341/42, f. 51v.

declaration, he found it based only on rumor and circumstance. Owen advised that *Vestiges* be reclassed as "anonymous," on the grounds that there were similar pointers to Combe, who had died in August 1858: the second edition contained corrections Owen had communicated only to the phrenologist.[72] This attribution was picked up by the *Critic*'s proprietor Edward William Cox, who did not name his source, but simply announced "upon evidence of the highest authority" that Combe had written the book. On this account, Chambers had merely been checking proof, and despite Page's assertion might not be the author after all.[73]

An attribution to Combe, made by someone more trustworthy than a disgruntled hack, spawned a flurry of letters to the newspapers. Nichol wrote to the Glasgow *North British Daily Mail* to state categorically "that Mr George Combe was NOT the author of that book, and had nothing to do with the preparation of it."[74] In Manchester, "J. S." supported Chambers as the author of "the obnoxious book" on stylistic grounds and his failure to disavow it after the Page affair. The editor of the *Manchester Guardian,* while not condemning "that famous production," pointed out that the events surrounding the lord provost election suggested that Chambers was indeed the author.[75] The *Critic* then repeated its assertion, this time identifying its source as "a witness, whose name in the world of science is second to none."[76]

Other suggestions flooded the papers. One of the most favored was made by "a well-known London author" in the *Newcastle Chronicle.* This shows the detail with which the problem was still being worked out:

> It is some years ago since we made a minute examination of the internal evidences of authorship to be found in the volume, and we came to the following conclusions: 1st, That no one person had written the whole. 2d, That there were at least two, if not three different styles employed. 3d, That here and there were passages interpolated in one portion of the work, written by another pen, the style of which was found in another part of the work. 4th, that this pen last mentioned acted as editor, while one or more writers were contributory. 5th, The editorial pen is found in the eloquence of the book, which sometimes bears evidence of manifest

72. R. Owen to J. Clark, [Apr. 1859], NHM, Owen Correspondence, vol. 7, f. 209 (draft), in which Owen claimed to have believed in Combe's authorship when the two men were indirectly in correspondence (e.g., R. Owen to J. Paget, 14 May 1855, NLS mss. 7350, f. 26, in the Combe archive). Owen's views between the spring of 1845 and 1855 are unclear.

73. "Literary News," *Critic,* 5 Mar. 1859, 236. This was repeated in most of the newspapers, notably *North British Daily Mail,* 7 Mar. 1859, 2. On Cox, see Watkins 1982, 89–90.

74. J. P. Nichol, "The Vestiges of Creation," *North British Daily Mail,* 9 Mar. 1859, 2 (see also *Critic,* 12 Mar. 1859, 245).

75. J. S., "The Authorship of the 'Vestiges of Creation,'" *Manchester Guardian,* 17 Mar. 1859, 4.

76. *Critic,* 19 Mar. 1859, 270.

> interpolations, and which are grand general inferences from particular facts. 6th, That the authorship was known to Mr. Robert Chambers, Professor Nichol, and to (at that time), a prominent member of the Manchester Literary and Philosophical Society. 7th, That the authorship could not be claimed by any one and could be denied by any one.
>
> Subsequent observation has confirmed these suspicions, and while we are prepared to state that George Combe was not the author of the "Vestiges," we are equally sure that he was one of the contributors to its pages.[77]

There was a further denial from Nichol, an emphatic restatement of Chambers's guilt from Page, and letters debating the merits and flaws of the book. (There were, in addition, newspaper reports that Chambers had disavowed the authorship at a public dinner.)[78] In the meantime, Combe's executors responded by denying that the phrenologist "never, by word, look, or silent acquiescence, knowingly gave the slightest countenance to the supposition that he was its author or had taken any part whatever in its production."[79] The *Critic*, somewhat abashed, identified its authority as Owen, who now said he did not know who the author was. Neither did others at the heart of the establishment, such as the royal physician Sir James Clark.[80] One writer went on to produce an elaborate manuscript essay, demonstrating that Chambers could not be the author. The net effect of years of discussion, accusation, and gossip had been to leave the authorship anonymous, as it henceforth was in the British Library catalogue.[81]

There was a final flurry of interest in the authorship when Alexander Ireland, the only surviving holder of the secret, formally announced that Chambers was the author. A new twelfth edition, with Chambers's name at last on the title page, spine, and pictorial front cover, was published by W. & R. Chambers in 1884. This posthumous edition cleared up the mystery—although Lyell is occasionally named as the author[82]—but could not affect the meanings

77. [A Man in the Streets], "Authorship of the 'Vestiges of Creation,'" *Newcastle Chronicle*, 19 Mar. 1859, 6; *Reasoner* 24 (3 Apr. 1859): 109; also in *North British Daily Mail*, 19 Mar. 1859, 2. J. Payn, a novelist and contributor to *Chambers's Journal* who knew Chambers well, also thought the work might have been written in collaboration (Payn 1884, 141–42).

78. J. P. Nichol, "The 'Vestiges of Creation,'" *North British Daily Mail*, 21 Mar. 1859, 2; D. Page, *Caledonian Mercury*, quoted in *Critic*, 26 Mar. 1859, 294. R. G., "The 'Vestiges of Creation,'" *Manchester Guardian*, 21 Mar. 1859, 4, notes the reported disavowal, but doubts its accuracy.

79. J. Coxe, "The 'Vestiges of Creation,'" *North British Daily Mail*, 25 Mar. 1859, 2.

80. *Critic*, 26 Mar. 1859, 294; R. Owen to J. Clark, [Apr. 1859], NHM, Owen Correspondence, vol. 7, f. 209 (draft); J. Clark to R. Owen, 11 Apr. 1859, NHM, Owen Correspondence, vol. 7, f. 208.

81. Ireland 1884, ix. According to the staff at the inquiry desk at the British Library, however, Combe's name did remain in the catalogue from 1860 until at least 1877 (Baxter 1993, 112 n. 62).

82. Probably inadvertently; for examples, see Church 1894, 63; Flint 1993, 233; and St. George 1993, xvi.

of the book for earlier readers. Even within the rumor mills of Fleet Street and Paternoster Row, only those who had been told by Chambers himself had ever known for sure. As the *Critic* had said in 1859, "the evidence of each week seems to make 'confusion worse confounded.' That there are persons who could at once settle the point, were they so minded, is not to be doubted. Why do they withhold their evidence? What room is there for mystery?"[83]

Wielding the Scalpel of Science

Among the diverse groups of reformers, the positive findings of science appeared to offer the foundations for agreement: nature's progressive history, expressed in relation to society, was the basis for free trade, national education, universal suffrage, and religious tolerance. As Marian Evans had said in her first article for the *Westminster*, "each age and each race has had a faith and a symbolism suited to its need and its stage of development."[84] For progress to continue, the latest science needed to be brought to readers. In November 1853, Chapman asked Spencer's unemployed friend Thomas Henry Huxley to write for the *Westminster*. He was, Evans told Combe, "a scientific man who is becoming celebrated in London."[85]

Like anyone forging a career, Huxley (fig. 14.9) had to make choices. The son of an impoverished mathematics teacher, he had served as surgeon naturalist on the *Rattlesnake* and, back in London, searched for a position in a scientific establishment. He published papers, was elected to the Royal Society, won medals, applied for professorships, and waited. Three years later he did not have a job, and his fiancée, Henrietta Heathorn, remained on the other side of the world. "A man of science may earn great distinction—great reputation," Huxley told her, "—but not bread. He will get invitations to all sorts of dinners & conversaziones, but not enough income to pay his cab hire."[86] The twenty-eight-year-old Huxley not only read *Pendennis,* he lived it.

One source of cab fares was journalism, and Huxley was glad to accept twelve guineas a sheet for reviewing in the *Westminster.* This was well-paying literary piecework, and Huxley was willing to act "as a kind of scientific jackal to the public" just as many had done before him.[87] Chapman's offices on the Strand were only a short walk from the "dingy but cosy dens" where men of science gathered for a drink.[88] But Huxley refused to make the traditional compromises of com-

83. *Critic,* 26 Mar. 1859, 294–95, at 295.

84. "Progress of the Intellect," *Westminster Review,* Jan. 1851, in Eliot 1990, 269.

85. G. Eliot to G. Combe, 25 Nov. 1853, in Haight 1954–78, 8:89. Huxley's reviewing, especially of *Vestiges,* is discussed in Desmond 1994, 38, 49, and Schwartz 1990, 147–52.

86. Quoted in Desmond 1994, 161.

87. [T. Huxley], "Science," *Westminster Review* 5 (Jan. 1854): 261.

88. Geikie 1895, 30, quoted in Desmond 1994, 152.

14.9 T. H. Huxley reading, his feet in the fire grate to keep warm; a self-portrait at age twenty. Huxley first read *Vestiges* at about this time, before leaving on the *Rattlesnake,* but claimed not to be able to remember what he thought of the work.

mercial science, even as he signed a contract with Churchill for a comparative anatomy textbook, translated German for "filthy lucre,"[89] and became friendly with Spencer, Lewes, and other active journalists. Fearing exile on Grub Street, he made public all the tensions in the pursuit of science for pay that Carpenter, Hunt, and especially Forbes—his ideal of the man of science—had kept hidden.

The new reviewer caused consternation at the *Westminster.* Huxley was willing to be impolite and impolitic, so that the friendships that tied scientific men into literary commerce might be broken. In notices consumed by sarcasm, he called a French work on human longevity "baseless and audacious"; "a more unsatisfactory production never came before us, from a man of real science." An English treatise on mollusks belonged to "this Rip van Winkle school of naturalists." Even the Germans, whose work Huxley typically praised, "have very good magnifying glasses, but they have yet to learn what a microscope in the English sense means."[90]

89. Ibid., 183.

90. [T. Huxley], "Science," *Westminster Review* 8 (July 1855): 240–63, at 241, 245; [T. Huxley], "Science," *Westminster Review* 5 (Apr. 1854): 580–95, at 591.

From the perspective of the Strand, such contemptuous intemperance could damage the cause. The line of conflict had been drawn when the cocky naturalist demolished Lewes's commentary on Auguste Comte and singled out the development hypothesis for special venom. Lewes's book, Huxley wrote, demonstrated "how impossible it is for even so acute a thinker as Mr. Lewes to succeed in scientific speculations, without the discipline and knowledge which result from being a worker also."[91] Such writings blurred the boundaries around Huxley's new and more rigid definition of "popular science." Evans was furious, not least because Lewes had read extensively in Continental science. Moreover, Comte's positive philosophy was central to the *Westminster*'s attempt to move beyond the transcendental mysticism of Carlyle and Emerson. Huxley's review, she fumed to Chapman, was "an utterly worthless & unworthy notice of a work by one of their own writers—a man of much longer & higher standing than Mr. Huxley."[92]

Huxley, not one to back down, continued the battle over development despite its centrality to the avant-garde agenda. He targeted the tenth edition of *Vestiges,* whose size, price, illustrations, binding, autobiographical preface, and appendix of authorities staked its claim as a classic. He savaged the book in the *Westminster,* but his longest attack appeared in the *British and Foreign Medico-Chirurgical Review,* the quarterly in which Carpenter had hailed the first edition as "a very beautiful and a very interesting book."[93] Churchill, still the *Review*'s publisher, probably expected a recognition of improvements, perhaps to repair any dents in his own reputation. Journals rarely drew attention to books in editions other than the first, and in such cases usually to praise.

Huxley later thought he must have read *Vestiges* before leaving England in 1846, but claimed the book made an impression only when after his return, when he found his *Westminster* friends taking it seriously.[94] His review, so vicious he later had to apologize, gave notice that there could be no more compromise with vague invocations of universal progress:

> We look for evidence of knowledge, and we find—what might be picked up by reading "Chambers's Journal" or the "Penny Magazine." We look for original research, and we find reason to doubt if the author ever performed an experiment or made an observation in any one branch of science. We seek for acuteness of

91. [T. Huxley], "Science," *Westminster Review* 5 (Jan. 1854): 254–70, at 255.

92. M. Evans to J. Chapman, Dec. 1853, quoted in Ashton 1996, 109; see also 145–47.

93. [W. Carpenter], "Natural History of Creation," *British and Foreign Medical Review* 19 (Jan. 1845): 155.

94. T. Huxley, "On the Reception of the 'Origin of Species,'" in Darwin 1887, 2:179–207, at 187. Huxley also read the first edition of *Vestiges,* and *Explanations,* carefully at some time after the *Origin,* as his notes include references to variation and natural selection. See "Vestiges of the Nat History of Creation 1844" and "Explanations," IC, Huxley Papers, vol. 41, ff. 57–60.

> thought, and we find nothing but confusion of ideas, and an ignorance of the first outlines of speculation. A spurious, glib eloquence, an affectation of reverence for truth and of scientific modesty, are not wanting to remind the curious observer all the more strongly of the total absence of that careful research and fair representation of both sides of a question, which should be the first-fruits of the latter qualities.[95]

The book offered the form of knowledge without its substance; the appearance of legitimacy masked a farrago of nonsense. Printing the new edition had been a mistake, as it was now "so much waste paper." The book, he noted sarcastically, was "very pretty and well got up," to appeal to "the ignorant of science," and "to the great glory and no small profit of the author."[96]

In *Vestiges,* Huxley thought, law became a thing to be bought and sold. The review argued that the book was inconsistent in defining natural law, a severe criticism of what was after all widely considered as its central concept. Sometimes law was "merely a term of human convenience" to express the orderly sequence by which divine will "worked out in external nature." This definition, as Huxley knew, had been developed to meet reviewers in the evangelical press. To criticize the book in this way was only to say that too much quarter had been given to the saints; in short, that it was controversial philosophizing, not real science. Worse still, *Vestiges* treated law as an entity; its "own abstraction" is confused with "objective fact." It was, as Huxley noted, a form of "fetish worship, where reverence is proportionate to the bigness of the idol." The objectifying of law, the review claimed, was the characteristic folly of scientific consumerism: "the great mass of those . . . who indulge at science at second-hand." Here the language of the undifferentiated "great mass" emphasizes the animal-like characteristics of the people, who buy up a "product of coarse feeling acting in a crude intellect" as though it were a work of a disciplined, original mind.[97]

To exhibit the character of real science, Huxley pointed to the study of fossil fish. Fish, as the first vertebrates in the history of life, were of great significance to the argument for progressive development. The relevant passages of *Vestiges* had already been revised in previous editions to conform with the best authorities, especially in response to the *Edinburgh*'s complaint that no reference had been made to the Swiss naturalist Louis Agassiz's expensive and difficult-to-obtain monograph on fishes. *Vestiges* made great play of Agassiz's

95. [T. Huxley], "The Vestiges of Creation," *British and Foreign Medico-Chirurgical Review* 13 (Apr. 1854): 425–39, at 438; reprinted in Huxley 1898–1903, 5:1–19. Huxley mentions his qualms about the review in "On the Reception," Darwin 1887, 2:188–89, and he spoke about it with one of Chambers's daughters (Priestley 1908, 41).

96. [Huxley], "Vestiges," 425, 426. Ironically, although Huxley could not have known it, the anonymous author had realized only a modest amount from *Vestiges,* and the tenth edition was the first in which all profits went to the publisher.

97. [Huxley], "Vestiges," 426–29.

claim that certain fishes of the Old Red Sandstone (fig. 14.7) were analogous to embryonic stages of the bony fishes of a later period. "These ancient fishes," Agassiz had said, "were not so fully developed as most of our fishes, being arrested, as it were, in their development."[98] This claim could buttress the developmental, embryological argument of *Vestiges,* although that was the last thing Agassiz had had in mind.

Huxley's discussion of fossil fish had many targets, especially Owen. The review noted that the early fishes were not "embryonic" or "arrested" in the great anatomist's Hunterian lectures on fishes, published in 1847; while in his (anonymous) *Quarterly* article of 1851, they were.[99] For Owen, this had surely been a question of tactics: his lectures targeted the developmental heresies of *Vestiges,* while the article stressed progression in the context of an attack on Lyell. Huxley was defending Lyell's doctrine of "absolute uniformity," buttressing the case against progress by arguing that the ganoids were highly organized—perhaps "*more highly organized*" than any other fish. This pro-Lyell, anti-Owen stance was shared by most of the Survey men, but Huxley went so far as to anticipate that potsherds would be found alongside trilobites in the oldest rocks.[100]

But more fundamental than such theoretical idiosyncrasies was the claim about expertise that underpinned the vendetta against "people's science." For Huxley based his comments on his own, original classification of fish, one he began to publish fully only in 1858. It drew upon work by the German naturalist Johannes Müller, who argued that the entire enterprise of classifying by fins, skin, and scales was mistaken. Internal organs provided the only diagnostic characters, and these were not available for the early fishes—such as *Cephalaspis* and *Pteraspis.* For this reason, the assumption that the earliest fishes were "low" in organization was untenable.[101] In effect, the expert on fossil fish now had nothing to say about the big issues of progress or development.

From Huxley's perspective, the earlier work upon which *Vestiges* had depended was itself misconceived through a failure of scientific discipline. Miller and Agassiz were just as mistaken as the hapless author of *Vestiges.* Agassiz had

98. Quoted in "Proofs, Illustrations, Authorities, etc.," *V*10, ix; for the revisions, see *E*1, 48–59. [A. Sedgwick], "Natural History of Creation," *Edinburgh Review* 82 (July 1845): 1–85, at 38–39, complained of the author's initial failure to read Agassiz.

99. [Huxley], "Vestiges," 433–36. On the early Owen-Huxley contest, see Desmond 1982 and Richards 1987, 168–71.

100. Desmond 1982, 37–41, 84–112; Bartholomew 1975; Di Gregorio 1984, 48–50. For anti-Lyellian views at the Survey, see Secord 1986b, 258–59.

101. T. H. Huxley, "On Cephalaspis and Pteraspis," *Quarterly Journal of the Geological Society of London* 14 (1858): 267–80; also in Huxley 1898–1903, 1:502–18, see esp. 517. Huxley's fish classification is discussed in Di Gregorio 1984, 69–74.

been led astray by "lively fancy," while Miller had been biased by religious polemic. "His geology is bad science," he told Ramsay privately, "but good Calvinism."[102] In place of accessible books like the *Old Red Sandstone,* Huxley wanted technical papers for the Geological Society and the Survey. As Lydia Miller noted in the preface to the posthumous edition of her husband's *Foot-prints of the Creator,* Huxley advocated "a more recondite and exclusive style of classification."[103] The difference was clear even when he stressed the force of practical reasoning in science. The review compares fish with tadpoles:

> The grounds of the comparison are worth noting; they are these: the large head, undistinguished from the thorax; the aggregation of the viscera anteriorly; the position of the anal fin and vent immediately behind the cephalo-thoractic expansion; and the appropriation of the rest of the trunk for locomotion. Any one who will go into the market and buy a sole, may satisfy himself that on these grounds that unhappy fish has been raised beyond his proper position, and is no better than an upstart tadpole.[104]

The crescendo of technical terminology is broken by the final sentence, with the everyday act of buying a fish; so that the technical language and structure of the passage belies the invitation to participate.

From the beginning, Huxley had no high opinion of what he condescendingly referred to as "the popular mind" or "the mob." He disdained the pioneering publications of the useful knowledge movement, the *Penny Magazine* and *Chambers's Journal,* implying that mass-circulation magazines had nothing to do with real science. Of course, Huxley famously believed—as he said in St. Martin's Hall in 1854—science is "nothing but *trained and organised common sense.*" But he went on to claim that "its methods differ from those of common sense only so far as the guardsman's cut and thrust differ from the manner in which a savage wields his club," differences that might be taken to imply a fairly substantial contrast in level of skill.[105] The same analogy was implicit in the review: the *Vestiges* author, attempting to legislate for nature, had cut "his fingers with the weapon he is unable to handle."[106]

In an extraordinary step, Huxley chose Sedgwick, the clerical bête noire of the progressive party, to exemplify the heroic character of the working man of science. Huxley failed to mention the theological animus that radical critics

102. T. H. Huxley to A. C. Ramsay, 29 Nov. [1854], IC, Ramsay Papers, KGA/Ramsay/8/102/3/1–2.

103. L. Miller, prefatory remarks to Miller 1861b, xlvi.

104. [Huxley], "Vestiges," 436.

105. "On the Educational Value of the Natural History Sciences," (1854), in Huxley 1893–94, 3:38–65, at 45.

106. [Huxley], "Vestiges," 438.

had found in the Cambridge geologist's attacks. Instead, the review portrayed Sedgwick as "a thorough, an earnest, and above all, a genial man, who has made truth the search of his life." If Sedgwick had lost his temper, it was because he was battling "the cool interposition of a mere sciolist with his 'hypotheses' in a neat pouncet box."[107] In the war between an anonymous charlatan and a working geologist—even one whose science was in the service of evangelical Christianity—Huxley chose his "horn" and showed which side he was on.

On Huxley's reading, the book was "a weed," "a lumber-room of second-hand scientific furniture," a "once attractive and still notorious work of fiction"—worth reviewing only to learn how it could have once been taken so seriously.[108] Huxley rewrote the *Vestiges* controversy as an object lesson in the dangers of mixing knowledge with commerce. He started with targets that would have been agreed by almost anyone who read a medical review, such as the table turning sensation in the newspapers. But in linking such issues to *Vestiges* Huxley lashed out at the entire scientific world, in which "true philosophy" was corrupted by the search for income and patronage. He implicitly condemned Carpenter for serving as a publisher's drudge, and Owen for selling his intellectual birthright. Underlying his demolition was a warning to literary friends who wrote cosmic philosophy. *Vestiges,* Huxley told them in the *Westminster,* was "pseudo-scientific" and heir to the antiquated ravings of scriptural geology.[109] The ethics of commercial bookmaking and reviewing would inevitably govern any attempt to create a "people's science"; in a market-based society, democratic epistemology would turn knowledge into a commodity.

"When the Brains Were Out, the Man Would Die"

For Huxley, war against the sciences of progress was total. Yet because his career was so vulnerable, his attacks could not escape the clutches of commerce. His review, for example, starts with a classic attention-grabbing opening from *Macbeth:* "'Time was, that when the brains were out, the man would die.'"[110] The same Shakespearean gambit had already been used in a review of *Explanations* in the *Torch* nearly a decade earlier.[111] Such recyclings were the stock-in-trade of Grub Street hacks. Huxley's cynical conviction that nothing could escape the marketplace comes out most strongly in his correspondence about potential jobs. He fully expected Chambers, whom he viewed as a second-rater and the certain author of *Vestiges,* to retaliate by blocking his career. When

107. Ibid., 438–39.

108. Ibid., 438.

109. [T. Huxley], "Science," *Westminster Review* 6 (Oct. 1854): 573.

110. [Huxley], "Vestiges," 425.

111. "Explanations—A Sequel to 'Vestiges of the Natural History of Creation,'" *Torch,* 10 Jan. 1846, 30–32, at 30.

Forbes died suddenly in November 1854, Huxley joined the queue of applicants for the lucrative natural history chair at the University of Edinburgh. By this time the ambitious young man had a Jermyn Street post, but the Scottish chair paid five times as much.[112] As Huxley told Ramsay, "Chambers, whatever he does, will do his best to oust me unless he possesses a generosity I give him no credit for—because I wrote that review of the *Vestiges*." Instead, Carpenter would be backed, "because the latter revised that broth of a book."[113] Although Huxley happened to be right about the authorship, he was wrong in thinking Chambers would be so naive as to retaliate. Huxley had no conception of the ostracism that anyone in Scotland would face if identified as "Mr. Vestiges." Much to his surprise, the young naturalist found himself invited to stay at Doune Terrace while canvassing for support.[114]

Other reviews show that Huxley's reading of *Vestiges* was the idiosyncratic product of the coterie of scientific careerists. In contrast, Lewes in the *Leader* had written the most enthusiastic notice of all, extending over four issues of the magazine. Lewes and Marian Evans did read Huxley's article in the *Medico-Chirurgical*—a copy is in Lewes's pamphlet collection—and were unconvinced. As Evans noted acerbically while taking over the *Leader*'s literary column during Lewe's absence, "A writer who evidently delights in wielding the scalpel in more senses than one has chosen the 'Vestiges' as a 'subject,' and dissects it with immense *gusto.*"[115] Both she and Lewes continued to speak favorably about *Vestiges* and its pioneering role in spurring debate, well after the appearance of the *Origin of Species.*

Darwin, another *Westminster* supporter, was ambivalent in his reaction to the review. He praised it in a letter to Huxley as by far the best on *Vestiges* ever written, for demolishing a vague concept of law, for the "exquisite & inimitable" way it handled Owen, and for making "mincemeat" with Agassiz's embryonic fish. Yet Darwin also thought it "rather hard on the poor author": this was not reviewing according to the rules of gentlemanly fair play. At the very least, "such a book, if it does no other good, spreads the taste for natural science." Darwin knew that Huxley would be unmoved, but was easing the way for his own work. This, he hinted to Huxley for the first time, would be "almost as unorthodox about species as the Vestiges itself, though I hope not *quite* so unphilosophical."

112. Desmond 1994, 206–7.

113. T. H. Huxley to A. C. Ramsay, 29 Nov. [1854], IC, Ramsay Papers, KGA/Ramsay/8/102/3/1–2.

114. RC to T. H. Huxley, 14 Dec. 1855, IC, Huxley Papers, vol. 12, f. 164. Another, less likely possibility, is that Chambers never read the review, as it does not appear to be referred to in his correspondence with Ireland.

115. [M. Evans], *Leader,* 15 Apr. 1854, 354. For Evans's authorship, see Postlethwaite 1984, 108 n. 103; and for the offprint, see Dr. Williams's Library (GHL), A.8.14(38).

To Hooker, Darwin worried about being attacked from the same quarter—that work on species would inevitably align him with populist cosmologizing. "What a deal I shall have to discuss with you," he told Hooker: "I shall have to look sharp that I do not 'progress' into one of the greatest bores in life to the few like you with lots of knowledge."[116]

Darwin's joke acknowledged that the meaning of expertise was changing. The campaign to create a paid clerisy had its roots in dissatisfaction with commercial science's subordinate place on the map of knowledge. Huxley later became this movement's most vigorous and skilled spokesman, so much so that there is a danger of reading the second half of the century through his eyes. But Huxley, Hooker, and their supporters were not the makers of the modern world. Neither were the gentlemanly specialists of the old regime, millenarian evangelicals, disillusioned Tractarians searching after a new creed—nor any of the other groups with visions of the role that science should occupy on the national stage. What mattered was the underlying structure of their debates, which were transforming the relations between knowledge, the market, and the reader.

Vestiges in the Origin

Toward the end of the 1850s, Churchill wanted to exploit the renewed interest with a new edition, but his still-shrouded author refused, unwilling to become immersed in "this horridly controverted subject"; others could best keep the subject before the public. One possibility was an essay contest, and Chambers drafted an advertisement:

> The Development Hypothesis as affected by Recent Science.
>
> ———
>
> Prize Essay.
>
> ———
>
> A prize of One Hundred Pounds is proposed for the Best Essay, 1, exhibiting the Development Hypothesis as affected by the discoveries and generally adopted views of the last ["ten or" *added*] twelve years in physiology and geology; 2, exposing the fallacies and misrepresentations of the Opponents of the Hypothesis during the same period. The essays to be lodged on or before the 1st of January 1858, when the best will be selected, and the prize immediately bestowed. The book will be published ["with the author's name" *added*], free of risk to him, and he will receive the half of any profits arising from it.[117]

116. C. Darwin to T. Huxley, 2 Sept. [1854], and C. Darwin to J. D. Hooker, 7 Sept. [1854], in Burkhardt et al. 1985–99, 5:212–15; see also Desmond and Moore 1991, 414, and Darwin's notes on the review, CUL, DAR 205.6.56–57.

117. Ms. memorandum, RC to AI, [1857], NLS Dep., 341/113/138–39. The idea of a contest was probably based on the Burnett Prize competition for the best essay on the evidences of Christian-

The contest was never held, but there are no prizes for guessing who would have won. On the other hand, Darwin might not have qualified at all. The draft advertisement sought out a book on the "development hypothesis" and an accompanying memorandum mentioned "the progressive history of life upon the globe"; Darwin was working on species and varieties. Chambers wanted a progressive story ranging across all the sciences; Darwin focused on a limited set of questions about a causal mechanism for species change.[118] In the very years that *Chambers's Journal* was complaining that specialists wasted their time determining species, Darwin commenced work on the barnacles.

The contrast is suggested by the people the two men suggested as appropriate to take up their respective projects. Chambers hoped that Lewes, Spencer, Powell, or the down-on-his-luck naturalist Edward Charlesworth would take on the job.[119] These men were all writers with links to scientific bohemia, favorable to the developmental philosophies discussed in the *Westminster* circle. Darwin, fearing his materials would have to be worked up after his death, suggested Lyell, Hooker, Owen, and Henslow as suitable candidates. These were either gentlemen of science, whom Darwin trusted to deal fairly with opinions they did not share, or researchers with undisputed credentials in specialist fields.[120]

When Darwin resumed full-time species work in 1854, he had to consider what kind of book he would write. His just-completed barnacle monographs, like his other writings, belonged to standard genres of specialist writing, but a treatise on species origins did not. He considered "a *very thin* & little volume," but worried that an unreferenced résumé of unpublished work would be "really dreadfully unphilosophical."[121] Finally, he commenced a multivolume treatise modeled on Lyell's *Principles,* a scale of work that could be contemplated only by the independently wealthy. As is well known, this plan was interrupted in June 1858 by a letter from Alfred Russel Wallace, who was collecting specimens in the Spice Islands for sale. As Darwin saw it, Wallace—who had been inspired by *Vestiges*—had solved the same problem he was examining, and in a broadly similar way.

---

ity; held every forty years, the contest in 1854–55 had attracted 208 entrants; Corsi 1988, 206. The two winning essays (Thompson 1855 and Tulloch 1855), although they attacked *Vestiges,* also illustrate the more measured attitude toward developmental evolution taking hold in moderate evangelical circles.

118. Hodge 1972b.

119. Memorandum, RC to Churchill, [1857], NLS Dep. 341/110/38; RC to AI, 30 Dec. 1859, NLS Dep. 341/112/158–59; memorandum, RC to Churchill, 23 Mar. 1860, NLS Dep. 341/112/174–75.

120. C. Darwin to E. Darwin, 5 July 1844, in Burkhardt et al. 1985–99, 3:43–44; Owen was later removed from the list when Darwin lost faith in his independence of judgment.

121. C. Darwin to J. D. Hooker, 9 May [1856], in Burkhardt et al. 1985–99, 6:106.

The timetables of commercial science had interrupted a gentlemanly scheme of literary production.[122] The need to publish quickly led Darwin to rethink what kind of work he was writing and who its readers would be. He considered papers in the Linnean Society *Transactions*, but worried that "religion would be brought in by men, whom I know"; instead he opted for the independence of a short book, an "abstract" that gradually grew in his hands to be several hundred manuscript pages.[123] Calling it an abstract (his first title was "An Abstract of an Essay on the Origin of Species and Varieties through Natural Selection") was a form of defense, a clear indication of difficulty in defining this project as "scientific" in terms that would distinguish it from *Vestiges* (fig. 14.10).

The *Origin* looked like the standard run of books published by John Murray, with green cloth casing, fifteen-shilling price, and octavo size; but it was a very unusual work, not at all typical of the period. No references, no illustrations, and a high proportion of speculation made Darwin's text much closer to works advocating the science of progress than is usually assumed. In many ways it recalled the reflective treatises of the Reform era that he had read on the *Beagle*, targeting "semi-scientific men" as well as men of science. But now it was much more difficult to reach these audiences simultaneously, and Darwin doubted that his literary talents were up to resolving their conflicting claims. "You say you dreamt that my Book was *entertaining*" he wrote Hooker; "that dream is pretty well over with me, & I begin to fear that the Public will find it intolerably dry & perplexing." Would so strange a book sell at all? Murray, a master at such things, estimated that only 500 copies would be needed, although eventually 1,250 were published in the first edition.[124]

Darwin knew that readers would approach his text in relation to *Vestiges*, which he characterized as a "popular" work, against which the *Origin* could be presented as "scientific." The years of following the reactions to the book paid off. In deprecating his authorial shortcomings, Darwin could define himself as someone who was not a journalist. This was not an intentional strategy on Darwin's part, but an outcome of his immersion in the literary forms and practices of specialist science. Skills in structuring arguments, practiced through writing on Glen Roy and other projects, were used to good effect.[125]

122. The so-called joint "papers" of Wallace and Darwin at the Linnean Society not only differed in their concepts of species formation, but in their modes of intervening in theoretical debate. Wallace's contribution, which would have been more appropriate in the commercial *Magazine of Natural History*, was at least straightforwardly a paper; Darwin's had to be assembled from letters and a privately circulated manuscript.

123. C. Darwin to J. D. Hooker, 24 Dec. [1858], in Burkhardt et al. 1985–99, 7:222; other letters in this volume also detail Darwin's publication plans.

124. C. Darwin to J. D. Hooker, 22 [June 1859], in Burkhardt et al. 1985–99, 7:308; on the print run, see Burkhardt et al. 1985–99, 7: 290 n. 1, 327 n. 1.

125. Rudwick 1974, 170–75.

14.10 Darwin, anxious to avoid being considered a speculator, sent Lyell this cautious title page for his proposed work in spring 1859. From C. Darwin to C. Lyell, 28 Mar. [1859].

An abstract of an Essay
on the
Origin
of
Species and Varieties
Through natural Selection
by
Charles Darwin M.A.
Fellow of the Royal, Geological & Linn. Socy.

London

1859

The *Origin* presented *Vestiges* as a failed attempt to establish a mechanism for species origins, rather than a general argument for a law-bound creation.[126] The first few pages ridiculed the earlier work—in mocking language echoing Huxley's and Sedgwick's reviews—as a means for hatching perfectly formed

126. The early chapters introduce natural selection through homely analogies with domesticated animals, especially the breeding of fancy pigeons. Lyell had found this strategy so successful that he encouraged Darwin to forget his "abstract of an essay" and instead issue the section on pigeons from his longer in-progress manuscript. This would demonstrate that Darwin had quantities of facts; it would get an overview of his theory in print; and, as the publisher's reader commented, "every body is interested in pigeons" (W. Elwin to J. Murray, 3 May 1859, in Burkhardt et al. 1985–99, 7:288–91). But Darwin rejected the pigeon plan.

mistletoe and woodpeckers.[127] (Like Huxley, Darwin later felt compelled to apologize for his cavalier dismissal.)[128] The balance of the argument had changed dramatically: the second and longer half of the essay of 1844, which had dealt with the general case for evolutionary change, occupied less space in the *Origin.* And this was despite Darwin's insistence that these issues—already treated in *Vestiges* and other works—mattered more than natural selection.[129]

Moreover, the *Origin* discussed neither Genesis nor human origins; as Darwin asked Lyell to reassure Murray, who believed in the biblical Flood, "my Book is not more *un*-orthodox than the subject makes inevitable." It could easily be read as sanctioning a miracle to bridge the gap between matter and "one primordial form, into which life was first breathed."[130] On these crucial issues, it gave much more ground to cautious readers than its anonymous predecessor. Above all, the *Origin* avoided narratives of successive events in nature. No one would accuse the *Origin* of reading like a novel.

The dryness of the *Origin* proved as important to its success as the eloquent extracts printed in the reviews. The book was just readable enough to sell, but unreadable enough not to be easily bracketed with journalism or cosmological potboilers. As the *Examiner* stressed, much of the *Origin* was "what ordinary readers would call 'tough reading;' that is, writing of which the full sense is only to be perceived by concentrated attention and some preparation for the task."[131] The work could be seen to be scientific, and aimed at specialist men of science, especially the new generation at the Survey and other government institutions, whose advocacy would make it difficult to dismiss the theory as imaginative fiction. Its presentation as a workmanlike piece of prose from a preeminently respectable publisher gave the book credibility to match Darwin's own reputation. The first words of the text ("When on board H.M.S. 'Beagle', as naturalist")[132] simultaneously reminded readers of his credentials as an observer and his authorship of a much-loved travel book. Without these associations and his name on the title page, the book would have been read in a very different way.

127. Darwin 1859, 3–4.

128. He made a more balanced assessment in the "historical sketch" added to later editions of the *Origin* (see Burkhardt et al. 1985–99, 8:572–76, at 574); and he apologized to the family after Chambers's death: "Several years ago I perceived that I had not done full justice to a scientific work which I believed and still believe he was intimately connected with, and few things have struck me with more admiration than the perfect temper and liberality with which he treated my conduct." C. Darwin to A. Dowie, 24 Mar. 1871, in Priestley 1908, 41–42.

129. *Athenaeum,* 9 May 1863, 617; Ospovat 1981, 89.

130. Darwin 1859, 484; C. Darwin to C. Lyell, 28 Mar. [1859], in Burkhardt et al. 1985–99, 7:270.

131. *Examiner,* 3 Dec. 1859, 772–73, at 773.

132. Darwin 1859, 1.

Only in the much-quoted final sentence were readers encouraged to rethink the text in terms of cosmic progression: "There is grandeur in this view of life, with its several powers, having been originally breathed into a few forms or into one; and that, whilst this planet has gone cycling on according to the fixed law of gravity, from so simple a beginning endless forms most beautiful and most wonderful have been, and are being, evolved."[133] This sentence, which had been part of the earliest drafts of the theory, gained new meaning at the close of a long and often difficult book, where it licensed readers to engage in the pleasure of speculation. The entire argument was reinscribed in a framework of cosmic progression, so that in the end, readers who understood natural selection were allowed to think back over the *Origin* in terms that resonated with *Vestiges*.

1859: Exiting an Age of Crisis

Although publication of the *Origin* is often portrayed as one of the great crises in intellectual history, the response was relatively muted. If critics had reacted with the fury often attributed to them, then Darwin would have counted his book a failure. The last thing he wanted was a holy war. Controversy in all areas of public life had moderated during the previous fifteen years. There was great interest; as the *Saturday Review* noted, debate had "passed beyond the bounds of the study and lecture-room into the drawing-room and the public street."[134] But the book and its author were difficult to dismiss even for the many reviewers who disagreed with what they read. There were exceptions, including the *Athenaeum* and an angry, aging Sedgwick in the *Spectator;* but most of the reviews were respectful. As Richard Church noted in a letter, the *Origin* raised far less of an outcry than "the once famous *Vestiges*."[135]

Uncontroversial respectability was the *Origin*'s great attraction for those eager to see science become a paid career. The book provided a powerful rallying point at a time when attempts to recruit a new generation of independent gentlemen had patently failed. Whewell, Sedgwick, Murchison, and Herschel were old men, honored but no longer at the cutting edge of intellectual argument. The change of tone was already evident in conversations at the Athenaeum, in reviews of books, in discussions at the Geological Society, and at British Association meetings. *Chambers's Journal* noted soon after publication that naturalists were increasingly inclined to treat the problem of species within

133. Ibid., 490. For Darwin's pleasure in writing, see Beer 1983, esp. 29–48.

134. Review of R. Owen, *Palaeontology, Saturday Review,* 5 May 1860, 573–74, at 573.

135. R. W. Church to A. Gray, 12 Mar. 1860, in Church 1895, 154; [J. Leifchild], review of *Origin of Species, Athenaeum,* 19 Nov. 1859, 659–60; [A. Sedgwick], "Objections to Mr. Darwin's Theory of the Origin of Species," *Spectator,* 24 Mar. 1860, 285–86; 7 Apr. 1860, 334–35 (revised). For the need to rethink the break around 1859, see esp. Hilton 2000, 179–83.

a framework of "order or fixed arrangement." It looked forward to observing "the effect, in the scientific world, of such views brought forward on scientific grounds by a naturalist of eminence."[136]

That effect was dramatic by any standard. Within a few months an important group of scientific men—Hooker, Huxley, Lyell, Owen, Ramsay, Carpenter, and many others—publicly stated their support for evolution. More important still, they did this both in newly dominant monthlies such as the *National* and *Macmillan's,* and in periodicals noted for having opposed earlier versions of progressive development.[137] The most significant volte-face was in the *Edinburgh.* Historians have always followed Darwin and his friends in seeing this as a theologically motivated betrayal by Owen, but contemporaries read through his spite and recognized a major shift by a leading quarterly. The *Edinburgh,* which had led the way in trashing *Vestiges,* now advocated creation as a continuous process and (in a rebuke to Sedgwick) condemned "sacerdotal revilers."[138] The biggest coup had come when Huxley was given the opportunity to review the *Origin* in the *Times.* The political journalist commissioned to do the job had felt unqualified to deal with a scientific book—something less likely to have happened a decade earlier—and asked Huxley to write on his behalf. As a result, the "Thunderer," the pioneer of mechanized newspaper printing nearly a half-century before, was hailing Darwin's book as an "ingenious hypothesis" that merited the "respect and attention of the scientific world."[139]

Among the advanced guard, the *Origin* was welcomed chiefly because of the alliance it offered with men of science. Marian Evans summed up the general view after reading it at home with Lewes:

> We have been reading Darwin's Book on the "Origin of Species" just now: it makes an epoch, as the expression of his thorough adhesion, after long years of study, to the Doctrine of Development—and not the adhesion of an anonym like

136. [Chambers], "Charles Darwin on the Origin of Species," *Chambers's Journal,* 17 Dec. 1859, 388–90; also in Chambers 1994, [209–10].

137. There is an extensive literature devoted to the reception of the *Origin* in Britain, including Bowler 1988, 1989; Desmond 1982; Hodge 1972a; Hull 1983; Moore 1979. The only account to deal substantively with the full range of responses is A. Ellegård's pioneering but now dated *Darwin and the General Reader* (1958).

138. [R. Owen], "Darwin on the Origin of Species" (1860) in Hull 1983, 175–213, at 192. See also S. P. Woodward to C. Lyell, 21 Feb. 1863, EUL Lyell 1/6479: "But knowing as I do the real sentiments of the Edinburgh Reviewer of April 1860—still less could I echo (Sir H. Holland[']s?) words in the 'Times' when he calls him the 'Champion of Orthodoxy'!"

139. "The Darwinian Hypothesis," *Times,* 26 Dec. 1859, 8; also in Huxley 1893–94, 2:1–21. On the background, C. Darwin to T. H. Huxley, 28 Dec. 1859, in Burkhardt et al. 1985–99, 7:458–59, and Desmond 1994, 263–64.

> the author of the "Vestiges," but of a long-celebrated naturalist. The book is ill-written and sadly wanting in illustrative facts. . . . This will prevent the work from becoming popular, as the "Vestiges" did, but it will have a great effect in the scientific world, causing a thorough and open discussion of a question about which people have hitherto felt timid. So the world gets on step by step towards brave clearness and honesty! [140]

From a very different perspective, the High Church *Guardian* agreed: "There are forms of speculation so wild and improbable, or, at any rate, so alien to our ordinary habits of thought, that they can only obtain a fair consideration under the protection of some illustrious name." What might be dismissed from an unknown, had to be taken seriously from "a man confessedly in the foremost ranks of natural philosophy." As Harriet Martineau enthused to Holyoake, with whom she was now on cigar-smoking terms, "The range & mass of knowledge take away one's breath." [141] She had none of Marian Evans's desire to understand the "mystery" behind the evolutionary process and was sorry only about Darwin's theological compromises. All these readers made little of natural selection, appreciating instead the way in which the *Origin* promised to transform the character of debate.

The changed situation was clear at the Oxford British Association in 1860, when Wilberforce launched into the *Origin* at a section meeting held in the newly built natural history museum. Thirteen years before, the bishop's sermon on pride had been hailed by scientific men and clerics alike. But the old consensus no longer held. Wilberforce was put down in speeches from Hooker and Lubbock, and by Huxley's angry remark that he would rather have a "miserable ape for a grandfather" than a man who misused his powers "for the mere purpose of introducing ridicule into a grave scientific discussion." [142] There had been expectations of a clash, although Huxley was present only because Chambers had begged him not to desert the cause. Afterward, many spectators felt the session was a draw; but the common view was that clerical interventions of this kind belonged to the past rather than the future.

This was a minor incident, later raised to mythical dimensions by the need to make the Darwinian debate look more heated than it actually was. The Reverend Frederick Temple, who preached the traditional Sunday sermon to the British Association, argued for a lawful creation and against arbitrary limits to science. Neither did Owen share Wilberforce's views; as president of the British Association at Leeds in 1858 he had famously advocated a continuous

140. G. Eliot to B. Bodichon, 5 Dec. 1859, in Haight 1954–78, 3:227.

141. *Guardian,* 8 Feb. 1860, 134–35, at 134; H. Martineau to G. Holyoake, [1859], BL Add. mss. 42726, f. 26.

142. T. H. Huxley to F. Dyster, 9 Sept. 1860, quoted in Desmond and Moore 1991, 497. Desmond 1994, 276–81; Jensen 1988; and Lucas 1979 provide the best accounts of the exchange.

creative law.[143] Bishop Wilberforce spoke only for a minority. Acland, unhappy at having his museum hijacked by a rogue cleric, wrote to Owen: "Whatever views Mr Huxley, or you, or Mr Darwin, or the Bishop of Oxford may have as to the essential Nature of Man, you all agree that however he so became, he is in some manner made in the image of God, by the ordinance of God."[144] Acland was being optimistic (especially about Huxley), but the tenor of controversy was shifting. Even Whewell agreed, though he thought Darwin's book "utterly unphilosophical." Wilberforce, he suspected, had been imprudent "to venture into a field where no eloquence can supersede the need of precise knowledge." This was no way to convince the younger generation of naturalists.[145] Holy wars were out of fashion.

The *Origin* was important in resolving a crisis, not in creating one. It offered the opportunity to reduce the tensions between those who advocated specialist ideals of research, and those who looked to developmental cosmology as the underpinning for a new kind of society. These distinctive visions could now be debated under the broad banner of "Darwinism"—a word coined in Huxley's *Westminster* review of the *Origin*.[146] The triumph of Darwinism was not one of doctrine—there was consensus neither about the meaning of evolution nor of the truth of natural selection. Rather, Darwinism was a convenient label for an arena of public discussion, structured by new relations between professional science and professional journalism. Gladstone and Huxley disagreed profoundly about evolution, but they could now do so in the pages of a single periodical. Reference to the *Origin* erased the troubled past out of which this new order had emerged, so that debate about the meaning of science could begin with a clean slate in 1859. It was, as George Eliot wrote in another context, "the make-believe of a beginning."[147] Darwin, whose character as a gentleman removed him from the corrupting influence of the market, became a hero of an aggressive, entrepreneurial, imperial science based on paid expertise and mass journalism. Darwinism was the science of the future.

143. For Temple's sermon, see Brooke 1991b, 274. For Owen, Rupke 1994, 325–28.

144. H. Acland to R. Owen, 4 Oct. 1862, cited in Yanni 1999, 89. N. Rupke (1994, 296–98) discusses relations between Acland and Owen at this time.

145. W. Whewell to J. D. Forbes, 24 July 1860, St. Andrews University Library, J. D. Forbes Incoming Letters, 1860, no. 145a.

146. [Huxley], "The Origin of Species," *Westminster Review*, n.s., 17 (Apr. 1860): 541–70, at 569, reprinted in Huxley 1893–94, 2:22–79, at 78; Desmond and Moore 1991, 481; Moore 1991.

147. G. Eliot, epigraph to *Daniel Deronda* (1876), chap. 1.

EPILOGUE

Lifting the Veil

Take the book . . . and read your eyes out,
you will never find what I find.

RALPH WALDO EMERSON, "Spiritual Laws," in *Essays* (1841)

AT FOUR O'CLOCK IN THE AFTERNOON, Wednesday, 19 April 1882, Charles Darwin received an enthusiastic greeting from the denizens of Hell. Contrary to rumor, Hell wasn't so bad—it was "a progressive institution," a place of good fellowship and sympathy rather than fire and brimstone. As a resident of No. 4,960,783 block explained to an undead friend, the arrival of the great evolutionist created more interest than any other of recent times. Everyone was amazed and amused that the body of their new inmate was to be buried in Westminster Abbey. The philosophers of Pandemonium flocked to hear the law of evolution at first hand from its discoverer. Democritus, Lucretius, Hobbes, Humboldt, Schopenhauer, all were there. Yet although the author of *Vestiges* had died some years before, he does not appear to have been among their number. Perhaps he was in another place—his last, unfinished work had been a Christian catechism. In Hell as on Earth, the *Origin of Species* had eclipsed *Vestiges* completely; at least that was the assumption of the freethinking secularists who published this report in a penny tract.[1]

Why do some books rather than others become part of our everyday experience? Every act of reading is an act of forgetting: the experience of reading is a palimpsest, in which each text partially covers those that came before. Those books that allow us to forget the most are accorded the authority of the classic. More than any other book of the modern era, the *Origin* has been endowed by successive generations of readers with the classic's timeless, transcendent power. The *Origin* is among the most pervasive remnants of the Victorian world in our culture, yet it simultaneously forces much of that world into oblivion. It is regularly referred to as the single work that transformed manners, the novel, the

1. Wheeler [1884], 7.

periodical press, the practice of the sciences. It shattered notions of humanity, morality, and truth. It removed at a single stroke any intellectual justification for God's role in nature. It is the secular Bible that made the modern world.[2]

Just stating these claims together shows how implausible they are. Repeated and reinforced throughout our culture, they attribute to a single book an intrinsic power accorded to no other force in history. This phenomenon is all the more striking in that the *Origin*'s main novelty, natural selection, was rejected by almost all readers for the first seventy-five years after publication.[3] The significance of the work cannot be attributed to the new truths that were revealed in its pages. Natural selection achieved major significance within science only in the 1930s, after population biologists and laboratory geneticists effectively "rediscovered" the *Origin*. This was a scientific world very different from the one in which the Victorians lived and worked. The situation is summed up by the photograph in figure E.1. What looks like Darwin is in fact a composite: the familiar bearded head has been superimposed upon the body of a twentieth-century laboratory researcher. This Darwin is a fake, wearing a modern suit, holding a modern microscope, and working in a modern laboratory.

The dominance of Darwin as heroic author and the *Origin* as a classic text is closely tied up with the division of learning into academic disciplines and the rise of the mass-circulation press in the late nineteenth century. Yet the apotheosis of the *Origin*, rather than being seen as an outcome of a restructuring of knowledge, is often claimed as its starting point. An ever-increasing number of fields, from philosophy to economic history, trace their history back to 1859. With modern sociobiology and Darwinian theories of the mind, the *Origin*'s status is now more firmly established than ever before. Moreover, as the classic of disciplined, science-based approaches to knowledge, the *Origin* is seen as one of the few genuinely interdisciplinary texts. This gives it a special status even in fields of the humanities that do not typically look to it for inspiration. One of the pathbreaking moves within literary criticism during the past two decades has been the broadening of the canon to include the *Origin* and a few other scientific works, yet in historical terms this has had the unintended effect of narrowing the agenda. Even those who contextualize and deconstruct Darwin's work are inevitably reinforcing its centrality.

The problem is not that we take meanings from the *Origin* that the Victorians could never have imagined. Like all readers, we are free to make what we can out of books in the context of our own interpretative communities. One of the marks of the classic is often taken to be its capacity for being read in differ-

2. Such claims are legion; for examples in historical work, see Cannon 1978, 275; Morgan 1994, 108.

3. Bowler 1988, Ruse 1996.

E.1 Darwin as a modern scientist.

ent languages, cultures, and historical periods.[4] It is remarkable how readers can still find useful insights in a book written so long ago—especially in the sciences, where most works end up on the rubbish heap after a few years. By all the standards used to define the classic, the *Origin* deserves its status.

As I have come to realize, the problem is more fundamental, involving the way that relations between texts, books, authors, and readers are dealt with in historical narratives. Take a simple example: it should be obvious by now that

4. Kermode 1983, 43–44.

the statement "Darwin published his *Origin of Species* in 1859" is false. Authors are not publishers, texts are not books, and books are not the interpretative property of authors.[5] Yet it is hard to avoid speaking as if all these things were true. By default "Darwin" functions as the basis for determining the meaning of a work that remains resolutely "his." Such statements define the terms on which concepts such as "the Darwinian Revolution" are based. Like all forms of hero-worship, this celebration of the author undermines possibilities for individual action, for none of us can be a Darwin, at least in the terms that the myth provides. It sets an unobtainable ideal—the genius revealing great discoveries—as the model of what a scientist should be. It obliterates decades of labor by teachers, theologians, technicians, printers, editors, and other researchers, whose work has made evolutionary debate so significant during the past two centuries.

THIS BOOK HAS BEEN AN EXPERIMENT in a different kind of history. It has explored the introduction of an evolutionary account of nature into public debate in order to see what happens when a major historical episode is approached from the perspective of reading. Its foundations are in the everyday practices of diary keeping, letter writing, debating, displaying, book production, lecturing, listening, and conversation. The tools I have employed are ready to hand in the writings of literary critics, cultural historians, and historians of the book.[6] They suggest that the notion of a "text" should be expanded to include pictures, maps, music, talk—any semantic system inviting interpretation and constructed of signs whose meaning is fixed by convention. Second, they stress that all texts are in the material form of a work, whether that be a published book, a periodical review, or an oil painting. Third, they show that material form is integral to the meaning of the work. Academics are so used to standard editions, microfilm, and on-line texts that they forget the central importance of format, price, paper, and typography in determining the audience for, and interpretation of, words and images. And finally, meaning is understood as the product of reading, undertaken in a context of struggles for authority over interpretation.[7]

Successful attempts to tackle analogous issues have come from a variety of fields, ranging from histories of medicine written from patients' point of view

5. This formulation is based on a comment by Robert E. Stoddard, quoted in Chartier 1994, 9.

6. McKenzie 1986; Chartier 1994; Bourdieu 1977, 1984; Darnton 1990, 107–35. For applications to the history of science, see Johns 1998b and Topham 2000.

7. Some historians, notably Rose (1992), have gone so far as to dismiss the insights of textual criticism outright, and as a result can have nothing to say about what their readers are responding to. From such a perspective, texts simply vanish, and anything goes. But texts do matter, once they are seen within shared contexts of interpretative codes and conventions.

to empirical studies of television audiences. Such accounts are helpful in considering the relationship between individual agency and the heterogeneous forces of social power.[8] They demonstrate that readers, watchers, and listeners actively engage with the media as a way of making sense of their own lives. They show that searching after "ordinary" or "common" readers, "typical" opinions, and truly "popular" culture—as many historians dream of doing—is fundamentally misconceived. Such fantasies aspire to a surveillance of vision akin to the "passive people meters" that monitor every glance of a statistical sample of television viewers.[9] To learn what is really important about reading, the limited and partial evidence of the situated case—very like that available for the past—remains vital even when audiences number in the millions.

The effectiveness of these tools demands the kind of rigorous attention to time and place I have attempted to provide here. Research over the past twenty years has revealed the richness and diversity of the settings for knowledge,[10] while analyses informed by gender and the social history of architecture suggest the need to look at more localized distinctions—to attend, for example, to variations in how books are read or television is watched in different parts of a house.[11] At every point the contexts in the present book have been selected to reveal networks of relations that make up the larger picture. The more completely a case can be situated, the more it reveals wider patterns and structures of response—of competing representations, appropriations, and contests over authority.

The sheer volume of sources for such a study is daunting, and anyone who has ventured beyond the familiar canon of Victorian authors has experienced the sense of being engulfed in a cacophony of conflicting voices. Attempts to bring order out of this seeming chaos go back to the origins of Victorian studies. The early literature tended to look for unity in dominant worldviews and ideologies. In 1957 Walter Houghton's *Victorian Frame of Mind* identified a way of thinking, a unitary temper, as a way of understanding the period. In the following decades Robert Young and others argued that religious, social, and scientific ideas were brought together in an ideological "common context."[12] This

8. Ang 1996; Morley 1992; Silverstone 1994; Radway 1991; Dayan and Katz 1992.

9. Ang 1996.

10. The need to attend to the geography of intellectual controversies is stressed, with specific reference to the Darwinian debate, in Livingstone 1992, 408–9. For references to a wide interdisciplinary literature, see Ophir and Shapin 1991 and Smith and Agar 1998.

11. Flint 1993; Davidoff and Hall 1987; Pearson 1999; Radway 1991; Winter 1995, 1998a.

12. Houghton 1957; Young 1985. Such views remain the stock-in-trade of most general surveys of Victorian Britain. Thus Hoppen (1998), in the volume in the *New Oxford History of England* that covers this period, devotes an entire chapter to evolutionary ideas, but without making any connection to the concerns with commerce, careers, and material circumstance that inform every other

analysis, which considered scientific ideas as forms of political and social representation, had as its centerpiece the demonstration that the discovery of natural selection had been informed by an ideology of Malthusian competitive capitalism. This was controversial at the time, but in retrospect this agenda can be seen to have been working within parameters set by the history of ideas and postwar evolutionary theory. As one commentator has noted, "only after biologists legitimated Darwin did historians rush to study him."[13]

Although its conclusions proved possible to assimilate within the traditional story, the contextual history pursued in the 1970s ultimately undermined the assumptions on which it was based. Like notions of a "Victorian world picture," the "common context" turned out to be a nostalgic invocation of a predisciplinary past in which everyone spoke the same language. Instead, historians have revealed a society riven by controversy. They have shown how alternative cosmologies thrived in the medical schools, how gentlemen of science managed the affairs of the British Association against outside threats, and how new sciences like mesmerism, weather prognostication, and geology competed with traditional ways of understanding.[14] Few would now say, as Young did, that popular phrenology and mechanics' institutes could be demarcated from the study of an "intelligentsia."[15]

The emphasis on diversity has been accompanied by a growing appreciation of the material and practical aspects of Victorian life. Cultural history has been transformed by being considered from local settings, from the Glasgow shipyards to the drawing rooms of the fashionable gentry. This has made it possible to escape the old image of science as dominated by a handful of great theorists and simultaneously to understand theory making as a form of practice. Intellectual history, which used to be written as a story of dramatic changes in worldview (the "Darwinian Revolution"), can be recast by looking at the basic material products of cultural life and drawing upon techniques developed for studying ordinary action.[16]

---

aspect of his account. Tellingly, a separate chapter deals with literature and the fine arts and is entitled "The Business of Culture."

13. Kohn 1985, 2.

14. E.g., Desmond 1989; Morrell and Thackray 1981; Winter 1998a; Anderson 1999; and Lightman 1997b.

15. Young 1985, 131–32. For the origins of the distinction between "lowbrow" and other forms of culture—a strongly marked characteristic in the writings of Ellegård, Young, and other authors of the 1950s and 1960s—see "On the History of the Middlebrow," in Radway 1997, 135–301, and Rubin 1992.

16. This trend, which is widespread in many historical fields, is sketched in Pickering 1992 and Golinski 1998. Its application is evident in many of the essays in Lightman 1997b, and in Smith and Agar 1998.

Yet there have been problems too. In history of science, a focus on the construction of knowledge has led to a proliferation of case studies, often centered on problems of epistemology.[17] In the humanities more generally, a critical emphasis on fragmentation and interpretative freedom has sometimes slipped into a celebration of Victorian values of liberal pluralism. As Adrian Desmond, who has done more than anyone else to open up the study of Victorian evolution, has said, knowledge becomes so contingent that it becomes difficult to draw conclusions beyond the level of the individual.[18] The consequences are most evident in biography, a genre whose popularity in the Victorian period is exceeded only in our own. No matter how strong their emphasis on "social explanation" or "context," biographies keep the individual at the center of readers' understanding.

From very different perspectives, much of the cultural and media studies literature has produced a parallel result: accounts of audience response illustrate diversity, but little else. All readings of a work become equally plausible in all circumstances.[19] These accounts often have the qualities of paintings by the Victorian artist W. P. Frith, such as *Derby Day* or the *Railway Station*, encompassing many sharply observed encounters and individuals within the context of an undifferentiated crowd. Frith's paintings (like Dickens's novels) are known for their reliance on a psychology of stereotypes. They teem with dozens of "characters" who must be scrutinized as individual specimens to be understood.[20] This, as I have stressed, is the language of "sensation" in its simultaneous attention to mass behavior and individualized response. To a surprising extent, accounts of the nineteenth century remain caught in the dichotomies of the Victorian social body, as tightly strapped as if wearing corsets.

This book has sought to escape these constricting frameworks. There are signs that older forms of intellectual and literary history are being recast as aspects of the history of communication. There have been outstanding studies of the reading habits of individuals, and general surveys of reading practices and images of the reader in particular historical periods. There is a long-standing tradition in literary studies of reader-response strategies, and a well-developed body of work analyzing the printing, publishing, and authorship of books and

17. Kohler 1999; see also Turner 1997 and Secord 1993.

18. Desmond 1997, 235–36.

19. This is a frequently made criticism of the work broadly characterized within literary studies as "new historicism." It is powerfully argued, in the context of a review of Mary Poovey's *Making of a Social Body,* in Gagnier 1999. The problem has also been extensively debated in the literature on modern communications, especially in relation to the work of John Fiske on the power of television audiences as makers of meaning. For critical discussions and references, see the introduction to Morley 1992, and Ang 1996, 178–79.

20. Cowling 1989, 232–316.

periodicals. What we have not had for the modern period is a full-length picture of how a substantial range of contemporary readers made meaning from a single work. That is what I have attempted to provide.

MY AIM HAS NOT BEEN to map scientific or philosophical doctrines onto political or religious doctrines, nor to portray texts as reflecting local contexts. As the literary historian Hans Robert Jauss has noted, "the specific achievement of literature in social existence is to be sought exactly where literature is not absorbed into the function of a *representational* art."[21] It is precisely when readings do *not* reinforce existing attitudes that they can change questions, upset the boundaries of disciplines, and confuse the conventions of genre. It is then that they can transform the practices of everyday life in a society and enter into wider processes of historical change.

Vestiges, as I have stressed throughout this book, was a literary hybrid—a Frankenstein, contemporaries said—whose status within accepted genres of fiction, science, and philosophy was indeterminate. A decade later it had become as much "familiar knowledge" as the simple facts with which *Vestiges* opened. The descent of man from the apes, spontaneous generation, the origins of the human mind through natural law, and cosmic progress were regular topics of conversation and controversy for hundreds of thousands of Victorian readers. Samuel Laing, who as a young man had compiled the *Expository Outline* of *Vestiges,* described the late 1850s as a saturated solution into which the *Origin* dropped like a crystal, around which all the diverse elements coalesced.[22] Historians used to say, vaguely, that evolution was "in the air," which is true enough for a book like *Vestiges* that was so much talked about; but it was also in drawing rooms, libraries, churches, pubs, clubs, and railway carriages.

"Science is no longer a lifeless abstraction floating above the heads of the multitude," one contemporary wrote. "It has descended to earth. It mingles with men. It penetrates our mines. It enters our workshops. It speeds along with the iron courser of the rail."[23] Many of the places for public discussion were new. The railways, the Penny Post, and the telegraph were transforming the face of the nation. The explosive growth of cities and the anonymity of urban life led to concerns for individual character. Periodical and newspaper reviews provided what were, in effect, closely circumstanced advice manuals for the use of books among differentiated groups of readers. A host of fine distinctions between public and private life were developed. Old forms of association,

21. "Literary History as a Challenge to Literary Theory," in Jauss 1982, 3–45, at 45; also Simpson 1988.

22. Laing 1889, 135.

23. Garvey 1852, 3.

especially the established churches, attempted to come to terms with the transformation. Conversation in the aristocratic and genteel salons was losing its centrality in defining intellectual fashion. Who would control the means of distribution and communication in the new age?

The industrialization of print culture has often been given a key role in these developments, for printing by machine is associated with the production of stable, permanent objects, which can serve as repositories of fixed meaning. The power of print lies in the assumption of its fixity. Although I have argued for the centrality of the early and mid–nineteenth century within the history of print, the explanation cannot rely primarily on specific technical innovations. In fact, mechanization initially produced a crisis of stability, as reading, authorship, printing, and publishing all came under close public scrutiny as part of the debate about the machine. Groups like the SDUK found "diffusing" uniform knowledge in print proved almost impossible: no one could agree on who would ensure uniformity, what useful knowledge was, who should have access to it, or how reading should take place. Attempts to create a "popular science" or "literature for the people" failed to achieve consensus. Books were less trusted to be what they claimed to be, and more open to alternative meanings, than at any time since the civil wars of the seventeenth century.[24]

Relative stability in print reemerged from the mid- and later 1840s, with the laying of a groundwork for a liberal nation-state, based on imperial free trade and an economic future clearly within the factory system.[25] The industrial revolution in communication resulted as much from changes in the forms of public debate as it was the consequence of technological innovation. The preceding chapters have sketched the beginnings of these transformations, which became firmly established in the second half of the nineteenth century. The structure of communication underwent further great changes, with repeal of the remaining taxes on knowledge, the introduction of the rotary press, and a huge expansion in newspaper and periodical publishing. Cheap national newspapers began to replace expensive provincial ones, and the heavyweight quarterlies, the forum of choice for gentlemanly debate, were superseded by fleet-footed monthlies. The "People," imagined by the entrepreneurs of useful knowledge in the 1830s, came into being as the market for the late Victorian mass-circulation press.[26]

24. For the creation of critical, engaged readerships in the mid–seventeenth century, see Raymond 1997 and Sharpe 2000. Johns 1998a, 628–32, uses the relative failure of the SDUK to downplay the effects of the industrial revolution in publishing in the nineteenth century. Although he is correct in recognizing that technology in itself did not produce stability, the alternative he offers (changes in the role of authorship) is too limited.

25. There is a large literature on this subject, surveyed from different perspectives in Price 1999, Parry 1993, Joyce 1991, Hoppen 1998, and Eley 1992.

26. Broks 1996, Brown 1985, and Curran and Seaton 1991.

The changed tone of discussion about creation and natural law became a significant factor in the emergence of liberal political institutions in British society. The connection between liberalism and evolutionary debate, usually traced to the post-Darwinian era, must be redated to the 1840s. The temper and tone of debate were changing. Whigs and liberal Tories, especially after the repeal of the Corn Laws in 1846, looked to rational economic relations that would mirror nature's progressive laws. "We have committed ourselves to the general laws of Providence," the *Times* wrote in 1852, "and Providence now rewards us with a vista of social improvements, and unexpected blessings, which men had not dreamt of ten years ago."[27] The changes are most evident in religious controversy, where the evolutionary discussion in the wake of *Vestiges* afforded opportunities for forging new alliances. It brought freethinkers closer to middle-class liberal debate, with a campaign tied to modern science rather than Enlightenment cosmology. Most important, perhaps, the *Vestiges* sensation contributed to the long-term weakening of organized religion, not least because the evangelicals were seen to have lost the war of manners in intellectual debate. A vengeful God who demanded retribution had less appeal for a cosmopolitan society with a rapidly advancing economy; speculation in science, as in the market, was no longer condemned as sinful. Even Hell was in decline; the secularists who reported Darwin's descent were not the only ones who doubted the reality of eternal punishment.[28]

Among the important changes was a new role for expertise. Specialist work was increasingly carried out behind laboratory doors, at field stations, and in technical journals, while cosmic evolutionary narratives were repeated and retold in mass-market books and magazines.[29] This was not the breakup of a common context, but a redrawing of boundaries around expertise consequent to the decline of aristocratic society and gentlemanly ideals of knowledge. Categories like "popular" science and "professional" science began to emerge as stable publishing genres through the debates traced in this book. Today, popular science has become everything that real "professional" science is not, and to call *Vestiges* "popular" (or "amateur") is effectively to dismiss it. Such a retrospective judgment has meaning only in the context of our own divide between "professional" and "popular."[30] Applying categories to the very debates that produced them clearly begs the question.

27. *Times,* Sept. 1852, quoted in Parry 1993, 168.

28. Rowell 1974; Hilton 1988, 255–97.

29. Broks 1996, Turner 1993.

30. For the redefining of the popular, see A. Secord 1994a, 299; Shiach 1989; "The Uses of the 'People,'" in Bourdieu 1990; Cooter and Pumfrey 1994. See also the account of "boundary-work" in Gieryn 1999.

The dismissive use of the term "popular," as it stabilized in the late nineteenth and twentieth centuries, was designed to render readers as invisible members of a mass audience. Just because books and other printed materials became factory-produced commodities, however, does not mean that readers agreed with what they found in them. Such a claim is in any event meaningless, as books are not checklists of messages against which responses can be compared.[31] The mass-communications industry never created a passive, homogeneous audience: it stereotyped books and newspapers, not readers. The power of scientists to dictate views on wider issues was (and remains) severely limited both by the institutions of the mass media and by the proliferation of alternative points of view. Even as the production of print became standardized, interpretation remained bound by local circumstance. Processes of national integration—like current trends toward globalization—enhanced awareness of difference.[32]

By the end of the nineteenth century, as the ambitious journalist Jaspar Milvain said in George Gissing's *New Grub Street* (1891), "'The struggle for existence among books is nowadays as severe as among men.'"[33] In fact, sales figures show that *Vestiges* continued to thrive long after the *Origin* is usually thought to have driven it out of the market (fig. E.2). Some fifteen thousand copies were published in the 1880s and 1890s, nearly two-fifths of the total published during the century. These posthumous editions show the place that evolution had come to occupy in the economy of print. They were cheap; their texts were unrevised (the text of the shilling version in Routledge's Universal Library was taken from the second edition, as copyright on it had expired in 1886); and from the twelfth edition of 1884 they were securely attributed to the cofounder of one of the largest mass publishing houses in the country.

Putting a name on the title page crystallized the book's place in history (fig. E.3). It was from this point that *Vestiges* began to be read as a "popularisation," as though there had already been some well-established body of evolutionary truth to diffuse. The author would henceforth be described as an "amateur," as though there had been a clear group of "professionals" in the 1840s who defined what science was. (Even Alfred Russel Wallace, while acknowledging his debt to the book, annotated his copy of the twelfth edition with dismissive comments: "Oh!" and "wretched logic.")[34] The intrinsic meaning of the book

31. It would, for this reason, be fruitless to boil down *Vestiges* to three or four isolated statements and evaluate their acceptance through a statistical analysis—the procedure Alvar Ellegård (1958) used to study the *Origin.*

32. These points are widely stressed in the literature on the mass media; see esp. Ang 1996, Morley 1992, and Radway 1991.

33. Gissing 1985, 493.

34. Copy in library of the Linnean Society, London, cited in Schwartz 1990, 146–47.

would be characterized as liberal and reformist, when the most plausible candidate for the authorship in the first year was a Tory aristocrat. And the book's readership would be defined as the commercial middle classes, although the early editions were too expensive for them to afford, and the sensation started in the salons of the aristocracy. Labeled as the work of a provincial popular author, *Vestiges* became unworthy of a permanent place in the library.

George Eliot, writing under a pseudonym and as an invented narrator in *Impressions of Theophrastus Such* (1879), had been astutely aware of the effect that revealing the authorship would have. Three years before Ireland made his announcement, she asked readers to think through the consequences:

> No doubt if it had been discovered who wrote the "Vestiges," many an ingenious structure of probabilities would have been spoiled, and some disgust might have been felt for a real author who made comparatively so shabby an appearance of likelihood. . . . Hardly any kind of false reasoning is more ludicrous than this on the probabilities of origination. It would be amusing to catechise the guessers as to their exact reasons for thinking their guess "likely:" why Hoopoe of John's has fixed on Toucan of Magdalen; why Shrike attributes its peculiar style to Buzzard, who

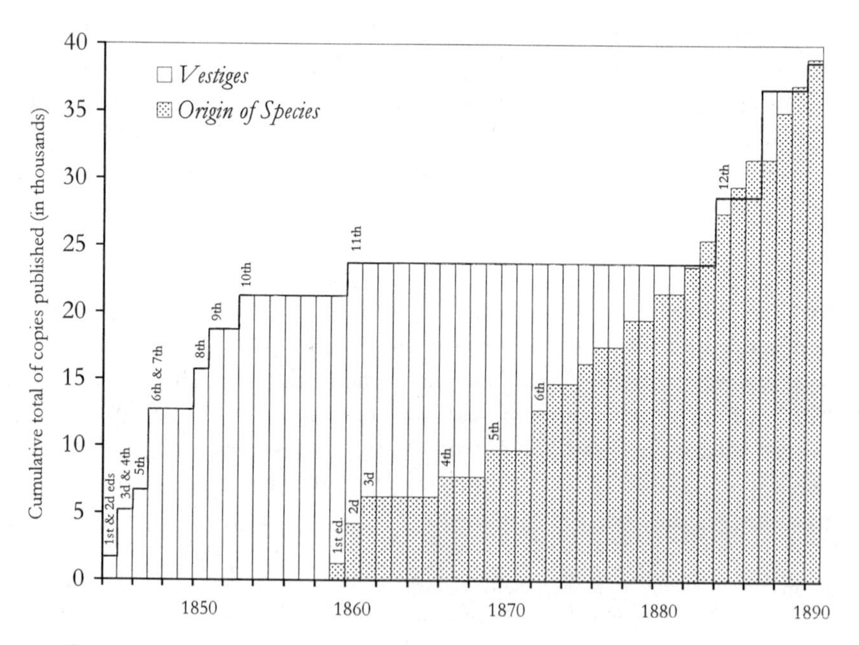

E.2 Comparative chart showing number of copies of *Vestiges* and the *Origin of Species* published in 1844–90. Ordinal numbers refer to editions. The *Origin* did not decisively overtake *Vestiges* until the twentieth century. At that point *Vestiges* was finally superseded by more up-to-date general evolutionary books, and the *Origin* went out of copyright. Data on the *Origin* is from Peckham 1959, appendix 2, 75–88.

E.3 The anonymity of *Vestiges,* as viewed by the graphic artist Borin Van Loon in the best-selling *Darwin for Beginners,* first published in 1982 and now retitled *Introducing Darwin and Evolution.* Miller and Van Loon 2000, 44.

> has not hitherto been known as a writer; why the fair Columba thinks it must belong to the reverend Merula; and why they are all alike disturbed in their previous judgment of its value by finding that it really came from Skunk, whom they had either not thought of at all, or thought of as belonging to a species excluded by the nature of the case.[35]

Written, as it was, by Skunk, the most discussed scientific book of the Victorian era became an embarrassment. Like phrenology, mesmerism, and the fashion for balloonlike crinolines, the *Vestiges* sensation was recalled as an endearing curiosity of a bygone era.

At the same time, publishers found new ways of marketing the book. Colorfully bound and printed on wood pulp paper, *Vestiges* could be assimilated to

35. G. Eliot 1994, 96.

E.4 Apes celebrating the gestation of humanity, by the light of Darwin's *Descent of Man.* Engraving by Ernst Moritz Geyger, published in 1892 by Charles Sedelmeyer of Paris.

the commodified surveys of science that became an important feature of late Victorian publishing. Its author could be remembered as precursor to the greater Darwin, who marched through the scientific jungles like an elephant. As George J. Romanes wrote in *Nature,* "it is difficult to realise the state of matters before the elephant appeared." Ireland, in introducing the posthumous edition of 1884, called *Vestiges* "the immediate forerunner" of Darwin.[36] The text remained the same, but for readers the book had a "Darwinian" look. These late editions became part of a flourishing genre of general evolutionary works; even the apes who watched at the birth of humanity did so in light of Darwin's *Descent of Man* (fig. E.4). At the beginning of the twentieth century, there were dozens of evolutionary titles, ranging from Arabella Buckley's childhood favorite *Winners in Life's Race,* through didactic writings of Richard Proctor and

36. G. J. Romanes, review of *Vestiges, Nature* 30 (22 May 1884): 73–74, at 74; Ireland, introduction to *V12*, xi. See also R. A. Proctor, "Vestiges of Creation," *Knowledge,* 9 May 1884, 329–30.

Robert Ball, to the *Origin of Species* itself, which appeared in a cheap edition from 1872 and in paperback early in the twentieth century.[37]

Controversy about the origin of the universe had erupted through the work of creating a progressive past for an industrial, imperial society. The Victorians read science in general, and *Vestiges* in particular, as part of their search for narratives of origins—to make sense of individual and collective experience.[38] They read to understand the seemingly infinite prospects for the future revealed by science—to wonder at new discoveries, new lands, new inventions, new forms of human relations. They erased old stories or remembered them in new ways. Take the memory recorded in Edward Maitland's pseudonymous novel *The Pilgrim and the Shrine* (1868). Having recently graduated from Cambridge, the hero is on a sugar plantation in Jamaica, lazing on the grass, watching the blacks work. A monkey rolling a bottle imitates a gardener rolling the path, and this leads to "a reverie" about *Vestiges*. "An ascending series indeed,—the monkey, the negro, and myself,—and complacently enough did I speculate as to where was the greatest interval, between the two former or the two latter."[39] The year is 1849, but in a novel published in 1868 readers would have read such a scene in light of the brutal suppression of a black rebellion three years before, in which 439 people were killed. An evolutionary hierarchy of progress, reinforced after 1859 by Darwinian ideas of struggle, was hardening into a racist belief in an imperial scale of civilization. Sanctified by practices in the laboratory, the university, and the publishing industry, evolution became what the *Westminster* had predicted, a "Whitworth gun in the armoury of liberalism," underwriting free trade, colonial expansion, the inferiority of women, and unrestrained market competition.[40]

Today, evolutionary controversies continue to dominate science in the mass media. Recycled fragments of the early Victorian debates have become part of the fabric of everyday life, read through their incorporation into other texts. They can be found not only in scientific reporting and documentaries but in the conventions of science fiction and the realist novel. They are present in our everyday talk about psychology, philosophy, and economics. More intangible are the ways that the *Vestiges* sensation entered into the making of basic features of our

37. For these works and on evolutionary journalism, see Lightman 1989, 1997c; Gates 1997; and Broks 1996, 83–97.

38. For narratives of nationhood, see Bhabha 1990; and for a helpful review of current problems of constructing a narrative for nineteenth-century British history, see Price 1999. On narratives of progress, Bowler 1989.

39. [E. Maitland] 1868, 1:52–55.

40. [T. H. Huxley], "The Origin of Species," *Westminster Review*, n.s. 17 (Apr. 1860): 541–70. The complex history of relations between liberalism and evolutionism is discussed in Freedan 1978; Hoppen 1998, 479–97; Desmond 1997; Crook 1994.

culture: the disciplinary structure of the sciences; the relations of science with the mass media and with religion; the boundaries between popular and expert knowledge; the limits of fiction and fact; the sense of an individual identity in urban society. All these were at a critical juncture in the 1840s, when the foundations for the modern world of industrial communications were being laid. Reading was like eating, the Victorians emphasized, and they read to feed the social body. To say that *Vestiges* enjoyed a "tea-table vogue" should be not to dismiss it,[41] but to recognize that the book became part of ordinary domestic rituals.

FOR THOUSANDS OF READERS, Tennyson's *In Memoriam* offered the most profound integration of the evolutionary narrative into everyday experience. The widowed Queen Victoria found solace in Tennyson's great elegy, as did soldiers in the trenches during the First World War. Most of the poem, although not published until a few years later, was already drafted when Tennyson encountered the *Examiner*'s review of *Vestiges* in November 1844.[42] Up to that point in his writing, science had been used to portray a disorienting cycle of extinction in which human beings seemed to occupy no particular place. Not only the "single life," but humanity itself (the "type") would disappear. The second half of *In Memoriam* begins a new cycle, drawing on a vision of union with God that leads toward an optimistic conclusion implicated in *Vestiges*.[43]

The poem ends with a wedding, looking beyond the perpetuation of the human race to a succession of more advanced forms. Nature's cosmic law is repeated in individual development as the moonlight reaches the wedding chamber:

> And touch with shade the bridal doors,
> With tender gloom the roof, the wall;
> And breaking let the splendour fall
> To spangle all the happy shores

41. Millhauser 1956, 213.

42. The poem made references to developmental progress, which after 1859 began to be read as Darwinian. "Nature red in tooth and claw" eventually united Tennyson and Darwin as the remembered authors of a common evolutionary faith. The republican poet Charles Algernon Swinburne waggishly suggested that Darwin had written not just *In Memoriam,* but all of Tennyson's poems. This was about as likely as Francis Bacon authoring Shakespeare, which was the point of the parody. See [Swinburne], "Dethroning Tennyson: A Contribution to the Tennyson-Darwin Controversy," *Nineteenth Century* 23 (Jan. 1888): 127–29.

43. As Sinfield (1971, 24–25) points out, the turning point comes when the poet reads the dead friend's letters in section 45; this subjective experience makes the forward-looking references to developmental progress more than wishful thinking.

By which they rest, and ocean sounds,
And, star and system rolling past,
A soul shall draw from out the vast
And strike his being into bounds,

And, moved through life of lower phase,
Result in man, be born and think,
And act and love, a closer link
Betwixt us and the crowning race. . . .[44]

These lines, which directly recall *Vestiges* ("Is our race but the initial of the grand crowning type?"), were read by Tennyson's audience in light of their shared experience of the book. As reviewers recognized, "the dreams of the author of 'Vestiges of Creation' seem to be realized and accepted by the poet."[45] Remembering our origins becomes a way of looking to the future, as nebular evolution—"star and system rolling past"—is united with the coming-into-being of a new soul.

In Memoriam's most significant incorporation of *Vestiges* is through the address to the reader in the final verses.[46] It is the reader, addressed as the groom, who is to be married, and the poet's sister is the bride. During the ceremony the address shifts briefly to the second-person plural, as the vows "have made you one"; thereafter the couple are spoken of in the third person ("We wish them store of happy days"). The reader observes the consummation of this cosmic marriage from outside and remains, simultaneously, a partner within it. By participating in this "thy marriage day," the reader of *In Memoriam* is invited to join in the forward advance of evolution. As in *Vestiges,* the act of reading thus becomes part of the progress toward "the crowning race":

Of those that, eye to eye, shall look
On knowledge; under whose command
Is Earth and Earth's, and in their hand
Is Nature like an open book. . . .

This is a biblical promise, recalling the potential for human perception in the garden of Eden. As *Vestiges* had said, the reader is to become that "nobler type of humanity, which shall . . . realize some of the dreams of the purest spirits of the present race."[47] In the future, the poem predicts, reading will be the model

44. *In Memoriam A. H. H.* (1850), in Ricks 1987, 2:456–57.

45. "New Poetry. Tennyson, Browning, and Taylor," *English Review* 14 (Sept. 1850): 65–92, at 76. For the references to *Vestiges,* see *V*1, 276, and esp. the discussion in Killham 1958, 252–66.

46. As far as I am aware, the complex role of the reader in the text of *In Memoriam* has yet to receive critical attention. For an analysis of Tennyson's related poem, *Lucretius,* from this perspective, see O'Neill 1991. For a sampling of recent approaches to Tennyson, see the essays in Stott 1996.

47. *V*1, 276.

for pure understanding: mastery over creation will be transcendent literacy. When nature can be read like a book, it will need no interpretation. The final verse looks to the union of knowledge and spiritual fulfillment in the reader who sees direct and whole.

Projections of this perfect reader are now most often found in the sciences. Books on the evolution of life, the solar system, and the human mind advertise the prospect of a key to universal knowledge. Specialist writing on science creates the same impression through a language that presupposes the transparency of the act of reading. Mathematics, of all forms of communication, is seen to approach transcendent truth. As Stephen Hawking wrote in the closing sentences of *A Brief History of Time*, a final theory in physics is likely to be so simple that everyone will be able to understand it and join in discussing the question of why we and the universe exist. To find the answer would be to "know the mind of God."[48] The dream of complete understanding is also held out by the electronic revolution, which promises the universal availability of information, perhaps a new stage in evolution. More ambiguously, the same hope for transcendence was expressed in Stanley Kubrick's celebrated film of the late sixties, *2001: A Space Odyssey*, which traces human progress from warring apes to the vision of an embryonic star-child floating in space. "One God, one law, one element, / And one far-off divine event, / To which the whole creation moves";[49] this is the image of universal gestation at the heart of the cosmological debates we have inherited from the Victorians.

The texts of science have no meaning apart from what readers make out of them, yet—ironically—they aspire to be a transcript of the truth of nature, needing no interpretation. Any attempt to recover their history is bound to project a very different image. If, as Michel de Certeau once wrote, history is cannibalism,[50] then the history of our understanding of nature has been among the most ferocious consumers. To recover past visions of nature is part of the process of reconstructing past readings of any kind: a fragmentary, inconsistent enterprise aimed at recalling memories that have disappeared. Few subjects other than science incorporate their past so completely into present experience, while throwing it so violently away. In remembering the *Origin* we forget *Vestiges;* in remembering *Vestiges* we forget the *Constitution of Man* and the bawdy philosophical books of the Enlightenment. In this way we have forged chronicles of progress, narratives that have prevented our seeing from any other perspective. As readers we can make a different history.

48. Hawking 1988, 175.
49. *In Memoriam A. H. H.* (1850), in Ricks 1987, 2:459.
50. De Certeau 1986, 3.

REFERENCES

Periodical and newspaper articles published before 1900 are cited fully in the notes and are not included in this list of references.

Acland, Henry Wentworth. 1848. *Remarks on the Extension of Education at the University of Oxford, in a Letter to the Rev. W. Jacobson.* Oxford: John Henry Parker.

Alborn, Timothy L. 1996. "The Business of Induction: Industry and Genius in the Language of British Scientific Reform, 1820–1840." *History of Science* 34:91–121.

Allen, David Elliston. 1985. "The Professionals in Natural History in the Early Nineteenth Century." In *From Linnaeus to Darwin: Commentaries on the History of Biology and Geology.* Ed. Alywne Wheeler and James H. Price, 1–12. London: Society for the History of Natural History.

———. 1996. "The Struggle for Specialist Journals: Natural History in the British Periodicals Market in the First Half of the Nineteenth Century." *Archives of Natural History* 23:107–23.

Allen, James Smith. 1991. *In the Public Eye: A History of Reading in Modern France, 1800–1940.* Princeton: Princeton University Press.

Altholz, Josef L. 1989. *The Religious Press in Britain, 1760–1900.* New York: Greenwood Press.

Altick, Richard D. 1957. *The English Common Reader: A Social History of the Mass Reading Public 1800–1900.* Chicago: University of Chicago Press.

———. 1962. "The Sociology of Authorship: The Social Origins, Education, and Occupations of 1100 British Writers, 1800–1935." *New York Public Library Bulletin* 66:389–404.

———. 1978. *The Shows of London.* Cambridge: Harvard University Press, Belknap Press.

———. 1991. *The Presence of the Present: Topics of the Day in the Victorian Novel.* Columbus: Ohio State University Press.

———. 1997. *Punch: The Lively Youth of a British Institution, 1841–1851.* Columbus: Ohio State University Press.

Anderson, John. 1850. *The Course of Creation.* London: Longman.

[Anderson, John, Jr.]. 1832. *Sketches of the Edinburgh Clergy of the Established Church of Scotland.* Edinburgh: John Anderson.

Anderson, Katharine. 1999. "The Weather Prophets: Science and Reputation in Victorian Meteorology." *History of Science* 37:179–216.

Anderson, Patricia J. 1991. *The Printed Image and the Transformation of Popular Culture.* Oxford: Clarendon Press.

Anderson, Patricia J., and Jonathan Rose, eds. 1991. *British Literary Publishing Houses, 1820–1880.* Detroit: Gale.

Ang, Ien. 1996. *Living Room Wars: Rethinking Media Audiences for a Postmodern World.* London: Routledge.

Ansted, D. T. 1847. *The Ancient World; Or, Picturesque Sketches of Creation.* London: John Van Voorst.

Appel, Toby. 1987. *The Cuvier-Geoffroy Debate: French Biology in the Decades before Darwin.* New York: Oxford University Press.

Archer, Thomas. 1848. "The Geological Evidences of the Existence of the Deity." In *YMCA Lectures, 1847–48,* 171–93. London: Benjamin L. Green.

———. 1864. "The Unity of the Species." In *YMCA Lectures, 1846–47,* 261–75. London: James Nisbet.

Ashton, Rosemary. 1991. *G. H. Lewes: A Life.* Oxford: Clarendon Press.

———. 1996. *George Eliot: A Life.* London: Hamish Hamilton.

Ashworth, William J. 1996. "Memory, Efficiency, and Symbolic Analysis: Charles Babbage, John Herschel, and the Industrial Mind." *Isis* 87:629–53.

Astore, William J. 1995. "Observing God: Thomas Dick (1774–1857), Evangelicalism and Popular Science in Victorian Britain and Antebellum America." Ph.D. diss., University of Oxford.

Atlay, J. B. 1903. *Sir Henry Wentworth Acland, Bart., K.C.B., F.R.S., Regius Professor of Medicine in the University of Oxford.* London: Smith, Elder.

Auerbach, Nina. 1990. *Private Theatricals: The Lives of the Victorians.* Cambridge: Harvard University Press.

Austin, S., J. Harwood, G. Pyne, and C. Pyne. 1831. *Lancashire Illustrated in a Series of Views.* London: H. Fisher.

Babbage, Charles. [1832] 1835. *On the Economy of Machinery and Manufactures.* 4th ed. London: Charles Knight.

———. 1838. *The Ninth Bridgewater Treatise: A Fragment.* 2d ed. London: John Murray.

Bailin, Miriam. 1994. *The Sickroom in Victorian Fiction: The Art of Being Ill.* Cambridge: Cambridge University Press.

Baker, William. 1992. *The Early History of the London Library.* Lewiston: Edwin Mellen Press.

Balfour, John Hutton. 1851. *Phyto-theology; Or, Botanical Sketches, Intended to Illustrate the Works of God in the Structure, Functions, and General Distribution of Plants.* London: Johnstone.

Bann, Stephen. 1984. *The Clothing of Clio: A Study of the Representation of History in Nineteenth-Century Britain and France.* Cambridge: Cambridge University Press.

Barker-Benfield, G. J. 1992. *The Culture of Sensibility: Sex and Society in Eighteenth-Century Britain.* Chicago: University of Chicago Press.

Barnes, Emm. 1995. "Fashioning a Natural Self: Guides to Self-presentation in Victorian England." Ph.D. diss., University of Cambridge.

Barnes, James J. 1974. *Authors, Publishers, and Politicians: The Quest for an Anglo-American Copyright Agreement.* London: Routledge.

Barrett P. H., P. J. Gautrey, S. Herbert, D. Kohn, and S. Smith, eds. 1987. *Charles Darwin's Notebooks.* Cambridge: British Museum (Natural History)/Cambridge University Press.

Bartholomew, Michael. 1973. "Lyell and Evolution: An Account of Lyell's Response to the Prospect of an Evolutionary Ancestry for Man." *BJHS* 6:261–303.

———. 1975. "Huxley's Defence of Darwin." *Annals of Science* 32:525–35.

Barton, Ruth. 1990. "'An Influential Set of Chaps': The X-Club and Royal Society Politics 1864–85." *BJHS* 23:53–81.

Bates, William. 1873. *The Maclise Portrait-gallery.* London: Chatto and Windus.

Baum, Joan. 1986. *The Calculating Passion of Ada Byron.* Hamden, Conn.: Archon Books.

Baxter, Paul. 1984. "Brewster, Evangelicalism, and the Disruption of the Church of Scotland." In *"Martyr of Science": Sir David Brewster 1781–1868.* Ed. A. D. Morrison-Low and J. R. R. Christie, 45–50. Edinburgh: Royal Scottish Museum Studies.

———. 1985. "Science and Belief in Scotland, 1805–1868: The Scottish Evangelicals." Ph.D. diss., University of Edinburgh.

———. 1993. "Deism and Development: Disruptive Forces in Scottish Natural Theology." In *Scotland in the Age of the Disruption.* Ed. Stuart J. Brown and Michael Fry, 98–112. Edinburgh: Edinburgh University Press.

Bayly, Thomas Haines. 1844. *Songs, Ballads, and Other Poems.* 2 vols. London: Richard Bentley.

Bayne, Peter. 1871. *The Life and Letters of Hugh Miller.* 2 vols. London: Strahan.

Beard, J. R., ed. 1846. *Unitarianism Exhibited in Its Actual Condition; Consisting of Essays by Several Unitarian Ministers and Others; Illustrative of the Rise, Progress, and Principles of Christian Anti-Trinitarianism in Different Parts of the World.* London.

Bebbington, David W. 1989. *Evangelicalism in Modern Britain: A History from the 1730s to the 1980s.* London: Unwin Hyman.

Becher, Harvey W. 1986. "Voluntary Science in Nineteenth Century Cambridge University to the 1850s." *BJHS* 19:57–87.

Bedford, James. 1854. *New Theories of the Universe, Explaining how Sun, Moon, Stars, etc. are Formed.* London: Simpkin.

Beer, Gillian. 1983. *Darwin's Plots: Evolutionary Narrative in Darwin, George Eliot and Nineteenth-Century Fiction.* London: Routledge.

Beetham, Margaret. 1990. "Towards a Theory of the Periodical as a Publishing Genre." In *Investigating Victorian Journalism.* Ed. Laurel Brake, Aled Jones, and Lionel Madden, 19–32. Basingstoke: Macmillan.

Benjamin, Marina. 1991. "Elbow Room: Women Writers on Science, 1790–1840." In *Science and Sensibility: Gender and Scientific Enquiry 1780–1945.* Ed. Marina Benjamin, 27–59. Oxford: Blackwell.

Bennett, Betty T., ed. 1980–88. *The Letters of Mary Wollstonecraft Shelley.* 3 vols. Baltimore: Johns Hopkins University Press.

Bennett, Daphne. 1977. *King without a Crown: Albert, Prince Consort of England, 1819–1861.* London: Heinemann.

Bennett, Scott. 1976. "John Murray's Family Library and the Cheapening of Books in Early Nineteenth Century Britain." *Studies in Bibliography* 29:139–66.

———. 1982. "Revolutions in Thought: Serial Publication and the Mass Market for Reading." In *The Victorian Periodical Press.* Ed. Joanne Shattock and Michael Wolff, 225–57. Leicester: Leicester University Press.

———. 1984. "The Editorial Character and Readership of *The Penny Magazine:* An Analysis." *Victorian Periodicals Review* 17:126–41.

Berg, Maxine. 1980. *The Machinery Question and the Making of Political Economy, 1815–1848.* Cambridge: Cambridge University Press.

———. 1985. *The Age of Manufactures: Industry, Innovation, and Work in Britain, 1700–1820.* Oxford: Basil Blackwell.

Berg, Maxine, and Pat Hudson. 1992. "Rehabilitating the Industrial Revolution." *Economic History Review* 45:24–50.

Berkeley, Edmund, and Dorothy Smith Berkeley. 1988. *George William Featherstonhaugh: The First U.S. Government Geologist.* Tuscaloosa: University of Alabama Press.

Bertram, James. 1893. *Some Memories of Books Authors and Events.* London: Archibald Constable.

Bhabha, Homi, ed. 1990. *Nation and Narration.* London: Routledge.

Bickersteth, Edward. 1846. *The Divine Warning to the Church, at This Time, of Our Enemies, Dangers and Duties, and as to Our Future Prospects.* 4th ed. London: Seeley.

Binfield, Clyde. 1973. *George Williams and the Y.M.C.A.: A Study in Victorian Social Attitudes.* London: Heinemann.

Binns, Edward. 1845. *The Anatomy of Sleep, or the Art of Procuring Sound and Refreshing Slumber at Will.* Annotations and additions by Earl Stanhope. London: John Churchill.

Birks, Thomas R. 1850. *Modern Astronomy.* London: Religious Tract Society.

Black, Alistair. 1997. "Lost Worlds of Culture: Victorian Libraries, Library History, and Prospects of a History of Information." *Journal of Victorian Culture* 2:95–112.

Blair, Ann. 1997. *The Theater of Nature: Jean Bodin and Renaissance Science.* Princeton: Princeton University Press.

Bloor, David. 1991. *Knowledge and Social Imagery.* Chicago: University of Chicago Press.

Blunt, Wilfrid. 1976. *The Ark in the Park: The Zoo in the Nineteenth Century.* London: Hamish Hamilton.

Boase, George Clement, and William Prideaux Courtney. 1878–82. *Bibliotheca Cornubiensis. A Catalogue of the Writings, both Manuscript and Printed, of Cornishmen, and of Works Relating to the County of Cornwall.* 3 vols. London: Longmans.

Bodenheimer, Rosemarie. 1994. *The Real Life of Mary Ann Evans: George Eliot, Her Letters and Fiction.* Ithaca: Cornell University Press.

Bonner, Hypatia Bradlaugh. 1891. *Catalogue of the Library of the Late Charles Bradlaugh.* London.

Bosanquet, Samuel Richard. 1843. *Principia: A Series of Essays on the Principles of Evil Manifesting Themselves in These Last Times in Religion, Philosophy, and Politics.* London: James Burns.

———. 1845. *"Vestiges of the Natural History of Creation:" Its Argument Examined and Exposed.* 2d ed. London: John Hatchard & Son.

Bourdieu, Pierre. 1977. *Outline of a Theory of Practice.* Trans. Richard Nice. Cambridge: Cambridge University Press.

———. 1984. *Distinction: A Social Critique of the Judgment of Taste.* Trans. Richard Nice. Cambridge: Harvard University Press.

———. 1990. *In Other Words: Essays Towards a Reflexive Sociology.* Stanford: Stanford University Press.

Bowler, Peter J. 1976. *Fossils and Progress: Paleontology and the Idea of Progressive Evolution in the Nineteenth Century.* New York: Science History Publications.

———. 1988. *The Non-Darwinian Revolution: Reinterpreting a Historical Myth.* Baltimore: Johns Hopkins University Press.

———. 1989. *Evolution: The History of an Idea.* Rev. ed. Berkeley and Los Angeles: University of California Press.

[Boyd, A. K. H.] 1890. *Twenty-Five Years of St. Andrews, September 1865 to September 1890.* 2 vols. London: Longman.

Bradfield, B. T. 1968. "Sir Richard Vyvyan and the Country Gentlemen, 1830–1834." *English Historical Review* 83:729–43.

Brantlinger, Patrick. 1998. *The Reading Lesson: The Threat of Mass Literacy in Nineteenth-Century British Fiction.* Bloomington: Indiana University Press.

Brewer, John. 1996. "Reconstructing the Reader: Prescriptions, Texts and Strategies in Anna Larpent's Reading." In *The Practice and Representation of Reading in England.* Ed. James Raven, Helen Small, and Naomi Tadmor, 226–45. Cambridge: Cambridge University Press.

Briggs, Asa, ed. 1959. *Chartist Studies.* London: Macmillan.

———. 1963. *Victorian Cities.* London: Odhams.

Brock, William H. 1980. "The Development of Commercial Science Journals in Victorian Britain." In *The Development of Science Publishing in Europe.* Ed. A. J. Meadows, 95–122. Amsterdam: Elsevier Science Publishers.

———. 1984. "Brewster as a Scientific Journalist." In *"Martyr of Science": Sir David Brewster 1781–1868.* Ed. A. D. Morrison-Low and J. R. R. Christie, 37–42. Edinburgh: Royal Scottish Museum Studies.

———. 1996. *Science for All: Studies in the History of Victorian Science and Education.* Aldershot, Hampshire: Ashgate.

Brock, William H., and Roy M. MacLeod, eds. 1980a. *Natural Knowledge in Social Context: The Journals of Thomas Archer Hirst FRS.* London: Mansell (microfiche ed., with index, of the typescript version).

———. 1980b. "Introduction: The Life of Thomas Hirst." In *Natural Knowledge in Social Context: The Journals of Thomas Archer Hirst FRS.* Ed. William H. Brock and Roy M. MacLeod, 5–37. London: Mansell.

Brock, William H., and A. J. Meadows. 1998. *The Lamp of Learning: Taylor & Francis and the Development of Science Publishing.* 2d ed. London: Taylor & Francis.

Brockman, John. 1995. *The Third Culture.* New York: Simon and Schuster.

Broks, Peter. 1996. *Media Science before the Great War.* Basingstoke: Macmillan.

Brooke, John Hedley. 1977. "Richard Owen, William Whewell, and the *Vestiges.*" *BJHS* 10:132–45.

———. 1979. "The Natural Theology of the Geologists: Some Theological Strata." In *Images of the Earth: Essays in the History of the Environmental Sciences.* Ed. L. J. Jordanova and Roy Porter, 39–64. Chalfont Saint Giles: British Society for the History of Science.

———. 1991a. "Indications of a Creator: Whewell as Apologist and Priest." In *William Whewell: A Composite Portrait.* Ed. Menachem Fisch and Simon Schaffer, 149–73. Oxford: Clarendon Press.

———. 1991b. *Science and Religion: Some Historical Perspectives.* Cambridge: Cambridge University Press.

———. 1996. "Like Minds: The God of Hugh Miller." In *Hugh Miller and the Controversies of Victorian Science.* Ed. Michael Shortland, 171–86. Oxford: Clarendon Press.

[Brougham, Henry]. 1825. *Practical Observations upon the Education of the People, Addressed to the Working Classes and Their Employers.* London: Longman.

[———]. 1827. *Objects, Advantages, and Pleasures of Science.* London: Baldwin and Cradock.

Brown, Callum G. 1987. *The Social History of Religion in Scotland since 1730.* London: Methuen.

Brown, David. 1979. *Walter Scott and the Historical Imagination.* London: Routledge and Kegan Paul.

Brown, J., and D. W. Forrest, eds. 1907. *Letters of Dr. John Brown: With Letters from Ruskin, Thackeray, and Others.* London: Adam and Charles Black.

Brown, Lucy. 1985. *Victorian News and Newspapers.* Oxford: Oxford University Press.

Brown, Philip A. H. 1982. *London Publishers and Printers, c. 1800–1870*. London: British Library.

Brown, Stewart J. 1982. *Thomas Chalmers and the Godly Commonwealth in Scotland.* Oxford: Oxford University Press.

———. 1997. "Religion and the Rise of Liberalism: The First Disestablishment Campaign in Scotland, 1829–1843." *Journal of Ecclesiastical History* 48:682–704.

Brown, Stewart J., and Michael Fry, eds. 1993. *Scotland in the Age of the Disruption.* Edinburgh: Edinburgh University Press.

Brown, Thomas. 1884. *Annals of the Disruption; with Extracts from the Narratives of Ministers who Left the Scottish Establishment in 1843*. Edinburgh.

Browne, Janet. 1995. *Charles Darwin: Voyaging*. London: Jonathan Cape.

———. 1998. "I Could Have Retched All Night: Charles Darwin and His Body." In *Science Incarnate: Historical Embodiments of Natural Knowledge.* Ed. Christopher Lawrence and Steven Shapin, 240–87. Chicago: University of Chicago Press.

Browning, Elizabeth Barrett. 1900. *The Poetical Works.* London: Smith, Elder.

Buchanan, James. 1855. *Faith in God and Modern Atheism Compared, in Their Essential Nature, Theoretic Grounds, and Practical Influence.* 2 vols. Edinburgh: James Buchanan, Junr.

Buck, Anne. 1984. *Victorian Costume and Costume Accessories.* 2d ed. Carlton, Bedford: Ruth Bean.

Buckland, William. 1836. *Geology and Mineralogy Considered with Reference to Natural Theology.* 2 vols. London: William Pickering.

Bud, Robert, and Gerrylynn K. Roberts. 1984. *Science versus Practice: Chemistry in Victorian Britain.* Manchester: Manchester University Press.

Budd, Susan. 1977. *Varieties of Unbelief: Atheists and Agnostics in English Society, 1830–1960*. London: Heinemann.

Bulwer-Lytton, Edward. [1833] 1970. *England and the English.* Chicago: University of Chicago Press.

Bunbury, Frances, ed. 1890–91. *Memorials of Sir C. J. F. Bunbury, Bart.* 9 vols. Mildenhall: privately printed.

Burke, J., and J. B. Burke. 1847. *A Genealogical and Heraldic Dictionary of the Peerage and Baronetage of the British Empire.* London: Henry Coburn.

Burke, Peter. 1993. *The Art of Conversation.* Cambridge: Polity Press.

———. 1995. *The Fortunes of the* Courtier. Cambridge: Polity Press.

Burkhardt, Frederick H. 1972. "England and Scotland: The Learned Societies." In *The Comparative Reception of Darwinism.* Ed. Thomas F. Glick, 32–74. Austin: University of Texas Press.

Burkhardt, Frederick H., et al., eds. 1985–99. *The Correspondence of Charles Darwin.* 11 vols. Cambridge: Cambridge University Press.

Burn, W. L. 1964. *The Age of Equipoise: A Study of the Mid-Victorian Generation.* London: George Allen and Unwin.

Bushnan, J. Stevenson. 1851. *Miss Martineau and Her Master.* London: John Churchill.

Butler, Samuel. [1903] 1993. *The Way of All Flesh.* Ed. Michael Mason. Oxford: Oxford University Press.

Bynum, W. F., Stephen Lock, and Roy Porter, eds. 1992. *Medical Journals and Medical Knowledge: Historical Essays.* London: Routledge.

Cairns, John. 1860. *Memoir of John Brown, D.D.* Edinburgh: Edmonston and Douglas.

Campbell, A. J. [n.d.] "Cupar: The Years of Controversy: A Study of Its Newspaper Press 1822–72." Unpublished typescript (St. Andrews University Library).

Campbell, Lewis, and William Garnett. 1882. *The Life of James Clerk Maxwell with a Selection from His Correspondence and Occasional Writings and a Sketch of His Contributions to Science.* London: Macmillan.

Candlish, Robert S. 1843. *Contributions Towards the Exposition of the Book of Genesis.* Edinburgh: John Johnstone.

———. 1859. *Reason and Revelation.* London: T. Nelson.

Cannon, Susan Faye. 1978. *Science in Culture: The Early Victorian Period.* New York: Science History Publications.

Carlile, Richard. [1821] 1972. "An Address to Men of Science." In *The Radical Tradition in Education.* Ed. Brian Simon, 91–137. London: Lawrence and Wishart.

Carpenter, William B. 1839. *Principles of General and Comparative Physiology, Intended as an Introduction to the Study of Human Physiology, and as a Guide to the Philosophical Pursuit of Natural History.* London: John Churchill.

———. 1840. *Remarks on Some Passages in the Review of "Principles of General and Comparative Physiology," in the Edinburgh Medical & Surgical Journal, January 1840.* Bristol: Philip and Evans.

———. 1841. *Principles of General and Comparative Physiology.* 2d ed. London: John Churchill.

———. 1888. *Nature and Man: Essays Scientific and Philosophical.* London: Kegan Paul, Trench & Co.

Carter, John W. 1935. *Publisher's Cloth: An Outline History of Publisher's Binding in England, 1820–1900.* London: Constable.

Carus, C. G. 1846. *The King of Saxony's Journey Through England and Scotland in the Year 1844.* Translated by S. C. Davison. London: Chapman and Hall.

Cashdollar, Charles D. 1989. *The Transformation of Theology, 1830–1890: Positivism and Protestant Thought in Britain and America.* Princeton: Princeton University Press.

Chadwick, Owen. 1971. *The Victorian Church: Part I.* 3d ed. London: Adam and Charles Black.

———. 1975. *The Secularization of the European Mind in the Nineteenth Century.* Cambridge: Cambridge University Press.

Chalmers, Thomas. 1835–42. *The Works.* 25 vols. Glasgow: William Collins.

———. 1847–49. *Posthumous Works.* Ed. William Hanna. 9 vols. Edinburgh: Thomas Constable.

Chambers, Robert. 1825. *Illustrations of the Author of Waverley: Being Notices of Real Characters, Scenes, and Incidents, Supposed to be Described in his Works.* 2d ed. Edinburgh: John Anderson, Junior.

———. 1828. *History of the Rebellions in Scotland, Under the Marquis of Montrose, and Others, from 1638 Till 1660.* 2 vols. Edinburgh: Constable.

[———]. [1833]. *Minor Antiquities of Edinburgh.* Edinburgh: William and Robert Chambers.

[———]. 1836. *Introduction to the Sciences. For Use in Schools and Private Instruction.* Edinburgh: William and Robert Chambers.

———, ed. 1843. *Cyclopædia of English Literature; a History, Critical and Biographical, of British Authors from the Earliest to the Present Times.* 2 vols. Edinburgh: William and Robert Chambers.

[———]. 1844a. *Vestiges of the Natural History of Creation.* London: John Churchill.

[———]. 1844b. *Vestiges of the Natural History of Creation.* 2d ed. London: John Churchill.

[———]. 1845a. *Vestiges of the Natural History of Creation.* 3d ed. London: John Churchill.

[———]. 1845b. *Vestiges of the Natural History of Creation.* 4th ed. London: John Churchill.

[———]. 1845c. *Vestiges of the Natural History of Creation.* 5th ed. London: John Churchill.

[———]. 1845d. *Explanations, A Sequel.* London: John Churchill.

[———]. 1846. *Explanations, A Sequel.* 2d ed. London: John Churchill.

———. 1847a. *Select Writings of Robert Chambers.* 7 vols. Edinburgh: W. and R. Chambers.

[———]. 1847b. *Vestiges of the Natural History of Creation.* 6th ed. London: John Churchill.

[———]. 1847c. *Vestiges of the Natural History of Creation.* 7th ed. London: John Churchill.

———. 1848. *Ancient Sea-margins, as Memorials of Changes in the Relative Level of Sea and Land.* Edinburgh: W. & R. Chambers.

[———]. 1850. *Vestiges of the Natural History of Creation.* 8th ed. London: John Churchill.

[———]. 1851. *Vestiges of the Natural History of Creation.* 9th ed. London: John Churchill.

[———]. 1853. *Vestiges of the Natural History of Creation.* 10th ed. London: John Churchill.

———. 1859. *Testimony: Its Posture in the Scientific World.* Edinburgh.

[———]. 1860. *Vestiges of the Natural History of Creation.* 11th ed. London: John Churchill.

———, ed. 1864. *The Book of Days: A Miscellany of Popular Antiquities in Connection with the Calendar Including Anecdote, Biography, & History, Curiosities of Literature and Oddities of Human Life and Character.* 2 vols. London: W. & R. Chambers.

———. 1884. *Vestiges of the Natural History of Creation.* 12th ed. Edinburgh: W. & R. Chambers.

———. 1887. *Vestiges of the Natural History of Creation.* With an introduction by Henry Morley. London: George Routledge.

———. 1994. *Vestiges of the Natural History of Creation and Other Evolutionary Writings.* Ed. James A. Secord. Chicago: University of Chicago Press.

Chambers, William, and Robert Chambers, eds. 1842. *Chambers's Information for the People.* New ed. 2 vols. Edinburgh: William and Robert Chambers.

Chandler, James. 1998. *England in 1819: The Politics of Literary Culture and the Case of Romantic Historicism.* Chicago: University of Chicago Press.

Chapman, Allan. 1998. *The Victorian Amateur Astronomer: Independent Astronomical Research in Britain, 1820–1920.* Chichester: Wiley.

Chapman, T. 1873. *Catalogue of the Very Valuable Library of the Late Robert Chambers, Esq. LL.D., F.R.S.E. . . . Which Will Be Sold by Auction.* Edinburgh: T. Chapman.

Chartier, Roger. 1988. *Cultural History: Between Practices and Representations.* Ithaca: Cornell University Press.

———. 1994. *The Order of Books: Readers, Authors, and Libraries in Europe between the Fourteenth and Eighteenth Centuries.* Cambridge: Polity.

Chorley, Henry, ed. 1872. *Letters of Mary Russell Mitford.* Series no. 2. 2 vols. London: Richard Bentley.

Christ, Carol T., and John O. Jordan, eds. 1995. *Victorian Literature and the Victorian Visual Imagination.* Berkeley and Los Angeles: University of California Press.

Christmas, Henry. 1850. *Echoes of the Universe: From the World of Matter and the World of Spirit.* London: Richard Bentley.

Church, Mary C., ed. 1894. *Life and Letters of Dean Church.* London: Macmillan.

Clark, John Willis, and Thomas McKenny Hughes. 1890. *The Life and Letters of the Reverend Adam Sedgwick.* 2 vols. Cambridge: Cambridge University Press.

Clark, Peter. 2000. *British Clubs and Societies, 1580–1800.* Oxford: Oxford University Press.

Clarke, W. K. Lowther. 1959. *A History of the S.P.C.K.* London: S.P.C.K.

Clement, Mark. 1996. "Sifting Science: Methodism and Natural Knowledge in Britain, 1815–70." Ph.D. diss., University of Oxford.

Clergy List. 1841–. *The Clergy List . . . Containing an Alphabetical List of the Clergy.* London: C. Cox.

Cobbe, Frances Power. 1894. *Life of Frances Power Cobbe. By Herself.* 2 vols. London: Richard Bentley.

Cochrane, John George. 1847. *Catalogue of the London Library.* London: M'Gowan.

Cockburn, William. 1804. *Remarks on a Publication of M. Volney, Called "The Ruins" &c.* Cambridge: Cambridge University Press.

———. 1845. *The Bible Defended Against the British Association: Being the Substance of a Paper Read in the Geological Section, at York, on the 27th of September, 1844.* 5th ed. London: Whittaker and Co.

Cockton, Henry. [1840] 1844a. *The Life and Adventures of Valentine Vox the Ventriloquist.* London: Willoughby.

———. 1844b. *Sylvester Sound the Somnambulist.* London: W. M. Clarke.

Colclough, Stephen. 1998. "Recovering the Reader: Commonplace Books and Diaries as Sources of Reading Experience." *Publishing History* 44:5–37.

Coleridge, Samuel Taylor. 1976. *On the Constitution of the Church and State.* Ed. John Colmer. *The Collected Works of Samuel Taylor Coleridge,* vol. 10. London: Routledge and Kegan Paul.

Collet, Collet D. 1899. *History of the Taxes on Knowledge: Their Origin and Repeal.* 2 vols. London: T. Fisher Unwin.

Collini, Stefan. 1991. *Public Moralists: Political Thought and Intellectual Life in Britain, 1850–1930.* Oxford: Oxford University Press.

Collins, Neil. 1994. *Politics and Elections in Nineteenth-Century Liverpool.* Aldershot, Hampshire: Scholar Press.

Combe, George. 1828. *The Constitution of Man Considered in Relation to External Objects.* Edinburgh: John Anderson.

———. 1835. *The Constitution of Man.* 2d ed. Edinburgh: John Anderson.

———. 1836a. *The Constitution of Man.* 4th ed. Edinburgh: William and Robert Chambers.

———. 1836b. *Testimonials on Behalf of George Combe as a Candidate for the Chair of Logic in the University of Edinburgh.* Edinburgh: John Anderson.

———. 1836c. *Outlines of Phrenology.* 6th ed. Edinburgh: Maclachlan.

———. 1843. *A System of Phrenology.* 5th ed. 2 vols. Edinburgh: Maclachlan.

———. 1847a. *The Constitution of Man Considered in Relation to External Objects.* 8th ed. Edinburgh: Maclachlan.

———. 1847b. "Kennedy's 'Nature and Revelation Harmonious.'" *Phrenological Journal* 20:425–35.

Cook, Edward. 1913. *The Life of Florence Nightingale.* 2 vols. London: Macmillan.

Cooney, Sondra Miley. 1970. "Publishers for the People: W. & R. Chambers: The Early Years, 1832–1850." Ph.D. diss., Ohio State University.

Cooper, Thomas. 1845. *The Purgatory of Suicides.* London: Jeremiah How.

———. 1872. *The Life of Thomas Cooper.* London: Hodder and Stoughton.

———. 1878. *Evolution, the Stone Book, and the Mosaic Record of Creation.* London: Hodder and Stoughton.

Cooter, Roger. 1984. *The Cultural Meaning of Popular Science: Phrenology and the Organization of Consent in Nineteenth-Century Britain.* Cambridge: Cambridge University Press.

Cooter, Roger, and Stephen Pumfrey. 1994. "Separate Spheres and Public Places: Reflections on the History of Science Popularization and Science in Popular Culture." *History of Science* 32:237–67.

Corsi, Pietro. 1988. *Science and Religion: Baden Powell and the Anglican Debate, 1800–1860.* Cambridge: Cambridge University Press.

Courtney, William Prideaux. 1908. *The Secrets of our National Literature: Chapters in the History of the Anonymous and Pseudonymous Writings of our Countrymen.* London: Archibald Constable.

Cowan, R. M. W. 1946. *The Newspaper in Scotland: A Study of Its First Expansion.* Glasgow: George Outram.

Cowie, Grace G., and Edward Royle. 1982. "Martin, Emma." In *Dictionary of Labour Biography.* Ed. Joyce M. Bellamy and John Saville, 6:188–91. London: Macmillan.

Cowling, Mary. 1989. *The Artist as Anthropologist: The Representation of Type and Character in Victorian Art.* Cambridge: Cambridge University Press.

Cox, Robert S. 1993. "A Spontaneous Flow: The Geological Contributions of Mary Griffith, 1772–1846." *Earth Sciences History* 12:187–95.

Crabbe, George. 1840. *An Outline of a System of Natural Theology.* London: William Pickering.

Crofton, Denis. 1855. *Genesis and Geology; Or, an Investigation into the Reconciliation of the Modern Doctrines of Geology with the Declarations of Scripture.* Glasgow: William Collins.

Crook, David Paul. 1994. *Darwinism, War, and History: The Debate over the Biology of War from the "Origin of Species" to the First World War.* Cambridge: Cambridge University Press.

Crosland, Camilla. 1893. *Landmarks of a Literary Life 1820–1892.* London: Sampson Low.

Crosland, Newton. 1898. *Rambles Round My Life: An Autobiography (1819–1896).* London: E. W. Allen.

Cross, Nigel. 1985. *The Common Writer: Life in Nineteenth-Century Grub Street.* Cambridge: Cambridge University Press.

Crowe, Michael J. 1986. *The Extraterrestrial Life Debate, 1750–1900: The Idea of a Plurality of Worlds from Kant to Lowell.* Cambridge: Cambridge University Press.

Cruikshank, George. 1870. *The Comic Almanack: An Ephemeris in Jest and Earnest, Containing Merry Tales, Humorous Poetry, Quips, and Oddities.* 2 vols. London: Chatto and Windus.

Cruse, Amy. 1935. *The Victorians and Their Books.* London: George Allen and Unwin.

Cumming, John. 1848. *Apocalyptic Sketches; or Lectures on the Book of Revelation.* London: Hall and Co.

———. 1851. "God in Science." In *YMCA Lectures, 1850–51,* 197–242. London: James Nisbet.

———. 1853a. *The Church Before the Flood.* London: Arthur Hall, Virtue, and Co.

———. 1853b. *Scripture Readings on the Book of Genesis.* London: John Farquahar Shaw.

Curran, James, and Jean Seaton. 1991. *Power without Responsibility: The Press and Broadcasting in Britain.* 4th ed. London: Routledge.

Curtin, M. 1987. *Propriety and Position: A Study of Victorian Manners.* New York: Garland.

Curwen, E. Cecil, ed. 1940. *The Journal of Gideon Mantell: Surgeon and Geologist.* London: Oxford University Press.

Curwen, Henry. 1873. *History of Booksellers: The Old and the New.* London: Chatto and Windus.

Darnton, Robert. 1979. *The Business of Enlightenment: A Publishing History of the* Encyclopédie, *1775–1800.* Cambridge: Harvard University Press, Belknap Press.

———. 1984. *The Great Cat Massacre and Other Episodes in French Cultural History.* New York: Vintage.

———. 1990. *The Kiss of Lamourette: Reflections in Cultural History.* New York: W. W. Norton.

———. 1996. *The Forbidden Best-Sellers of Pre-revolutionary France.* London: HarperCollins.

Darwin, Charles. 1859. *On the Origin of Species by Means of Natural Selection, or the Preservation of Favoured Races in the Struggle for Life.* London: John Murray.

———. 1868. *The Variation of Animals and Plants Under Domestication.* 2 vols. London: John Murray.

———. 1909. *The Foundations of the Origin of Species: Two Essays Written in 1842 and 1844.* Cambridge: Cambridge University Press.

Darwin, Francis, ed. 1887. *The Life and Letters of Charles Darwin, Including an Autobiographical Chapter.* 3 vols. London: John Murray.

Daunton, M. J. 1995. *Progress and Poverty: An Economic and Social History of Britain 1700–1850.* Oxford: Oxford University Press.

David, Deirdre. 1987. *Intellectual Women and Victorian Patriarchy: Harriet Martineau, Elizabeth Barrett Browning, George Eliot.* Ithaca: Cornell University Press.

Davidoff, Leonore. 1973. *The Best Circles: Society, Etiquette, and the Season.* London: Croom Helm.

Davidoff, Leonore, and Catherine Hall. 1987. *Family Fortunes: Men and Women of the English Middle Class, 1780–1850.* Chicago: University of Chicago Press.

Davidson, Cathy N. 1988. *Revolution and the Word: The Rise of the Novel in America.* New York: Oxford University Press.

D[avidson], W[illiam]. 1851. *Letter to the Rev. John Cumming, D.D., on the Subject of his Lecture Entitled God in Science.* London: H. Baillière.

Davy, Humphry. 1830. *Consolations in Travel, or the Last Days of a Philosopher.* London: John Murray.

Dawson, Carl. 1979. *Victorian Noon: English Literature in 1850.* Baltimore, Johns Hopkins University Press.

Dayan, Daniel, and Elihu Katz. 1992. *Media Events: The Live Broadcasting of History.* Cambridge: Harvard University Press.

Dean, Dennis R. 1999. *Gideon Mantell and the Discovery of Dinosaurs.* Cambridge: Cambridge University Press.

De Certeau, Michel. 1986. *Heterologies: Discourse on the Other.* Trans. Brian Massumi. Manchester: Manchester University Press.

De Giustino, David. 1975. *Conquest of Mind: Phrenology and Victorian Social Thought.* London: Croom Helm.

De Morgan, Augustus. 1872. *A Budget of Paradoxes.* London: Longmans.

Desmond, Adrian. 1982. *Archetypes and Ancestors: Palaeontology in Victorian London, 1850–1875.* London: Blond and Briggs.

———. 1985. "The Making of Institutional Zoology in London, 1822–1836." *History of Science* 23:153–85, 223–50.

———. 1987. "Artisan Resistance and Evolution in Britain, 1819–1848." *Osiris,* 2d ser., 3:77–110.

———. 1989. *The Politics of Evolution: Morphology, Medicine, and Reform in Radical London.* Chicago: University of Chicago Press.

———. 1994. *Huxley: The Devil's Disciple.* London: Michael Joseph.

———. 1997. *Huxley: Evolution's High Priest.* London: Michael Joseph.

Desmond, Adrian, and James Moore. 1991. *Darwin.* London: Michael Joseph.

D'Holbach, Paul Thyry, Baron. [M. de Mirabaud, pseud.]. 1834. *Nature, and her Laws: As Applicable to the Happiness of Man, Living in Society; Contrasted with Superstition and Imaginary Systems.* 2 vols. London: James Watson.

[Dibdin, Thomas Frognall]. 1832. *Bibliophobia. Remarks on the Present Languid and Depressed State of Literature and the Book Trade. In a Letter Addressed to the Author of the Bibliomania.* London: Henry Bohn.

Di Gregorio, Mario A. 1984. *T. H. Huxley's Place in Natural Science.* New Haven: Yale University Press.

Di Gregorio, Mario A., and N. W. Gill. 1990. *Charles Darwin's Marginalia.* Vol. 1. New York: Garland.

Dickson, David. 1814. *The Influence of Learning on Religion: A Sermon, Preached Before the Society in Scotland, (Incorporated by Royal Charter,) for Propagating Christian Knowledge.* Edinburgh: W. Creech.

Disraeli, Benjamin. [1847] 1927. *Tancred or the New Crusade.* London: Peter Davies.

[Disraeli, Isaac]. 1807. *Curiosities of Literature.* 5th ed. 2 vols. London: John Murray.

Dixon, Edmund Saul. 1848. *Ornamental and Domestic Poultry: Their History and Management.* London: Gardeners' Chronicle.

Dodd, George. 1843. *Days at the Factories; Or, the Manufacturing Industry of Great Britain Described, and Illustrated by Numerous Engravings of Machines and Processes.* London: Charles Knight.

Dodds, John W. 1953. *The Age of Paradox: A Biography of England 1841–1851.* London: Victor Gollancz.

Dooley, Allan C. 1992. *Author and Printer in Victorian England.* Charlottesville: University Press of Virginia.

Driver, Felix. 1988. "Moral Geographies: Social Science and the Urban Environment in Mid-Nineteenth Century England." *Transactions, Institute of British Geographers* 13:257–87.

Drummond, Andrew L., and James Bulloch. 1975. *The Church in Victorian Scotland, 1843–1874.* Edinburgh: Saint Andrew Press.

Dunn, Waldo Hilary Dunn. 1961. *James Anthony Fronde: A Biography.* 2 vols. Oxford: Clarendon Press.

Durant, John R. 1979. "Scientific Naturalism and Social Reform in the Thought of Alfred Russel Wallace." *BJHS* 12:31–58.

Dyos, H. J., and Michael Wolff, eds. 1973. *The Victorian City: Images and Realities.* 2 vols. London: Routledge.

Eger, Martin. 1993. "Hermeneutics and the New Epic of Science." In *The Literature of Science: Perspectives on Popular Science Writing.* Ed. Murdo William McRae. Athens: University of Georgia Press.

Egerton, Frank N. 1970. "Refutation and Conjecture: Darwin's Response to Sedgwick's Attack on Chambers." *Studies in History and Philosophy of Science* 1:176–83.

Eley, Geoff. 1992. "Nations, Publics, and Political Cultures: Placing Habermas in the Nineteenth Century." In *Habermas and the Public Sphere.* Ed. Craig Calhoun, 289–339. Cambridge: MIT Press.

Eliot, George. [1876] 1988. *Daniel Deronda.* Oxford: Oxford University Press.

———. 1990. *Selected Essays, Poems, and Other Writings.* Ed. A. S. Byatt and Nicholas Warren. London: Penguin Books.

———. [1879] 1994. *Impressions of Theophrastus Such.* Ed. Nancy Henry. London: William Pickering.

Eliot, Simon. 1994. *Some Patterns and Trends in British Publishing, 1800–1919.* Occasional Papers Number 8. London: Bibliographical Society.

———. 1997–98. "*Patterns and Trends* and the *NSTC:* Some Initial Observations." *Publishing History* 42:79–104; 43:71–112.

Ellegård, Alvar. 1958. *Darwin and the General Reader: The Reception of Darwin's Theory of Evolution in the British Periodical Press, 1859–1872.* Göteborg: Göteborgs Universitets Arsskrift.

Ellis, Sarah. 1839. *The Women of England, Their Social Duties, and Domestic Habits.* London: Fisher.

———. 1842. *The Daughters of England, Their Position in Society, Character & Responsibility.* London: Fisher.

Emerson, Ralph Waldo. 1841. *Essays.* With a preface by Thomas Carlyle. London: James Fraser.

Engel, A. J. 1983. *From Clergyman to Don: The Rise of the Academic Profession in Nineteenth-Century Oxford.* Oxford: Clarendon Press.

English, Mary P. 1990. *Victorian Values: The Life and Times of Dr. Edwin Lankester, M.D., F.R.S.* Bristol: Biopress.

Erickson, Lee. 1995. *The Economy of Literary Form: English Literature and the Industrialization of Publishing, 1800–1830.* Baltimore: Johns Hopkins University Press.

Espinasse, Francis. 1893. *Literary Recollections and Sketches.* London: Hodder and Stoughton.

Eve, A. S., and C. H. Creasey. 1945. *Life and Work of John Tyndall.* London: Macmillan.

Feather, John. 1988. *A History of British Publishing.* London: Croom Helm.

———. 1994. *Publishing, Piracy, and Politics: An Historical Study of Copyright in Britain.* London: Mansell.

Feltes, N. N. 1986. *Modes of Production of Victorian Novels.* Chicago: University of Chicago Press.

Ferris, Ina. 1991. *The Achievement of Literary Authority: Gender, History, and the Waverley Novels.* Ithaca: Cornell University Press.

Fichte, Johann Gottlieb. 1848. *The Vocation of Man.* Trans. William Smith. London: John Chapman.

Finney, Colin. 1993. *Paradise Revealed: Natural History in Nineteenth-Century Australia.* Melbourne: Museum of Victoria.

Fisch, Menachem, and Simon Schaffer, eds. 1991. *William Whewell: A Composite Portrait.* Oxford: Clarendon Press.

Fish, Stanley. 1980. *Is There a Text in This Class? The Authority of Interpretative Communities.* Cambridge: Harvard University Press.

Fleming, John. 1851. *The Temperature of the Seasons, and Its Influence on Inorganic Objects, and on Plants and Animals.* Edinburgh: Johnstone and Hunter.

Fletcher, John. 1835–37. *Rudiments of Physiology.* Edinburgh: John Carfrae.

———. 1836. *Discourse on the Importance of the Study of Physiology as a Branch of Popular Education.* Edinburgh: Black.

Flint, Kate. 1993. *The Woman Reader, 1837–1914.* Oxford: Clarendon Press.

———. 1996. "Women, Men, and the Reading of *Vanity Fair.*" In *The Practice and Representation of Reading in England.* Ed. James Raven, Helen Small, and Naomi Tadmor, 246–62. Cambridge: Cambridge University Press.

Foot, M. R. D., and H. C. G. Matthew, eds. 1968–94. *The Gladstone Diaries*. 14 vols. Oxford: Clarendon Press.

Forbes, Duncan. 1953. "The Rationalism of Sir Walter Scott." *The Cambridge Journal* 7:20–35.

Forgan, Sophie. 1999. "Bricks and Bones: Architecture and Science in Victorian Britain." In *The Architecture of Science*. Ed. Peter Galison and Emily Thompson, 181–208. Cambridge: MIT Press.

Forster, Margaret. 1988. *Elizabeth Barrett Browning: A Biography*. London: Chatto & Windus.

Fox, Celina. 1988. *Graphic Journalism in England during the 1830s and 1840s*. New York: Garland.

Fox, Robert. 1997. "The University Museum and Oxford Science, 1850–1880." In *The History of the University of Oxford*. Vol. 6: *Nineteenth-Century Oxford, Part I*. Ed. M. G. Brock and M. C. Curthoys, 641–91. Oxford: Clarendon Press.

Frasca-Spada, Marina, and Nicholas Jardine, eds. 2000. *Books and the Sciences in History*. Cambridge: Cambridge University Press.

Fraser, Angus. 1997. "John Murray's Colonial and Home Library." *Papers of the Bibliographical Society of America* 91:339–408.

Fraser, Derek, ed. 1982. *Municipal Reform and the Industrial City*. Leicester: Leicester University Press.

———. 1985. "The Editor as Activist: Editors and Urban Politics in Early Victorian England." Ed. Joel H. Wiener. In *Innovators and Preachers: The Role of the Editor in Victorian England*. Westport, Conn.: Greenwood Press.

Freedan, Michael. 1978. *The New Liberalism: An Ideology of Social Reform*. Oxford: Oxford University Press.

Freeman, Michael. 1999. *Railways and the Victorian Imagination*. New Haven: Yale University Press.

Freeman, Michael, and Derek Aldcroft. 1985. *The Atlas of British Railway History*. London: Croom Helm.

Freeman, Richard Broke, and Douglas Wertheimer. 1980. *Philip Henry Gosse: A Bibliography*. Folkestone: Dawson.

Fullom, S. W. 1852. *The Marvels of Science, and their Testimony to Holy Writ*. London: Longman.

Fyfe, Aileen. 1997. "The Reception of William Paley's *Natural Theology* in the University of Cambridge." *BJHS* 30:321–35.

———. 2000. "Reading Children's Books in Eighteenth-Century Dissenting Families." *Historical Journal* 43:453–74.

Gadamer, Hans-Georg. 1975. *Truth and Method*. Ed. Garrett Barden and John Cumming. London: Sheed and Ward.

Gagnier, Regenia. 1999. "Methodology and the New Historicism." *Journal of Victorian Culture* 4:116–22.

Gallagher, Catherine. 1994. *Nobody's Story: The Vanishing Acts of Women Writers in the Marketplace, 1670–1820.* Berkeley and Los Angeles: University of California Press.

Gardiner, Brian G. 1993. "Edward Forbes, Richard Owen, and the Red Lions." *Archives of Natural History* 20:349–72.

Gardner, J. Helen, and Robin J. Wilson. 1993. "Thomas Archer Hirst: Mathematician Xtravagant." *American Mathematical Monthly* 100:435–41, 531–38, 723–31, 827–34, 907–15.

Garland, Martha MacMackin. 1980. *Cambridge before Darwin: The Ideal of a Liberal Education, 1800–1860.* Cambridge: Cambridge University Press.

Garrard, John. 1983. *Leadership and Power in Victorian Industrial Towns, 1830–80.* Manchester: Manchester University Press.

Garside, Peter D. 1975. "Scott and the 'Philosophical' Historians." *Journal of the History of Ideas* 36:497–512.

Garvey, Michael Angelo. 1852. *The Silent Revolution: Or the Future Effects of Steam and Electricity Upon the Condition of Mankind.* London: William and Frederick G. Cash.

Gaskell, Philip. [1972] 1985. *A New Introduction to Bibliography.* Oxford: Clarendon Press.

Gates, Barbara T. 1997. "Revisioning Darwin with Sympathy: Arabella Buckley." In *Natural Eloquence: Women Reinscribe Science.* Ed. Barbara T. Gates and Ann B. Shteir, 164–76. Madison: University of Wisconsin Press.

Gates, Barbara T., and Ann B. Shteir, eds. 1997. *Natural Eloquence: Women Reinscribe Science.* Madison: University of Wisconsin Press.

Gaury, Gerald de. 1972. *Travelling Gent: The Life of Alexander Kinglake (1809–1891).* London: Routledge & Kegan Paul.

Gay, Hannah, and John W. Gay. 1997. "Brothers in Science: Science and Fraternal Culture in Nineteenth-Century Britain." *History of Science* 35:425–53.

Geikie, Archibald. 1875. *Life of Sir Roderick I. Murchison.* 2 vols. London: John Murray.

———. 1895. *Memoir of Sir Andrew Crombie Ramsay.* London: Macmillan.

Geison, Gerald L. 1978. *Michael Foster and the Cambridge School of Physiology: The Scientific Enterprise in Victorian Society.* Princeton: Princeton University Press.

Gibbon, Charles. 1878. *The Life of George Combe, Author of "The Constitution of Man."* 2 vols. London: Macmillan.

Gieryn, Thomas F. 1999. *Cultural Boundaries of Science: Credibility on the Line.* Chicago: University of Chicago Press.

Gilbert, Alan D. 1976. *Religion and Society in Industrial England: Church, Chapel and Social Change, 1740–1914.* London: Longman.

Gillispie, Charles Coulston. 1951. *Genesis and Geology: A Study in the Relations of Scientific Thought, Natural Theology, and Social Opinion in Great Britain, 1790–1850.* Cambridge: Harvard University Press.

Gilmartin, Kevin. 1997. *Print Politics: The Press and Radical Opposition in Early Nineteenth-Century England.* Cambridge: Cambridge University Press.

Ginswick, J. 1983. *Labour and the Poor in England and Wales 1849–1851.* 8 vols. London: Frank Cass.

Gissing, George. [1891]. 1985. *New Grub Street.* Harmondsworth: Penguin Books.

Godwin, Benjamin. 1853. *The Philosophy of Atheism Examined and Compared with Christianity.* 3d ed. London: Arthur Hall.

Goldgar, Anne. 1995. *Impolite Learning: Conduct and Community in the Republic of Letters, 1680–1750.* New Haven: Yale University Press.

Goldie, Sue. 1983. *A Calendar of the Letters of Florence Nightingale.* Oxford: Oxford Microform Publications.

Golinski, Jan. 1998. *Making Natural Knowledge: Constructivism and the History of Science.* Cambridge: Cambridge University Press.

Gosse, Edmund. 1890. *The Life of Philip Henry Gosse, F.R.S.* London: Kegan Paul.

Gosse, Philip Henry. 1857. *Omphalos: An Attempt to Untie the Geological Knot.* London: John Van Voorst.

Gough, John B. 1882. *Sunlight and Shadow; Or, Gleanings from My Life-work.* 2d ed. London: Hodder and Stoughton.

Gould, Stephen Jay. 1977. *Ontogeny and Phylogeny.* Cambridge: Harvard University Press, Belknap Press.

Grandville [Jean-Ignace-Isidore Gérard]. 1844. *Un autre monde.* Paris: H. Fournier.

Grant, Julia M., Katharine H. McCutcheon, and Ethel F. Sanders, eds. 1927. *St. Leonards School 1877–1927.* Oxford: Oxford University Press.

Grant, Robert Edmond. 1828. *An Essay on the Study of the Animal Kingdom.* London: John Taylor.

Green, S. J. D. 1990. "Religion and the Rise of the Common Man: Mutual Improvement Societies, Religous Associations and Popular Education in Three Industrial Towns in the West Riding of Yorkshire c. 1850–1900." In *Cities, Class and Communication: Essays in Honour of Asa Briggs.* Ed. Derek Fraser, 25–43. New York: Harvester Wheatsheaf.

Griest, Guienevere L. 1970. *Mudie's Circulating Library and the Victorian Novel.* Newton Abbott, Devon: David and Charles.

Griffin, Robert J. 1999. "Anonymity and Authorship." *New Literary History* 30:877–95.

Gross, John. 1969. *The Rise and Fall of the Man of Letters: Aspects of English Literary Life since 1800.* London: Weidenfeld and Nicholson.

Gunn, J. A. W., et al., eds. 1982–. *Benjamin Disraeli Letters.* Toronto: University of Toronto Press.

Gutteridge, Joseph. 1893. *Lights and Shadows in the Life of an Artisan.* Coventry: Curtis and Beamish.

Hadley, Elaine. 1995. *Melodramatic Tactics: Theatricalized Dissent in the English Marketplace, 1800–1885.* Stanford: Stanford University Press.

Haight, Gordon S., ed. 1954–78. *The George Eliot Letters.* 9 vols. New Haven: Yale University Press.

Halifax Mechanics' Institution. 1851. *Alphabetical and Classified Catalogue of the Library of the Halifax Mechanics' Institution and Mutual Improvement Society.* Halifax: printed by N. Burrows, 1851.

Hall, Marie Boas. 1984. *All Scientists Now: The Royal Society in the Nineteenth Century.* Cambridge: Cambridge University Press.

Hall, S. C. 1883. *Retrospect of a Long Life: From 1815 to 1883.* London: Richard Bentley.

Halttunen, Karen. 1982. *Confidence Men and Painted Women: A Study of Middle-Class Culture in America, 1830–1870.* New Haven: Yale University Press.

Harden, Edgar F., ed. 1994. *The Letters and Private Papers of William Makepeace Thackeray: A Supplement.* 2 vols. New York: Garland.

Harman, P. M., ed. 1990–95. *The Scientific Letters and Papers of James Clerk Maxwell.* 2 vols. Cambridge: Cambridge University Press.

Harris, John. 1846. *The Pre-Adamite Earth: Contributions to Theological Science.* London: Ward.

Harris, P. R. 1998. *A History of the British Museum Library, 1753–1973.* London: British Library.

Harrison, Edward R. 1928. *Harrison of Ightham: A Book about Benjamin Harrison, of Ightham, Kent, Made up Principally of Extracts from His Notebooks and Correspondence.* Oxford: Oxford University Press.

Harrison, J. F. C. 1969. *Robert Owen and the Owenites in Britain and America: The Quest for the New Moral World.* London: Routledge.

———. 1979. *The Second Coming: Popular Millenarianism 1780–1850.* London: Routledge.

Harvey, John. 1995. *Men in Black.* Chicago: University of Chicago Press.

Hawking, Stephen W. 1988. *A Brief History of Time: From the Big Bang to Black Holes.* London: Bantam Press.

Hawkins, Thomas. 1840. *The Book of the Great Sea-dragons, Ichthyosaura and Plesiosauri, Gedolim Taninim, of Moses. Extinct Monsters of the Ancient Earth.* London: William Pickering.

Hays, J. N. 1981. "The Rise and Fall of Dionysius Lardner." *Annals of Science* 38:527–42.

———. 1983. "The London Lecturing Empire, 1800–50." In *Metropolis and Province: Science in British Culture, 1780–1850.* Ed. Ian Inkster and Jack Morrell, 91–119. London: Hutchinson.

Hayter, Alethea. 1965. *A Sultry Month: Scenes of London Literary Life in 1846.* London: Faber.

Helmstadter, Richard J., and Bernard Lightman, eds. 1990. *Victorian Faith in Crisis: Essays on Continuity and Change in Nineteenth-Century Religious Belief.* Basingstoke: Macmillan.

Hempton, David. 1996. *Religion and Political Culture in Britain and Ireland: From the Glorious Revolution to the Decline of Empire.* Cambridge: Cambridge University Press.

Henderson, Heather. 1989. *The Victorian Self: Autobiography and Biblical Narrative.* Ithaca: Cornell University Press.

Henry, John. 1996. "Palaeontology and Theodicy: Religion, Politics, and the *Asterolepis* of Stromness." In *Hugh Miller and the Controversies of Victorian Science.* Ed. Michael Shortland, 151–70. Oxford: Clarendon Press.

Herndon, William H., and Jesse W. Weik. 1893. *Abraham Lincoln: The Story of a Great Life.* 2 vols. London: Sampson Low.

Herschel, John F. W. 1830. *A Preliminary Discourse on the Study of Natural Philosophy.* London: Longman.

———. 1833. *A Treatise on Astronomy.* London: Longman.

———. 1857. *Essays from the Edinburgh and London Quarterly Reviews, with Addresses and Other Pieces.* London: Longman.

Heyck, T. W. 1982. *The Transformation of Intellectual Life in Victorian England.* London: Croom Helm.

Heywood, James, ed. 1843. *Illustrations of the Manchester Meeting of the British Association for the Advancement of Science, June 1842.* Manchester: Thomas Forrest.

Hiller, Mary Ruth. 1978. "The Identification of Authors: The Great Victorian Enigma." In *Victorian Periodicals.* Ed. J. Donn Vann and Rosemary T. Van Arsdel, 123–48. New York: MLA.

Hilton, Boyd. 1988. *The Age of Atonement: The Influence of Evangelicalism on Social and Economic Thought, 1795–1865.* Oxford: Clarendon Press.

———. 1994. "Whiggery, Religion, and Social Reform: The Case of Lord Morpeth." *Historical Journal* 37:829–59.

———. 2000. "The Politics of Anatomy and an Anatomy of Politics c. 1825–1850." In *History, Religion, and Culture: Essays in British Intellectual History, 1750–1950.* Ed. Stefan Collini, Richard Whatmore, and Brian Young, 179–97. Cambridge: Cambridge University Press.

Hinton, D. A. 1979. "Popular Science in England, 1830–1870." Ph.D. diss., Bath University.

Hitchcock, Edward. 1851. *The Religion of Geology and Its Connected Sciences.* London: David Bogue.

Hobhouse, John Cam. 1911. *Recollections of a Long Life . . . with Additional Extracts from his Diaries.* 6 vols. London: John Murray.

Hodge, M. J. S. 1972a. "England." In *The Comparative Reception of Darwinism.* Ed. Thomas F. Glick, 3–31. Austin: University of Texas Press.

———. 1972b. "The Universal Gestation of Nature: Chambers' 'Vestiges' and 'Explanations.'" *Journal of the History of Biology* 5:127–51.

———. 1991. "The History of the Earth, Life, and Man: Whewell and Palaetiological Science." In *William Whewell: A Composite Portrait.* Ed. Menachem Fisch and Simon Schaffer, 255–88. Oxford: Clarendon Press.

Hodgson, William Ballantyne. [1839]. *Testimonials in Favour of Mr. William Ballantyne Hodgson, of Edinburgh.* Glasgow: for the author.

———. 1845. *Address Delivered to the Mental Improvement Society of the Liverpool Mechanics' Institution at the Opening of the Session 1845.* Liverpool: Smith.

Hoeveler, J. David, Jr. 1981. *James McCosh and the Scottish Intellectual Tradition: From Glasgow to Princeton.* Princeton: Princeton University Press.

Hollis, Patricia. 1970. *The Pauper Press: A Study in Working-Class Radicalism of the 1830s.* Oxford: Oxford University Press.

Holyoake, George Jacob. [1847]. *Paley Refuted in His Own Words.* 2d ed. London: Hetherington.

———. 1852. *The Last Days of Mrs. Emma Martin, Advocate of Free Thought.* London: J. Watson.

———. 1892. *Sixty Years of an Agitator's Life.* 2 vols. London: T. F. Unwin.

Hooker, Joseph. 1844–47. *Flora Antarctica.* London: Reeve.

Hope, Thomas. 1831. *An Essay on the Origin and Prospects of Man.* 3 vols. London: John Murray.

Hoppen, K. Theodore. 1998. *The Mid-Victorian Generation: 1846–1886.* Oxford: Clarendon Press.

Hoskin, Michael. 1987. "John Herschel's Cosmology." *Journal for the History of Astronomy* 18:1–34.

———. 1990. "Rosse, Robinson, and the Resolution of the Nebulae." *Journal for the History of Astronomy* 21:331–44.

Houghton, Walter E. 1957. *The Victorian Frame of Mind, 1830–1870.* New Haven: Yale University Press.

Houghton, Walter E., et al., eds. 1966–89. *The Wellesley Index to Victorian Periodicals, 1824–1900.* 5 vols. Toronto: University of Toronto Press.

Howe, Ellic, and Harold E. Waite. 1948. *The London Society of Compositors.* London: Cassell.

Howitt, Mary. 1889. *Autobiography.* Ed. Margaret Howitt. 2 vols. London: Wm. Isbister.

Howsam, Leslie. 1991. *Cheap Bibles: Nineteenth-Century Publishing and the British and Foreign Bible Society.* Cambridge: Cambridge University Press.

———. 1993. "Forgotten Victorians: Contracts with Authors in the Publication Books of Henry S. King and Kegan Paul, Trench 1871–89." *Publishing History* 34:51–70.

Hughes, Kathryn. 1998. *George Eliot: The Last Victorian.* London: Fourth Estate.

Hughes, Linda K., and Michael Lund. 1991. *The Victorian Serial.* Charlottesville: University Press of Virginia.

Hughes, R. Elwyn. 1989. "Alfred Russel Wallace; Some Notes on the Welsh Connection." *BJHS* 22:401–18.

Hughes, Winifred. 1980. *The Maniac in the Cellar: Sensation Novels of the 1860s.* Princeton: Princeton University Press.

Hull, David L. 1983. *Darwin and His Critics: The Reception of Darwin's Theory of Evolution by the Scientific Community.* Chicago: University of Chicago Press.

Hume, Abraham. 1845. *Examination of the Theory Contained in* "Vestiges of the Natural History of Creation." Liverpool: M. J. Whitty.

———. 1851. *Suggestions for the Advancement of Literature and Learning in Liverpool.* Liverpool: Deighton and Laughton.

———. 1858. *Condition of Liverpool, Religious and Social; Including Notices of the State of Education, Morals, Pauperism, and Crime.* 2d ed. Liverpool: printed by T. Brakell.

Hunt, Robert. 1848. *The Poetry of Science, or Studies of the Physical Phenomena of Nature.* London: Reeve.

———. 1849. *Panthea, the Spirit of Nature.* London: Reeve.

Hutchinson, Gov. 1980. "Robert Chambers's Vision of Science: The Diffusion of Scientific Ideas to the General Reader in Early-Victorian Britain." Ph.D. diss., Temple University.

Huxley, Leonard. 1900. *Life and Letters of Thomas Henry Huxley.* 2 vols. London: Macmillan.

———. 1918. *Life and Letters of Sir Joseph Dalton Hooker.* 2 vols. London: John Murray.

Huxley, Thomas H. 1893–94. *Collected Essays.* 9 vols. London: Macmillan.

———. 1898–1903. *Scientific Memoirs.* Ed. Michael Foster and E. R. Lankester. 5 vols. London: Macmillan.

Hyman, Anthony. 1984. *Charles Babbage: Pioneer of the Computer.* Oxford: Oxford University Press.

Inkster, Ian. 1979. "London Science and the Seditious Meetings Act of 1817." *BJHS* 12:192–96.

———. 1981. "Seditious Science: A Reply to Paul Weindling." *BJHS* 14:181–87.

———. 1985. "Polarised Culture and Steam Intellect: A Case Study of Liverpool and Its Region, Circa 1820–1850's." In *The Steam Intellect Societies: Essays on Culture, Education and Industry Circa 1820–1914.* Ed. Ian Inkster, 44–59. Nottingham: Department of Adult Education, University of Nottingham.

Ireland, Alexander. 1884. *The Book-lover's Enchiridion: Thoughts on the Solace and Companionship of Books, and Topics Incidental Thereto; Gathered from the Best Writers of Every Age, and Arranged in Chronological Order.* 4th ed. London: Simpkin, Marshall, & Co.

Iser, Wolfgang. 1978. *The Act of Reading: A Theory of Aesthetic Response.* Baltimore: Johns Hopkins University Press.

Jackson, Leon. 1999. "The Reader Retailored: Thomas Carlyle, His American Audiences, and the Politics of Evidence." *Book History* 2:146–72.

Jacyna, L. S. 1983. "Immanence or Transcendence: Theories of Life and Organization in Britain, 1790–1835." *Isis* 74:311–29.

———. 1984. "Principles of General Physiology: The Comparative Dimension to British Neuroscience in the 1830s and 1840s." *Studies in History of Biology* 7:47–92.

James, Frank A. J. L. 2000. "Books on the Natural Sciences in the Nineteenth Century." In *Thornton and Tully's Scientific Books, Libraries, and Collectors: A Study of Bibliography and the Book Trade in Relation to the History of Science.* Ed. Andrew Hunter, 258–71. 4th ed. Aldershot: Ashgate.

James, John Angell. 1849. "The Possession of Spiritual Religion the Surest Preservative from the Snares of Infidelity and the Seductions of False Philosophy." In *YMCA Lectures, 1848–49,* 185–224. London: William Jones.

James, Louis. [1963] 1974. *Fiction for the Working Man, 1830–50.* Harmondsworth: Penguin University Books.

———, ed. 1976. *English Popular Literature 1819–1851.* New York: Columbia University Press.

Jardine, Lisa, and Anthony Grafton. 1990. "'Studied for Action': How Gabriel Harvey Read His Livy." *Past and Present,* no. 129:30–78.

Jauss, Hans Robert. 1982. *Toward an Aesthetic of Reception.* Trans. Timothy Bahti. Minneapolis: University of Minnesota Press.

Jensen, J. Vernon. 1988. "Return to the Wilberforce-Huxley Debate." *BJHS* 21:161–79.

Johns, Adrian. 1996. "The Physiology of Reading in Restoration England." In *The Practice and Representation of Reading in England.* Ed. James Raven, Helen Small, and Naomi Tadmor, 138–61. Cambridge: Cambridge University Press.

———. 1998a. *The Nature of the Book: Print and Knowledge in the Making.* Chicago: University of Chicago Press.

———. 1998b. "Science and the Book in Modern Cultural Historiography." *Studies in History and Philosophy of Science* 29A: 167–94.

Johnson, Richard. 1979. "'Really Useful Knowledge': Radical Education and Working-Class Culture, 1790–1848." In *Working-Class Culture: Studies in History and Theory.* Ed. J. Clarke, C. Critcher, and R. Johnson, 75–102. London: Hutchinson.

Johnson, Samuel. 1755. *A Dictionary of the English Language.* 2 vols. London: J. and P. Knapton.

Jones, Aled. 1996. *Powers of the Press: Newspapers, Power, and the Public in Nineteenth-Century England.* Aldershot: Scholar Press.

Jones, William. 1850. *The Jubilee Memorial of the Religious Tract Society: Containing a Record of Its Origin, Proceedings, and Results.* London: Religious Tract Society.

Jordan, John O., and R. L. Patten, eds. 1995. *Literature in the Marketplace: Nineteenth-Century British Publishing and Reading Practices.* Cambridge: Cambridge University Press.

Joyce, Patrick. 1991. *Visions of the People: Industrial England and the Question of Class, 1848–1914.* Cambridge: Cambridge University Press.

———. 1994. *Democratic Subjects: The Self and the Social in Nineteenth-Century England.* Cambridge: Cambridge University Press.

Judd, Catherine A. 1995. "Male Pseudonyms and Female Authority in Victorian England." In *Literature in the Marketplace.* Ed. John O. Jordan and Robert L. Patten, 250–68. Cambridge: Cambridge University Press.

Jukes Browne, C. A., ed. 1871. *Letters and Extracts from the Addresses and Occasional Writings of J. Beete Jukes, M.A. F.R.S. F.G.S.* London: Chapman and Hall.

Kain, Roger J., and Hugh C. Prince. 1985. *The Tithe Surveys of England and Wales.* Cambridge: Cambridge University Press.

Kelley, Philip, et al., eds. 1984–98. *The Brownings' Correspondence.* 14 vols. Winfield, Kans.: Wedgestone Press.

Kemble, Frances Anne. 1878. *Records of a Girlhood.* 2 vols. London: Richard Bentley.

———. 1882. *Records of Later Life.* 3 vols. London: Richard Bentley.

Kemp, T. Lindley. 1852. *The Natural History of Creation.* London: Longman.

———. 1854. *Indications of Instinct: A Sequel to "The Natural History of Creation."* London: Longman.

Kennedy, C. J. 1846. *Nature and Revelation Harmonious: A Defence of Scriptural Truths Assailed in Mr George Combe's Work on "The Constitution of Man, Considered in Relation to External Objects."* Edinburgh: William Oliphant.

Kennedy, John. 1851. *The Natural History of Man; Or, Popular Chapters on Ethnography.* 2 vols. London: John Cassell.

Kermode, Frank. 1983. *The Classic: Literary Images of Permanence and Change.* Cambridge: Harvard University Press.

Kernan, Alvin. [1987] 1989. *Samuel Johnson and the Impact of Print.* Princeton: Princeton University Press.

Killham, John. 1958. *Tennyson and* The Princess: *Reflections of an Age.* London: Athlone.

King, David. 1850. *The Principles of Geology Explained, and Viewed in their Relation to Revealed and Natural Religion.* Edinburgh: Johnstone.

———. 1853. *The Principles of Geology Explained. . . .* 4th ed. Edinburgh: Johnstone.

Kingsley, Charles. [1850] 1983. *Alton Locke: Tailor and Poet: An Autobiography.* Ed. Elizabeth A. Cripps. Oxford: Oxford University Press.

Kirby, William. 1835. *On the Power, Wisdom and Goodness of God, as Manifested in the Creation of Animals and in their History, Habits and Instincts.* 2 vols. London: William Pickering.

Kitteringham, Guy. 1982. "Science in Provincial Society: The Case of Liverpool in the Early Nineteenth Century." *Annals of Science* 39:329–48.

Klancher, Jon P. 1987. *The Making of English Reading Audiences, 1790–1832.* Madison: University of Wisconsin Press.

Knell, Simon J. 2000. *The Culture of English Geology, 1815–1851: A Science Revealed through its Collecting.* Aldershot: Ashgate.

Knies, Earl A. 1984. *Tennyson at Aldworth: The Diary of James Henry Mangles.* Athens: Ohio University Press.

Knight, Charles. 1854. *The Old Printer and the Modern Press.* London: John Murray.

———. 1864. *Passages of a Working Life During Half a Century: With a Prelude of Early Reminiscences.* 3 vols. London: Bradbury and Evans.

Knight, Frances. 1995. *The Nineteenth-Century Church and English Society.* Cambridge: Cambridge University Press.

Knight, William. 1907. *Colloquia Peripatetica: Deep-Sea Soundings: Being Notes of Conversations with the Late John Duncan, LL.D.* 6th ed. Edinburgh: Oliphant.

Knox, Robert. 1850. *The Races of Men: A Fragment.* London: Henry Renshaw.

Kohler, Robert E. 1999. "The Constructivists' Tool Kit." *Isis* 90:329–31.

Kohn, David. 1985. "Introduction: A High Regard for Darwin." In *The Darwinian Heritage.* Ed. David Kohn, 1–5. Princeton: Princeton University Press.

[Laing, Samuel]. 1846. *An Expository Outline of the "Vestiges of the Natural History of Creation;" with a Comprehensive and Critical Analysis of the Arguments by Which the Extraordinary Hypotheses of the Author Are Supported and Have Been Impugned, with Their Bearing Upon the Religious and Moral Interests of the Community. With a Notice of the Author's "Explanations:" a Sequel to the Vestiges.* London: Effingham Wilson.

———. 1889. *Problems of the Future and Essays.* London: Chapman and Hall.

Lane, Adrian. 1995. "Renovations of the Stars: John Pringle Nichol and the Architecture of the Heavens." Master's essay, Department of History and Philosophy of Science, University of Cambridge.

Lang, Cecil Y., and Edgar F. Shannon, eds. 1982–90. *The Letters of Alfred Tennyson.* 3 vols. Oxford: Clarendon Press.

Lankester, Edwin. 1848. "The Natural History of Creation." In *YMCA Lectures, 1847–48,* 1–32. London: Benjamin L. Green.

Lathbury, Daniel Connor. 1910. *Correspondence on Church and Religion of William Ewart Gladstone.* 2 vols. London: John Murray.

Laudan, Rachel. 1982. "The Role of Methodology in Lyell's Science." *Studies in History and Philosophy of Science* 13:215–49.

Layman, C. H., ed. 1990. *Man of Letters: The Early Life and Love Letters of Robert Chambers.* Edinburgh: Edinburgh University Press.

Lehmann, R. C. 1908. *Memories of Half a Century.* London: Smith, Elder, and Co.

Lenoir, Timothy. 1989. *The Strategy of Life: Teleology and Mechanics in Nineteenth Century German Biology.* Chicago: University of Chicago Press.

Léonard, Adrien. 1842. *Essai sur l'education des animaux. Le chien, pris pour type.* Lille.

Levine, George. 1988. *Darwin and the Novelists: Patterns of Science in Victorian Fiction.* Chicago: University of Chicago Press.

Lewis, Donald M. 1986. *Lighten Their Darkness: The Evangelical Mission to Working-Class London, 1828–1860.* New York: Greenwood Press.

Lightman, Bernard. 1989. "Ideology, Evolution, and Late-Victorian Agnostic Popularizers." In *History, Humanity, and Evolution.* Ed. James R. Moore, 285–309. Cambridge: Cambridge University Press.

———. 1997a. "Constructing Victorian Heavens: Agnes Clerke and the 'New Astronomy.'" In *Natural Eloquence: Women Reinscribe Science.* Ed. Barbara T. Gates and Ann B. Shteir, 61–75. Madison: University of Wisconsin Press.

———, ed. 1997b. *Victorian Science in Context.* Chicago: University of Chicago Press.

———. 1997c. "'The Voices of Nature': Popularizing Victorian Science." In *Victorian Science in Context.* Ed. Bernard Lightman, 187–211. Chicago: University of Chicago Press.

[Linton, Eliza Lynn]. 1885. *The Autobiography of Christopher Kirkland.* 3 vols. London: Richard Bentley.

———. 1899. *My Literary Life.* London: Hodder and Stoughton.

Liverpool Library. 1850. *Catalogue of the Liverpool Library.* Liverpool: T. Brakell.

Livingstone, David N. 1987. *Darwin's Forgotten Defenders: The Encounter between Evangelical Theology and Evolutionary Thought.* Grand Rapids, Mich.: William B. Eerdmans.

———. 1992. "Darwinism and Calvinism: The Belfast-Princeton Connection." *Isis* 83:408–28.

Lockhart, John Gibson. 1837–38. *Memoirs of the Life of Sir Walter Scott, Bart.* 7 vols. Edinburgh: Robert Cadell.

Logan, Peter Melville. 1997. *Nerves and Narratives: A Cultural History of Hysteria in Nineteenth-Century British Prose.* Berkeley and Los Angeles: University of California Press.

Lonsdale, Henry. 1870. *A Sketch of the Life and Writings of Robert Knox the Anatomist.* London: Macmillan.

Lord, Perceval B. 1834. *Popular Physiology; Being a Familiar Explanation of the Most Interesting Facts Connected with the Structure and Functions of Animals, and Particularly of Man. Adapted for General Readers.* London: John W. Parker.

Lovejoy, Arthur O. 1959. "The Argument for Organic Evolution before the *Origin of Species,* 1830–1858." In *Forerunners of Darwin: 1745–1859.* Ed. Bentley Glass, Oswei Temkin, and William L. Strauss, Jr, 356–414. Baltimore: Johns Hopkins University Press.

Low, Sampson. 1864. *The English Catalogue of Books Published from Jan. 1835, to Jan. 1863.* London.

Lowry, H. F., A. L. P. Norrington, and F. L. Mulhauser, eds. 1951. *The Poems of Arthur Hugh Clough.* Oxford: Clarendon Press.

Lucas, J. R. 1979. "Wilberforce and Huxley: A Legendary Encounter." *Historical Journal* 22:313–30.

Lukács, Georg. 1962. *The Historical Novel.* Trans. Hannah and Stanley Mitchell. London: Merlin.

Lyell, Katherine M., ed. 1881. *Life Letters and Journals of Sir Charles Lyell, Bart.* 2 vols. London: John Murray.

Machin, G. I. T. 1977. *Politics and the Churches in Great Britain 1832 to 1868.* Oxford: Clarendon Press.

Mackay, Charles. [1841]. 1852. *Memoirs of Extraordinary Popular Delusions and the Madness of Crowds.* 2d ed. 2 vols. London: Office of the National Illustrated Library.

———. 1887. *Through the Long Day, Or, Memorials of a Literary Life During Half a Century.* 2 vols. London: W. H. Allen.

[MacKenzie, George S., prob.]. 1846. *Vestiges of Error in Religious Doctrine.* London: John Mardon.

Mackie, Alexander. [1845]. *A Word to the Dupes of Mrs Martin, Messrs. Buchanan, Southwell & Co., of the Hall of Science, Manchester. And a Receipt for Making Money Without Working.* Manchester: David Brown.

Mackintosh, Thomas Simmons. [1841]. *The "Electrical Theory" of the Universe. Or the Elements of Physical and Moral Philosophy.* London: Simpkin.

Macleod, Donald. 1996. "Hugh Miller, the Disruption, and the Free Church of Scotland." In *Hugh Miller and the Controversies of Victorian Science.* Ed. Michael Shortland, 187–205. Oxford: Clarendon Press.

Maguire, Robert, ed. 1858. *Scenes from My Life, by a Working Man.* London: Seeleys.

Maidment, Brian E. 1984. "Magazines of Popular Progress and the Artisans." *Victorian Periodicals Review* 17:82–94.

[Maitland, Edward]. 1868. *The Pilgrim and the Shrine; Or, Passages from the Life and Correspondence of Herbert Ainslie, B.A., Cantab.* 3 vols. London: Tinsley Brothers.

[Malkin, Benjamin Heath]. [1829]. *Astronomy.* In *Library of Useful Knowledge: Natural Philosophy.* London: Baldwin and Cradock.

Mallock, W. H., and Gwendden Ramsden, eds. 1893. *Letters, Remains, and Memoirs of Edward Adolphus Seymour Twelfth Duke of Somerset K. G.* London: Richard Bentley.

Mandler, Peter. 1984. "Cain and Abel: Two Aristocrats and the Early Victorian Factory Acts." *Historical Journal* 27:83–109.

———. 1990. *Aristocratic Government in the Age of Reform.* Oxford: Clarendon Press.

Manguel, Alberto. 1996. *A History of Reading.* London: HarperCollins.

Mantell, Gideon. 1839. *Wonders of Geology.* 2 vols. Relfe and Fletcher.

———. 1846. *Thoughts on Animalcules; Or, a Glimpse of the Invisible World Revealed by the Microscope.* London: John Murray.

———. 1848. *Wonders of Geology.* 2 vols. Henry G. Bohn.

Martin, A. Patchett. 1893. *Life and Letters of the Right Honourable Robert Lowe, Viscount Sherbrooke.* 2 vols. London: Longmans.

Martin, Emma. [1844a]. *A Few Reasons for Renouncing Christianity, and Professing Infidel Opinions.* London: Watson.

———. [1844b]. *First Conversation on the Being of God.*

Martineau, Harriet. 1877. *Autobiography.* With memorials by Maria Weston Chapman. 3 vols. London: Smith, Elder, & Co.

Martineau, James. 1839. *The Bible: What It Is, and What It is Not: A Lecture, Delivered in Paradise Street Chapel, Liverpool, on Tuesday, February 19, 1839.* Liverpool: Willmer and Smith.

Mason, Michael. 1994a. *The Making of Victorian Sexuality.* Oxford: Oxford University Press.

———. 1994b. *The Making of Victorian Sexual Attitudes.* Oxford: Oxford University Press.

Mason, Thomas Monck. 1845. *Creation by the Immediate Agency of God, as Opposed to Creation by Natural Law; Being a Refutation of the Work Entitled* Vestiges of the Natural History of Creation. London: John W. Parker.

Masson, David. 1908. *Memories of London in the 'Forties.* Edinburgh: William Blackwood.

Maurice, Frederick. 1884. *The Life of Frederick Denison Maurice Chiefly Told in His Own Letters.* 2d ed. 2 vols. London: Macmillan.

Mayhew, Henry. 1861–62. *London Labour and the London Poor.* 4 vols. London: Griffin.

Mays, Kelly J. 1995. "The Disease of Reading and Victorian Periodicals." In *Literature in the Marketplace.* Ed. John O. Jordan and Robert L. Patten, 165–94. Cambridge: Cambridge University Press.

McCabe, Joseph. 1908. *Life and Letters of George Jacob Holyoake.* 2 vols. London: Watts.

McCalman, Iain. 1984. "Unrespectable Radicalism: Infidels and Pornography in Early Nineteenth-Century London." *Past and Present,* no. 104:74–110.

———. 1988. *Radical Underworld: Prophets, Revolutionaries, and Pornographers in London, 1795–1840.* Cambridge: Cambridge University Press.

———. 1992. "Popular Irreligion in Early Victorian England: Infidel Preachers and Radical Theatricality in 1830s London." In *Religion and Irreligion in Victorian Society: Essays in Honor of R. K. Webb.* Ed. R. W. Davis and R. J. Helmstadter, 51–67. London: Routledge.

McCartney, Paul J. 1977. *Henry De la Beche: Observations on an Observer.* Cardiff: Friends of the National Museum of Wales.

McGann, Jerome. 1983. *The Romantic Ideology.* Chicago: University of Chicago Press.

McKenzie, Donald Francis. 1960. "Printers' Perks: Paper Windows and Copy Money." *Library,* 5th ser., 15:288–91.

———. 1986. *Bibliography and the Sociology of Texts: The Panizzi Lectures 1985.* London: British Library.

McKinney, H. Lewis. 1969. "Wallace's Earliest Observations on Evolution: 28 December 1845." *Isis* 60:370–73.

———. 1972. *Wallace and Natural Selection.* New Haven: Yale University Press.

McKitterick, David. 1998. *A History of Cambridge University Press.* Vol. 2, *Scholarship and Commerce, 1698–1872.* Cambridge: Cambridge University Press.

Meiklejohn, J. M. D., ed. 1883. *Life and Letters of William Ballantyne Hodgson.* Edinburgh: David Douglas.

Mill, John Stuart. 1973–74. *A System of Logic.* Vols. 7 and 8 of *Collected Works of John Stuart Mill.* Ed. J. M. Robson. With an introduction by R. F. McRae. Toronto: University of Toronto Press.

———. 1986. "The Spirit of the Age." In *Newspaper Writings.* Vol. 22 of *Collected Works of John Stuart Mill.* Ed. A. P. Robson and R. M. Robson, 227–34, 238–45, 252–58, 289–95, 312–16. Toronto: University of Toronto Press.

Miller, Edward. 1973. *That Noble Cabinet: A History of the British Museum.* London: André Deutsch.

Miller, Hugh. 1841. *The Old Red Sandstone; Or, New Walks in an Old Field.* Edinburgh: John Johnstone.

———. 1849. *Foot-prints of the Creator: Or, the Asterolepis of Stromness.* London: Johnstone.

———. 1854. "The Two Records: Mosaic and Geological." In *YMCA Lectures, 1853–54,* 377–407. London: James Nisbet.

———. 1861a. *The Headship of Christ, and the Rights of the Christian People.* Edinburgh: Adam and Charles Black.

———. 1861b. *Footprints of the Creator: Or, the Asterolepis of Stromness.* Edinburgh: Adam and Charles Black.

Miller, Jonathan, and Borin Van Loon. 2000. *Introducing Darwin and Evolution.* Cambridge: Icon Books.

[Miller, William Haig]. [1850]. *The Problem of Life; Or the Three Questions, What am I? Whence Came I? Wither Do I Go?* London: Religious Tract Society.

Millgate, Jane. 1987. *Scott's Last Edition: A Study in Publishing History.* Edinburgh: University of Edinburgh Press.

Millhauser, Milton. 1956. "The Literary Impact of *Vestiges of Creation.*" *Modern Language Quarterly* 17:213–26.

———. 1959. *Just before Darwin: Robert Chambers and* Vestiges. Middletown, Conn.: Wesleyan University Press.

Mills, Eric I. 1984. "A View of Edward Forbes, Naturalist." *Archives of Natural History* 11:365–93.

Mills, John. 1899. *From Tinder-box to the "Larger" Light. Threads from the Life of John Mills, Banker (Author of "Vox Humana"). Interwoven with some Early Century Recollections by his Wife.* Manchester: Sherratt & Hughes.

Milner, Thomas. 1846. *The Gallery of Nature: A Pictorial and Descriptive Tour Through Creation, Illustrative of the Wonders of Astronomy, Physical Geography, and Geology.* London: Wm. S. Orr.

Mitchell, Charles. 1846. *Newspaper Press Directory.* London: C. Mitchell.

Mitchell, Graham. 1848. *The Young Man's Guide Against Infidelity: Embracing New Arguments, Arising from Recent Investigations, in Favour of the Religion of Jesus.* Edinburgh: William Whyte.

M'Neile, Hugh. 1850. "The Bible: Its Provision and Adaptation for the Moral Necessities of Fallen Man." In *YMCA Lectures, 1849–50*, 251–86. London: James Nisbet.

Moore, Doris Langley. 1977. *Ada, Countess of Lovelace: Byron's Legitimate Daughter.* London: John Murray.

Moore, James R. 1979. *The Post-Darwinian Controversies: A Study of the Protestant Struggles to Come to Terms with Darwin in Great Britain and America 1870–1900.* Cambridge: Cambridge University Press.

———, ed. 1988. *Religion in Victorian Britain.* Vol. 3: *Sources.* Manchester: Manchester University Press.

———, ed. 1989. *History, Humanity, and Evolution: Essays for John C. Greene.* Cambridge: Cambridge University Press.

———. 1990. "Theodicy and Society: The Crisis of the Intelligentsia." In *Victorian Faith in Crisis: Essays on Continuity and Change in Nineteenth-Century Religious Belief.* Ed. Richard J. Helmstadter and Bernard Lightman, 153–86. Stanford: Stanford University Press.

———. 1991. "Deconstructing Darwinism: The Politics of Evolution in the 1860s." *Journal of the History of Biology* 24:353–408.

———. 1997. "Wallace's Malthusian Moment: The Common Context Revisited." In *Victorian Science in Context.* Ed. Bernard Lightman, 290–311. Chicago: University of Chicago Press.

Moore, Kevin. 1992. "'This Whig and Tory Ridden Town': Popular Politics in Liverpool in the Chartist Era." In *Popular Politics, Riot and Labour: Essays in Liverpool History 1790–1940.* Ed. John Belchem, 38–67. Liverpool: Liverpool University Press.

Moran, James. 1965. *The Composition of Reading Matter: A History from Case to Computer.* London: Wace.

———. 1973. *Printing Presses: History and Development from the Fifteenth Century to Modern Times.* London: Faber.

Morgan, Marjorie. 1994. *Manners, Morals, and Class in England, 1774–1858.* New York: St. Martin's Press.

Morley, David. 1992. *Television, Audiences, and Cultural Studies.* London: Routledge.

Morley, Edith J., ed. 1938. *Henry Crabb Robinson on Books and Their Writers.* 3 vols. London: J. M. Dent.

Morley, John. 1903. *The Life of William Ewart Gladstone.* 3 vols. London: Macmillan.

Morley, John Cooper. 1887. *A Brief Memoir of the Rev. Abraham Hume.* Liverpool: the author.

Morrell, J. B. 1971. "Individualism and the Structure of British Science in 1830." *Historical Studies in the Physical Sciences* 3:183–204.

———. 1985. "Wissenschaft in Worstedopolis: Public Science in Bradford, 1800–1850." *BJHS* 18:1–23.

———. 1990. "Professionalisation." In *Companion to the History of Modern Science.* Ed. R. C. Olby, G. N. Cantor, J. R. R. Christie, and M. J. S. Hodge, 980–89. London: Routledge.

Morrell, J. B., and A. Thackray. 1981. *Gentlemen of Science: Early Years of the British Association for the Advancement of Science.* Clarendon Press: Oxford.

Morrison-Low, A. D., and J. R. R. Christie, eds. 1984. *"Martyr of Science": Sir David Brewster 1781–1868.* Edinburgh: Royal Scottish Museum.

Morus, Iwan Rhys. 1998. *Frankenstein's Children: Electricity, Exhibition, and Experiment in Early-Nineteenth-Century London.* Princeton: Princeton University Press.

Morus, Iwan Rhys, Simon Schaffer, and James A. Secord. 1992. "Scientific London." In *London: World City 1800–1840.* Ed. Celina Fox, 129–42. New Haven: Yale University Press.

Mozley, H., ed. 1885. *Letters of the Rev. J. B. Mozley, D.D.* London: Rivingtons.

[Mudie, Robert]. 1825. *Babylon the Great: A Dissection and Demonstration of Men and Things in the British Capital.* 2 vols. London: Charles Knight.

Mulhauser, Frederick L., ed. 1957. *The Correspondence of Arthur Hugh Clough.* Oxford: Clarendon Press.

Murchison, Roderick Impey. 1854. *Siluria: The History of the Oldest Known Rocks Containing Organic Remains, with a Brief Sketch of the Distribution of Gold Over the Earth.* London: John Murray.

Murchison, Roderick Impey, Edouard de Verneuil, and Alexander von Keyserling. 1845. *The Geology of Russia in Europe and the Ural Mountains.* 2 vols. London: John Murray.

Murphy, James. 1959. *The Religious Problem in English Education: The Crucial Experiment.* Liverpool: Liverpool University Press.

Murray, John. 1837. *Considerations on the Vital Principle; with a Description of Mr. Crosse's Experiments.* London: E. Wilson.

———. 1845. *Strictures on Morphology: Its Unwarrantable Assumptions, and Atheistical Tendency.* London: Hamilton.

Myers, Frederic. 1852. *Sermons Preached Before the University of Cambridge.* London: John W. Parker.

Myers, Greg. 1990. *Writing Biology: Texts in the Social Construction of Scientific Knowledge.* Madison: University of Wisconsin Press.

Napier, Macvey, ed. 1879. *Selections from the Correspondence of the Late Macvey Napier.* London: Macmillan.

Neal, Frank. 1988. *Sectarian Violence: The Liverpool Experience, 1819–1914: An Aspect of Anglo-Irish History.* Manchester: Manchester University Press.

Needler, G. H. 1939. *Letters of Anna Jameson to Ottilie von Goethe.* London: Oxford University Press.

Neuberg, Victor E. 1977. *Popular Literature: A History and Guide from the Beginning of Printing to the Year 1897.* Harmondsworth: Penguin.

Nichol, John Pringle. 1837. *Views of the Architecture of the Heavens: In a Series of Letters to a Lady.* Edinburgh: William Tait.

———. 1838. *Views of the Architecture of the Heavens: In a Series of Letters to a Lady.* 2d ed. Edinburgh: William Tait.

———. 1848. *Thoughts on Some Important Points Relating to the System of the World.* Edinburgh: John Johnstone.

Nockles, P. B. 1997. "'Lost Causes and . . . Impossible Loyalties': The Oxford Movement and the University." In *The History of the University of Oxford.* Vol. 6: *Nineteenth-Century Oxford, Part I.* Ed. M. G. Brock and M. C. Curthoys, 195–267. Oxford: Clarendon Press.

North, John S. 1989. *The Waterloo Directory of Scottish Newspapers and Periodicals, 1800–1900.* 2 vols. Waterloo, Canada: North Waterloo Academic Press.

Ogilvie, Marilyn Bailey. 1973. "Robert Chambers and the Successive Revisions of the *Vestiges of the Natural History of Creation.*" Ph.D. diss., University of Oklahoma.

———. 1975. "Robert Chambers and the Nebular Hypothesis." *BJHS* 8:214–32.

O'Neill, Patricia. 1991. "An Aesthetic of Reception and the Productive Reader of Victorian Poetry: Tennyson's 'Lucretius' among Victorian and Contemporary Critics." *Victorian Poetry* 29:385–400.

Ophir, Adi, and Steven Shapin. 1991. "The Place of Knowledge: A Methodological Survey." *Science in Context* 4:3–21.

Oppenheim, Janet. 1985. *The Other World: Spiritualism and Psychical Research in England, 1850–1914.* Cambridge: Cambridge University Press.

Ospovat, Dov. 1976. "The Influence of K. E. von Baer's Embryology, 1828–1859." *Journal of the History of Biology* 9:1–28.

———. 1981. *The Development of Darwin's Theory: Natural History, Natural Theology, and Natural Selection, 1838–1859.* Cambridge: Cambridge University Press.

Outram, Dorinda. 1984. *Georges Cuvier: Vocation, Science, and Authority in Post-revolutionary France.* Manchester: Manchester University Press.

Owen, Richard. 1894. *The Life of Richard Owen by His Grandson.* 2 vols. London: John Murray.

Packer, Brian A. 1984. "The Founding of the Liverpool Domestic Mission and Its Development under the Ministry of John Johns." *Transactions of the Unitarian Historical Society* 18, no. 2: 39–53.

Page, David. 1861. *The Past and Present Life of the Globe: Being a Sketch in Outline of the World's Life-system.* Edinburgh: William Blackwood.

Paradis, James G. 1997. "Satire and Science in Victorian Culture." In *Victorian Science in Context.* Ed. Bernard Lightman, 143–75. Chicago: University of Chicago Press.

Parry, Jonathan. 1993. *The Rise and Fall of Liberal Government in Victorian Britain.* New Haven: Yale University Press.

Parsons, Gerald, ed. 1988. *Religion in Victorian Britain.* 4 vols. Manchester: Manchester University Press.

Paston, George. 1932. *At John Murray's: Records of a Literary Circle 1843–1892.* London: John Murray.

Patten, Robert L. 1978. *Charles Dickens and His Publishers.* Oxford: Clarendon Press.

Paxton, Nancy L. 1991. *George Eliot and Herbert Spencer: Feminism, Evolutionism, and the Reconstruction of Gender.* Princeton: Princeton University Press.

Payn, James. 1884. *Some Literary Recollections.* London: Smith, Elder.

Pearson, Jacqueline. 1999. *Women's Reading in Britain, 1750–1835: A Dangerous Recreation.* Cambridge: Cambridge University Press.

Pearson, Thomas. 1853. *Infidelity; Its Aspects, Causes, and Agencies: Being the Prize Essay of the British Organization of the Evangelical Alliance.* London: Partridge.

Peckham, Morse. 1951. "Dr. Lardner's *Cabinet Cyclopaedia.*" *Papers of the Bibliographical Society of America* 45:37–58.

———, ed. 1959. *The Origin of Species by Charles Darwin: A Variorum Text.* Philadelphia: University of Pennsylvania Press.

Perkin, Harold. 1969. *The Origins of Modern English Society, 1780–1880.* London: Routledge.

Perkin, Michael, ed. 1987. *The Book Trade in Liverpool 1806–50: A Directory.* Liverpool: Liverpool Bibliographical Society.

———. 1990. "Egerton Smith and the Early 19th-Century Book Trade in Liverpool." Ed. Robin Myers and Michael Harris. In *Spreading the Word: The Distribution Networks of Print 1550–1850,* 151–64. Winchester: St. Paul's Bibliographies.

Peterson, M. Jeanne. 1984. "No Angels in the House: The Victorian Myth and the Paget Women." *American Historical Review* 89:677–798.

Pickering, Andrew, ed. 1992. *Science as Practice and Culture.* Chicago: University of Chicago Press.

Place, G. W. 1983. "John Brindley (1811–1873), Cheshire Schoolmaster and Opponent of Atheism." *Transactions of the Historic Society of Lancashire and Cheshire* 133:113–32.

Pollins, Harold. 1971. *Britain's Railways: An Industrial History.* Newton Abbot: David and Charles.

Poovey, Mary. 1984. *The Proper Lady and the Woman Writer: Ideology as Style in the Works of Mary Wollstonecraft, Mary Shelley, and Jane Austen.* Chicago: University of Chicago Press.

———. 1988. *Uneven Developments: The Ideological Work of Gender in Mid-Victorian England.* Chicago: University of Chicago Press.

———. 1995. *Making a Social Body: British Cultural Formation.* Chicago: University of Chicago Press.

Pope, Willard Bissell, ed. 1963. *The Diary of Benjamin Robert Haydon.* 5 vols. Cambridge: Harvard University Press.

Porter, Roy. 1978a. "Gentlemen and Geology: The Emergence of a Scientific Career, 1860–1920." *Historical Journal* 21:809–36.

———. 1978b. "Philosophy and Politics of a Geologist: G. H. Toulmin, (1754–1817)." *Journal of the History of Ideas* 39:435–50.

———. 1994. *London: A Social History.* London: Hamish Hamilton.

Postlethwaite, Diana. 1984. *Making It Whole: A Victorian Circle and the Shape of Their World.* Columbus: Ohio State University Press.

Powell, Baden. 1838. *The Connexion of Natural and Divine Truth; Or, the Study of the Inductive Philosophy Considered as Subservient to Theology.* London: John W. Parker.

———. 1855. *Essays on the Spirit of the Inductive Philosophy, the Unity of Worlds, and the Philosophy of Creation.* London: Longman.

Prescott, Gertrude M. 1985. "Faraday: Image of the Man and the Collector." In *Faraday Rediscovered: Essays on the Life and Work of Michael Faraday.* Ed. Frank A. J. L. James, 15–31. Basingstoke: Macmillan.

Price, Leah. 1997. "George Eliot and the Production of Consumers." *Novel* 30:145–69.

Price, Richard. 1999. *British Society, 1680–1880: Dynamism, Containment, and Change.* Cambridge: Cambridge University Press.

Priestley, Elizabeth. 1908. *The Story of a Lifetime.* London: Kegan Paul.

Pykett, Lyn. 1990. "Reading the Periodical Press: Text and Context." In *Investigating Victorian Journalism.* Ed. Laurel Brake, Aled Jones, and Lionel Madden, 3–18. Basingstoke: Macmillan.

Pym, Horace N., ed. 1882. *Memories of Old Friends: Being Extracts from the Journals and Letters of Caroline Fox of Penjerrick Cornwall, from 1835 to 1871.* 2d ed. 2 vols. London: Smith, Elder.

Radway, Janice A. 1986. "Reading Is Not Eating: Mass-Produced Literature and the Theoretical, Methodological, and Political Consequences of a Metaphor." *Book Research Quarterly* 2:7–29.

———. 1991. *Reading the Romance: Women, Patriarchy, and Popular Literature.* 2d ed. Chapel Hill: University of North Carolina Press.

———. 1997. *A Feeling for Books: The Book-of-the-Month Club, Literary Taste, and Middle-Class Desire.* Chapel Hill: University of North Carolina Press.

Rainy, Robert, and James Mackenzie. 1871. *Life of William Cunningham, D.D.* London: T. Nelson and Sons.

Raven, James, Helen Small, and Naomi Tadmor, eds. 1996. *The Practice and Representation of Reading in England.* Cambridge: Cambridge University Press.

Ray, Gordon N., ed. 1945–46. *The Letters and Private Papers of William Makepeace Thackeray.* 4 vols. London: Oxford University Press.

Raymond, Joad. 1996. *The Invention of the Newspaper: English Newsbooks, 1641–1649.* Oxford: Clarendon Press.

Reach, Angus B. [1847]. *The Natural History of "Bores."* London: Kent.

Rehbock, Philip F. 1979. "Edward Forbes (1815–1854): An Annotated List of Published and Unpublished Writings." *Journal of the Society for the Bibliography of Natural History* 9:171–218.

———. 1983. *The Philosophical Naturalists: Themes in Early Nineteenth-Century British Biology.* Madison: University of Wisconsin Press.

Reid, Wemyss. 1899. *Memoirs and Correspondence of Lyon Playfair.* London: Cassell.

[Rennie, James]. 1828. *Conversations on Geology.* London: Samuel Maunder.

Rice, Daniel P. 1971. "Natural Theology and the Scottish Philosophy in the Thought of Thomas Chalmers." *Scottish Journal of Theology* 24:23–46.

Richards, Evelleen. 1987. "A Question of Property Rights: Richard Owen's Evolutionism Reassessed." *BJHS* 20:129–71.

———. 1989a. "Huxley and Woman's Place in Science: The 'Woman Question' and the Control of Victorian Anthropology." In *History, Humanity, and Evolution.* Ed. James R. Moore, 253–84. Cambridge: Cambridge University Press.

———. 1989b. "The 'Moral Anatomy' of Robert Knox: The Interplay between Biological and Social Thought in Victorian Scientific Naturalism." *Journal of the History of Biology* 22:373–436.

———. 1990. "'Metaphorical Mystifications': The Romantic Gestation of Nature in British Biology." In *Romanticism and the Sciences.* Ed. Andrew Cunningham and Nicholas Jardine, 130–43. Cambridge: Cambridge University Press.

———. 1994. "A Political Anatomy of Monsters, Hopeful and Otherwise: Teratogeny, Transcendentalism, and Evolutionary Theorizing." *Isis* 85:377–411.

Richards, Joan. 1988. *Mathematical Visions: The Pursuit of Geometry in Victorian England.* London: Academic Press.

Richards, Robert J. 1987. *Darwin and the Emergence of Evolutionary Theories of Mind and Behavior.* Chicago: University of Chicago Press.

———. 1992. *The Meaning of Evolution: The Morphological Construction and Ideological Reconstruction of Darwin's Theory.* Chicago: University of Chicago Press.

Richards, Thomas. 1991. *The Commodity Culture of Victorian England: Advertising and Spectacle, 1851–1914.* London: Verso.

Richardson, Ruth. 1987. *Death, Dissection, and the Destitute.* London: Routledge.

[Richmond, Legh]. [1814]. *The Negro Servant.* London: Religious Tract Society.

———. 1843. *Domestic Portraiture.* 6th ed. London: R. B. Seeley.

Ricks, Christopher, ed. 1987. *The Poems of Tennyson.* 2d ed. 3 vols. London: Longman.

Robb-Smith, A. H. T. 1997. "Medical Education." In *The History of the University of Oxford, Volume 6. Nineteenth-Century Oxford, Part I.* Ed. M. G. Brock and M. C. Curthoys, 563–82. Oxford: Clarendon Press.

Roberts, Jon H. 1988. *Darwinism and the Divine in America: Protestant Intellectuals and Organic Evolution, 1859–1900.* Madison: University of Wisconsin Press.

Robertson, Fiona. 1994. *Legitimate Histories: Scott, Gothic, and the Authorities of Fiction.* Oxford: Clarendon Press.

Robinson, Howard. 1948. *The British Post Office: A History.* Princeton: Princeton University Press.

Robson, John M. 1990. "The Fiat and Finger of God: The Bridgewater Treatises." In *Victorian Faith in Crisis.* Ed. Richard J. Helmstadter and Bernard Lightman, 71–125. Basingstoke: Macmillan.

Roderick, Gordon W., and Michael D. Stephens. 1978. *Education and Industry in the Nineteenth Century: The English Disease?* London: Longman.

Roget, Peter Mark. 1834. *Animal and Vegetable Physiology Considered with Reference to Natural Theology.* 2 vols. London: William Pickering.

Rose, Jonathan. 1992. "Rereading the English Common Reader: A Preface to a History of Audiences." *Journal of the History of Ideas* 53:47–70.

———. 1995. "How Historians Study Reader Response: Or, What Did Jo Think of *Bleak House?*" In *Literature in the Marketplace.* Ed. John O. Jordon and Robert L. Patten, 195–212. Cambridge: Cambridge University Press.

Rose, Mark. 1993. *Authors and Owners: The Invention of Copyright.* Cambridge: Harvard University Press.

Rose, Michael B. 1998. *Darwin's Spectre: Evolutionary Biology in the Modern World.* Princeton: Princeton University Press.

Rose, R. B. 1957. "John Finch, 1784–1857: A Liverpool Disciple of Robert Owen." *Transactions of the Historic Society of Lancashire and Cheshire* 109:157–84.

Rosenberg, Sheila. 1982. "The Financing of Radical Opinion: John Chapman and the *Westminster Review.*" In *The Victorian Periodical Press.* Ed. Joanne Shattock and Michael Wolff, 167–92. Leicester: Leicester University Press.

Rosman, Doreen M. 1984. *Evangelicals and Culture.* London: Croom Helm.

Rothblatt, Sheldon. 1968. *The Revolution of the Dons: Cambridge and Society in Victorian England.* New York: Basic Books.

Rowell, Geoffrey. 1974. *Hell and the Victorians: A Study of the Nineteenth-Century Theological Controversies Concerning Eternal Punishment and the Future Life.* Oxford: Clarendon Press.

Royal Society of London. 1867–1925. *Catalogue of Scientific Papers (1800–1900).* 19 vols. London: Eyre and Spottiswood.

Royle, Edward. 1974. *Victorian Infidels: The Origins of the British Secularist Movement.* Manchester: Manchester University Press.

———. 1976. *The Infidel Tradition from Paine to Bradlaugh.* London: Macmillan.

Rubin, Joan Shelley. 1992. *The Making of Middlebrow Culture.* Chapel Hill: University of North Carolina Press.

Rudwick, Martin J. S. 1974. "Darwin and Glen Roy: A 'Great Failure' in Scientific Method?" *Studies in History and Philosophy of Science* 5:97–185.

———. 1976. *The Meaning of Fossils: Episodes in the History of Palaeontology.* 2d ed. New York: Science History Publications.

———. 1982. "Charles Darwin in London: The Integration of Public and Private Science." *Isis* 73:186–206.

———. 1985. *The Great Devonian Controversy: The Shaping of Scientific Knowledge among Gentlemanly Specialists.* Chicago: University of Chicago Press.

———. 1986. "The Shape and Meaning of Earth History." in *God and Nature: Historical Essays on the Encounter between Christianity and Science.* Ed. David C. Lindberg and Ronald L. Numbers, 296–321. Berkeley and Los Angeles: University of California Press.

Rupke, Nicolaas A. 1983. *The Great Chain of History: William Buckland and the English School of Geology (1814–1849).* Oxford: Oxford University Press, Clarendon Press.

———. 1993. "Richard Owen's Vertebrate Archetype." *Isis* 84:231–51.

———. 1994. *Richard Owen: Victorian Naturalist.* New Haven: Yale University Press.

———. 1997. "Oxford's Scientific Awakening and the Role of Geology." In *The History of the University of Oxford.* Vol. 6: *Nineteenth-Century Oxford, Part I.* Ed. M. G. Brock and M. C. Curthoys, 543–62. Oxford: Clarendon Press.

———. 2000. "Translation Studies in the History of Science: The Example of *Vestiges.*" *BJHS* 33:209–22.

Ruse, Michael. 1979. *The Darwinian Revolution.* Chicago: University of Chicago Press.

———. 1996. *Monad to Man: The Concept of Progress in Evolutionary Biology.* Cambridge: Harvard University Press.

Rusk, Ralp L., ed. 1939. *The Letters of Ralph Waldo Emerson.* 6 vols. New York: Columbia University Press.

Russell, E. S. 1916. *Form and Function: A Contribution to the History of Animal Morphology.* London: John Murray.

Ryals, Clyde de L., et al., eds. 1993. *The Collected Letters of Thomas and Jane Welsh Carlyle.* Vol. 19. Durham, North Carolina: Duke University Press.

Sabertash, Orlando [John Mitchell]. 1842. *The Art of Conversation, with Remarks on Fashion and Address.* London: G. W. Nickisson.

Sadleir, Michael. 1930. *The Evolution of Publishers' Binding Styles, 1770–1900.* London: Constable.

Saenger, Paul. 1982. "Silent Reading: Its Impact on Late Medieval Script and Society." *Viator* 13:367–414.

St. George, Andrew. 1993. *The Descent of Manners: Etiquette, Rules, and the Victorians.* London: Chatto & Windus.

Sala, George Augusta. 1859. *Twice Round the Clock.* London: Houlston and Wright.

Samuel, Raphael. 1977. "Workshop of the World: Steam Power and Hand Technology in Mid-Victorian Britain." *History Workshop,* no. 3: 6–72.

[Sargent, George Etell]. [1859]. *The Story of a Pocket Bible.* London: Religious Tract Society.

Saville, John. 1987. *1848: The British State and the Chartist Movement.* Cambridge: Cambridge University Press.

Schaffer, Simon. 1988. "Astronomers Mark Time: Discipline and the Personal Equation." *Science in Context* 2:115–45.

———. 1989. "The Nebular Hypothesis and the Science of Progress." In *History, Humanity and Evolution.* Ed. James R. Moore, 131–64. Cambridge: Cambridge University Press.

———. 1991. "The History and Geography of the Intellectual World: Whewell's Politics of Language." In *William Whewell: A Composite Portrait.* Ed. Menachem Fisch and Simon Schaffer, 201–31. Oxford: Clarendon Press.

———. 1994. "Babbage's Intelligence: Calculating Engines and the Factory System." *Critical Inquiry* 21:203–27.

———. 1996. "Babbage's Dancer and the Impresarios of Mechanism." In *Cultural Babbage: Technology, Time and Invention.* Ed. Francis Spufford and Jenny Uglow, 53–80. London: Faber and Faber.

———. 1998. "On Astronomical Drawing." In *Picturing Science, Producing Art.* Ed. Caroline A. Jones and Peter Galison, 441–74. New York: Routledge.

Schivelbusch, Wolfgang. 1986. *The Railway Journey: The Industrialization of Time and Space in the Nineteenth Century.* Berkeley and Los Angeles: University of California Press.

Scholnick, Robert. 1999. "'The Fiery Cross of Knowledge': *Chambers's Edinburgh Journal,* 1832–1843." *Victorian Periodicals Review* 32:324–58.

Schwartz, Joel S. 1990. "Darwin, Wallace, and Huxley, and *Vestiges of the Natural History of Creation.*" *Journal of the History of Biology* 23:127–53.

Scott, Walter. [1814] 1986. *Waverley; Or, 'tis Sixty Years Since.* Ed. Claire Lamont. Oxford: Oxford University Press.

Scott, William. 1836. *The Harmony of Phrenology with Scripture, Shewn in a Refutation of the Philosophical Errors Contained in Mr. Combe's "Constitution of Man."* Edinburgh: Fraser.

Scottish Association for Opposing Prevalent Errors. 1847. *Report of the Proceedings of the First Public Meeting of the Scottish Association for Opposing Prevalent Errors, Held in the Saloon of Gibb's Royal Hotel, Princes Street, Edinburgh, on Tuesday 9th March, 1847.* Edinburgh: Alexander Padon.

Sealts, Merton M., ed. 1973. *The Journals and Miscellaneous Notebooks of Ralph Waldo Emerson.* Vol. 10. Cambridge: Harvard University Press, Belknap Press.

Secord, Anne. 1994a. "Science in the Pub: Artisan Botanists in Early Nineteenth-Century Lancashire." *History of Science* 32:269–315.

———. 1994b. "Corresponding Interests: Artisans and Gentlemen in Nineteenth-Century Natural History." *BJHS* 27:383–408.

Secord, James A. 1981. "Nature's Fancy: Charles Darwin and the Breeding of Pigeons." *Isis* 72:162–86.

———. 1985. "John W. Salter: The Rise and Fall of a Victorian Palaeontological Career." In *From Linnaeus to Darwin: Commentaries on the History of Biology and Geology.* Ed. Alywne Wheeler and James H. Price, 61–75. London: Society for the History of Natural History.

———. 1986a. *Controversy in Victorian Geology: The Cambrian-Silurian Dispute.* Princeton: Princeton University Press.

———. 1986b. "The Geological Survey of Great Britain as a Research School, 1839–1855." *History of Science* 24:223–75.

———. 1989a. "Behind the Veil: Robert Chambers and *Vestiges." In History, Humanity, and Evolution.* Ed. James R. Moore, 165–94. Cambridge: Cambridge University Press.

———. 1989b. "Extraordinary Experiment: Electricity and the Creation of Life in Victorian England." In *The Uses of Experiment: Studies in the Natural Sciences.* Ed. David Gooding, Trevor Pinch, and Simon Schaffer, 337–83. Cambridge: Cambridge University Press.

———, ed. 1993. "The Big Picture." *BJHS* 26:387–483.

———. 1994. Introduction to Robert Chambers, *Vestiges of the Natural History of Creation and Other Evolutionary Writings,* [vii–xlviii]. Chicago: University of Chicago Press.

———. 1997. Introduction to Charles Lyell, *Principles of Geology,* ix–xliii. London: Penguin Books.

Sedgwick, Adam. 1850. *A Discourse on the Studies of the University of Cambridge.* 5th ed. London: John W. Parker.

Seed, John. 1982. "Unitarianism, Political Economy, and the Antinomies of Liberal Culture in Manchester, 1830–50." *Social History* 7:1–25.

Sellers, Ian. 1969. "Liverpool Nonconformity (1786–1914)." Ph.D. diss., Keele University.

Sennett, Richard. 1992. *The Fall of Public Man.* New York: W. W. Norton.

Sewell, William. 1861. *Christian Vestiges of Creation.* Oxford: J. H. and Jas. Parker.

Shapin, Steven. 1975. "Phrenological Knowledge and the Social Structure of Early Nineteenth-Century Edinburgh." *Annals of Science* 32:219–43.

———. 1991. "'A Scholar and a Gentleman': The Problematic Identity of the Scientific Practitioner in Early Modern England." *History of Science* 29:279–327.

———. 1994. *A Social History of Truth: Civility and Science in Seventeenth-Century England.* Chicago: University of Chicago Press.

Shapin, Steven, and Barry Barnes. 1977. "Science, Nature, and Control: Interpreting Mechanics' Institutes." *Social Studies of Science* 7:31–74.

Sharpe, Kevin. 2000. *Reading Revolutions: The Politics of Reading in Early Modern England.* New Haven: Yale University Press.

Shattock, Joanne. 1982. "Problems of Parentage: The *North British Review* and the Free Church of Scotland." In *The Victorian Periodical Press: Samplings and Soundings*. Ed. Joanne Shattock and Michael Wolff, 145–66. Leicester: Leicester University Press.

———. 1989. *Politics and Reviewers: The* Edinburgh *and the* Quarterly *in the Early Victorian Age*. Leicester: Leicester University Press.

Shaw, Harry E. 1983. *The Forms of Historical Fiction: Sir Walter Scott and His Successors*. Ithaca: Cornell University Press.

[Shaw, Jonathan]. 1869. *Recollections of Liverpool Cotton Brokers by One of Themselves*. Liverpool.

Sheepshanks, Richard. 1845. *Correspondence Respecting the Liverpool Observatory, Between Mr. John Taylor and the Rev. R. Sheepshanks*. London: George Barclay.

Sheets-Pyenson, Susan. 1981a. "Darwin's Data: His Reading of Natural History Journals, 1837–1842." *Journal of the History of Biology* 14:231–48.

———. 1981b. "A Measure of Success: The Publication of Natural History Journals in Early Victorian Britain." *Publishing History* 9:21–36.

———. 1981c. "War and Peace in Natural History Publishing: *The Naturalist's Library*, 1833–1844." *Isis* 72:50–72.

———. 1985. "Popular Science Periodicals in Paris and London: The Emergence of a Low Scientific Culture, 1820–1875." *Annals of Science* 42:549–72.

Sheppard, John. 1845. *A Lecture on the Arguments for Christian Theism, From Organized Life and Fossil Osteology; Containing Remarks on a Work Entitled "Vestiges of the Natural History of Creation." Delivered Before the Frome Literary and Scientific Institution, Feb. 7, 1845*. London: Jackson and Walford.

Sherman, William H. 1994. *John Dee: The Politics of Reading and Writing in the English Renaissance*. Amherst: University of Massachusetts Press.

Shiach, Morag. 1989. *Discourse on Popular Culture: Class, Gender, and History in Cultural Analysis, 1730 to the Present*. Cambridge: Polity.

Shillingsburg, Peter. 1992. *Pegasus in Harness: Victorian Publishing and W. M. Thackeray*. Charlottesville: University Press of Virginia.

Shipley, A. E. 1913. *"J.": A Memoir of John Willis Clark*. London: Smith, Elder.

Shortland, Michael. 1994. "Robert Southey's *The Doctor, &c:* Anonymity and Authorship." *English Language Notes*. 31:54–63.

———. 1996a. "Bonneted Mechanic and Narrative Hero: The Self-modelling of Hugh Miller." In *Hugh Miller and the Controversies of Victorian Science*. Ed. Michael Shortland, 14–75. Oxford: Clarendon Press.

———, ed. 1996b. *Hugh Miller and the Controversies of Victorian Science*. Oxford: Clarendon Press.

Shteir, Ann B. 1996. *Cultivating Women, Cultivating Science: Flora's Daughters and Botany in England, 1760 to 1860*. Baltimore: Johns Hopkins University Press.

———. 1997. "Elegant Recreations? Configuring Science Writing for Women." In

Victorian Science in Context. Ed. Bernard Lightman, 236–55. Chicago: University of Chicago Press.

Shuttleworth, Sally. 1984. *George Eliot and Nineteenth-Century Science: The Make-Believe of a Beginning.* Cambridge: Cambridge University Press.

Silverstone, Roger. 1994. *Television and Everyday Life.* London: Routledge.

Simon, Brian, ed. 1972. *The Radical Tradition in Education in Britain.* London: Lawrence and Wishart.

Simpson, David. 1988. "Literary Criticism and the Return to 'History.'" *Critical Inquiry* 14:721–47.

Sinfield, Alan. 1971. *The Language of Tennyson's "In Memoriam."* Oxford: Basil Blackwell.

Sinnema, Peter W. 1998. *Dynamics of the Pictured Page: Representing the Nation in the* Illustrated London News. Aldershot: Ashgate.

Small, Helen. 1996. "A Pulse of 124: Charles Dickens and a Pathology of the Mid-Victorian Reading Public." In *The Practice and Representation of Reading in England.* Ed. James Raven, Helen Small, and Naomi Tadmor, 263–90. Cambridge: Cambridge University Press.

Smedley, Frank E. 1855. *Harry Coverdale's Courtship, and All That Came of It.* London: Virtue.

Smiles, Samuel. 1891. *A Publisher and His Friends: Memoir and Correspondence of the Late John Murray, with an Account of the Origin and Progress of the House, 1768–1843.* 2 vols. London: John Murray.

Smith, Crosbie. 1998. *The Science of Energy: A Cultural History of Energy Physics in Victorian Britain.* London: Athlone.

Smith, Crosbie, and Jon Agar, eds. 1998. *Making Space for Science: Territorial Themes in the Shaping of Knowledge.* Basingstoke: Macmillan.

Smith, Crosbie, and M. Norton Wise. 1989. *Energy and Empire: A Biographical Study of Lord Kelvin.* Cambridge: Cambridge University Press.

Smith, Jeremiah Finch. 1868. *The Admission Register of the Manchester School with Some Notices of the More Distinguished Scholars.* 3 vols. Manchester: Chetham Society.

Smith, John Pye. 1839. *On the Relation between the Holy Scriptures and Some Parts of Geological Science.* London: Jackson and Walford.

———. 1852. *The Relation between the Holy Scriptures and Some Parts of Geological Science.* 5th ed. London: Henry G. Bohn.

Smith, Mary. 1892a. *The Autobiography of Mary Smith, Schoolmistress and Nonconformist, a Fragment of a Life.* London: Bemrose.

———. 1892b. *Miscellaneous Poems.* London: Bemrose.

Smith, William Henry. 1857. *Thorndale or the Conflict of Opinions.* Edinburgh: William Blackwood.

Somerville, Martha. 1873. *Personal Recollections, from Early Life to Old Age, of Mary Somerville.* London: John Murray.

Somerville, Mary. 1834. *On the Connexion of the Physical Sciences.* London: John Murray.

Spacks, Patricia Meyer. 1986. *Gossip.* Chicago: University of Chicago Press.

———. 1995. *Boredom: The Literary History of a State of Mind.* Chicago: University of Chicago Press.

Spencer, Herbert. 1904. *An Autobiography.* 2 vols. London: Williams and Norgate.

Spencer, Thomas. 1885. *Of the Origin and Reproduction of Animal and Vegetable Life on the Globe.* London: Effingham Wilson.

Spielmann, M. H. 1895. *The History of "Punch."* London: Cassell.

Stafford, Robert A. 1989. *Scientist of Empire: Sir Roderick Murchison, Scientific Exploration and Victorian Imperialism.* Cambridge: Cambridge University Press.

Stair-Douglas, J. 1882. *The Life and Selections from the Correspondence of William Whewell, D.D., Late Master of Trinity College Cambridge.* 2d ed. London: Kegan Paul.

Stedman Jones, Gareth. 1983. *Languages of Class: Studies in English Working Class History, 1832–1982.* Cambridge: Cambridge University Press.

Stein, Dorothy. 1985. *Ada: A Life and a Legacy.* Cambridge: MIT Press.

Stein, Richard L. 1987. *Victoria's Year: English Literature and Culture, 1837–1838.* New York: Oxford University Press.

Stephens, Michael D., and Gordon W. Roderick. 1972. "Nineteenth Century Ventures in Liverpool's Scientific Education." *Annals of Science* 28:61–86.

Steuart, T. [1848]. *Edinburgh: A Complete View of the City & Environs, as Seen in a Walk Round the Calton Hill.* London: Ackermann.

Stewart, Garrett. 1996. *Dear Reader: The Conscripted Audience in Nineteenth-Century British Fiction.* Baltimore: Johns Hopkins University Press.

Stirling Maxwell, William. 1872. *The Scott Exhibition: MDCCCLXXI. Catalogue of the Exhibition Held at Edinburgh, in July and August 1871.* Edinburgh.

[Stonehouse, James]. [1846]. *Pictorial Liverpool: Its Annals; Commerce; Shipping; Institutions; Public Buildings; Sights; Excursions; &c., &c. A New and Complete Hand-book for Resident, Visitor, and Tourist.* 2d ed. Liverpool: Henry Lacey.

Stott, Rebecca, ed. 1996. *Tennyson.* London: Longman.

Stoughton, John. 1864. "Biblical Statements in Harmony with Scientific Discoveries." In *YMCA Lectures, 1845–46,* 1–32. London: James Nisbet.

Stowell, Hugh. 1849. "Modern Infidel Philosophy." In *YMCA Lectures, 1848–49,* 145–84. London: William Jones.

———. 1852. *Religion: What It Is, and What It is Not. A Lecture, Delivered in the Concert Hall, Liverpool, on Sunday, August 15th, 1852.* Liverpool: Egerton Smith.

Suleiman, Susan R., and Inge Crosman, eds. 1980. *The Reader in the Text: Essays on Audience and Interpretation.* Princeton: Princeton University Press.

Sutherland, John. 1976. *Victorian Novelists and Publishers.* Chicago: University of Chicago Press.

———. 1987. "The British Book Trade and the Crash of 1826." *Library,* 6th ser., 9:148–61.

———. 1988. "Publishing History: A Hole at the Centre of Literary Sociology." *Critical Inquiry* 14:574–89.

———. 1995. *The Life of Sir Walter Scott: A Critical Biography.* Oxford: Blackwell.

Swan, John. 1873. *The Valuable Library of Scientific & Miscellaneous Books of the Late Rev. Adam Sedgwick, LL.D., F.R.S. . . . to Be Sold by Auction.* Cambridge: John Swan.

Taylor, Andrew. 1996. "Into His Secret Chamber: Reading and Privacy in Late Medieval England." In *The Practice and Representation of Reading in England.* Ed. James Raven, Helen Small, and Naomi Tadmor, 41–61. Cambridge: Cambridge University Press.

Taylor, Barbara. 1983. *Eve and the New Jerusalem: Socialism and Feminism in the Nineteenth Century.* London: Virago.

Taylor, George. 1855. *The Indications of the Creator; Or, the Natural Evidences of Final Cause.* Glasgow: William Collins.

Terhune, Alfred McKinley, and Annabelle Burdick Terhune, eds. 1980. *The Letters of Edward FitzGerald.* 4 vols. Princeton: Princeton University Press.

Terrall, Mary. 1996. "Salon, Academy, and Boudoir: Generation and Desire in Maupertuis's Science of Life." *Isis* 87:217–29.

Thackeray, William Makepeace. 1847–48. *Vanity Fair: A Novel without a Hero.* London: Bradbury and Evans.

———. [1848–50] 1994. *The History of Pendennis: His Fortunes and Misfortunes, His Friends and His Greatest Enemy.* Ed. John Sutherland. Oxford: Oxford University Press.

Thackray, John C. 1979. "R. I. Murchison's *Geology of Russia* (1845)." *Journal of the Society for the Bibliography of Natural History* 8:421–33.

Thompson, E. P. 1968. *The Making of the English Working Class.* Harmondsworth: Penguin.

Thompson, Robert Anchor. 1855. *Christian Theism: The Testimony of Reason and Revelation to the Existence and Character of the Supreme Being.* 2 vols. London: Rivingtons.

Tiedemann, Friedrich. 1834. *A Systematic Treatise on Comparative Physiology, Introductory to the Physiology of Man.* Trans. J. M. Gully and J. H. Lane. Vol. 1 (all published). London: John Churchill.

Tiffen, Herbert J. 1935. *A History of the Liverpool Institute Schools, 1825 to 1935.* Liverpool: Liverpool Institute Old Boy's Association.

[Timbs, John]. 1846. *The Year-book of Facts in Science and Art.* London: David Bogue.

———. 1866. *Club Life of London with Anecdotes of the Clubs, Coffee-houses and Taverns of the Metropolis During the 17th, 18th, and 19th Centuries.* 2 vols. London.

Tinsley, William. 1900. *Random Recollections of an Old Publisher.* London: Simpkin.

[Todd, John]. [1848]. *Self-Improvement: Chiefly Addressed to the Young.* London: Religious Tract Society.

Todd, William B. 1972. *A Directory of Printers and Others in Allied Trades: London and Vicinity, 1800–1840.* London: Printing Historical Society.

Todhunter, I. 1876. *William Whewell, D.D., Master of Trinity College, Cambridge. An Account of His Writings with Selections from His Literary and Scientific Correspondence.* 2 vols. London: Macmillan.

Tollemache, Lionel A. 1885. *Recollections of Pattison.* London: C. F. Hodgson.

Tomkins, Jane P., ed. 1980. *Reader-Response Criticism: From Formalism to Post-structuralism.* Baltimore: Johns Hopkins University Press.

Tomlinson, Charles, ed. 1854. *Cyclopaedia of Useful Arts, Mechanical and Chemical, Manufactures, Mining, and Engineering.* 2 vols. London: George Virtue.

Topham, Jonathan. 1992. "Science and Popular Education in the 1830s: The Role of the *Bridgewater Treatises.*" *BJHS* 25:397–430.

———. 1993. "'An Infinite Variety of Arguments': The *Bridgewater Treatises* and British Natural Theology in the 1830s." Ph.D. diss., University of Lancaster.

———. 1998. "Beyond the 'Common Context': The Production and Reading of the Bridgewater Treatises." *Isis* 89:233–62.

———. 1999. "Science, Natural Theology, and Evangelicalism in Early Nineteenth-Century Scotland: Thomas Chalmers and the *Evidence* Controversy." In *Evangelicals and Science in Historical Perspective.* Ed. David N. Livingstone, N. G. Hart, and Mark A. Noll. New York: Oxford University Press.

———. 2000. "Scientific Publishing and the Reading of Science in Early Nineteenth-Century Britain: An Historical Survey and Guide to Sources." *Studies in History and Philosophy of Science,* 31A.

Tosh, John. 1999. *A Man's Place: Masculinity and the Middle-Class Home in Victorian England.* New Haven: Yale University Press.

Townley, Henry, and George Jacob Holyoake. 1852. *Atheistic Controversy. A Public Discussion on the Being of a God.* London: Ward and Co.

Tresise, Geoffrey R. 1989. "Chirotherium: The First Finds at Storeton Quarry and the Role of the Liverpool Natural History Society." *Geological Curator* 5:135–51.

———. 1991. "The Storeton Quarry Discoveries of Triassic Vertebrate Footprints, 1838: John Cunningham's Account." *Geological Curator* 5:225–30.

Tuchman, Gaye. 1989. *Edging Women Out: Victorian Novelists, Publishers, and Social Change.* London: Routledge.

Tulloch, John. 1855. *Theism: The Witness of Reason and Nature to an All-wise and Beneficent Creator.* Edinburgh: William Blackwood.

Turner, Frank M. 1981. "John Tyndall and Victorian Scientific Naturalism." In *John Tyndall: Essays on a Natural Philosopher.* Ed. W. H. Brock, N. D. McMillan, and R. C. Mollan, 169–80. Dublin: Royal Dublin Society.

———. 1993. *Contesting Cultural Authority: Essays in Victorian Intellectual Life.* Cambridge: Cambridge University Press.

———. 1997. "Practicing Science: An Introduction." In *Victorian Science in Context.* Ed. Bernard Lightman, 283–89. Chicago: University of Chicago Press.

Twyman, Michael. 1998. *Printing 1770–1970: An Illustrated History of Its Development and Uses in England.* New ed. London: British Library.

Tylecote, Mabel. 1957. *The Mechanics' Institutes of Lancashire and Yorkshire before 1851.* Manchester: Manchester University Press.

Unitarianism Confuted: A Series of Lectures Delivered in Christ Church, Liverpool, in MDCCCXXXIX. 1839. Liverpool: Henry Perris.

United Kingdom. Parliament. 1852. *Report of Her Majesty's Commissioners Appointed to Inquire into the State, Discipline, Studies, and Revenues of the University and Colleges of Cambridge: Together with an Appendix.* London: Her Majesty's Stationery Office.

[Urquhart, William Pollard]. 1849. *Some Thoughts on Natural Theology, Suggested by a Work, Entitled "Vestiges of the Natural History of Creation."* London: Longman.

Vanderkiste, Robert W. 1852. *Notes and Narratives of a Six Years' Mission, Principally Among the Dens of London.* London: James Nisbet.

Vernon, James. 1993. *Politics and the People: A Study in English Political Culture, 1815–1867.* Cambridge: Cambridge University Press.

Vickery, Amanda. 1993. "Golden Age to Separate Spheres: A Review of the Categories and Chronology of English Women's History." *Historical Journal* 36:384–414.

Vincent, David. 1981. *Bread, Knowledge, and Freedom: A Study of Nineteenth-Century Working Class Autobiography.* London: Europa.

———. 1989. *Literacy and Popular Culture: England 1750–1914.* Cambridge: Cambridge University Press.

———. 1999. *The Culture of Secrecy: Britain, 1832–1998.* Oxford: Oxford University Press.

Vyvyan, Richard. 1845. *On the Harmony of the Comprehensible World.* London: privately printed.

Wahrman, Dror. 1995. *Imagining the Middle Class: The Political Representation of Class in Britain, c. 1780–1840.* Cambridge: Cambridge University Press.

Wainwright, David. 1960. *Liverpool Gentlemen: A History of Liverpool College, an Independent Day School, from 1840.* London: Faber.

Walker, Archibald Stodart, ed. 1909. *The Letters of John Stuart Blackie to His Wife, with a Few Earlier Ones to His Parents.* Edinburgh: William Blackwood.

Walker, J. U. 1845. *Walker's Directory of the Parish of Halifax; to Which Is Appended, a Variety of Useful Statistical Information.* Halifax: J. U. Walker.

Walker, R. B. 1968. "Religious Changes in Liverpool in the Nineteenth Century." *Journal of Ecclesiastical History* 19:195–211.

Wallace, Alfred Russel. 1905. *My Life: A Record of Events and Opinions.* 2 vols. London: Chapman and Hall.

Waller, P. J. 1981. *Democracy and Sectarianism: A Political and Social History of Liverpool 1868–1939.* Liverpool: Liverpool University Press.

Wallis, John. 1845. *A Brief Examination of the Nebulous Hypothesis, with Strictures on a Work Entitled Vestiges of the Natural History of Creation.* London: R. Groombridge & Sons.

Walters, Alice N. 1997. "Conversation Pieces: Science and Politeness in Eighteenth-Century England." *History of Science* 35:121–54.

Ward, W. R. 1973. *Religion and Society in England 1790–1850.* New York: Schocken Books.

Watkins, Charlotte C. 1982. "Edward William Cox and the Rise of 'Class Journalism.'" *Victorian Periodicals Review* 15:86–93.

Watson, Hewett C. 1836. *An Examination of Mr. Scott's Attack Upon Mr. Combe's "Constitution of Man."* London: printed for the author.

Weatherall, Mark. 2000. *Gentlemen, Scientists, and Doctors: Medicine at Cambridge 1800–1940.* Woodbridge, Suffolk: Boydell Press.

Webb, R. K. 1955. *The British Working Class Reader, 1790–1848: Literacy and Social Tension.* London: George Allen & Unwin.

———. 1960. *Harriet Martineau: A Radical Victorian.* London: Heinemann.

———. 1978. "John Hamilton Thom: Intellect and Conscience in Liverpool." In *The View from the Pulpit: Victorian Ministers and Society.* Ed. P. T. Phillips, 211–43. Toronto: Macmillan.

———. 1990. "The Faith of Nineteenth-Century Unitarians: A Curious Incident." In *Victorian Faith in Crisis.* Ed. Richard J. Helmstadter and Bernard Lightman, 126–49. Basingstoke: Macmillan.

Wees, J. Dustin, in collaboration with Michael J. Campbell. 1986. *Darkness Visible: The Prints of John Martin.* Williamstown, Mass.: Sterling and Francine Clark Art Institute.

Weindling, Paul. 1980. "How Effective Were the Acts Licensing Lectures and Meetings, 1795–1819?" *BJHS* 13:139–53.

Weld, Charles. 1848. *A History of the Royal Society with Memoirs of the Presidents.* 2 vols. London: John W. Parker.

Welsh, Alexander. 1985. *George Eliot and Blackmail.* Cambridge: Harvard University Press.

Wheeler, J. M., trans. [1884]. *Letters from Hell.* London: Progressive Publishing Company.

Whewell, William. 1833. *Astronomy and General Physics Considered with Reference to Natural Theology.* London: William Pickering.

———. 1845a. *The Elements of Morality, Including Polity.* 2 vols. London: John W. Parker.

———. 1845b. *Indications of the Creator. Extracts, Bearing Upon Theology, from the History and the Philosophy of the Inductive Sciences.* London: John W. Parker.

———. 1846. *Indications of the Creator . . .* 2d ed. London: John W. Parker.

White, Paul. Forthcoming. *Thomas Huxley: Making the "Man of Science."* Cambridge: Cambridge University Press.

White, Walter. 1898. *The Journals of Walter White, Assistant Secretary of the Royal Society.* London: Chapman and Hall.

Wiener, Joel H. 1969. *The War of the Unstamped: The Movement to Repeal the British Newspaper Tax, 1830–1836.* Ithaca: Cornell University Press.

———. 1983. *Radicalism and Freethought in Nineteenth-Century Britain: The Life of Richard Carlile.* Westport, Conn.: Greenwood Press.

Wilberforce, Samuel. 1847. *Pride a Hindrance to True Knowledge: A Sermon Preached in the Church of St. Mary the Virgin, Oxford, before the University, on Sunday, June 27, 1847.* London: Rivington.

[Williams, Charles]. [1852]. *Caxton and the Art of Printing.* London: Religious Tract Society.

Williams, Frederick S. 1852. *The Wonders of the Heavens.* London: John Cassell.

Wilson, Frederick J. F. 1879. *Typographic Printing Machines and Machine Printing.* London: Wyman and Sons.

Wilson, George. 1852. *Electricity and the Electric Telegraph: Together with the Chemistry of the Stars; An Argument Touching the Stars and their Inhabitants.* London: Longman.

———. 1862. *Religio Chemici: Essays.* London: Macmillan.

Wilson, George, and Archibald Geikie. 1861. *Memoir of Edward Forbes, F.R.S.* London: Macmillan.

Wilson, Jessie Aitken. 1860. *Memoir of George Wilson.* Edinburgh: Edmonston and Douglas.

Wilson, Leonard G., ed. 1970. *Sir Charles Lyell's Scientific Journals on the Species Question.* New Haven: Yale University Press.

———. 1998. *Lyell in America: Transatlantic Geology, 1841–1853.* Baltimore: Johns Hopkins University Press.

Wilson, R. Jackson. 1989. *Figures of Speech: American Writers and the Literary Marketplace, from Benjamin Franklin to Emily Dickinson.* Baltimore: Johns Hopkins University Press.

Winter, Alison. 1995. "Harriet Martineau and the Reform of the Invalid in Victorian Britain." *Historical Journal* 38:597–616.

———. 1997. "The Construction of Orthodoxies and Heterodoxies in the Early Victorian Life Sciences." In *Victorian Science in Context.* Ed. Bernard Lightman, 24–50. Chicago: University of Chicago Press.

———. 1998a. *Mesmerized: Powers of Mind in Victorian Britain.* Chicago: University of Chicago Press.

———. 1998b. "A Calculus of Suffering: Ada Lovelace and the Bodily Constraints on Women's Knowledge in Early Victorian England." In *Science Incarnate: Historical Embodiments of Natural Knowledge.* Ed. Christopher Lawrence and Steven Shapin, 202–39. Chicago: Unviersity of Chicago Press.

Winter, James. 1993. *London's Teeming Streets: 1830–1914.* London: Routledge.

Withrington, Donald J. 1993. "The Disruption: A Century and a Half of Historical Interpretation." *Records of the Scottish Church History Society* 25:118–53.

Wood, Marcus. 1994. *Radical Satire and Print Culture, 1790–1822.* Oxford: Clarendon Press.

Woodmansee, Martha. 1994. *The Author, Art, and the Market: Rereading the History of Aesthetics.* New York: Columbia University Press.

Woolley, Benjamin. 1999. *The Bride of Science: Romance, Reason, and Byron's Daughter.* London: Macmillan.

Worsley, Thomas. 1845–49. *The Province of the Intellect in Religion Deduced from Our Lord's Sermon on the Mount, and Considered with Reference to Prevalent Errors.* 2 vols. London: John W. Parker.

Wyatt, John. 1995. *Wordsworth and the Geologists.* Cambridge: Cambridge University Press.

Yanni, Carla. 1999. *Nature's Museums: Victorian Science and the Architecture of Display.* London: Athlone.

Yeo, Eileen. 1981. "Christianity in Chartist Struggle, 1838–1842." *Past and Present,* no. 91: 109–39.

Yeo, Richard. 1984. "Science and Intellectual Authority in Mid-Nineteenth-Century Britain: Robert Chambers and *Vestiges of the Natural History of Creation.*" *Victorian Studies* 28:5–31.

———. 1991. "Reading Encyclopaedias: Science and the Organisation of Knowledge in British Dictionaries of Arts and Sciences." *Isis* 82:24–49.

———. 1993. *Defining Science: William Whewell, Natural Knowledge, and Public Debate in Early Victorian Britain.* Cambridge: Cambridge University Press.

Youatt, William. 1855. *The Horse: With a Treatise of Draught.* New Edition. London: Longman.

Young Man's Best Companion. 1831. London: J. Smith.

Young, Robert M. 1985. *Darwin's Metaphor: Nature's Place in Victorian Culture.* Cambridge: Cambridge University Press.

Yule, John David. 1976. "The Impact of Science on British Religious Thought in the Second Quarter of the Nineteenth Century." Ph.D. diss., Cambridge University.

Zboray, Ronald J. 1993. *A Fictive People: Antebellum Economic Development and the American Reading Public.* New York: Oxford University Press.

CREDITS

Grateful acknowledgment is made to those who have kindly granted permission to quote from their manuscripts: Her Majesty the Queen; Bodleian Library, University of Oxford; British Geological Survey; the British Library; Cambridge Philosophical Society; A. S. Chambers; the Hon. Simon Howard; Edinburgh University Library; Archives of Imperial College of Science, Technology and Medicine, London; Trustees of the National Library of Scotland; The Natural History Museum, London; Harry Ransom Humanities Research Center, the University of Texas at Austin; The Library of the University of St. Andrews; the Department of Geology at the National Museums and Galleries of Wales; the Royal Institution of Great Britain; the Royal Literary Fund; the President and Council of the Royal Society; Master and Fellows of Trinity College, Cambridge; the Director of Dr Williams's Library on behalf of the Trustees (extract from the H. C. Robinson papers).

Grateful acknowledgment is made to those who have kindly granted permission to reproduce copyright images in their possession. Figure 1.1: Birmingham Museums and Art Gallery; 1.2, 3.3, 8.3, and 8.5: Trustees of the National Library of Scotland; 1.4, 1.7, 1.10, 1.11, 2.6, 2.8, 2.9, 2.10, 4.1, 4.2, 4.3, 4.5, 4.6, 4.8, 4.9, 5.3, 6.1, 6.2, 6.3, 6.4, 6.5, 7.2, 7.11, 8.4, 8.8, 8.9, 9.3, 9.8, 9.10, 9.11, 9.12, 9.13, 11.1, 11.2, 12.1, 12.3, 12.4, 12.5, 12.8, 4.13, 12.9, 12.10, 12.12, 12.13, 13.1, 13.2, 13.4, 13.8, 13.9, 14.2, 14.3, 14.4, 14.6: Syndics of Cambridge University Library; 1.6, J. R. Topham; 1.8: Bell and Howell Information and Learning Company; 1.9, 2.3, 2.5, 2.11, 2.12, 4.10, 4.14, 4.15, 8.1, 11.6, 14.5: author's collection; 2.1: The Natural History Museum, London; 2.2: Wellcome Library, London; 2.4, 7.6, 9.1, 12.11: The British Museum; 2.13, 5.2, 5.7, 6.9, 7.8, 9.4, 9.6, 9.7, 10.4, 10.5, 12.2, 12.7, 13.5, 13.7, 13.12: The British Library; 3.6 and 14.8: A. S. Chambers; 4.10: Director of Dr. Williams's Library on behalf of the Trustees; 4.12: Pearson Education Ltd.; 4.15: Director of Dr. Williams's Library on behalf of the Trustees and J. R. Moore; 5.5: Lord Howick; 5.6, 7.9: Whipple Museum for the History of Science, Cambridge; 5.9: Sir Ferrars Vyvyan; 5.10: Laurence Pollinger Limited and the Earl of Lytton/Science and Society Picture Library; 6.7: Local Studies Unit, Manchester Central Library; 7.3: Sedgwick Museum, Department of Earth Sciences, University of Cambridge; 7.4 and 7.5: Master and Fellows of Trinity College, Cambridge; 7.7: Museum of the History of Science, Oxford; 7.9: Whipple Library, Department of the History and Philosophy of Science, University of Cambridge; 7.10: National Museum of Photography/Science and Society Picture Library; 8.2: Scottish National Portrait Gallery, photograph by D. O. Hill and R. Adamson; 10.1, 10.3: Calderdale District Archives, West Yorkshire Archive Service, Halifax; 11.3: crown copyright, Royal

Commission on the Ancient and Historical Monuments of Scotland; 11.5: Edinburgh University Library; 12.6: Geological Society of London; 13.3: Guildhall Library, Corporation of London; 14.1 and 14.9: Archives of Imperial College of Science, Technology and Medicine, London; 14.10: American Philosophical Society; E.1: Brown Brothers; E.3: Icon Books Ltd.; E.4: Mary Evans Picture Library.

INDEX

Page numbers in italics refer to figures.